Methods in Microbiology
Volume 29

Recent titles in the series

Methods in Microbiology

Volume 29
Genetic Methods for Diverse Prokaryotes

Edited by

Margaret CM Smith

Institute of Genetics
University of Nottingham
Queens Medical Centre
Nottingham, UK

and

R Elizabeth Sockett

Institute of Genetics
University of Nottingham
Queens Medical Centre
Nottingham, UK

ACADEMIC PRESS

San Diego London Boston
New York Sydney Tokyo Toronto

Academic Press
24–28 Oval Road, London NW1 7DX, UK
http://www.hbuk.co.uk/ap/

Academic Press
a division of Harcourt Brace & Company
525 B Street, Suite 1900, San Diego, California 92101-4495, USA
http://www.apnet.com

A catalogue record for this book is available from the British Library

ISBN 0–12–521529–0 (Hardback)
ISBN 0–12–652340–1 (Comb bound)

Typeset by Phoenix Photosetting, Chatham, Kent
Printed in Great Britain by Redwood Books, Trowbridge, Wiltshire
99 00 01 02 03 04 RB 9 8 7 6 5 4 3 2 1

Contents

JUL 2 1 1999

Series Advisors

Gordon Dougan   Department of Biochemistry, Wolfson Laboratories, Imperial College of Science, Technology and Medicine, London, UK

Graham J Boulnois   Zeneca Pharmaceuticals, Mereside, Alderley Park, Macclesfield, Cheshire, UK

Jim Prosser   Department of Molecular and Cell Biology, Marischal College, University of Aberdeen, Aberdeen, UK

Ian R Booth   Department of Molecular and Cell Biology, Marischal College, University of Aberdeen, Aberdeen, UK

David A Hodgson   Department of Biological Sciences, University of Warwick, Coventry, UK

David H Boxer   Department of Biochemistry, Medical Sciences Institute, The University, Dundee, UK

Contributors

E Allan Microbial Pathogenicity Research Group, Department of Medical Microbiology, St Bartholomew's and the Royal London School of Medicine and Dentistry, West Smithfield, London EC1A 7BE, UK

A Cooper Department of Microbiology, University of Illinois, 407 S Goodwin Avenue, Urbana, IL 61801, USA

SM Cutting School of Biological Sciences, University of London Royal Holloway and Bedford New College, Egham, Surrey TW20 0EX, UK

J D'Elia Department of Microbiology, University of Illinois, 407 S Goodwin Avenue, Urbana, IL 61801, USA

N Dorrell Microbial Pathogenicity Research Group, Department of Medical Microbiology, St Bartholomew's and the Royal London School of Medicine and Dentistry, West Smithfield, London EC1A 7BE, UK

PJ Dyson School of Biological Sciences, University of Wales – Swansea, Singleton Park, Swansea SA2 8PP, UK

VJ Evans Institute of Biological Sciences, University of Wales Aberystwyth, Cledwyn Building, Aberystwyth, Ceredigion SY23 3DA, UK

S Foynes Microbial Pathogenicity Research Group, Department of Medical Microbiology, St Bartholomew's and the Royal London School of Medicine and Dentistry, West Smithfield, London EC1A 7BE, UK

DH Green School of Biological Sciences, University of London Royal Holloway and Bedford New College, Egham, Surrey TW20 0EX, UK

GF Hatfull Department of Biological Sciences, University of Pittsburgh, Pittsburgh, Pennsylvania 15260, USA

JR Jefferies Institute of Biological Sciences, University of Wales Aberystwyth, Cledwyn Building, Aberystwyth, Ceredigion SY23 3DA, UK

KCB Jennert Institute of Biological Sciences, University of Wales Aberystwyth, Cledwyn Building, Aberystwyth, Ceredigion SY23 3DA, UK

WW Metcalf Department of Microbiology, University of Illinois, 407 S Goodwin Avenue, Urbana, IL 61801, USA

TF Meyer Max Planck Institut für Biologie, Abteilung Infektionsbiologie, Spemannstrasse 34, D-72076 Tübingen, Germany and Max Planck Institut für Infektionsbiologie, Abteilung Molekulare Biologie, Monbijoustrasse 2, 10117 Berlin, Germany

MD Page   Department of Biochemistry, Oxford University, South Parks Road, Oxford OX1 3QU, UK

ZEV Phillips   Institute of Biological Sciences, University of Wales Aberystwyth, Cledwyn Building, Aberystwyth, Ceredigion SY23 3DD, UK

L Radnedge   Environmental Molecular Microbiology Laboratory, Biology and Biotechnology Research Program, Lawrence Livermore National Laboratory, PO Box 808, Livermore, CA 94551, USA

A Ravagnani   Institute of Biological Sciences, University of Wales Aberystwyth, Cledwyn Building, Aberystwyth, Ceredigion SY23 3DD, UK

CED Rees   Division of Food Sciences, University of Nottingham, Nottingham NG7 2UH, UK

H Richards   Department of Biology, Darwin Building, University College London, Gower Street, London WC1E 6BT, UK

P Rosa   Laboratory of Microbial Structure and Function, Rocky Mountain Laboratories, NIAID, NIH, 903 South Fourth Street, Hamilton, Montana 59840, USA

T Rudel   Max Planck Institut für Biologie, Abteilung Infektionsbiologie, Spemannstrasse 34, D-72076 Tübingen, Germany

GPC Salmond   Department of Biochemistry, University of Cambridge, Tennis Court Road, The Downing Site, Cambridge CB2 1QW, UK

A Salyers   Department of Microbiology, University of Illinois, 407 S Goodwin Avenue, Urbana, IL 61801, USA

JR Saunders   School of Biological Sciences, Life Sciences Building, University of Liverpool, Liverpool L69 7BZ, UK

VA Saunders   School of Biomolecular Sciences, Liverpool John Moores University, Liverpool L3 3AF, UK

T Schwan   Byk Gulden Pharmazeutika, FB3, Byk Gulden Str. 2, 78467 Konstanz, Germany

J Shipman   Department of Microbiology, University of Illinois, 407 S Goodwin Avenue, Urbana, IL 61801, USA

N Shoemaker   Department of Microbiology, University of Illinois, 407 S Goodwin Avenue, Urbana, IL 61801, USA

MCM Smith   Institute of Genetics, University of Nottingham, Queens Medical Centre, Nottingham NG7 2UH, UK

RE Sockett Institute of Genetics, University of Nottingham, Queen's Medical Centre, Nottingham NG7 2UH, UK

B Stevenson Laboratory of Microbial Structure and Function, Rocky Mountain Laboratories, NIAID, NIH, 903 South Fourth Street, Hamilton, Montana 59840, USA

JD Thomas Department of Biochemistry, University of Cambridge, Tennis Court Road, The Downing Site, Cambridge CB2 1QW, UK

NR Thomson Department of Biochemistry, University of Cambridge, Tennis Court Road, The Downing Site, Cambridge CB2 1QW, UK

K Tilly Laboratory of Microbial Structure and Function, Rocky Mountain Laboratories, NIAID, NIH, 903 South Fourth Street, Hamilton, Montana 59840, USA

B Wren Microbial Pathogenicity Research Group, Department of Medical Microbiology, St Bartholomew's and the Royal London School of Medicine and Dentistry, West Smithfield, London EC1A 7BE, UK

DI Young Institute of Biological Sciences, University of Wales Aberystwyth, Cledwyn Building, Aberystwyth, Ceredigion SY23 3DD, UK

M Young Institute of Biological Sciences, University of Wales Aberystwyth, Cledwyn Building, Aberystwyth, Ceredigion SY23 3DD, UK

Contributors

Preface

Research of prokaryotes has undergone a renaissance in recent times. The completed genome sequences of many diverse prokaryotes, including archael genomes, have greatly encouraged researchers to study organisms other than *E. coli* or *Salmonella*. Furthermore, as pointed out by Professor Hopwood and Professor Chater in the Preface to the excellent volume 'Genetics of Bacterial Diversity' edited by DA Hopwood and KF Chater, *E. coli* and *Salmonella* simply do not display important complex phenotypes associated with other bacteria. Therefore if one wishes to study photosynthesis a non-photosynthetic organism cannot be used. Whilst the aforementioned volume contains a wide range of chapters describing the genetics of various bacterially-encoded phenomena, we felt that there was a need for technical reviews of the methods for genetic research in diverse organisms.

The aim of this book is to produce a guide for genetic techniques in bacteria other than *E. coli* or *Salmonella* or other close relatives. It is meant as a primer for researchers who, as the result of genome sequencing information, may wish to use genetic methods to study cellular processes in a bacterium which is new to them. The book presents technical reviews on molecular genetic and genetic methods and approaches in three sections:

The first section describes general genetic techniques that will increase awareness of methods and allow workers to see how they may apply methodologies from well-studied bacteria to new organisms. These chapters bring together procedures from many different bacteria and occasionally refer to systems that have been adapted from *E. coli*, e.g. Tn5 derivatives, promiscuous plasmids. The second section provides case studies of genetic techniques involving important bacteria (Mycobacteria, Clostridia) which may also have unusual genetic properties (e.g. *Bacteroides*) or in which genetic techniques are newly emerging (*Borrelia*). For completeness a survey of Archael methods are included in Chapter 10. The third section presents analyses of genetic methodology that have gone into the elucidation of complex phenomena in specific organisms. These chapters aim to give a historical perspective as well as a description of modern techniques, and will illustrate how a diversity of methodologies complement each other in order to analyse problems in different organisms. Hopefully these chapters show how new approaches are invented, sometimes because of the novel physiology of the organism and sometimes because of the biochemical properties of the phenomenon under study.

A single volume of this size cannot be a fully comprehensive guide. We have chosen organisms in which the genetic systems are novel, or in which unusual, complex cellular processes occur. *In vitro* methods for gene manipulation have largely overtaken classical genetic methods and this volume reflects this situation. For several organisms the genetic methods are largely adapted from those used in enteric bacteria. For instance, the genetics of electron transport systems in *Rhodobacter* and

Paracoccus, and secretion in *Erwinia* have been included even though some of the study methods have much in common with those for *E. coli*. It should be noted that, solely due to the similarity of the genetic methods that are employed, we have been forced to omit studies on many important and interesting complex phenomena in other Gram-negative bacteria. Whilst we chose to include *Rhodobacter*, *Paracoccus* and *Erwinia*, we could equally have picked *Agrobacterium*, *Rhizobium*, *Yersinia*, Myxobacteria and others.

We hope that this volume will be useful to undergraduate and postgraduate students and research scientists. The aim is to educate on the technical feasibility of genetic experiments in many diverse prokaryotes and to point the way to the uninitiated. We hope also that the volume will be of use to people embarking on genetic studies in novel prokaryotes and novel phenomena where a genetic methodology has not previously been applied.

PART I

Essential Techniques for Genetic Analysis in Bacteria

◆◆◆

1 Introduction of DNA into Bacteria

Jon R. Saunders[1] **and Venetia A. Saunders**[2]

[1]School of Biological Sciences, University of Liverpool, Liverpool, UK; [2]School of Biomolecular Sciences, Liverpool John Moores University, Liverpool, UK

◆◆

CONTENTS

◆◆◆◆◆◆ I. INTRODUCTION

The introduction of purified DNA into bacteria is an almost universally applicable means of transferring genes between organisms, and has the major advantage that no vector is required: the nucleic acid itself mediates information transfer. The term transformation was originally applied to gene transfer between bacteria that did not require cell-to-cell contact (which distinguishes it from conjugation) and that was sensitive to deoxyribonuclease (which distinguishes it from transduction). Transformation involves the uptake of exogenous plasmid or chromosomal DNA and its subsequent incorporation into the genome so that the cell acquires an altered genotype. The uptake of naked bacteriophage DNA and subsequent infection of a host cell is often referred to as transfection (Benzinger, 1978), but the mechanisms of DNA uptake involved are essentially the same as those in transformation. Therefore, the term transformation will be used throughout this work, whatever the nature of the transforming/transfecting DNA involved. Transformation can be distinguished from other natural mechanisms of gene transfer in bacteria by sensitivity of the process to added deoxyribonuclease. While the mainstay of many gene transfer and gene cloning procedures, transformation can also be used in direct assay of biological activity, using genetic markers carried on specific DNA fragments or replicons such as plasmids (see, for example, Romanowski *et al.*, 1992). Transformation is frequently the only method of gene transfer available for genetic analysis of bacteria. An inability to transform a particular microorganism, especially with plasmid DNA, may therefore be a significant factor in preventing the application of genetic techniques to previously unexplored species. This chapter will consider the variety of molecular mechanisms involved in transformation phenomena and how they may be exploited in genetic manipulation of bacteria.

In order to be transformed, bacteria must be able to become competent, that is, to take up DNA into a deoxyribonuclease-resistant form, and thus internal to the cell envelope. Transformation systems can be divided into two classes, natural (physiological) and artificial (Stewart, 1989; Solomon and Grossman, 1996). Based on laboratory culture techniques, this classification makes certain, perhaps unwarranted, assumptions concerning the vast range of physical and chemical regimes that bacteria might conceivably encounter, albeit occasionally, in various natural environments. However, the distinction between "natural" and "artificial" is a useful working categorization that will be applied here. In naturally transformable bacterial species, individual cells are either always competent or, more commonly, become competent at particular growth phases, or following shifts in nutritional status. Many bacterial species are not intrinsically competent under normal growth conditions and must be artificially rendered permeable to DNA by chemical or physical means, while retaining viability. Such "artificially induced" permeability can be achieved chemically, in some cases simply by treating whole bacterial cells with high concentrations of divalent cations, Tris(hydroxymethyl)aminomethane (Tris) or other chemical treatments that transiently alter the cell surface. Alternatively, all or part of the cell envelope components that lie outside the cytoplasmic membrane can be removed to form protoplasts, or spheroplasts respectively. Protoplasts or spheroplasts can subsequently be induced to take up DNA in the presence of divalent cations and/or polyethylene glycol (PEG). More convenient in some ways, and certainly applicable on a wide scale, is the permeabilization of bacteria to DNA by subjection to a strong electrical field in the process of electroporation (Miller, 1994). This form of artificial transformation appears to be effective for a very wide range of organisms, including higher eukaryotes.

◆◆◆◆◆◆ II. TRANSFORMATION MECHANISMS

A. Transformation as a Gene Transfer Mechanism

1. The functional stages of transformation

Transformation can be operationally divided into three stages: (i) binding of exogenous DNA to the outside of the cell; (ii) transport of all or a proportion of that DNA across the cell envelope; and (iii) establishment of the incoming DNA either as a replicon, or, in the case of a non-replicating fragment, by recombination with a resident replicon. The efficiency of the first two stages is largely dependent on the natural properties of the cell envelope of the recipient bacterial species and its response to nutritional and other signals or artificial treatments required to achieve competence. There is considerable inter- and even intra-species variation in the biochemical and molecular mechanisms involved in permitting DNA molecules to cross the bacterial cell envelope. The efficiency of the third stage depends partly on the topological status of the transforming DNA on entry into the cytoplasm, the host's repertoire of enzymes mediating DNA transactions, and the ability of the incoming DNA to replicate independently or to undergo recombination with resident DNA sequences.

2. Size and topology of transforming DNA

More often than not, bacteria that can be transformed by homologous chromosomal DNA can also be transformed with DNA of appropriate plasmids or phages. There are, however, marked differences in the efficiency with which plasmid, phage and chromosomal markers can be used to transform competent cells. This is usually the result of differential processing of topological forms of DNA and the ability of surviving DNA molecules to self-replicate or to recombine with a resident replicon. Large fragments of chromosomal DNA are fragile and susceptible to shear forces, which reduce the efficiency of transformation. Furthermore, the termini of linear DNA molecules (as in the genomes of some bacteriophages and chromosomal fragments prepared by most DNA extraction methods) present targets for exonucleases and other enzymes involved in housekeeping functions for nucleic acids in bacteria. Circular genomes are intrinsically refractory to many such enzymes, having no free ends. On the other hand, linear molecules may be preferred substrates for recombination with host DNA sequences. In some species, circular double-stranded plasmid DNA is less good as a substrate for uptake than enzymatically linearized plasmid DNA (Biswas *et al.*, 1986; Stein, 1991). Furthermore, the process of DNA uptake may itself lead to linearization in some naturally transformable bacteria. Transformation of naturally competent bacterial cells by a single plasmid DNA molecule may therefore actually be uncommon. Interaction with another replicon may be required to repair damage arising from the DNA processing involved in uptake and establishment of a plasmid.

Transformation with plasmids is also limited in most cases by low efficiency of the process and by reduced efficiency of uptake and establishment as the length of DNA increases. In general, transformation efficiency is inversely related to the molecular mass of the transforming plasmid DNA (Hanahan, 1983). For example, frequencies of transformation obtained with plasmids of >75 kbp may be prohibitively low (Hanahan, 1987). Low levels of transformation are normally achieved with larger plasmids, due to the time taken for a complete DNA molecule to traverse the cell envelope and increased potential for DNA damage prior to establishment of the incoming replicon. During recombinant DNA procedures, there is likely to be bias against recombinant plasmids containing larger DNA inserts during transformation of *in vitro* ligation mixtures. This may cause under-representation of some gene sequences during construction of gene libraries. This property of transformation in *Escherichia coli* provided much impetus to develop phage and hybrid vectors that circumvent a transformation stage by *in vitro* packaging (see Chapter 3).

B. Transformation Kinetics

1. Dose–response relationships

The DNA uptake systems in transformable bacteria are normally saturated by high concentrations of exogenous DNA. As a result, characteristic dose–response relationships are observed when fixed numbers of competent bacterial cells or protoplasts are exposed to increasing amounts of transforming DNA (Figure 1). At non-saturating DNA concentrations, the slope of dose–response curves reflects the number of DNA molecules required to produce a single transformant. Typically, for single or closely linked chromosomal genetic markers this slope is 1, and for two unlinked markers it is 2 (Trautner and Spatz, 1973). If the slope is 1 with homogeneous populations of plasmid (or phage) DNA molecules, a single molecule is evidently sufficient to produce one transformant. If, on the other hand, the dose–response slope is 2, then two separate DNA molecules must react co-operatively in order to produce a single transformant (Saunders and Guild, 1981a; Weston *et al.*, 1981). For a theoretical discussion of the relationship between the number of transformed cells and the concentration of transforming DNA in artificially induced competence, see Rittich and Spanova (1996). It should also be realized that, where homologous recombination is required to rescue incoming DNA molecules, e.g. in *Bacillus subtilis*, kinetics may be affected by the length of homology between participating sequences (Michel and Ehrlich, 1983).

Knowledge of dose–response kinetics is of great practical significance in the development of new transformation systems. Firstly, the presence of excessive concentrations of DNA will increase the chances of forming transformants receiving unwanted multiple fragments of DNA molecules, which may complicate the subsequent unravelling of genotypes. Secondly, any assays of biological activity of DNA molecules made by transformation must be performed at sub-saturating DNA concentrations, or there will be potential underestimation. Furthermore, if transformation is to be used quantitatively in an assay for the biological activity of DNA preparations, the dose–response relationship must be defined beforehand.

2. Frequency and efficiency parameters

The frequency of transformation is normally expressed as the number of transformants per viable recipient cell. With saturating concentrations of an homologous population of transforming DNA molecules, which is the case with plasmid or phage replicons, transformation frequency is a measure of the number of transformable bacteria in a population (Humphreys *et al.*, 1979). In contrast, the total number of transformable cells in a population of competent bacteria is not calculable directly in chromosomal

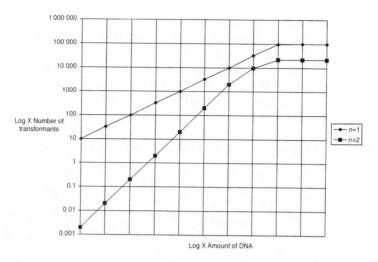

Figure 1. Dose–response relationship for transformation using a single or double unlinked markers. Dose–response curve for a single plasmid molecule or genetic marker on a chromosomal fragment ♦ ($n = 1$). Dose–response curve for two separate plasmids or chromosomal fragments bearing unlinked markers simultaneously transforming the same population of bacteria ■ ($n = 2$).

transformations, where DNA fragments carrying one selected marker compete with a random population of DNA fragments that do not bear the marker concerned. Therefore, only a proportion of the competent population undergoes transformation measured with any given marker. Fortunately, it is possible to estimate the number of competent cells by comparing the observed number of cells that become doubly transformed for two unlinked markers with the number that would have been predicted to occur on the basis of probability theory. If all the cells in the population are competent, then the number of transformants acquiring both markers should equal the product of the numbers obtained when selection is made independently for the two markers. A higher than expected frequency of double transformants can be explained only if a sub-fraction of the population is competent (Goodgal and Herriot, 1961). The proportion of competent cells can be estimated from the formula:

$$f = N_1 N_2 / N_D B$$

where f = the fraction of competent cells; N_1, N_2 and N_D are the number of cells transformed with markers 1, 2, singly and both 1 and 2 together, respectively, and B is the total number of bacteria.

An additional and important parameter to consider in transformation systems is the efficiency with which added DNA is taken up and incorporated into the transformed cells. Transformation efficiency is normally expressed as transformants per microgram of DNA (Saunders *et al.*, 1987) when measured at sub-saturating concentrations of DNA (Figure 1).

When attempting to develop transformation systems for previously untransformed bacteria, the niceties of transformation kinetics may initially be ignored. It may be that all that is required is to achieve a means of constructing a single strain, containing, for example, a single plasmid. In such cases, a high efficiency of transformation is not necessary. In contrast, construction of gene libraries will require refinement of the system to give high efficiencies and frequencies of transformation.

3. Congression as a means of introducing unselected markers

Co-transformation (congression) is a useful procedure when faced with the need to introduce into a cell DNA molecules that lack readily identifiable or selectable phenotypes. It allows indirect selection, by exploiting the ability of a competent cell/protoplast/spheroplast or electroporated cell to take up more than one DNA molecule. For example, when *E. coli* cells were transformed with a small amount of the DNA of the genetically-marked indicator plasmid pSC201 (whose replication is temperature sensitive) together with an excess (10^3–10^5-fold) of DNA of a different plasmid lacking an identifiable marker, Kretschmer and co-workers (1975) demonstrated that 50–85% of transformants selected for acquisition of the marker plasmid also acquired the unmarked plasmid. Cells carrying the unmarked plasmid by itself were then isolated by growing double transformants at the replication-restrictive temperature of 42°C and subsequently eliminating pSC201 by segregation. A similar approach has been used with chromosomal DNA as indicator and a chromosomally-marked hyper-transformable recipient *E. coli* strain (Bergmans *et al.*, 1980). Selection for transformation by a specific chromosomal marker (for example, by complementation of auxotrophic markers on minimal salts agar), after transformation with chromosomal DNA in the presence of a 10–20-fold excess of unmarked plasmid DNA, produces about 5–10% co-inheritance of the plasmid. In this method, there is no need to remove the indicator plasmid from the co-transformed strain.

C. Natural Transformation

I. The role of natural transformation in bacterial populations

The ability to bind and take up exogenous DNA is widespread in bacteria, suggesting that it is an important mechanism for horizontal gene transfer in natural bacterial populations (Stewart, 1989; Lorenz and Wacknagel, 1994; Soloman and Grossman, 1996). Indeed, the ease of gene transfer mediated by transformation may provide certain bacteria with some of the benefits provided for eukaryotes by sexual reproduction (Redfield, 1988, 1993; Redfield *et al.*, 1997). The process of transformation is not only a frequently employed laboratory technique, but is also known to occur in many bacteria occupying natural environments and under conditions far from those experienced during exponential growth in rich laboratory media (see, for example, Rochelle *et al.*, 1988; Stewart, 1989; Khanna and Stotzky, 1992; Romanowski *et al.*, 1993;

Brautigam *et al.*, 1997; Sikorski *et al.*, 1998). In some species, for example the pathogenic *Neisseria* (Redfield, 1993; Facius *et al.*, 1996), gene transfer is extremely frequent and analysis of population genetics demonstrates free exchange of chromosomal genes, in contrast to the more familiar clonal structure exhibited by *Enterobacteriaceae* (Smith *et al.*, 1991, 1993; Bygraves and Maiden, 1992; Feavers *et al.*, 1992).

The mechanisms involved in natural transformation are complex and varied, being regulated by nutritional conditions and/or responses to cellular signalling events. It is by no means clear that the primary impetus for the evolution of natural transformation systems has been as a gene transfer mechanism. Various features of DNA uptake systems in transformable bacteria indicate that utilization of nucleic acids as a nutritional source may have been equally as important, and probably pre-dated a role in genetic exchange. Because of the highly variable nature of transformation mechanisms, there is no guarantee that individual strains of a given species will be transformable naturally, even if a type strain is so able. Moreover, natural competence has been well characterized in only a few organisms, primarily the Gram-positive bacteria *Bacillus subtilis* and *Streptococcus pneumoniae*, and the Gram-negative *Haemophilus influenzae* and *Neisseria* species.

2. Mechanisms of transformation in naturally competent bacteria

(a) *Gram-positive bacteria*

Many Gram-positive bacteria, for example, *Streptococcus mutans*, *Streptococcus pneumoniae*, *Streptococcus sanguis* and *Bacillus subtilis*, as well as Archaea such as *Methanococcus voltae* (Patel *et al.*, 1994), become competent naturally under normal growth conditions. The competent state in streptococci is transitory, develops in exponential phase cultures and is regulated by extracellular protein competence factors (CFs) of about 10 kDa. Competence occurs at a critical cell density depending on medium composition and the strain involved. The population density for competence development is markedly affected by starting pH (over the range 6.8–8.0). At lower initial pH, a single wave of competence occurs, but a series of competence cycles is observed in more alkaline media (Chen and Morrison, 1987). Competence factor accumulation is apparently dependent on pH and CF seems to govern the critical density for competence. Competence can be produced in otherwise non-competent streptococci by addition of purified CF (Yother *et al.*, 1986; Chen and Morrison, 1987). CFs from different strains of *S. sanguis* are biochemically similar, yet exhibit differing specificities in ability to induce competence in other strains (Gaustad, 1985). Addition of CF to cultures of *S. pneumoniae* induces increases in intracellular pH and Na^+ content, without changing either K^+ content or membrane potential (Lopez *et al.*, 1989). Such changes are accompanied by stimulation of glycolysis, leading to an increase in the ATP pool within cells.

Development of competence in streptococci is accompanied by the appearance of a number of new proteins (competence-induced proteins

(CIPs)) and an efficient DNA processing system for uptake and recombination (Vijayakumar and Morrison, 1986). The half-life of intracellular CIPs is short and their selective removal suggests the involvement of a specific protease within streptococci.

Competence of *B. subtilis* for transformation by either plasmid or chromosomal DNA normally develops slowly in batch culture, during transition from exponential to stationary phase (Contente and Dubnau, 1979a). However, it is possible to transform *B. subtilis* on agar plates, rather than in liquid culture, if desired (Hauser and Karamata, 1994). Competence is accompanied by the appearance of new proteins and reduction in the rate of RNA and DNA synthesis. Com⁻ mutants of *B. subtilis* are poorly transformable by both chromosomal and plasmid DNA and are defective in DNA uptake. Some mutants fail to bind significant amounts of DNA. Expression of *com* genes appears to be dependent on both growth conditions and genotype (Espinosa *et al.*, 1980). Competence develops late in growth and is maximal in simple media, such as salts–glucose minimal broth. In contrast, cells grown in complex media do not exhibit competence. Development of competence is dependent on *spoOH* (σ^{30}) and *spoOA* (required for control of sporulation and synthesis of products elaborated in stationary phase). The *com* genes are expressed, in a temporal sequence that precedes maximal competence, under transcriptional control (Smith *et al.*, 1985).

The induction of competence for transformation in exponentially growing cultures and subsequent autolysis in late exponential phase may be regarded as a stress response (Trombe *et al.*, 1994). All regulatory routes in the competence–signal–transduction network converge at the level of expression of the positive autoregulatory gene *comK*, which is identical to Competence Transcription Factor (CTF) (Van Sinderen *et al.*, 1995). ComK is required for transcriptional induction of *comK* itself, and also for late competence genes. These specify morphogenetic and structural proteins necessary for manufacture of the DNA-binding/uptake apparatus.

DNA-damage-inducible (*din*) genes are activated during the induction of competence, possibly implying a certain degree of overlap with induction of the SOS system (Love *et al.*, 1985). However, no induction or enhancement of competence by application of DNA damaging agents has been observed in either *B. subtilis* or *H. influenzae*, representatives of Gram-positive and Gram-negative bacterial species respectively, strongly suggesting that the primary function of bacterial transformation is not as an aid to DNA repair (Redfield, 1993, Redfield *et al.*, 1997).

The expression of competence genes in *B. subtilis* is controlled by a signal transduction cascade that increases expression of *comK* during transition from exponential growth to stationary phase (Kong and Dubnau, 1991; Soloman and Grossman, 1996). Transcription of *comK* is decreased by the product of the *mecA* gene, which interacts directly with ComK and whose inhibitory effect is relieved in response to a signal received from upstream in the regulatory pathway. Inactivation of *mecA*, or a further gene *mecB* (encoding MecB which functions prior to MecA in the signalling pathway) causes overproduction of ComK. The concentration of

MecA protein does not vary significantly as a function of growth, or in competent and non-competent cells. The model of Kong and Dubnau (1991) suggests sequestration of ComK by MecA-binding and the release of the transcription factor when the appropriate signal is relayed to MecA by MecB.

Mutants of the *comA* gene in *Streptococcus gordonii* Challis are completely deficient in transformation, exhibiting decreased levels of DNA-binding and hydrolysis (Lunsford and Roble, 1997). Synthesis of the single 6.0 kb *comA*-containing transcript occurs in cells grown under conditions promoting maximal competence, and its production is strictly dependent on exogenous competence factor for expression in ComA1⁻ cells. The deduced 319 amino acid ComA polypeptide exhibits high similarity to the transporter protein of *B. subtilis,* and to general secretory pathway components in Gram-negative bacteria. Putative ComC and ComY peptides also have leader sequences similar to those of type IV (*N*-methylphenylalanine) pilins of Gram-negative bacteria (Lunsford and Roble, 1997).

The regulation of competence for genetic transformation in *S. pneumoniae* depends on a quorum-sensing system comprising an extracellular peptide signal and an ABC-transporter required for its export (Pestova *et al.,* 1996). Competence for genetic transformation in *S. pneumoniae* arises in growing cultures at a critical cell density, in response to a secreted protease-sensitive signal. *S. pneumoniae* produces a 17-residue ribosomally – manufactured peptide (NH₂-Glu-Met-Arg-Leu-Ser-Lys-Phe-Phe-Arg-Asp-Phe-Ile-Leu-Gln-Arg-Lys-Lys-COOH) called ComC, that induces cells of this species to develop competence (Havarstein *et al.,* 1995). A synthetic peptide of the same sequence is biologically active in small quantities and can be used to extend the range of conditions suitable for development of competence in cultures. The *comC* gene encodes a pre-peptide containing the Gly-Gly consensus processing site, found in peptide bacteriocins, and the peptide is therefore likely to be exported by a specialized ATP-binding cassette transport protein characteristic of these bacteriocins. The allele, *comC2,* whose predicted mature product is a slightly different heptadecapeptide has been found (Pozzi *et al.,* 1996). Transcription of *comC* increases *ca.* 40-fold above basal expression level in response to exogenous synthetic activator. This is consistent with the activator acting autocatalytically (Pestova *et al.,* 1996). Two new genes, *comD* and *comE,* which encode members of the histidine protein kinase and response-regulator families, and that are linked to *comC* are also required for both response to synthetic activator peptide and endogenous competence induction.

There is no specificity for DNA uptake in transformation of *Bacillus* and *Streptococcus,* which incorporate DNA from their own species (homospecific transformation) and that from unrelated species (heterospecific transformation). Uptake of DNA in such genera involves membrane-bound nucleases that degrade one strand of double-stranded DNA during concomitant passage of the complementary strand across the cell envelope (Figure 2) (Lacks, 1977). (The presence of surface-located nucleases that degrade DNA as part of the uptake process is common in

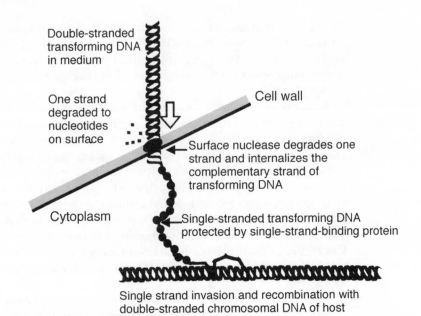

Double-stranded transforming DNA in medium

One strand degraded to nucleotides on surface

Cell wall

Surface nuclease degrades one strand and internalizes the complementary strand of transforming DNA

Cytoplasm

Single-stranded transforming DNA protected by single-strand-binding protein

Single strand invasion and recombination with double-stranded chromosomal DNA of host

Figure 2. Single-stranded DNA uptake following surface processing of double-stranded DNA in naturally competent Gram-positive bacteria.

naturally transformable bacteria. For example, during transformation of the cyanobacterium *Synechocystis*, DNA is converted from double-stranded to single-stranded, probably by a Ca^{2+}-dependent nuclease, located in the cytoplasmic membrane (Barten and Lill, 1995).) By the end of exponential growth, about one in ten cells of *B. subtilis* become committed to entering the pathway leading to competence. This commitment is the result of a signal transduction pathway involving a phosphorylation cascade, and the integration of signals reflecting the nutritional environment, population density, and growth stage (Van Sinderen *et al.*, 1995). However, the cardinal step in determining commitment to competence is transcription of *comK* (Kong and Dubnau, 1991; Hahn *et al.*, 1995).

A protein family has been identified and characterized that is responsible for elaborating the cell-surface-located DNA transport apparatus in *B. subtilis*, probably involving an aqueous pore formed by protein E3, through which DNA is transported by an ATP-driven DNA translocase (F1) (Inamine and Dubnau, 1995). Other proteins are probably involved in DNA-binding and its presentation to the uptake apparatus (Chung and Dubnau, 1995). A membrane-bound protein complex of 75 kDa is required for binding and entry of donor DNA in *B. subtilis* (Vosman *et al.*, 1988). The entry nuclease activity resides in a polypeptide that produces single- and double-stranded endonucleolytic breaks in plasmid DNA *in vitro*, provided Mg^{2+} ions are present. In contrast, total breakdown of DNA occurs in the presence of Mn^{2+} ions (Vosman *et al.*, 1988). DNA-binding and entry probably involves nicking of one strand of double-stranded donor DNA at, or close to, the site(s) of binding. An initial single-stranded nick is followed by a second break in the other strand at an opposite, or nearly opposite, position. Subsequently, the entry nuclease produces acid-soluble oligonucleotides by degradation of one of the strands, while the complementary strand enters the cell.

A membrane-bound polypeptide has also been implicated in DNA-binding during transformation of *S. pneumoniae*. The major endonuclease of *S. pneumoniae* is also required for DNA entry and seems to produce breaks opposite initial nicks that are introduced into donor DNA during binding. The resulting double-stranded breaks probably initiate the process of DNA entry *per se*. The *B. subtilis* nuclease is specific for double-stranded DNA, but in contrast, the entry nuclease of *S. pneumoniae* acts on both single- and double-stranded DNA. As a consequence, single-stranded DNA entering *B. subtilis* becomes refractory to nuclease activity. However, single-stranded DNA in streptococci is likely to remain susceptible to the entry nuclease unless protected in some way, possibly by complexing with single-strand binding proteins. Single-stranded DNA molecules are not capable of transformation if re-extracted from transformed cells at this stage, which has been referred to as eclipse phase. Single-stranded DNA bound to proteins forms an eclipse complex in which the DNA is protected against nucleolytic degradation. Following uptake, single strands of chromosomal DNA are inserted into homologous regions of the recipient chromosome in both streptococci and *Bacillus* (Lacks, 1962; Davidoff-Abelson and Dubnau, 1973; de Vos and Venema, 1982). The integration process, which is similar to that mediated by the *E.coli* RecA protein, begins within 1 min of the DNA entering the cytoplasm and is essentially complete within 12 min.

Covalently closed circular (CCC), but not open circular (OC) or linear plasmid DNA can be used to transform *B. subtilis*. Monomeric CCC forms of plasmids are inactive in transforming intact cells of *B. subtilis*. (Monomeric plasmid DNA can, however, transform protoplasts of *B. subtilis* (see below).) In contrast, plasmid oligomers (dimers, trimers, etc.) transform efficiently (Mottes *et al.*, 1979). A single DNA molecule of a plasmid oligomer is sufficient to produce a transformant (Contente and Dubnau, 1979a; de Vos and Venema, 1981; de Vos *et al.*, 1981). Plasmid DNA is processed extensively by *B. subtilis* during and immediately after uptake. Only single-stranded plasmid DNA is found inside cells immediately following exposure to monomeric plasmid DNA, but fully and partially double-stranded DNA is found additionally with oligomeric transforming DNA (de Vos *et al.*, 1981). Re-circularization of normal length plasmid molecules internal to the cell by re-annealing of partially homologous single strands generated during processing of surface-bound oligomers, probably accounts for plasmid transformation observed in competent *B. subtilis* cells (Michel *et al.*, 1982; Michel and Ehrlich, 1983; see also Figure 3).

Genetic markers on plasmid monomers can also be rescued by homologous recombination (*recE*-dependent) (de Vos and Venema, 1983) in competent *B. subtilis* cells that already carry a plasmid that exhibits sequence homology with the incoming plasmid (Contente and Dubnau, 1979b; Docherty *et al.*, 1981). The efficiency of such marker rescue is dependent on the extent of homologous sequences on the rescuing

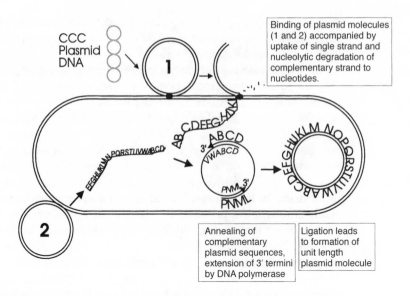

CCC Plasmid DNA

1 Binding of plasmid molecules (1 and 2) accompanied by uptake of single strand and nucleolytic degradation of complementary strand to nucleotides.

2 Annealing of complementary plasmid sequences, extension of 3' termini by DNA polymerase

Ligation leads to formation of unit length plasmid molecule

Figure 3. Restoration of functional plasmid molecules from re-annealed single-stranded molecules following transformation of *Bacillis subtilis*.

plasmid (Michel *et al.*, 1982). Monomeric CCC forms of plasmid DNA bearing cloned segments of *B. subtilis* chromosome, or (in a lysogen) fragments of bacteriophage DNA (Bensi *et al.*, 1981), are also effective substrates in transformation. The resulting homology can be used to direct the integration of recombinant plasmids into specific sites in the recipient chromosome (Iglesias *et al.*, 1981)

Transformation of streptococci with plasmid DNA shows similar low efficiency (10^{-3} to 10^{-4} transformants per viable cell at saturation) to that observed in intact *B. subtilis* (Barany and Tomasz, 1980). However, compared with streptococci, 10–50 times more plasmid DNA is required to saturate competent *B. subtilis*. CCC plasmid oligomers efficiently transform *S. pneumoniae* (Barany and Tomasz, 1980; Saunders and Guild, 1981a) and *S. sanguis* (Macrina *et al.*, 1981) with single-hit kinetics. In contrast to *B. subtilis*, CCC monomers are effective in transformation for both streptococcal species, but dose–response data indicate two-hit kinetics. Plasmid molecules linearized at a single restriction site are inactive in transformation of *S. pneumoniae*, whereas mixtures of molecules linearized at different sites are active (Saunders and Guild, 1981b). This indicates that two physically independent plasmid monomers that have become damaged during uptake, co-operate to produce a functional unit length plasmid molecule in every transformant. This rescue process is presumed to require annealing of overlapping single strands, followed by repair and ligation to produce a circular plasmid replicon. Damage to the incoming plasmid DNA may also promote recombination with the chromosome if there is sufficient homology (Pozzi and Guild, 1985).

Regulation of competence for genetic transformation in *S. pneumoniae* involves the *comAB* locus and a small extracellular protein, the competence factor (CF). Mutations in *comA* or *comB* block both spontaneous induction of competence and the elaboration of CF, yet permit competence induction by exogenous CF and subsequent normal transformation. ComA and the proteins, PdcD and LcnC, which act in the secretion of pediocin A-1 and lactococcin A, respectively, form a sub-family within the large ABC-transporter protein family. ComB is homologous to LcnD, which is required for secretion of the peptide antibiotic, lactococcin A. The *comAB* locus therefore displays homology to two lactococcin A secretion genes, but is devoid of additional linked competence genes. The mechanism of CF production is similar to that in the small peptide bacteriocins, lactococcin A and pediocin A-1, but its genetic organization is unusual in being split into at least two separate operons. In view of such findings, the presence of bacteriocin production might be a useful indicator of the likelihood of natural competence in strains of Gram-positive bacteria.

Competent cells must be sufficently metabolically active to produce the motive force for DNA uptake (Van Nieuwenhoven *et al.*, 1982). DNA transport in competent cells can be regarded as a homeostatic response that prevents autolysis and is linked to Ca^{2+} transport (Trombe *et al.*, 1994). Electrogenic and co-operative Ca^{2+} transport appears to mediate this Ca^{2+} response in *S. pneumoniae*. Some of these bacterial mutants had reduced co-operativity of Ca^{2+} uptake which was associated with an absolute requirement for added CF to develop competence and with resistance to autolysis. In other mutants, enhanced co-operativity of Ca^{2+} uptake was associated with facilitated competence and hypersensitivity to autolysis or with increases in CF response by competence-defective mutants.

(b) Gram-negative bacteria

Various Gram-negative bacteria are naturally transformable, often with very high efficiencies, for example, *Acinetobacter calcoaceticus* (Chamier *et al.*, 1993; Palmen *et al.*, 1993), *Helicobacter pylori* (Nedenskovsorensen *et al.*, 1990), *Campylobacter coli* (Haas *et al.*, 1993; Richardson and Park 1997), *Pseudomonas stutzeri* (Lorenz and Wackernagel, 1991; Sikorski *et al.*, 1998) and *Synechocystis* sp. (Barten and Lill, 1995). Competence in *Haemophilus influenzae* and *Haemophilus parainfluenzae* is a stable state that is induced following the transfer of exponentially growing cultures to medium that does not support growth. Competence is lost rapidly following the return of competent cells to a rich medium. In *H. influenzae*, competence for transformation is stimulated by cyclic AMP (cAMP) and requires cAMP-dependent catabolite regulatory protein CRP (Dorocicz *et al.*, 1993). Like *E. coli*, *H. influenzae* uses cAMP and CRP to regulate uptake and utilization of nutrients, lending support to the notion that uptake of DNA may primarily be a mechanism for nutrient acquisition. *H. influenzae cya⁻* mutants fail to develop competence either spontaneously, or after transfer to competence-inducing media, but become as competent as the wild type in the presence of exogenous cAMP. This suggests that failure of exogenously added cAMP to induce optimum competence in wild-type cells is not due to a limitation to entry of the metabolite into cells, but rather favours a model where competence induction requires both an increase in intracellular cAMP and a further, as yet unidentified, regulatory step.

The immediate benefits of DNA uptake are believed to be primarily nutritional. Competence may serve primarily to provide *H. influenzae* with the nucleotides that are released when the DNA fragments are hydrolysed within the cell, and that DNA repair and genetic recombination are possible secondary benefits (Redfield, 1993). Competence is probably controlled by two signals: a general signal, in the form of a rise in intracellular cAMP, occurring when adenylate cyclase is stimulated by the phosphotransferase system (PTS) in response to absence of the PTS-transported sugar fructose; and a specific signal arising when nucleotide pools become depleted. Both signals increase expression of the competence-regulating gene *sxy*: the general signal, by binding to the transcriptional regulator CRP and activating *sxy* transcription; the specific signal by limiting secondary structure in *sxy* mRNA, thus permitting its translation. According to this hypothesis, when *H. influenzae* cannot scavenge sufficient free nucleotides from the environment nor spare the energy to synthesize them *de novo*, exogenous DNA is taken up by the cell as a source of nucleotides. This view of competence predicts that recombination proficiency will not be directly co-regulated with competence. cAMP permits transcription of the competence activator gene *sxy* (Williams *et al.*, 1994) (also called *tfoX* (Zulty and Barcak, 1995), which encodes a 19 kDa basic protein with no evident processing signals or transmembrane domains, or homology to known proteins. Null mutations in *sxy* eliminate all competence (Williams *et al.*, 1994), and reduce competence-induced transcription of DNA uptake genes *com101A* and *dprA* (Zulty and Barcak, 1995). Point mutations in *sxy*, cause hyper-competence phenotypes, all by single-base substitutions that potentially disrupt secondary structure in *sxy* mRNA. The best characterized of these mutations is *sxy-1*, which causes moderate competence to develop under conditions that are not inducing for wild-type cells, and full competence under conditions that normally cause only partial competence (Redfield, 1991). The *sxy-1* mutation causes a 100–1000-fold increase in spontaneous natural competence (Williams *et al.*, 1994). However, elevated competence is absolutely dependent on cAMP and CRP. The *sxy-1* mutation increases DNA uptake and transformation and increases transcription of *com101A*, but does not affect growth or DNA repair (Redfield, 1991). The first step in competence induction may be a rise in intracellular cAMP, caused by PTS-dependent activation of adenylate cyclase when fructose is absent (Dorocicz *et al.*, 1993; Redfield, 1993; MacFadyen *et al.*, 1996). In the presence of cAMP, CRP allows initiation of *sxy* transcription, but Sxy is not expressed unless a second signal, the depletion of nucleotide pools, is present. By limiting/preventing formation of secondary structure in *sxy* mRNA, nucleotide depletion may allow transcription to produce functional mRNA. Sxy protein, in turn, may then activate transcription of DNA uptake genes, and the resulting uptake and degradation of DNA would relieve pressure on nucleotide pools (Redfield, 1993, Redfield *et al.*, 1997). A periplasmic oxidoreductase (product of *por*) is required for the correct assembly and/or folding of one or more disulphide-containing cell envelope proteins involved either in development of competence, or in the DNA-binding and uptake machinery (Tomb, 1992). The 20.6 kDa Por protein is highly similar to two other polypeptides, DsbA from *E. coli* and TcpG from *Vibrio cholerae*, which promote disulphide bond formation in periplasmic proteins, and are required for pilus biogenesis.

No extracellular competence factor has been detected in *Haemophilus* (Tomb *et al.*, 1989, 1991). Furthermore, the presence of pili (fimbriae), which are filamentous surface appendages of bacteria, seems to enhance transformability of *Haemophilus* spp. For example, piliated cells of *H. parainfluenzae* are transformed by chromosomal and plasmid DNA at frequencies that are 20 and 100 times greater, respectively, than non-piliated cells. Frequencies of transformation with plasmid DNA in both *H. parainfluenzae* and *H. influenzae* are in any case generally lower by several orders of magnitude than can be achieved with chromosomal DNA (Gromkova and Goodgal, 1979). Furthermore, in contrast to chromosomal DNA, transformations with circular or linearized plasmid DNA preparations are stimulated 50-fold by the addition of Ca^{2+} and/or Mg^{2+} ions (Gromkova and Goodgal, 1979). Further evidence for a separate system involved in chromosomal and plasmid transformation has come from mutant studies. A gene, *dprA*, has been identified that encodes a 41.6 kDa inner membrane protein required for efficient chromosomal, but not plasmid DNA transformation (Karudapuram *et al.*, 1995)

In contrast to normal transformation in *Haemophilus*, which exhibits marked specificity for the origin of the transforming DNA (see below), Ca^{2+}-stimulated transformation occurs with heterologous as well as homologous DNA. Stimulation of transformability in *H. parainfluenzae* occurs only in cells that are already competent for transformation with chromosomal DNA (Gromkova and Goodgal, 1981). This is in marked contrast to *E. coli* transformation in which only a small proportion of an otherwise non-competent population is rendered competent (see below). The capsular polysaccharide of *H. influenzae* seems to have no inhibitory effect on transformation (Rowji *et al.*, 1989). This is in contrast to *S. pneumoniae* where capsular polysaccharide inhibits the release of CF and the transformation efficiency of encapsulated strains is about fourfold lower than with unencapsulated variants (Yother *et al.*, 1986).

In the obligate human pathogens *Neisseria gonorrhoeae* (the gonococcus) and *Neisseria meningitidis* (the meningococcus) (Facius *et al.*, 1996), cells are highly competent at all stages of growth and can be transformed efficiently by homologous chromosomal DNA, but less efficiently with plasmid DNA (Biswas *et al.*, 1986) (see also Chapter 5,III.C.). DNA is taken up in double-stranded form (Biswas and Sparling, 1981; Scocca, 1990). Competence for transformation in *Neisseria* spp. is associated with the production of proteinaceous filaments (type IV pili), which extend from the cell surface. Expression of type IV pili is a prerequisite for infectivity in *N. gonorrhoeae* and *N. meningitidis* (Fussenegger *et al.*, 1997). Piliation is metastable in pathogenic *Neisseria*, with cells undergoing a phase change from piliated (P⁺) to non-piliated (P⁻) and vice versa. In order to be competent, gonococci must be piliated: non-piliated derivatives that cannot revert to produce pili are not transformable (Seifert *et al.*, 1988, 1990).

Most transformable bacteria will take up DNA from other species as well as DNA from their own species. For example, the naturally competent cyanobacterium, *Synechocystis* sp. is easily transformed by both homologous and heterologous DNA (Barten and Lill, 1995). Remarkably, *H. influenzae* (Scocca, 1990), *N. gonorrhoeae* (Dougherty *et al.*, 1979), *N. meningitidis* (Jyssum *et al.*, 1971) and *Campylobacter* spp. (Wang and Taylor, 1990) actively discriminate against heterospecific DNA and can naturally only take up DNA of the same or closely related species. DNA uptake by naturally competent cells of *H. influenzae* requires the presence of a specific base sequence (uptake site) on double-stranded DNA. Competent cells of *H influenzae* can bind up to 300 kb of double-stranded DNA. The uptake sequence (USS) required for transformation in *H.influenzae* is 5'-AAGTGCGGTCA-3' (Danner *et al.*, 1980; Pifer and Smith, 1985), and occurs about once for each 4 kilobase pairs (kbp) of *Haemophilus* DNA and 1465 times in the 1830 kb *H. influenzae* genome (Smith *et al.*, 1995). Fragments containing one or more copies of the core USS (5'-AAGTGCGGT-3') are taken up 10–100 times more efficiently than fragments without a USS (Goodgal and Mitchell, 1990). The sequence occurs randomly in heterologous DNA, about once every 300 kbp, which renders uptake of foreign DNA by *Haemophilus* unlikely. *Haemophilus* preferentially binds DNA fragments carrying the "core" uptake signal sequence (Goodgal and Mitchell, 1990), which is highly over-represented in the *H. influenzae* genome (Smith *et al.*, 1995). DNA uptake in *Haemophilus* involves recognition of the uptake sequence by surface components on competent cells (Danner *et al.*, 1980; Fitzmaurice *et al.*, 1984). Soon after binding, double-stranded DNA becomes incorporated into membrane vesicles called transformasomes and then becomes refractory to both DNase (Kahn *et al.*, 1983) and restriction endonucleases (Barany and Kahn, 1985). One strand of the DNA sequestered in the vesicle is degraded to produce a single strand, which is probably then transported across the membrane to pair with homologous regions of the recipient genome (Barany *et al.*, 1983). During translocation of DNA into the cytoplasm, which does not occur with circular DNA, the nucleotides resulting from degradation are reused for *de novo* DNA synthesis (Pifer and Smith, 1985).

Specificity for uptake in *Neisseria* spp. seems to be less stringent than in *Haemophilus* (Dougherty *et al.*, 1979) and a different uptake sequence of 10 bp (5'-GCCGTCTGAA-3') is required (Burnstein *et al.*, 1988; Goodman and Scocca, 1988). Furthermore, DNA uptake independent of sequence motifs is also observed (Boyle-Vavra and Seifert, 1996). The specific sequence for gonococcal DNA uptake is a frequent component of transcription terminator sequences in *N. gonorrhoeae* (Goodman and Scocca, 1988). This suggests that uptake sequences may have evolved as regulatory systems whose frequent occurrence was exploited to confer specificity in transformation. The presence of uptake signals in transcription sequences also allows the specific sequence to occur frequently without interfering with coding capacity. Furthermore, the location of an uptake site at gene boundaries would ensure that coding information was retained following uptake. However, this does not appear to be the case in

H. influenzae, where most USS do not occur in inverted-repeat pairs at the ends of genes (Smith *et al.,* 1995).

Neisseria spp. are not only highly competent, but are also very efficient at recombination using homologous chromosomal DNA. Indeed, there is good evidence that chromosomal transformation plays a very important role in the population biology of gonococci. Certainly, genetic exchange by transformation can be achieved in mixed cultures of *N. gonorrhoeae* (Frosch and Meyer, 1992; Rudel *et al.,* 1995; Facius *et al.,* 1996). Horizontal transmission of chromosomal DNA between gonococci appears to favour the spread of intact alleles, as opposed to expanding the allelic repertoire through the formation of gene mosaics (Hill, 1996). Transformation efficiency is drastically reduced in *recA⁻*gonococci (Koomey and Falkow, 1987; Biswas and Sparling, 1989).

The overall transformation process in *N. gonorrhoeae* can be divided into three distinct steps: (i) sequence-specific uptake of transforming DNA into a DNase-resistant state; (ii) the transfer of DNA to the cytosol; (iii) processing and recombination of incoming and resident DNA (Facius *et al.,* 1996). Two phase-variable pilus-associated proteins, the major pilus subunit (pilin – PilE), and PilC1/PilC2 (proteins involved in assembly and adherence of neisserial pili) are essential for transformation competence (Rudel *et al.,* 1995; Ryll *et al.,* 1997). The two pilus structural components appear to act on the bacterial cell surface and co-operate in DNA recognition and/or outer membrane translocation. PilE and PilC proteins are necessary for the conversion of linearized plasmid DNA carrying the *Neisseria*-specific DNA uptake signal into a DNase-resistant form, but production of typical pilus fibres is neither essential nor sufficient for the process. Mutants defective in *pilG,* a pilus-assembly gene, which is adjacent to the gonococcal *pilD* gene encoding the pre-pilin leader peptidase, are devoid of pili and display drastically reduced competence for transformation (Tonjum *et al.,* 1995). These findings cannot be accounted for by pilin-gene alterations, or by polarity exerted on *pilD* expression, suggesting a role for a further pilus gene in transformation of *Neisseria* spp.

Competence factors unrelated to pilus biogenesis, ComA, ComL and Tpc, are also required. These are not essential for DNA uptake, but act in a later stage. However, mutants in any of the three genes encoding these non-pilus products, lack the characteristic nucleolytic processing observed with incoming DNA in both wild-type and non-transformable *recA*-deficient *N. gonorrhoeae,* indicating a blockade in processing and/or the delivery of DNA to the cytoplasm. In view of the instability of piliation, and hence competence, of *Neisseria* spp., it is important to determine whether individual strains are piliated. For gonococci, and somewhat less evidently meningococci, it is possible to determine piliation phenotype by the appearance of colonies growing on clear typing agar (Mathis and Scocca, 1984). Piliated colonies tend to be smaller, dense and rounded, with characteristic highlights when illuminated from an angle above, whereas non-piliated colonies are larger, more diffuse and lack highlights. These differences may be partly obscured in capsulate organisms such as *N. meningitidis* (Blake *et al.,* 1989).

Transformation of *N. gonorrhoeae* or *Haemophilus* spp. with plasmid

DNA (even if containing the appropriate USS) is less efficient than with chromosomal DNA. This is partly due to restriction of plasmid DNA, even though the same plasmids may not be subject to restriction when entering by conjugation (Stein *et al.*, 1988; Stein, 1991). A high proportion of trans-formed bacteria contain deleted plasmids (Stuy, 1980; Saunders *et al.*, 1986). However, transformation frequencies can be increased if the recipient cell already contains a plasmid that is at least partly homologous to the incoming transforming plasmid DNA. This reflects a requirement for recombination to rescue plasmid molecules that become damaged during DNA uptake. Damage occurs in *N. gonorrhoeae* because circular plasmid molecules are apparently linearized by non-site-specific endonuclease action, during or immediately after uptake (Biswas *et al.*, 1986). Incoming plasmid molecules are presumably then degraded by exonucleases, possibly stimulating recombinational rescue by resident plasmids carrying homologous sequences. A less efficient process, occurring regardless of the presence of an indigenous homologous plasmid, involves perfect recircularization of incoming plasmid molecules that have escaped exonucleolytic processing (Saunders *et al.*, 1986). Nucleolytic erosion of linear plasmid molecules followed by recombinational recyclization is also responsible for the formation of plasmid deletants (see below).

D. Artificial Competence in Intact Cells

1. Competence induced by alkali metal ions

Much genetic manipulation in diverse prokaryotes requires intermediate cloning in *E. coli*. It is important to consider efficiency of *E. coli* transformation here in its own right and to inform work in a few diverse prokaryotes that can be transformed by CACl$_2$ methods. *E. coli* populations do not become naturally competent at any stage in growth. However, intact cells that are treated with CaCl$_2$ at around 0°C become competent for transformation by the DNA of lambda and other bacteriophages (Benzinger, 1978), by chromosomal (Cosloy and Oishi, 1973; Wackernagel, 1973) or plasmid DNA (Cohen *et al.*, 1972). Modifications of procedures employed for *E. coli* have been devised for transformation of a wide range of Gram-negative bacteria.

The generation of artificial competence in *E. coli* generally involves exposing early to mid-exponential phase cultures to 50–100 mM CaCl$_2$ at 0–4°C. The resulting competent cells (which, if stored at 4°C for several hours or even days, may actually demonstrate increased efficiencies of competence (Jones *et al.*, 1981; Chung *et al.*, 1989)) are exposed to the transforming DNA at 4°C for 30–45 min, followed by a short (1–2 min) heat pulse at 42°C. Transformed cells are then generally returned to growth medium for a short period to allow expression of selective markers (Saunders and Saunders, 1988). The CaCl$_2$ heat shock procedure causes losses in cellular viability of between 10 and 50%. Increasing the concentration of CaCl$_2$ normally increases transformation frequencies achieved, but high concentrations (>200 mM) cause increasing losses in viability (Weston *et al.*, 1981). Preferences for different divalent cations and optimal concentrations required vary considerably between species and strains of bacteria due to differences in cell envelope chemistry. Increased transformation frequencies can be obtained with combinations of $CaCl_2$ and $MgCl_2$ in *E. coli* (Bergmans *et al.*, 1980), and treatment with $MgCl_2$ prior to $CaCl_2$ results in improved transformation in both *Pseudomonas aeruginosa* and *Salmonella typhimurium* (Lederberg and Cohen, 1974). Pre-treatment with enzymes, such as α-amylase and protease, has been found to promote $CaCl_2$-mediated transformation of some bacteria, such as *Flavobacterium* sp., probably by removing surface components that would mask the outer membrane (Negoro *et al.*, 1980). Rubidium ions also have a stimulatory effect in divalent cation-mediated transformation (Kushner, 1978; Bagdasarian and Timmis, 1981). A complex procedure that involves treatment of *E.coli* with Ca^{2+}, Mn^{2+}, Rb^+, hexamine cobalt(III), dimethyl sulphoxide and dithiothreitol has been found to give reproducibly high efficiencies of transformation (>10^9 transformants per microgram plasmid DNA) (Hanahan, 1983), and is the basis of preparation of some commercially available competent cells of *E. coli*.

Transformation frequencies vary over a range of 100 during batch culture of *E. coli* (Brown *et al.*, 1979; Hanahan, 1983; Saunders *et al.*, 1987). Transformation frequencies rise to a maximum in early exponential phase, and then decline dramatically. Only about 5–12% of the bacteria in a population of *E. coli* can be transformed under ideal conditions (Hanahan *et al.*, 1991). The $CaCl_2$ treatment procedure is inefficient at inducing competence by the standards of many natural transformation systems. Even under optimal conditions, only 10^{-2} to 10^{-3} of plasmid DNA molecules in a transformation mixture actually give rise to a transformant (Hanahan, 1983). Nevertheless, individual competent cells may be capable of taking up more than one molecule. There is no evidence for a protein competence factor in *E. coli*, but maximal competence occurs in batch growth when the mean cell volume of the culture is at its highest (Saunders *et al.*, 1987), and cells are, incidentally, most susceptible to killing by divalent cations at concentrations used to induce competence (Brown *et al.*, 1979). Maximal growth rates in batch or continuous cultures usually produce bacterial populations containing highest numbers of competent cells (Jones *et al.*, 1981; Hanahan, 1983; Saunders *et al.*, 1987). There is a direct relationship between modal cell volume (which is maximal at higher growth rates) and maximum transformability, which may result both from stress of the cell envelope and the increased surface area available for binding and uptake of DNA (Saunders *et al.*, 1987).

Divalent cations may be required to promote DNA binding to cell membranes by neutralization of negative charges prior to DNA transport (Weston *et al.*, 1981). Ca^{2+} produces rearrangements of the lipopolysaccharide and outer membrane that may promote binding and subsequent transport of DNA molecules to the interior of the cell. Ca^{2+} treatment at low temperature also causes freezing of membrane lipids. Sudden warming during the heat pulse stage could cause unfreezing of the lipids and subsequent ingestion of any DNA molecules that happen to be bound to the cell envelope. Ca^{2+}–DNA complexes may also be transported down a Ca^{2+} gradient extending across the inner membrane to the cell interior (Grinius, 1980). Plasmid DNA responsible for the production of transformants is known not to enter the bacterial cell until the heat pulse has occurred (Weston *et al.*, 1981). Nevertheless, a heat pulse carried out in the absence of DNA does render *E. coli* transformable by chromosomal DNA, a phenomenon that suggests that the induction of competence is distinct from the process of DNA uptake *per se* (Bergmans *et al.*, 1980). A complex of polyhydroxybutyrate (PHB), Ca^{2+} and inorganic polysulphate (polyP) has been proposed as the membrane component responsible for competence for DNA entry in *E. coli* (Reusch and Sadoff, 1988). A novel form of polyP comprising about 60–70 residues has been identified in chloroform extracts of competent cell membranes, in a stoichiometric ratio of PHB to polyP of 2 : 1 (Castuma *et al.*, 1995). Fluorescent lipid probes demonstrate that there are extensive rigid domains in the membranes of competent cells, leading to the proposal by Castuma and co-workers (1995) that a complex of PHB, Ca^{2+} and polyP perturbs the conformation of the lipid matrix, making it more permeable to charged molecules, thus allowing DNA entry.

Covalently closed circular, open circular and linear plasmid DNA molecules are taken up by *E. coli* with approximately equal efficiency and with single-hit kinetics (Cohen *et al.*, 1972; Conley and Saunders, 1984). Thus, one plasmid molecule is sufficient to transform a single cell. DNA is taken up in the form in which it is presented: therefore, double-stranded circular DNA is taken up without being converted to a single-stranded form, which is in contrast to mechanisms encountered in some naturally transformable bacteria. There is no sequence specificity for DNA uptake and heterologous DNA molecules are taken up without discrimination (Brown *et al.*, 1981; Hanahan, 1983). Also, around 10–90% of transformable cells, depending on the strain and regime used to induce competence, can become doubly transformed by two independently marked plasmids (Weston *et al.*, 1979; Hanahan, 1983).

Transformation of wild-type *E. coli* with chromosomal DNA fragments is prevented by the exonucleolytic activities of exonuclease V, the *recBCD* product (Cosloy and Oishi, 1973; Wackernagel, 1973; Hoekstra *et al.*, 1980; Rinken and Wackernagel, 1992). This presents a severe limitation to chromosomal transformation in this species, but it can be overcome by utilizing *recBC-* or *recBCD*-deficient strains (lacking exonuclease V activity), that also contain *sbcB* or *sbcA* mutations opening up alternative recombination pathways (*recF*, or *recF* and *recE* systems, respectively) to ensure recombinational integration of donor chromosomal DNA. Transformation of wild-type *E. coli* with linearized plasmid DNA occurs at frequencies that are around 100 times lower than those observed with CCC or OC molecules of the same plasmid (Cohen *et al.*, 1972; Conley and Saunders, 1984). In contrast to chromosomal transformation, absence of *recBC* reduces the efficiency of transformation of *E. coli* with linear plasmid molecules. This is probably because the exonucleolytic action of the enzyme may be required for the *in vivo* recircularization of plasmid molecules (Conley and Saunders, 1984; Conley *et al.*, 1986a). Many other recombination functions are required for efficient recyclization of linearized plasmid DNA following transformation (McFarlane and Saunders, 1996)

Treatment of naturally competent species, for example, *Acinetobacter calcoaceticus* (Lorenz *et al.*, 1992), with elevated concentrations of divalent cations may dramatically stimulate competence levels. In *A. calcoaceticus*, uptake of DNA into a DNase-I-resistant state, but not binding of DNA to cells, is strongly stimulated by divalent cations: an increase of nearly three orders of magnitude in transformation frequency was obtained in response to 0.25 mM Ca^{2+} (Lorenz *et al.*, 1992). DNA competition experiments indicate that *A. calcoaceticus* does not discriminate between homologous and heterologous DNA, nor between linear and circular DNA molecules. High efficiency plasmid transformation was obtained with *A. calcoaceticus* in non-sterile natural groundwater or extracts of fresh or air-dried soil. Competent cells of *A. calcoaceticus* can also take up any free DNA without discrimination, including that of plasmids from natural environments, such as soil, sediment and groundwater (Lorenz *et al.*, 1992).

2. Polyethylene glycol treatment of intact cells

Intact cells of *E. coli* may be transformed after treatment with polyethylene glycol and concentrations of Mg^{2+} or $Ca^{2+} < 30$ mM at comparable efficiencies to those obtained with $CaCl_2$ treatment alone (Klebe *et al.*, 1983). PEG-induced transformation occurs optimally at room temperature and is inhibited by the low temperatures used in the $CaCl_2$ method.

PEG acts to concentrate DNA by excluding water and may therefore increase the chances of interactions between DNA molecules and the cell surface. It promotes plasmid transformation in *Mycoplasma mycoides* (King and Dybvig, 1991), a bacterium lacking a conventional rigid cell wall. High concentrations of Tris (500 mM) with $CaCl_2$ and PEG also induce competence in intact cells of the photosynthetic bacterium *Rhodobacter sphaeroides* (Fornari and Kaplan, 1982). Tris disrupts the lipopolysaccharide outer membrane of Gram-negative bacteria and may assist the action of Ca^{2+} and PEG in rearranging the structure of the cell envelope. Competence can also be induced using Tris in Gram-positive bacteria, for example *Bacillus anthracis* (Quinn and Dancer, 1990), *Bacillus brevis* and *Bacillus thuringiensis* (Heierson *et al.*, 1987). Addition of $MgCl_2$ or $CaCl_2$ is inhibitory for *B. brevis* transformation and reduces transformation frequency in *B. thuringiensis*. CCC DNA is more efficient than linear DNA in transforming Tris-treated *B. thuringiensis*, but oligomers give decreased transformation frequencies compared with CCC monomers. This is in contrast to natural transformation of *B. subtilis*, where CCC monomers are very poor at producing transformants (see section II. C.2 (a)).

3. Transformation with liposome-entrapped plasmid DNA

Biologically active plasmid DNA entrapped in lipid vesicles (liposomes) can be used to transform intact Ca^{2+}-treated *E. coli* cells (Fraley *et al.*, 1979). However, efficiencies of transformation are only about 1% of those achieved with free DNA. High frequencies of transformation of *Streptomyces* protoplasts (see below) can be achieved by encasing chromosomal DNA in liposomes (Makins and Holt, 1981). DNA-free liposomes also stimulate transformation in *Streptomyces* (Rodicio and Chater, 1982) and *Streptococcus lactis* (Van der Vossen *et al.*, 1988) protoplasts. Liposome transformation systems have not proved to be widely applicable, but have been useful in bacterial species that are normally non-transformable due to the production of extracellular nucleases (see, for example, Fromm *et al.*, 1995; Metcalf *et al.*, 1997; Chapter 10,III.B).

4. Transformation of frozen and thawed bacteria

E. coli can be transformed without added Ca^{2+} by freezing the cells with DNA at $-70°C$ or $-196°C$, and then thawing at $42°C$ (Dityatkin and Iliyashenko, 1979). However, this regime reproduces many features of Ca^{2+} treatment, notably the sequential freezing and unfreezing of membrane lipids, producing sufficient stress on the cell envelope to allow penetration of DNA. In *E. coli*, transformation frequencies obtained by freeze–thaw are much lower than those achieved with $CaCl_2$ heat shock. However, freeze–thaw methods have proved valuable when alternatives were not available, for example, in *Agrobacterium tumefaciens* (Holsters *et al.*, 1978) and *Sinorhizobium* (*Rhizobium*) *meliloti* (Selvaraj and Iyer, 1981). Freeze–thawing following treatment with glycine, which reduces the cross-linking of peptidoglycan, has also permitted transformation of *B. anthracis* (Stepanov *et al.*, 1990).

5. Electroporation

(a) Electroporation procedures

Electroporation is the process of subjecting cells to a rapidly changing electrical field of high voltage, which transiently generates pores in cell membranes. This method has proved highly effective in inducing plant and animal cells to take up DNA (see, for example, Fromm *et al.*, 1985; Toneguzzo and Keating, 1986). However, the electrical conditions used so successfully for higher eukaryotes produced relatively inefficient transformation of bacteria, when compared for example with divalent cation treatment. This was due largely to the smaller size of bacteria, which means that they require a much larger electrical field, of several kilovolts per centimetre, to induce poration. The first application of electroporation to bacteria was of protoplasts from *Bacillus cereus* (Shivarova *et al.*, 1983). Transformation of intact bacterial cells by electroporation was initially reported in *Lactococcus lactis* (Harlander, 1987). A wide variety of Gram-positive and Gram-negative bacteria has subsequently been transformed in this way, including many organisms that are not naturally competent for DNA uptake. In many cases, electroporation is much more efficient than alternative methods. For example, under optimal conditions, transformation efficiency by electroporation is five orders of magnitude higher than the conventional $CaCl_2$ treatment procedure for *Serratia marcescens* (Sakurai and Komatsubara, 1996).

The efficiency of transformation by electroporation depends on various factors, including the status of the recipient cells and the quality of the transforming DNA (see Table 1). Generally, there is no necessity to remove the cell wall completely prior to electroporation. However, growth conditions that result in increasing fragility of the cell wall, or gentle enzymatic digestion enhance DNA uptake in Gram-positive bacteria. For example, *B. subtilis* 168 *trp* was found to be transformable with the tetracycline-resistance plasmid pAB124 at very low frequencies (McDonald *et al.*, 1995). However, supplementing the growth medium with glycine, or especially DL-threonine, allowed much more efficient electrotransformation at frequencies of up to 2.5×10^3 transformants per microgram of plasmid DNA. Transformation was optimal with cells grown in medium containing a racemic mixture of the D- and L-threonine isomers; no transformants were obtained when pure D- or L-threonine was used. The same procedure works for other *Bacillus* strains, but with variable success.

The time of harvesting of bacteria is critical for achieving maximal transformation efficiency (see, for example, Tyurin *et al.*, 1996), as is the case for many other transformation procedures. For *E. coli* the precise time varies with strain and growth medium, but cells harvested in exponential phase seem to give the best efficiencies. Cells that are harvested at other times may show efficiencies that are reduced by as much as three orders of magnitude (Calvin and Hanawalt, 1988). Optimal conditions for electroporation of competent cells depend on the organism involved and can vary between strains of the same species, depending on the genetic diversity involved. The cell density or growth phase is a significant factor for successful electrotransformation (Shigekawa and Dower, 1988; Dower *et al.*, 1998, 1992; Nickoloff, 1995) and in natural or other forms of artificially induced competence. For example, a linear relationship was observed between growth phase and transformation efficiency up to OD_{600} of 2.0 when electroporating *Pseudomonas putida* cells (Cho *et al.*, 1995). *Borrelia burgdorferi* (Samuels and Garon, 1997) will either not transform efficiently or not plate efficiently if the cell density is too high (Samuels, 1995). At lower cell densities ($1–2 \times 10^7$ cells cm^{-3}), pelleting the cells at a fairly high g force (up to $5000 \times g$) and adjusting the final volume of the cell suspension may assist in raising the frequencies obtained. However, higher cell densities can be used successfully for preparing competent cells of *Borrelia* (Rosa *et al.*, 1996). Thorough washing is important to remove components of the medium. Maintaining the competent cells at 4°C generally yields optimal efficiencies (Shigekawa and Dower, 1988; Dower *et al.*, 1992; Nickoloff, 1995). Electroporation in the presence of high ionic strength solutions causes electrical arcing (and a lowered time constant) which may kill all of the cells.

The electroporation medium often contains sucrose or glycerol and the procedure is generally performed at 0–4°C. However, in the case of mycobacteria elevated temperature (37°C) is required for efficient electrotransformation of slow-growing strains, contrasting with the low temperature (0°C) for fast-growing strains (Ward and Collins, 1996). This implies differences in cell wall structure between such strains which affect DNA uptake. A range of electrical field strengths is used for electroporation,

Table 1 Selected factors affecting transformation by electroporation

Factor	Comment	Reference
Status of recipient cells		
● Growth medium	Solid medium more effective for *Clavibacter michiganensis*.	Laine *et al.*, 1996
	Avoid extensive pellet formation for streptomycetes.	Pigac and Schrempf, 1995
	Glycine requirement, e.g.	
	Streptococcus spp.	Dunny *et al.*, 1991
	Rhodococcus spp.	Singer and Finnerty, 1988
	Corynebacterium glutamicum	Haynes and Britz, 1990
	Slow-growing mycobacteria	Ward and Collins, 1996
	Glycine independent, e.g.	
	Streptomyces spp.	Pigac and Schrempf, 1995
	Bacillus spp.	Mahillon *et al.*, 1989
	Acetobacter woodii	Stratz *et al.*, 1994
	Isonicotinic acid hydrazide, e.g.	Kawagishi *et al.*, 1994
	Thermoanaerobacterium sp.	Mai *et al.*, 1997
● Growth phase	Early exponential, e.g.	
	E. coli	Calvin and Hanawalt, 1988
	Bifidobacterium spp.	Rossi *et al.*,1997
	Late exponential, e.g.	
	Corynebacterium glutamicum	Haynes and Britz, 1990
	Rhodococcus spp.	Singer and Finnerty, 1988
	Streptomyces spp.	Pigac and Schrempf, 1995
	Early stationary, e.g.	
	Marine *Vibrio* spp.	Kawagishi *et al.*, 1994
Pretreatment of recipient cells		
● Agents interfering with murein synthesis, e.g. lysozyme	For several Gm +ve bacteria, e.g.	
	Streptomyces spp.	Pigac and Schrempf, 1995
	Brevibacterium lactofermentum	Bonnassie *et al.*, 1990
● Polyethylene glycol	For increased survival, interaction with cell membrane, volume exclusion, e.g.	Hui *et al.*, 1996
	Streptomyces spp.	Pigac and Schrempf, 1995
	Rhodococcus fascians	Desomer *et al.*, 1990
	Lactobacillus hilgardii	Josson *et al.*, 1989
	Bacillus thuringiensis	Mahillon *et al.*, 1989
● Low temperature	Freezing in glycerol, storage at $-70°C$, e.g.	
	Bifidobacterium longum	Argnani *et al.*, 1996
● Osmotic shock	Permeabilize outer membrane and removal of DNase in marine *Vibrio* spp.	Kawagishi *et al.*, 1994
Treatment post-electroporation		
● Medium	Rapid regeneration of damaged cells required, e.g. Mg^{2+} and	

generally from 5 to 16.7 kV cm^{-1}. However, *E. coli* can be transformed, although with reduced efficiencies, at lower field strengths (3.5–4.0 kV cm^{-1}), as can *Streptococcus* spp.

The formation of pores by electroporation is reversible, with the membrane resealing to restore membrane integrity within seconds, minutes or hours, depending on the temperature. However, a proportion of the cells is killed by the process. This depends on the electrical parameters and procedures used, following electroporation, to regenerate the damaged cells. This is manifested as a reduction in colony-forming ability, which seems to occur as a consequence of cellular filamentation. Inclusion of pantoyl lactone, an inhibitor of filamentation, into the selective agar plates used results in a threefold increase in the recovery of transformants. Competent cell volume together with the presence of polyethylene glycol 6000 and single-stranded DNA during electrical pulsing were found to be critical for producing high frequencies of 10^7–10^8 transformants per microgram of plasmid DNA in *B. brevis* (Okamoto *et al.*, 1997).

Various protocols for electrotransformation of bacteria are given by Weaver (1995). Efficiencies vary greatly between species and strains, ranging from about 10^{10} transformants per microgram of DNA for *E.coli* to 20 transformants per microgram in certain staphylococci. However, most values are in the range 10^4–10^8 transformants per microgram. Electroporation efficiency generally increases with DNA concentration and the topology of the DNA influences uptake. For example, during electroporation of *B. subtilis*, transformation by circular, supercoiled plasmid DNA is approximately an order of magnitude more efficient than circular DNA that had been relaxed with topoisomerase I, and no transformants were obtained with linearized plasmid DNA (Ohse *et al.*, 1997).

(b) Mechanism of DNA uptake

The interaction of DNA molecules with differing cell wall components plays an important role in determining successful penetration of DNA through the electroporated cell envelope (Rittich and Spanova, 1996). The mechanism of entry of DNA by electrotransformation is not completely clear. The application of a brief, high voltage pulse appears to cause structural rearrangements of the membrane by local disruption of the lipid bilayer. This creates aqueous pathways, resulting in enhanced permeability, particularly to charged ions and molecules. The transmembrane voltage may also provide a local driving force for transport across the membrane. Membrane breakdown can occur at electrical field strengths of as low as about 3.5 kV cm^{-1}, as field strength increases, transformation efficiency increases, with pore formation occurring at multiple sites on the cell membrane.

The first demonstration of the physical effects of electroporation on bacterial membranes is shown in Figure 4 for *Borrelia burgdorferi* (Samuels and Garon, 1997) (See also Chapter 7, III.B. in this volume. Electron microscopy reveals dark staining regions on the surface of the spirochaete, when subjected to electroporation, but not on the controls. These regions may represent the pores, induced by the electrical pulse, for entry of the DNA.

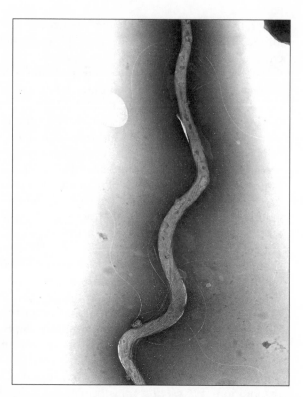

Figure 4. Surface perturbations on *Borrelia burgdorferi* following electroporation. Bacteria that had been subjected to electroporation were negatively stained with ammonium molybdate and subjected to transmission electron microscopy. Note the darkly stained regions which correspond to electrically induced damage, and the flagella which are often liberated by the electroporation procedure. (*Courtesy of D. Scott Samuels and Claude Garon.*)

Electroporation provides a means of avoiding the sequence specificity inherent in some natural transformation systems, for example *H. influenzae* (Setlow and Albritton, 1992). Thus, plasmid DNA can be introduced by electroporation into the competence-proficient *H. influenzae* strain Rd and its competence-deficient uptake mutants, as well as strains Ra, Rc, Re and Rf which are not naturally transformable. Plasmid DNA without USS can also be electroporated into both competent and non-competent cells of naturally transformable bacteria. Indeed, competent cells are several orders of magnitude less efficient than non-competent *H.influenzae* cells for electroporation of plasmid DNA. Frozen "electrocompetent" cells of *Streptococcus agalactiae* (group B) can be prepared by repeated washes in 10% (v/v) glycerol (Ricci *et al.*, 1994). Freezing *S. marcescens* cells in glycerol (100 g L^{-1}) and storage at –80°C has been shown not to affect transformation efficiency adversely (Sakurai and Komatsubara, 1996).

The activities of restriction enzymes and DNase could represent a significant barrier to successful transformation by electroporation in some organisms such as *S. pneumoniae*, *Corynebacterium glutamicum*, *Brevibacterium flavum* and *E. coli*. While in others, including *Lactococcus lactis, Brevibacterium lactofermentum* and *Yersinia* spp. the transforming DNA appears to escape restriction (Lefrancois and Sicard, 1997). This may be due to inactivation of the restriction system during electric shock. Alternatively, a mechanism may operate that protects the incoming DNA from degradation in such species. In *S. pneumoniae* the restriction-modification systems, *Dpn*I and *Dpn*II, degrade methylated and non-methylated DNA respectively, acting only on double-stranded DNA. Plasmid DNA entering by electroporation is fully restricted in restriction-proficient cells, implying that the DNA enters as presented, in double-stranded form, without being converted to single strands. In this respect electroporation appears to be like transformation of Ca^{2+}-treated *E. coli* and in contrast to natural transformation of *S. pneumoniae* or *B subtilis* (see sections II.C.2 (a) and II.D.1).

By increasing the strength of electrical field (to 14 000 V cm^{-1}) and the duration of pulses, it has proved possible to transform *E. coli* with efficiencies of up to 5×10^9 per microgram of plasmid DNA, which exceeds those routinely achieved by treatment with Ca^{2+} ions (Calvin and Hanawalt, 1988). Perhaps not surprisingly, the generation of pores in the bacterial membrane leads to the release of plasmid DNA molecules already present in cells. This can be exploited to extract plasmid DNA, relatively free of RNA and other cellular components, from electroporated cells.

(c) Electroporation in gene manipulation

Electroporation has become the mechanism of choice for introducing recombinant DNA into a wide diversity of bacteria. Few naturally transformable species have been described among commercially important bacteria. Electroporation therefore provides a rapid, efficient and reproducible method for introducing DNA into many organisms, and has been instrumental in rendering a wide range of bacteria amenable to genetic investigation. The relatively high transformation efficiencies achieved by electroporation compared with other transformation procedures make the technique particularly useful for the performance of various gene manipulations, and it can serve as an alternative to transduction or conjugation for transfer of genomic DNA. The use of homologous recombination strategies for making chromosomal insertions demands high transformation efficiencies, particularly where high integration efficiency, high specificity and high stability of integrants are required. Electrotransformation has proved applicable to site-specific integration of genes and plasmids into the chromosome in organisms such as *Lactobacillus sake* (Berthier *et al.*, 1996). Additionally, site-specific mutations have been introduced directly by recombination of oligonucleotides with the host genome in *Borrelia burgdorferi* (Samuels and Garon, 1997). Single-stranded antisense oligonucleotides were more efficient in transformation than sense or double-stranded forms. However, different strand preferences may be found in different organisms.

In electroporation of *E. coli*, single-stranded circular phage DNA is less infective than double-stranded replicative DNA (Kimoto and Taketo,

1996). Linearized double-stranded DNA of phages or plasmids exhibits reduced efficiency in electroporation of *E. coli* (Kimoto and Taketo, 1996; McFarlane and Saunders, 1996). The use of electroporation has been shown to reduce the frequency of unwanted deletion mutations encountered using conventional transformation in some organisms, for example *Shigella flexneri* (Porter and Dorman, 1997).

6. Transformation of spheroplasts

Spheroplasts are bacterial cells that have been treated to remove part of the cell envelope and expose areas of cytoplasmic membrane. *E. coli* spheroplasts can be transformed when osmotically buffered in the presence of low concentrations of divalent cations (Benzinger, 1978). However, spheroplasts generated conventionally by lysozyme treatment cannot regenerate functional cell envelopes to allow sufficient bacterial growth and division to form colonies. Consequently, such spheroplasts cannot be used for transformation by chromosomal or plasmid DNA, where a colony is required for selection. Certain mutants of *E. coli*, with temperature-sensitive defects in peptidoglycan biosynthesis, produce spheroplasts that can regenerate to form colonies on the surface of agar when returned to permissive temperatures (Suzuki and Szalay, 1979). Such spheroplasts can be transformed with similar efficiencies to those obtained following Ca^{2+} treatment of intact cells. Similar approaches are also applicable to other bacteria into which suitable temperature-sensitive mutations can be introduced such as *Aerobacter aerogenes, H. influenzae, H. parainfluenzae, Klebsiella pneumoniae, Micrococcus luteus S. typhimurium* (Suzuki and Szalay, 1979; Mycobacteria Chapter 9, III.B in this volume).

7. Transformation of protoplasts

Protoplasting is very effective in removing barriers to transformation and other forms of genetic exchange, many of which reside in the surface layers of the cell (see, for example, Regue *et al.*, 1992; Chapter 6, III.B and for Archaea Chapter 10, IV.C). The development of protoplast-based systems has proved successful in producing efficient transformation in bacterial species that could not be made competent by alternative methods. Protoplast transformation procedures have often been derived from protoplast fusion protocols designed to permit gene transfer by cell fusion. In protoplasts, all of the cell wall material is removed, thus exposing the cell membrane to direct access by added DNA. Uptake of plasmid DNA by bacterial protoplasts was initially described in *Streptomyces coelicolor* and *Streptomyces parvulus* (Hopwood, 1981). PEG-protoplast transformation procedures for *B. subtilis* (Chang and Cohen, 1979) have been modified for the introduction of plasmid DNA into various Gram-positive bacteria, notably, *Bacillus thuringiensis* and *Bacillus megaterium* (Brown and Carlton, 1980), and *Lactobacillus* spp. (Morelli *et al.*, 1987).

Protoplasts are normally generated by enzymatic degradation of the cell wall in osmotically buffered media. This involves the use of lysozyme or similar enzymes that degrade the peptidoglycan layer of the bacterial cell. Once produced, protoplasts must be protected from lysis by including osmotic stabilizers, such as mannitol, sorbitol or NaCl. Stabilizer use is dependent on absence of toxicity and/or inability of the bacterial strain concerned to metabolize the relevant carbon compound. Some stabilizers also affect the expression of selectable resistance genes, for example those to aminoglycosides. Protoplasts are subsequently exposed to transforming DNA in the presence of PEG. Requirements for ideal polymer chain length and concentration of PEG vary between different bacteria, and some period of experimentation and optimization will normally be required when devising a new system (see Chapter 10, IV.C). The PEG is finally washed out, and the protoplasts incubated to allow expression of transformed markers, and then inoculated on to a complex, osmotically stabilized solid growth medium that favours regeneration of a functional cell wall.

Many factors influence the successful transformation of protoplasts, notably growth phase of starting cells or mycelium at the time of protoplasting, the incubation temperature for regeneration of protoplasts, and the composition and degree of dryness of the regeneration medium. Protoplasts of *B. subtilis* are transformed at the same efficiency by plasmid monomers as by dimers or higher oligomers (de Vos and Venema, 1981). Linear monomers, open circular and non-supercoiled DNA molecules produce transformants, but at frequencies of about 10–1000 times lower than observed with native plasmid DNA molecules (Chang and Cohen, 1979). Plasmid DNA enters protoplasts in a double-stranded and largely CCC form, with each molecule entering being sufficient to produce one transformant.

◆◆◆◆◆◆ III. FORMATION OF *IN VIVO* DELETIONS

A. Deletion following Transformation

In some bacteria, largely those that are naturally competent such as *N. gonorrhoeae* (Biswas *et al.*, 1986) and *B. subtilis* (Bron *et al.*, 1988), restriction and/or deletion may occur spontaneously during or immediately following DNA uptake. Deletions and other rearrangements of the transforming DNA are frequently encountered following any transformation technique, or indeed other forms of gene transfer (see, for example, Porter and Dorman, 1997). This appears to result from DNA processing as a consequence of the DNA uptake mechanism, or from the induction of DNA repair processes as the host cell recognizes incoming DNA as worthy of attention. In many cases, deletions of transformed DNA are a nuisance that must be tolerated. The best solution to the problem is to check all constructs, whether plasmidic or chromosomal, for integrity before making use of them in subsequent experiments. However, the tendency of transformed DNA to become deleted, may in some cases be exploited as a tool in genetic manipulation and strain construction.

B. *In vivo* Deletion in *Escherichia coli*

Deletion of circular transforming DNA does not occur in intact Ca^{2+}-treated or electroporated cells of *E. coli*, or following protoplast or spheroplast transformation in most species (Conley *et al.*, 1986a; McFarlane and Saunders, 1996). To obtain deletants deliberately in *E. coli*, it is necessary to use plasmid DNA linearized by limited mechanical shearing or by cleavage with restriction endonucleases. Nucleolytic processing of linear DNA molecules occurs *in vivo* following entry into transformed cells. Most of the entering DNA molecules are degraded and do not produce transformants. However, if only limited erosion of termini occurs, then *in vivo* circularization of a small proportion of molecules can generate functional plasmid molecules that contain deletions of varying size that extend on either side or both sides of the site used for linearization. Deletants may also be produced *in vitro* by exonucleolytic digestion and ligation prior to transformation.

Transformation of bacteria with linearized plasmid DNA molecules occurs accidentally during transformation with *in vitro* recombinant DNA ligation mixtures. Transformation of *E.coli* with linearized plasmid DNA is reduced 100–1000 times compared with equivalent circular molecules (Conley and Saunders, 1984). Most resulting transformants are perfectly recircularized, losing no nucleotides from their exposed termini. However, about 10% contain deletions that extend varying distances from one or both sides of the site used for linearization. Linear plasmid molecules carrying cohesive termini produce more transformants and less deletants than blunt-ended molecules, where >85% of the few survivor plasmids in the transformants contain deletions (Conley *et al.*, 1986a). Deletion occurs by exonucleolytic erosion of termini, followed by recombination at positions on the two arms of the linear molecule apparently possessing short (as low as 3 bp) homology matches (Conley *et al.*, 1986b; McFarlane and Saunders, 1996). Most of the deleted plasmids produced have a size that is less than monomeric and exhibit simple uni- or bi-directional erosion from the linearization termini as a result of *in vivo* recyclization. A small proportion of deletant plasmids are more than monomeric, but less than dimeric, and contain plasmid sequence deletions and duplications in either direct or indirect repeat. Yet more rarely, plasmids arise in transformants that are more than dimeric but less than trimeric (Conley *et al.*, 1986a), probably as a consequence of recombinational events involving two or more plasmid molecules, stimulated by the presence of linear DNA molecules in recipients (Conley *et al.*, 1986b). Such deletions and rearrangements, which may be regarded as a pathological response of the host recombination system, account for many of the aberrant recombinants that may be generated in transformation and gene manipulation experiments.

C. Exploitation of *in vivo* Deletion To Make Genetic Constructs

The deletions that occur following transformation of some bacteria such as *E. coli*, are essentially predictable. Hence the deliberate use of linearized plasmid DNA to transform the bacterium can be exploited to make deletions of varying size, extending away from the termini. Transformation of *E. coli* with linearized recombinant plasmid DNA has thus been utilized to introduce essentially random deletions, extending from restriction sites within the *tra* operon of the F plasmid (Thompson and Achtman, 1979). Recombination *in vivo* between partially homologous genetic sequences carried on either terminus of a linear recombinant plasmid molecule can also be used to generate novel chimeric genes, by driving recombination between partially homologous terminal sequences on a linear DNA substrate. The directed *in vivo* recombination approach has proved successful, for example, in the construction of various novel interferon genes by exploiting the essentially random nature of recombination and deletion end-points (Weber and Weissman, 1983).

◆◆◆◆◆◆ IV. STRATEGIES FOR DEVELOPING NEW TRANSFORMATION PROTOCOLS

A. General Approaches

The approach to devising transformation methods in previously untransformed species is dependent largely on the efficiency that will be demanded of the protocol. If all that is required is to introduce one or perhaps a few genetic constructs into the strain, then an efficient system is not required: in essence all that will be necessary is a very limited number of transformed colonies on selection plates. If, on the other hand, the system is to be used semi-quantitatively, for example in the construction of gene libraries or in a biological assay for DNA, considerable effort may have to be expended to develop a highly efficient system that will deliver large numbers of transformants.

B. Choice of Transforming DNA Molecule and its Preparation

The best choice of transformation substrate to test in a novel system is one that can be prepared in homogeneous form, to high purity, and which carries an easily selectable genetic marker, For this reason plasmids, or to a lesser extent bacteriophage DNA molecules, are to be preferred; the use of chromosomal DNA fragments will reduce the efficiency of selection for individual markers and present difficulties in preparation of fragments of uniform size. It is advisable to use the smallest feasible plasmid or other DNA molecule for experiments designed to develop a novel transformation system. This is because larger DNA molecules are more likely to suffer damage during uptake, which will compromise subsequent establishment and replication or recombination with resident replicons. This is borne out by studies of the efficiency of transformation in electroporation of *Xanthomonas campestris* (Ferreira *et al.*, 1995) or in Ca^{2+}-treated *E. coli* (Hanahan, 1983), which in both cases is inversely related to the size of the plasmid employed.

For some species, a well-characterized plasmid of sufficiently broad host range, for example, IncQ non-conjugative plasmids or IncP conjugative plasmids (Bingle, 1988; Metzler *et al.*, 1992), may provide a suitable substrate for testing transformability. If possible, it is advisable to ensure that such plasmids can actually replicate and survive in the intended recipient (see Chapter 2). This can be accomplished by mobilizing the test plasmid into the new host using a suitable conjugative plasmid. Having established that the test plasmid can replicate stably in the new host, any subsequent failures are likely to be due to inappropriate transformation conditions, rather than the use of an unsuitable or doomed test substrate. However, the broad host range plasmid approach is not guaranteed to be succcessful, due to peculiarities in the establishment and stability of different replicons when arriving in "virgin" hosts. Thus, vectors based on the broad-host-range plasmid, pBBR1, could be transferred by electroporation into strains of the methylotrophic bacteria *Hyphomicrobium denitrificans*, *Hyphomicrobium facilis*, *Methylobacterium extorquens*, *Methylobacillus glycogenes* and *Methylophilus methylotrophus* (Gliesche, 1997) (see also Chapter 2). However, other broad host range plasmids, such as pLA2917, pKT231, pSUP2021, pRZ705, or bacteriophage DNA, could not be transferred to *Hyphomicrobium* spp. using the same electroporation regime.

If the broad host range plasmid strategy fails, then cryptic plasmid replicons, indigenous to the target species, may first have to be identified and characterized at molecular and functional levels in order to provide test systems. For example, successful electroporation was achieved with dairy strains of *Lactobacillus casei* and *Lactobacillus delbrueckii* ssp. *lactis* once suitable cryptic plasmids had been identified in dairy starter cultures (Klein *et al.*, 1996). Even so, not all plasmids appear suitable as electroporation substrates. For example, of three plasmids tested in attempting to electroporate *Leuconostoc oenos*, only one was expressed in transformed cells (Dicks, 1994). Even among suitable plasmids there may be differences in effectiveness during application. For example, the rolling circle replication (RCR) plasmid, pGKV21, was found to be more stable in *Lactobacillus lactis* than the *theta* replication plasmid pIL253, and therefore more suitable as a substrate for developing a workable electroporation protocol (Kim *et al.*, 1996).

Transformation may apparently fail at the selection stage if sufficient time is not allowed for establishment and expression of resistance or other marker phenotype before selection is imposed. Peculiarities of gene expression in the host may also affect the outcome of transformation experiments. For example, using electrotransformation with the kanamycin-resistance plasmid pIKM1 of *Thermoanaerobacterium* sp., the levels of antibiotic used for selection had to be varied with growth temperature from 400 µg cm^{-3} at 48°C to 200 µg cm^{-3} at 60°C (Mai *et al.*, 1997).

The immediate source of test DNA may also be a significant factor in determining success. For example, electrotransformation efficiencies for the thermophile *Clostridium thermosaccharolyticum* ATCC 31960, were an order of magnitude higher with plasmid DNA prepared from *C. thermosaccharolyticum* than when prepared from *E. coli* (Klapatch *et al.*, 1996).

Much of this problem arises due to differential host modification signatures from the different DNA methylases in the hosts used for preparing the transforming DNA, and/or the proclivities of restriction systems in the intended recipient strain. These factors are sometimes difficult to predict, since restriction systems may be methylation dependent or methylation protective. Hemi-methylated DNA may also be restricted (see, for example, Bron *et al.*, 1984). Some success in heterologous transformations can be achieved with donor DNA produced in *dam⁻* or *dcm⁻* mutant strains (which lack DNA adenine- or cytosine-dependent methylase activities, respectively), or *dam⁻*, *dcm⁻* double mutant strains of *E. coli*. Following preshock incubation, the highest number of *Clostridium perfringens* transformants was obtained when the transforming plasmid was both Dam⁺- and Dcm⁺-modified, while no transformants were obtained when the plasmid DNA was not methylated or solely Dcm-methylated (Chen *et al.*, 1996). This is consistent with the observation that DNA of the plasmid used was protected against digestion by *C. perfringens* cell-associated nucleases for up to 3 min, when methylated by both methylases. DNA containing methylated adenine residues is restricted in some *B. anthracis* strains, resulting in decreased plasmid DNA-mediated transformation frequencies (Marrero and Welkos, 1995). Such inhibition can again be alleviated by propagating plasmids in DNA methyltransferase (MTase)-deficient *E. coli* strains such as GM2929 (*dam⁻*, *dcm⁻*) or *B. subtilis* 168, which lacks DNA adenine-methylation activity, before their introduction into problematical strains of *B. anthracis*. DNA derived from MTase-deficient *E. coli* strains transforms *Streptomyces coelicolor* protoplasts with high efficiency and permits efficient plasmid transfer by intergeneric conjugation (MacNeil *et al.*, 1992; Flett *et al.*, 1997).

Production of transforming DNA in a non-methylating host may remove a potentially "foreign" methylation signature in DNA, but only to a limited extent, since the DNA could still be restricted by a non-methylation-dependent restriction endonuclease that might be present in uncharacterized bacterial recipients. Without a pre-emptive analysis of restriction-modification patterns in donor DNA host and recipient strain, all that can be done is to try different combinations of DNA source and recipient until a workable system is seen. As an example of the inherent variability experienced, differences of up to 33 000-fold in electrotransformation of highly DNA restrictive corynebacteria are observed if the transforming DNA is a shuttle plasmid extracted from *E. coli* propagated in different nutritional regimes (Ankri *et al.*, 1996b). Growth of the plasmid DNA donor host in minimal medium increased plasmid transformability, whereas growth on rich media decreased it. This starvation-dependent increase in transformability is reverted by supplementing the growth medium with methionine, suggesting that a nutritionally modulated SAM-dependent DNA-methyltransferase of *E. coli* may be involved.

One way to overcome restriction problems has been to use PCR-produced DNA as the transforming substrate. For example, highly restricting corynebacteria can be transformed with DNA made *in vitro* by PCR amplification of a sequence that contains the replication origin of a plasmid common to many corynebacteria (Ankri *et al.*, 1996a). In all recipient strains tested, transformation efficiencies of PCR-synthesized DNA are equal to, or better, than those obtained with heterologous DNA prepared biologically from wild-type or *dam⁻*, *dcm⁻* strains of *E. coli*.

C. Choice of Recipient

One perfectly valid approach to developing a transformation regime for a new strain or species is to adapt a method previously developed for a close relative. For example, optimization of an electroporation protocol performed for one strain of *Streptococcus thermophilus* proved to be very effective for all other strains of the same species that were tested (Marciset and Mollet, 1994). Similarly, conditions optimal for *Xanthobacter autotrophicus* GJ10 were used to electrotransform four out of nine other *Xanthobacter* strains with high efficiencies (Swaving *et al.*, 1996). There are, however, no guarantees that the establishment of a workable transformation protocol for one strain can be employed without alteration for a close relative. Thus, electroporation protocols devised for one species may not work even with other closely related bacteria. For example, the method described for *Streptomyces parvulus* and *Streptomyces vinaceus* is not applicable to other *Streptomyces* species such as *S. lividans or S. coelicolor* (MazyServais *et al.*, 1997). Moreover, even when generic electroporation protocols can be applied to a wide variety of streptomycetes, because of the diversity of this group, conditions have to be optimized for individual species and strains (Pigac and Schrempf, 1995). Even very closely related bacteria may present different problems when devising new transformation protocols. For example, two strains of *Lactobacillus acidophilus* Group A1, ATCC 4356 and NCFM-N2, a human isolate used as a dietary supplement, exhibit markedly different transformation characteristics. This is so even though the two strains appear very similar genomically in terms of their *Sma*I-restriction fragment patterns (Walker *et al.*, 1996). See Chapter 7, III.D and Chapter 8, II.A for other examples of recipient variability in *Borrelia* and *Bacteroides* respectively.

In practice, a transformation system will more often than not be desired for a specific strain that has been selected for its other attributes, such as industrial or other use. There may be no close relative on which to base a working protocol for transformation. In such cases, the deficiencies of the host as a transformation recipient just have to be tolerated. Desirable features in recipient strains would appear to be the absence of capsular polysaccharide, and the presence of pili. Extracellular nuclease production is likely to be extremely prejudicial to transformation. It may therefore be worth while checking for the presence of such nucleases before embarking on system development (see, for example, Ahrenholtz *et al.*, 1994) and, if necessary, removing any activities by mutation. Similarly, a restriction-deficient phenotype is desirable. This may be achieved again by mutation, or, if applicable, restriction systems may sometimes be annulled temporarily by transient exposure of the recipient to elevated, but non lethal temperatures (Irani and Rowe, 1997). Electroporation has the distinct advantage that the electric field may temporarily disable restriction systems (Lefrancois and Sicard, 1997).

One strategy to improve an existing, workable transformation system is to isolate mutants with increased transformability. These may result from mutations in competence development and/or recombination/establishment of the test transformation substrate (Zulty and Barcak, 1995). However, such an approach is only warranted if a robust, high efficiency transformation system is the desired aim.

◆◆◆◆◆◆ V. CONCLUDING REMARKS

The possibilities for introducing DNA into bacteria are numerous. Although various transformation systems are potentially available, there is no doubt that electroporation provides the most widely applicable approach to manipulating previously unexplored organisms. The method is more robust and less subject to the vagaries of cell wall chemistry and nucleic acid metabolism. On balance, electroporation is currently therefore the best place to start when attempting to introduce DNA into a desired organism where there is no existing precedent.

References

Ahrenholtz, I., Lorenz, M.G. and Wackernagel, W. (1994). The extracellular nuclease of *Serratia marcescens* – studies on the activity *in vitro* and effect on transforming DNA in a groundwater aquifer microcosm. *Arch. Microbiol.* **161**, 176–183.

Ankri, S., Reyes, O. and Leblon, G. (1996a). Electrotransformation of highly DNA-restrictive Corynebacteria with synthetic DNA. *Plasmid* **35**, 62–66.

Ankri, S., Reyes, O. and Leblon, G. (1996b). Improved electrotransformation of highly DNA-restrictive Corynebacteria with DNA extracted from starved *Escherichia coli. FEMS Microbiol. Lett.* **140**, 247–251.

Argnani, A., Leer, R.J., Van Luijk, N. and Pouwels, P.H. (1996). A convenient and reproducible method to genetically transform bacteria of the genus *Bifidobacterium. Microbiology* **142**, 109–114.

Bagdasarian, M. and Timmis, K.N. (1981). Host–vector systems for gene-cloning in *Pseudomonas. Curr. Top. Microbiol. Immunol.* **1**, 47–67.

Barany, F. and Kahn, M.E. (1985). Comparison of transformation mechanisms of *Haemophilus parainfluenzae* and *Haemophilus influenzae. J. Bacteriol.* **161**, 72–79.

Barany, F. and Tomasz, A. (1980). Genetic transformation of *Streptococcus pneumoniae* by heterologous plasmid deoxyribonucleic acid. *J. Bacteriol.* **144**, 698–709.

Barany, F., Kahn, M.E. and Smith, H.O. (1983). Directional transport and integration of donor DNA in *Haemophilus influenzae* transformation. *Proc. Natl. Acad. Sci. USA* **80**, 7274–7278.

Barten, R. and Lill, H. (1995). DNA uptake in the naturally competent cyanobacterium, *Synechocystis. FEMS Microbiol. Lett.* **129**, 83–88.

Bensi, G., Iglesias, A., Canosi, U. and Trautner, T.A. (1981). Plasmid transformation in *Bacillus subtilis* – the significance of partial homology between plasmid and recipient cell DNAs. *Mol. Gen. Genet.* **184**, 400–404.

Benzinger, R. (1978). Transfection of Enterobacteriaceae and its applications. *Microbiol. Rev.* **42**, 194–236.

Bergmans, H.E.N., Kooijman, D.H. and Hoekstra, W.P.M. (1980). Cotransformation of linear chromosomal DNA and plasmid DNA in *Escherichia coli. FEMS Microbiol. Lett.* **9**, 211–214.

Berthier, F., Zagorec, M., Champomierverges, M., Ehrlich, S.D. and Moreldeville, F. (1996). Efficient transformation of *Lactobacillus-sake* by electroporation. *Microbiology* **142**, 1273–1279.

Bingle, W.H. (1988). Transformation of *Azotobacter-vinelandii* OP with a broad host range plasmid containing a cloned chromosomal *nif*-DNA marker. *Plasmid* **19**, 242–250.

Biswas, G.D. and Sparling, P.F. (1981). Entry of double-stranded deoxyribonucleic acid during transformation of *Neisseria gonorrhoeae*. *J. Bacteriol.* **129**, 638–640.

Biswas, G.D. and Sparling, P.F. (1989). Transformation-deficient mutants of piliated *Neisseria gonorrhoeae*. *J. Bacteriol.* **171**, 6576–6664.

Biswas, G.D., Burnstein, K.L. and Sparling, P.F. (1986). Linearization of donor DNA during plasmid transformation in *Neisseria gonorrhoeae*. *J. Bacteriol.* **168**, 756–761.

Blake, M.S., Macdonald, C.M. and Klugman, K.P. (1989). Colony morphology of piliated *Neisseria meningitidis*. *J. Exp. Med.* **170**, 1727–1736.

Bonnassie, S., Burini, J.F., Oreglia, J., Trautwetter, A., Patte, J.C. and Sicaed, A.M. (1990). Transfer of plasmid DNA to *Brevibacterium lactofermentum* by electroporation, *J. Gen. Microbiol.* **136**, 2107–2112.

Boyle-Vavra, S. and Seifert, H.S. (1996). Uptake-sequence-independent DNA transformation exists in *Neisseria gonorrhoeae*. *Microbiology* **142**, 2839–2845.

Brautigam, M., Hertel, C. and Hammes, W.P. (1997). Evidence for natural transformation of *Bacillus subtilis* in foodstuffs. *FEMS Microbiol. Lett.* **155**, 93–98.

Bron, S., Luxen, E. and Venema, G. (1984). Restriction of hemimethylated DNA by the *Bacillus subtilis*-r system. *Mol. Gen. Genet.* **195**, 370–373.

Bron, S., Janniere, L. and Ehrlich, S.D. (1988). Restriction and modification in *Bacillus subtilis* marburg-168 – target sites and effects on plasmid transformation. *Mol. Gen. Genet.* **211**, 186–189.

Brown, B.J. and Carlton, B.C. (1980). Plasmid-mediated transformation in *Bacillus megaterium*. *J. Bacteriol.* **142**, 508–512.

Brown, M.G.M., Weston, A., Saunders, J.R. and Humphreys, G.O. (1979). Transformation of *Escherichia coli* by plasmid DNA at different phases of growth. *FEMS Microbiol. Lett.* **5**, 219–222.

Brown, M.G.M., Saunders, J.R. and Humphreys, G.O. (1981). Lack of specificity in DNA-binding and uptake during transformation of *Escherichia coli*. *FEMS Microbiol. Lett.* **11**, 97–100.

Burnstein, K.L., Dyer, D.W. and Sparling, P.F. (1988). Preferential uptake of restriction fragments from a gonococcal cryptic plasmid by competent *Neisseria gonorrhoeae*. *J. Gen. Microbiol.* **134**, 547–557.

Bygraves, J.A. and Maiden, M.C.J. (1992). Analysis of the clonal relationships between strains of *Neisseria meningitidis* by pulsed field gel-electrophoresis. *J. Gen. Microbiol.* **138**, 523–531.

Calvin, N.M. and Hanawalt, P.C. (1988). High-efficiency transformation of bacterial cells by electroporation. *J. Bacteriol.* **170**, 2796–2801.

Castuma, C.E., Huang, R.P., Kornberg, A. and Reusch, R.N. (1995). Inorganic polyphosphates in the acquisition of competence in *Escherichia coli*. *J. Biol. Chem.* **270**, 12980–12983.

Chamier, B., Lorenz, M.G. and Wackernagel, W. (1993). Natural transformation of *Acinetobacter calcoaceticus* by plasmid DNA adsorbed on sand and groundwater aquifer material. *Appl. Environ. Microbiol.* **59**, 1662–1667.

Chang, S. and Cohen, S.N. (1979). High-frequency transformation of *Bacillus subtilis* protoplasts by plasmid DNA. *Mol. Gen. Genet.* **168**, 111–115.

Chen, J.D. and Morrison, D.A. (1987). Modulation of competence for genetic-transformation in *Streptococcus pneumoniae*. *J. Gen. Microbiol.* **133**, 1959–1967.

Chen, C.K., Boucle, C.M. and Blaschek, H.P. (1996). Factors involved in the transformation of previously non-transformable *Clostridium-perfringens* type-b. *FEMS Microbiol. Lett.* **140**, 185–191.

Cho, J.H., Kim, E.K. and So, J.S. (1995). Improved transformation of *Pseudomonas putida* KT2440 by electroporation. *Biotech. Techn.* **9**, 41–44.

Chung, C.T., Niemela, S.L. and Miller, R.H. (1989). One-step preparation of competent *Escherichia coli*: transformation and storage of bacterial cells in the same solution. *Proc. Natl. Acad. Sci. USA.* **86**, 2172–2176.

Chung, Y.S. and Dubnau, D. (1995). ComC is required for the processing and translocation of ComGC, a pilin-like competence protein of *Bacillus subtilis. Mol. Microbiol.* **15**, 543–551.

Cohen, S.N., Chang, A.C.Y. and Hsu, C.L. (1972). Non-chromosomal antibiotic resistance in bacteria: genetic transformation of *Escherichia coli* by R factor DNA. *Proc. Natl. Acad. Sci. USA* **69**, 2110–2114.

Conley, E.C. and Saunders, J.R. (1984). Recombination-dependent recircularization of linearized pBR322 plasmid DNA following transformation of *Escherichia coli. Mol. Gen. Genet.* **194**, 211–218.

Conley, E.C., Saunders, V.A., Jackson, V. and Saunders, J.R. (1986a). Mechanism of intramolecular recyclization and deletion formation following transformation of *Escherichia coli* with linearized plasmid DNA. *Nucl. Acids Res.* **14**, 8919–8932.

Conley, E.C., Saunders, V.A. and Saunders, J.R. (1986b). Deletion and rearrangement of plasmid DNA during transformation of *Escherichia coli* with linear plasmid molecules. *Nucl. Acids Res.* **14**, 8905–8917.

Contente, S. and Dubnau, D. (1979a). Characterization of plasmid transformation in *Bacillus subtilis*: kinetic properties and the effect of DNA concentration. *Mol. Gen. Genet.* **167**, 215–258.

Contente, S. and Dubnau, D. (1979b). Marker rescue by linear plasmid DNA. *Plasmid* **2**, 555–571.

Cosloy, S.D. and Oishi, M. (1973). Genetic transformation in *Escherichia coli. Proc. Natl. Acad. Sci. USA* **69**, 2110–2114.

Danner, D.B. Deich, R.A., Sisco, K.L. and Smith, H.O. (1980). An 11-base pair sequence determines the specificity of DNA uptake in *Haemophilus* transformation. *Gene* **11**, 311–318.

Davidoff-Abelson, R. and Dubnau, D. (1973). Conditions affecting the isolation from transformed cells of *Bacillus subtilis* of high-molecular weight single stranded deoxyribonucleic acid of donor DNA. *J. Bacteriol.* **116**, 146–153.

Desomer, J., Dhaese, P. and Van Montagu, M. (1990). Transformation of *Rhodococcus fascians* by high-voltage electroporation and development of *Rhodococcus fascians* cloning vectors. *Appl. Environ. Microbiol.* **56**, 2818–2825.

de Vos, W.M. and Venema, G. (1981). Fate of plasmid DNA in transformation of *Bacillus subtilis* protoplasts. *Mol. Gen. Genet.* **182**, 39–43.

de Vos, W.M. and Venema, G. (1982). Transformation of *Bacillus subtilis* competent cells – identification of a protein involved in recombination. *Mol. Gen. Genet.* **187**, 439–445.

de Vos, W.M. and Venema, G. (1983). Transformation of *Bacillus subtilis* competent cells – identification and regulation of the *recE* gene-product. *Mol. Gen. Genet.* **190**, 56–64.

de Vos, W.M., Venema, G., Canosi, U. and Trautner, T.A. (1981). Plasmid transformation in *Bacillus subtilis* – fate of plasmid DNA. *Mol. Gen. Genet.* 181, 424–433.

Dicks, L.M.T. (1994). Transformation of *Leuconostoc oenos* by electroporation. *Biotech. Techn.* **8**, 901–904.

Dityatkin, S.Y.A. and Iliyashenko, B.N. (1979). Plasmid transformation of frozen-thawed bacteria. *Genetika* **15**, 220–225.

Docherty, A., Grandi, G., Grandi, R., Gryczan, T.J., Shivakumar, A.G. and Dubnau, D. (1981). Naturally occurring macrolide-lincosamide-streptogramin-b resistance in *Bacillus licheniformis. J. Bacteriol.* **145**, 129–137.

Dorocicz, I., Williams, P. and Redfield, R.J. (1993). The *Haemophilus influenzae* adenylate cyclase gene: cloning, sequence and essential role in competence. *J. Bacteriol.* **175**, 7142–7149.

Dougherty, T.J., Asmus, A. and Tomasz, A. (1979). Specificity of DNA uptake in genetic transformation of gonococi. *Biochem. Biophys. Res. Comm.* **86**, 97–104.

Dower, W.J., Miller, J.F. and Ragsdale, C.W. (1988). High efficiency transformation of *E. coli* by high voltage electroporation. *Nucl. Acids Res.* **16**, 6127–6145.

Dower, W.J., Chassy, B.M., Trevors, J.T. and Blaschek, H.P. (1992). Protocols for the transformation of bacteria by electroporation. In *Guide to electroporation and electrofusion*, pp. 485–499. Academic Press, San Diego.

Dunny, G.M., Lee, L.N. and Leblanc, D.J. (1991). Improved electroporation and cloning vector system for Gram-positive bacteria. *Appl. Environ. Microbiol.* **57**, 1194–1198.

Espinosa, M., Joenje, H. and Venema, G. (1980). DNA-binding and deoxyribonuclease activity in *Bacillus subtilis* during temperature-induced competence development. *J. Gen. Microbiol.* **121**, 79–84.

Facius, D., Fussenegger, M. and Meyer, T.F. (1996). Sequential action of factors involved in natural competence for transformation of *Neisseria gonorrhoeae*. *FEMS Microbiol. Lett.* **137**, 159–164.

Feavers, I.M., Heath, A.B., Bygraves, J.A. and Maiden, M.C.J. (1992). Role of horizontal genetic exchange in the antigenic variation of the class-1 outer-membrane protein of *Neisseria meningitidis. Mol. Microbiol.* **6**, 489–495.

Ferreira, H., Barrientos, F.J.A., Baldini, R.L., Rosato, Y.B. (1995). Electro-transformation of 3 pathovars of *Xanthomonas campestris. Appl. Microbiol. Biotech.* **43**, 651–655.

Fitzmaurice, W.P., Benjamin, R.C., Huang, P.C. and Scocca, J.J. (1984). Characterization of recognition sites on bacteriophage HP1c1 DNA which interact with the DNA uptake system of *Haemophilus influenzae* Rd. *Gene* **31**, 187–196.

Flett, F., Mersinias, V. and Smith, C.P. (1997). High efficiency intergeneric conjugal transfer of plasmid DNA from *Escherichia coli* to methyl DNA-restricting streptomycetes. *FEMS Microbiol. Lett.* **155**, 223–229.

Fornari, C.S. and Kaplan, S. (1982). Genetic-transformation of *Rhodopseudomonas sphaeroides* by plasmid DNA. *J. Bacteriol.* **152**, 89–97.

Fraley, R.T., Fornari, C.S. and Kaplan, S. (1979). Entrapment of a bacterial plasmid in phospholipid membranes. *Proc. Natl. Acad. Sci. USA* **76**, 3348–3352.

Fromm, M., Taylor, L. and Walbot, V. (1985). Expression of genes transferred into monocot and dicot plants by electroporation. *Proc. Natl. Acad. Sci. USA* **82**, 5824–5828.

Frosch, M. and Meyer, T.F. (1992). Transformation-mediated exchange of virulence determinants by cocultivation of pathogenic *Neisseriae. FEMS Microbiol. Lett.* **100**, 345–349.

Fussenegger, M., Rudel, T., Barten, R., Ryll, R. and Meyer, T.F. (1997). Transformation competence and type-4 pilus biogenesis in *Neisseria gonorrhoeae* – a review. *Gene* **192**, 125–134.

Gaustad, P. (1985). Genetic transformation of *Streptococcus sanguis*. Effects on genetic transformation by culture filtrates of *Streptococcus sanguis* (serogroups H and W) and *Streptococcus mitis* (mitior) with reference to identification. *Acta Path. Microbiol. Scand.* Sect.B93, 283–287.

Gliesche, C.G. (1997). Transformation of methylotrophic bacteria by electroporation. *Can. J. Microbiol.* **43**, 197–201.

Goodgal, S.H. and Herriot, R.M. (1961). Studies on transformation of *Haemophilus influenzae* I. Competence. *J. Gen. Physiol.* **44**, 1201–1227.

Goodgal, S.H. and Mitchell, M.A. (1990). Sequence and uptake specificity of cloned sonicated fragments of *Haemophilus influenzae* DNA. *J. Bacteriol.* **172**, 5924–5928.

Goodman, S.D. and Scocca, J.J. (1988). Identification and arrangement of the DNA sequence recognized in specific transformation of *Neisseria gonorrhoeae*. *Proc. Natl. Acad. Sci. USA* **85**, 6982–6986.

Grinius, L. (1980). Nucleic acid transport driven by ion gradient across cell membrane. *FEBS Lett.* **113**, 1–10.

Gromkova, R. and Goodgal, S.H. (1979). Transformation by plasmid and chromosomal DNA in *Haemophilus influenzae*. *Biochem. Biophys. Res. Comm.* **88**, 1428–1434.

Gromkova, R. and Goodgal, S.H. (1981). Uptake of plasmid deoxyribonucleic acid by *Haemophilus*. *J. Bacteriol.* **146**, 79–84.

Haas, R., Meyer, T.F. and Vanputten, J.P.M. (1993). Aflagellated mutants of *Helicobacter pylori* generated by genetic-transformation of naturally competent strains using transposon shuttle mutagenesis. *Mol. Microbiol.* **8**, 753–760.

Hahn, J., Bylund, J., Haines, M., Higgins, M. and Dubnau, D. (1995). Inactivation of MecA prevents recovery from the competent state and interferes with cell division and the partitioning of nucleoids in *Bacillus subtilis*. *Mol. Microbiol.* **18**, 755–767.

Hanahan, D. (1983). Studies on transformation of *Escherichia coli* with plasmids. *J. Mol. Biol.* **166**, 557–580.

Hanahan, D. (1987). Mechanisms of DNA transformation. In *Escherichia coli* and *Salmonella typhimurium. Cellular and Molecular Biology*, Vol 2, pp. 1177–1183. American Society for Microbiology, Washington.

Hanahan, D., Jessee, J. and Bloom, F.R. (1991). Plasmid transformation of *Escherichia coli* and other bacteria. *Meth. Enzymol.* **204**, 63–113.

Harlander, S. (1987). Gene transfer systems in lactic streptococci. In *Streptococcal Genetics*. (J.J.Ferretti and R.C. Curtiss, eds), pp. 229–233. American Society for Microbiology, Washington.

Hauser, P.M. and Karamata, D. (1994). A rapid and simple method for *Bacillus subtilis* transformation on solid media. *Microbiology* **140**, 1613–1617.

Havarstein, L.S., Coomaraswamy, G. and Morrison, D.A. (1995). An unmodified heptadecapeptide pheromone induces competence for genetic-transformation in *Streptococcus pneumoniae*. *Proc. Natl. Acad. Sci. USA* **92**, 11140–11144.

Haynes, J.A. and Britz, M.L. (1990). The effect of growth-conditions of *Corynebacterium glutamicum* on the transformation frequency obtained by electroporation. *J. Gen. Microbiol.* **136**, 255–263.

Heierson, A., Landen, R., Lovgren, A., Dalhammar, G. and Boman, H.G. (1987). Transformation of vegetative cells of *Bacillus thuringiensis* by plasmid DNA. *J. Bacteriol.* **169**, 1147–1152.

Hill, S.A. (1996). Limited variation and maintenance of tight genetic-linkage characterize heteroallelic *pilE* recombination following DNA transformation of *Neisseria gonorrhoeae*. *Mol. Microbiol.* **20**, 507–518.

Hoekstra, W.P.M., Bergamans, J.E.N. and Zuidweg, E.M. (1980). Role of *recBC* nuclease in *Escherichia coli* transformation. *J. Bacteriol.* **143**, 1031–1032.

Holsters, M., De Waele, D., Messens, E., van Montagu, M. and Schell, J. (1978). Transfection and transformation of *Agrobacterium tumefaciens*. *Mol. Gen. Genet.* **163**, 181–187.

Hopwood, D. (1981). Genetic studies with bacterial protoplasts. *Annu. Rev. Microbiol.* **35**, 237–272.

Hui, S.W., Stoicheva, N. and Zhao, Y.L. (1996). High efficiency loading, transfection, and fusion of cells by electroporation in 2-phase polymer systems. *Biophys. J.* **71**, 147–150.

Humphreys, G.O., Weston, A., Brown, M.G.M. and Saunders, J.R. (1979). Plasmid transformation of *Escherichia coli*. In *Transformation 1978* (S.W. Glover and L.O. Butler, eds), pp. 254–279. Cotswold Press, Oxford.

Iglesias, A., Bensi, G., Canosi, U. and Trautner, T.A. (1981). Plasmid transformation in *Bacillus subtilis* – alterations introduced into the recipient homologous DNA of hybrid plasmids can be corrected in transformation. *Mol. Gen. Genet.* **184**, 405–409.

Inamine, G. and Dubnau, D. (1995). ComEA, a *Bacillus subtilis* integral membrane protein required for genetic transformation, is needed for both DNA binding and transport. *J. Bacteriol.* **177**, 3045–3051.

Irani, V.R. and Rowe, J.J. (1997). Enhancement of transformation in *Pseudomonas aeruginosa* PAO1 by Mg^{2+}. *Biotechniques* **22**, 54.

Jones, I.M., Primrose, S.B., Robinson, A. and Ellwood, D.C. (1981). Effect of growth-rate and nutrient limitation on the transformability of *Escherichia coli* with plasmid deoxyribonucleic-acid. *J. Bacteriol.* **146**, 841–846.

Josson, K., Scheirlinck, T., Michiels, F., Platteeuw, C., Stanssens, P., Joos, H., Dhaese, P., Zabeau, M. and Mahillon, J. (1989). Characterization of a Gram-positive broad-host-range plasmid isolated from *Lactobacillus hilgardii*. *Plasmid*, **21**, 9–20.

Jyssum, K., Jyssum, S. and Gundersen, W.B. (1971). Sorption of DNA and RNA during transformation of *Neisseria meningitidis*. *Acta Path. Microbiol. Scand.* Sect B. **79**, 563–571.

Kahn, M.E., Barany, F. and Smith, H.O. (1983). Transformasomes: specialized membranous structures which protect DNA during transformation. *Proc. Natl. Acad. Sci. USA* **80**, 6927–6931.

Karudapuram, S., Zhao, X.M. and Barcak, G.J. (1995). DNA-sequence and characterization of *Haemophilus influenzae dprA*(+), a gene required for chromosomal but not plasmid DNA transformation. *J. Bacteriol.* **177**, 3235–3240.

Kawagishi, I., Okunishi, I., Homma, M. and Imae, Y. (1994). Removal of periplasmic DNase before electroporation enhances efficiency of transformation in the marine bacterium *Vibrio alginolyticus*. *Microbiology* **140**, 2355–2361.

Khanna, M. and Stotzky, G. (1992). Transformation of *Bacillus subtilis* by DNA bound on montmorillonite and effect of DNAse on the transforming ability of bound DNA. *Appl. Environ. Microbiol.* **58**, 1930–1939.

Kim, W.J., Heo, T.R. and So, J.S. (1996). Optimized transformation by electroporation of *Lactococcus-lactis*. *Biotech. Techn.* **10**, 553–558.

Kimoto, H. and Taketo, A. (1996) Studies on electrotransfer of DNA into *Escherichia coli* – effect of molecular-form of DNA. *Biochim. et Biophys. Acta* **1307**, 325–330.

King, K.W. and Dybvig, K. (1991). Plasmid transformation of *Mycoplasma mycoides subspecies mycoides* is promoted by high-concentrations of polyethylene-glycol. *Plasmid* **26**, 108–115.

Klapatch, T.R., Guerinot, M.L. and Lynd, L.R. (1996). Electrotransformation of *Clostridium thermosaccharolyticum*. *J. Indust. Microbiol.* **16**, 342–347.

Klebe, R.J., Harriss, J.V., Sharp, Z.D. and Douglas, M.G. (1983). A general-method for polyethylene-glycol-induced genetic-transformation of bacteria and yeast. *Gene* **25**, 333–341.

Klein, J.R., Ulrich, C., Wegmann, U., Meyerbarton, E., Plapp, R. and Henrich, B. (1996). Molecular tools for the genetic-modification of dairy Lactobacilli. *System. Appl. Microbiol.* **18**, 493–503.

Kong, L. and Dubnau, D. (1991). Regulation of competence-specific gene expression by Mec-mediated protein–protein interaction in *Bacillus subtilis. Proc. Natl. Acad. Sci. USA* **91**, 5793–5797.

Koomey, J.M. and Falkow, S. (1987). Cloning of the *recA* gene of *Neisseria gonorrhoeae* and construction of gonococcal *recA* mutants. *J. Bacteriol.* **169**, 790–795.

Kretschmer, P.J., Chang, A.C.Y. and Cohen, S.N. (1975). Indirect selection of bacterial plasmids lacking identifiable phenotypes. *J. Bacteriol.* **124**, 225–231.

Kushner, S.R. (1978). An improved method for transformation of *Escherichia coli* with ColE1-derived plasmids. In *Genetic Engineering* (H.W. Boyer and S. Nicosa, eds), pp. 17–23, Elsevier, Amsterdam.

Lacks, S. (1962). Molecular fate of DNA in genetic transformation of pneumococcus. *J. Mol. Biol.* **5**, 119–131.

Lacks, S.A. (1977). Binding and entry of DNA in bacterial transformation. In *Microbial Interactions; Receptors and Recognition. Series B* (J.L. Reissig, ed.), pp. 179–232, Chapman & Hall, London.

Laine, M.J., Nakhei, H., Dreier, J., Lehtila, K., Meletzus, D., Eichenlaub, R. and Metzler, M.C. (1996). Stable transformation of the Gram-positive phytopathogenic bacterium *Clavibacter michiganensis* subsp. *sepedonicus* with several cloning vectors. *Appl. Environ. Microbiol.* **62**, 1500–1506.

Lederberg, E.M. and Cohen, S.N. (1974). Transformation of *Salmonella typhimurium* by plasmid deoxyribonucleic acid. *J. Bacteriol.* **119**, 1072–1074.

Lefrancois, J. and Sicard, A.M. (1997). Electrotransformation of *Streptococcus pneumoniae*: evidence for restriction of DNA on entry. *Microbiology* **143**, 523–526.

Lopez, A., Clave, C., Capeyrou, R., Lafontan, V. and Trombe, M.-C. (1989). Ionic and energetic changes at competence in the naturally transformable bacterium *Streptococcus pneumoniae. J. Gen. Microbiol.* **135**, 2189–2197.

Lorenz, M.G. and Wackernagel, W. (1991). High-frequency of natural genetic-transformation of *Pseudomonas stutzeri* in soil extract supplemented with a carbon energy and phosphorus source. *Appl. Environ. Microbiol.* **57**, 1246–1251.

Lorenz, M.G. and Wackernagel, W. (1994). Bacterial gene transfer by natural genetic transformation in the environment. *Microb. Rev.* **58**, 563–602.

Lorenz, M.G., Reipschlager, K. and Wackernagel, W. (1992). Plasmid transformation of naturally competent *Acinetobacter calcoaceticus* in nonsterile soil extract and groundwater. *Arch. Microbiol.* **157**, 355–360.

Love, P.E., Lyle, M.J. and Yasbin, R.E. (1985). DNA-damage-inducible (*din*) loci are transcriptionally activated in competent *Bacillus subtilis. Proc. Natl. Acad. Sci. USA* **82**, 6201–6205.

Lunsford, R.D. and Roble, A.G. (1997). *ComYA*, a gene similar to *comGA* of *Bacillus subtilis*, is essential for competence-factor-dependent DNA transformation in *Streptococcus gordonii. J. Bacteriol.* **179**, 3122–3126.

McDonald, I.R., Riley, P.W., Sharp, R.J. and McCarthy, A.J. (1995) Factors affecting the electroporation of *Bacillus subtilis. J. Appl. Bacteriol.* **79**, 213–218.

MacFadyen, L.P., Dorocicz, I.R., Reizer, J., Saier, M.H. and Redfield, R.J. (1996). Regulation of competence development and sugar utilization in *Haemophilus influenzae* rd by a phosphoenolpyruvate-fructose phosphotransferase system. *Mol. Microbiol.* **21**, 941–952.

McFarlane, R.J. and Saunders, J.R. (1996). Molecular mechanisms of intramolecular recircularization of linearized plasmid DNA in *Escherichia coli*: requirements for the *ruvA*, *ruvB*, *recG*, *recF* and *recR* gene products. *Gene* **177**, 209–216.

MacNeil, D.J., Gewain, K.M., Ruby, C.L., Dezeny, G., Gibbons, P.H. and MacNeil, T. (1992). Analysis of *Streptomyces avernutilis* genes required for avermectin biosynthesis utilizing a novel integration vector. *Gene* **111**, 61–68.

Macrina, F.L., Jones, K.R. and Welch, R.A. (1981). Transformation of *Streptococcus sanguis* with monomeric pVA736 plasmid deoxyribonucleic-acid. *J. Bacteriol.* **146**, 826–830.

Mahillon, J., Chungjatupornchai, W., Decock, J., Dierickx, S., Michiels, F., Peferoen, M. and Joos, H. (1989). Transformation of *Bacillus thuringiensis* by electroporation. *FEMS Microbiol. Lett.* **60**, 205–210.

Mai, V., Lorenz, W.W. and Wiegel, J. (1997). Transformation of *Thermoanaerobacterium* sp. strain JW/SL-YS485 with plasmid pIKM1 conferring kanamycin resistance. *FEMS Microbiol. Lett.* **148**, 163–167.

Makins, J.F. and Holt, G. (1981). Liposome-mediated transformation of streptomycetes by chromosomal DNA. *Nature* **293**, 671–673.

Marciset, O. and Mollet, B. (1994). Multifactorial experimental-designs for optimizing transformation – electroporation of *Streptococcus-thermophilus*. *Biotech. Bioeng.* **43**, 490–496.

Marrero, R. and Welkos, S.L. (1995). The transformation frequency of plasmids into *Bacillus anthracis* is affected by adenine methylation. *Gene* **152**, 75–78.

Mathis, L.S. and Scocca, J.J. (1984). On the role of pili in transformation of *Neisseria gonorrhoeae*. *J. Gen. Microbiol.* **130**, 3165–3173.

MazyServais, C., Baczkowski, D. and Dusart, J. (1997). Electroporation of intact cells of *Streptomyces parvulus* and *Streptomyces vinaceus*. *FEMS Microbiol. Lett.* **151**, 135–138.

Metcalf, W.W., Zhang, J.K., Apolinario, E., Sowers, K.R. and Wolfe, R.S. (1997). A genetic system for *Archaea* of the genus *Methanosarcina*: liposome-mediated transformation and construction of shuttle vectors. *Proc. Natl. Acad. Sci. USA* **94**, 2626–2631.

Metzler, M.C., Zhang, Y.P. and Chen, T.A. (1992). Transformation of the Gram-positive bacterium *Clavibacter xyli* subsp *cynodontis* by electroporation with plasmids from the IncP incompatibility group. *J. Bacteriol.* **174**, 4500–4503.

Michel, B. and Ehrlich, S.D. (1983). Recombination efficiency is a quadratic function of the length of homology during plasmid transformation of *Bacillus subtilis* protoplasts and *Escherichia coli* competent cells. *EMBO J.* **3**, 2879–2884.

Michel, B., Niaudet, B. and Ehrlich, S.D. (1982). Intramolecular recombination during plasmid transformation of *Bacillus subtilis* competent cells. *EMBO J.* **1**, 1565–1571.

Miller, J.F. (1994). Bacterial transformation by electroporation. *Meth. Enzymol.* **235**, 375–385.

Morelli, L., Cocconcelli, P.S., Bottazzi, V., Damiani, G., Ferretti, L. and Sgaramella, V. (1987). *Lactobacillus* protoplast transformation. *Plasmid* **17**, 73–75.

Mottes, M., Grandi, G., Sgaramella, V., Canosi, U., Morelli, G. and Trautner, T. A. (1979), Different specific activities of the monomeric and oligomeric forms of plasmid DNA in transformation of *B. subtilis* and *E. coli*. *Mol. Gen. Genet.* **174**, 281–286.

Nedenskovsorensen, P., Bukholm, G. and Bovre, K. (1990). Natural competence for genetic-transformation in *Campylobacter pylori*. *J. Infect. Dis.* **161**, 365–366.

Negoro, S., Shinagawa, H., Nagata, A., Kinoshita, S., Hatozaki, T. and Okada, H. (1980). Plasmid control of 6-aminohexanoic acid cyclic dimer degradation enzymes by *Flavobacterium* sp. K172. *J. Bacteriol.* **143**, 248–245.

Nickoloff, J.A. (ed.) (1995). *Electroporation Protocols for Microorganisms. Methods in Molecular Biology*. Humana Press, Totowa, New Jersey.

Ohse, M. , Kawade, K. and Kusaoke, H. (1997). Effects of DNA topology on transformation efficiency of *Bacillus subtilis* ISW1214 by electroporation. *Biosci. Biotech. Biochem.* **61**, 1019–1021.

Okamoto, A., Kosugi, A., Koizumi, Y., Yanagida, F. and Udaka, S. (1997). High efficiency transformation of *Bacillus brevis* by electroporation. *Biosci. Biotech. Biochem.* **61**, 202–203.

Palmen, R., Vosman, B., Buijsman, P., Breek, C.K.D. and Hellingwerf, K.J. (1993). Physiological characterization of natural transformation in *Acinetobacter calcoaceticus*. *J. Gen. Microbiol.* **139**, 295–305.

Patel, G.B., Nash, J.H.E., Agnew, B.J. and Sprott, G.D. (1994). Natural and electro-poration-mediated transformation of *Methanococcus voltae* protoplasts. *Appl. Environ. Microbiol.* **60**, 903–907.

Pestova, E.V., Havarstein, L.S. and Morrison, D.A. (1996). Regulation of competence for genetic-transformation in *Streptococcus pneumoniae* by an auto-induced peptide pheromone and a 2-component regulatory system. *Mol. Microbiol.* **21**, 853–862.

Pifer, M.L. and Smith, H.O. (1985). Processing of donor DNA during *Haemophilus influenzae* transformation: analysis using a model plasmid system. *Proc. Natl. Acad. Sci. USA* **82**, 3731–3735.

Pigac, J. and Schrempf, H. (1995). A simple and rapid method of transformation of *Streptomyces rimosus* R6 and other streptomycetes by electroporation. *Appl. Environ. Microbiol.* **61**, 352–356.

Porter, M.E. and Dorman, C.J. (1997). Virulence gene deletion frequency is increased in *Shigella flexneri* following conjugation, transduction, and transformation. *FEMS Microbiol. Lett.* **147**, 163–172.

Pozzi, G. and Guild, W.R. (1985). Modes of integration of heterologous plasmid DNA into the chromosome of *Streptococcus pneumoniae*. *J. Bacteriol.* **161**, 909–912.

Pozzi, G., Masala, L., Iannelli, F., Manganelli, R., Havarstein, L.S., Piccoli, L., Simon, D. and Morrison, D.A. (1996). Competence for genetic-transformation in encapsulated strains of *Streptococcus pneumoniae* – 2 allelic variants of the peptide pheromone. *J. Bacteriol.* **178**, 6087–6090.

Quinn, C.P. and Dancer, B.N. (1990). Transformation of vegetative cells of *Bacillus anthracis* with plasmid DNA. *J. Gen. Microbiol.* **136**, 1211–1215.

Redfield, R.J. (1988) Evolution of bacterial transformation: is sex with dead cells ever better than no sex at all? *Genetics* **119**, 213–221.

Redfield, R.J. (1991) *sxy-1*, a *Haemophilus influenzae* mutation causing greatly enhanced competence. *J. Bacteriol.* **173**, 5612–5618.

Redfield, R.J. (1993). Evolution of natural transformation: testing the DNA repair hypothesis in *Bacillus subtilis* and *Haemophilus influenzae*. *Genetics* **133**, 755–761.

Redfield, R.J., Schrag, M.R. and Dean, A.M. (1997). The evolution of bacterial transformation: sex with poor relations. *Genetics* **146**, 27–38.

Regue, M., Enfedaque, J., Camprubi, S. and Tomas, J.M. (1992). The O-antigen lipopolysaccharide is the major barrier to plasmid DNA uptake by *Klebsiella pneumoniae* during transformation by electroporation and osmotic shock. *J. Microbiol. Meth.* **15**, 129–134.

Reusch, R.N. and Sadoff, H.L. (1988). Putative structure and functions of a poly-beta-hydroxybutyrate calcium polyphosphate channel in bacterial plasma-membranes. *Proc. Natl. Acad. Sci. USA* **85**, 4176–4180.

Ricci, M.L., Manganelli, R., Berneri, C., Orefici, G. and Pozzi, G. (1994). Electrotransformation of *Streptococcus agalactiae* with plasmid DNA. *FEMS Microbiol. Lett.* **119**, 47–52.

Richardson, P.T. and Park, S.F. (1997). Integration of heterologous plasmid DNA into multiple sites on the genome of *Campylobacter coli* following natural transformation. *J. Bacteriol.* **179**, 1809–1812.

Rinken, R. and Wackernagel, W. (1992). Inhibition of the RecBCD dependent activation of *chi* recombinational hot-spots in SOS-induced cells of *Escherichia coli*. *J. Bacteriol.* **174**, 1172–1178.

Rittich, B. and Spanova, A. (1996). Electrotransformation of bacteria by plasmid DNAs – Statistical evaluation of a model quantitatively describing the relationship between the number of electrotransformants and DNA concentration. *Bioelectrochem. Bioenerget.* **40**, 233–238.

Rochelle, P.A., Day, M.J. and Fry, J.C. (1988). Occurrence, transfer and mobilization in epilithic strains of *Acinetobacter* of mercury-resistance plasmids capable of transformation. *J. Gen. Microbiol.* **134**, 2933–2941.

Rodicio, M.R. and Chater, K.F. (1982). Small DNA-free liposomes stimulate transfection of *Streptomyces* protoplasts. *J. Bacteriol.* **151**, 1078–1085.

Romanowski, G., Lorenz, M.G., Sayler, G. and Wackernagel, W. (1992). Persistence of free plasmid DNA in soil monitored by various methods, including a transformation assay. *Appl. Environ. Microbiol.* **58**, 3012–3019.

Romanowski, G., Lorenz, M.G. and Wackernagel, W. (1993). Plasmid DNA in a groundwater aquifer microcosm-adsorption, DNAase resistance and natural genetic-transformation of *Bacillus subtilis. Molec. Ecol.* **2**, 171–181.

Rosa, P., Samuels, D.S., Hogan, D., Stevenson, B., Casjens, S. and Tilly, K. (1996). Directed insertion of a selectable marker into a circular plasmid of *Borrelia burgdorferi. J. Bacteriol.* **178**, 5946–5953.

Rossi, M., Brigidi, P. and Matteuzzi, D. (1997). An efficient transformation system for *Bifidobacterium* spp. *Lett. Appl. Microbiol.* **24**, 33–36.

Rowji, P., Gromkova, R. and Koornhof, H. (1989). Genetic transformation in encapsulated clinical isolates of *Haemophilus influenzae* type b. *J. Gen. Microbiol.* **135**, 2775–2782.

Rudel, T., Facius, D., Barten, R., Scheuerpflug, I., Nonnenmacher, E. and Meyer, T.F. (1995). Role of pili and the phase-variable PilC protein in natural competence for transformation of *Neisseria gonorrhoeae. Proc. Natl. Acad. Sci. USA* **92**, 7986–7990.

Ryll, R.R., Rudel, T., Scheuerpflug, I., Barten, R. and Meyer, T.F. (1997). PilC of *Neisseria meningitidis* is involved in class II pilus formation and restores pilus assembly, natural transformation competence and adherence to epithelial cells in *pilC*-deficient gonococci. *Mol. Microbiol.* **23**, 879–892.

Sakurai, N. and Komatsubara, S. (1996). Simple and versatile electrotransformation of *Serratia-marcescens* SR41. *Lett. Appl. Microbiol.* **23**, 23–26.

Samuels, D.S. (1995). Electrotransformation of the spirochete *Borrelia burgdorferi.* In *Electroporation Protocols for Microorganisms*, pp. 253–259. Humana Press, Totowa, New Jersey.

Samuels, D.S. and Garon, C.F. (1997). Oligonucleotide-mediated genetic transformation of *Borrelia burgdorferi. Microbiology* **143**, 519–522.

Saunders, C.W. and Guild, W.R. (1981a). Pathway of plasmid transformation in pneumococcus – open circular and linear-molecules are active. *J. Bacteriol.* **146**, 517–526.

Saunders, C.W. and Guild, W.R. (1981b). Monomer plasmid DNA transforms *Streptococcus pneumoniae. Mol. Gen. Genet.* **181**, 57–62.

Saunders, J.R. and Saunders, V.A. (1988). Bacterial transformation with plasmid DNA. *Meth. Microbiol.* **21**, 79–128.

Saunders, J.R., Hart, C.A. and Saunders, V.A. (1986). Plasmid-mediated resistance to β-lactam antibiotics in Gram-negative bacteria: the role of in vivo recyclization reactions in plasmid evolution. *J. Antimicrob. Chemother.* **18**, Suppl. C, 57–66.

Saunders J.R., Docherty, A. and Humphreys, G.O. (1987). Transformation of bacteria by plasmid DNA. *Meth. Microbiol.* **17**, 61–95.

Scocca, J.J. (1990). The role of transformation in the variability of the *Neisseria gonorrhoeae* cell-surface. *Mol. Microbiol.* **4**, 321–327.

Seifert, H.S., Ajioka, R.S., Marchal, C., Sparling, P.F. and So, M. (1988). DNA transformation leads to pilin antigenic variation in *Neisseria gonorrhoeae. Nature* **336**, 392–395.

Seifert, H.S., Ajioka, R.S., Paruchuri, D., Heffron, F. and So, M. (1990). Shuttle mutagenesis of *Neisseria gonorrhoeae* – pilin null mutations lower DNA transformation competence. *J. Bacteriol.* **172**, 40–46.

Selvaraj, G. and Iyer, V.N. (1981). Genetic-transformation of *Rhizobium meliloti* by plasmid DNA. *Gene* **15**, 279–283.

Setlow, J.K. and Albritton, W.L. (1992). Transformation of *Haemophilus influenzae* following electroporation with plasmid and chromosomal DNA. *Curr. Microbiol.* **24**, 97–100.

Shigekawa, K. and Dower, W.J. (1988). Electroporation of eukaryotes and prokaryotes: a general approach to the introduction of macromolecules into cells. *BioTechniques* **6**, 742–751.

Shivarova, N., Forster, W., Jacob, H.E. and Grigorova, R. (1983). Microbiological implications of electric-field effects .7. Stimulation of plasmid transformation of *Bacillus cereus* protoplasts by electric-field pulses. *Zeitsch. All. Mikrobiol.* **23**, 595–599.

Sikosrski, J., Graupner, S., Lorenz, M.G. and Wackernagel, W. (1998). Natural genetic transformation of *Pseudomonas stutzeri* in a non-sterile soil. *Microbiology* **144**, 569–576.

Singer, M.E.V. and Finnerty, W.R. (1988). Construction of an *Escherichia coli-Rhodococcus* shuttle vector and plasmid transformation in *Rhodococcus*-spp. *J. Bacteriol.* **170**, 638–645.

Smith, H., Wiersma, K., Venema, G. and Bron, S. (1985). Transformation in *Bacillus subtilis* – further characterization of a 75,000-dalton protein complex involved in binding and entry of donor DNA. *J. Bacteriol.* **164**, 201–206.

Smith, J.M., Dowson, C.G. and Spratt, B.G. (1991). Localized sex in bacteria. *Nature* **349**, 29–31.

Smith, J.M., Smith, N.H., O'Rourke, M. and Spratt, B.G. (1993). How clonal are bacteria. *Proc. Natl. Acad. Sci. USA* **90**, 4384–4388.

Smith, H.O., Tomb, J.-F., Dougherty, B.A., Fleischmann, R.D. and Venter, J.C. (1995). Frequency and distribution of DNA uptake signal sequences in the *Haemophilus influenzae* Rd genome. *Science* **269**, 538–540.

Solomon, J.M. and Grossman, A.D. (1996) Who's competent and when – regulation of natural genetic competence in bacteria. *Trends Genet.* **12**, 150–155.

Stein, D.C. (1991). Transformation of *Neisseria gonorrhoeae* – physical requirements of the transforming DNA. *Can. J. Microbiol.* **37**, 345–349.

Stein, D.C., Gregoire, S. and Piekarowicz, A. (1988). Restriction of plasmid DNA during transformation but not conjugation in *Neisseria gonorrhoeae*. *Infect. Immun.* **56**, 112–116.

Stepanov, A.S., Puzanova, O.G., Dityatkin, S.Y., Loginova, O.G. and Ilyashenko, B.N. (1990). Glycine-induced cryotransformation of plasmids into *Bacillus anthracis*. *J. Gen. Microbiol.* **136**, 1217–1221.

Stewart, G.J. (1989). The mechanism of natural transformation. In *Gene transfer in the environment.* (S.B. Levy and R.V. Miller, eds), pp. 139–164. McGraw-Hill, New York.

Stratz, M., Sauer, U., Kuhn, A. and Durre, P. (1994). Plasmid transfer into the homoacetogen *Acetobacterium woodii* by electroporation and conjugation. *Appl. Environ. Microbiol.* **60**, 1033–1037.

Stuy, J.H. (1980). Mechanism of additive genetic transformation in *Haemophilus influenzae*. *J. Bacteriol.* **144**, 999–1002.

Suzuki, M. and Szalay, A.A. (1979). Bacterial transformation using temperature-sensitive mutants deficient in peptidoglycan synthesis. In *Methods in Microbiology* (S.W. Glover and D.A. Hopwood, eds), pp. 49–109. Cambridge University Press, London.

Swaving, J., Van Leest, W., Van Ooyen, A.J.J. and Debont, J. (1996). Electro-transformation of *Xanthobacter autotrophicus* GJ10 and other *Xanthobacter* strains. *J. Microbiol. Meth.* **25**, 343–348.

Thompson, R. and Achtman, M. (1979). The control region of the F sex factor DNA transfer cistrons: physical mapping by deletion analysis. *Mol. Gen. Genet.* **169**, 49–57.

Tomb, J.F. (1992). A periplasmic protein disulfide oxidoreductase is required for transformation of *Haemophilus influenzae* RD. *Proc. Natl. Acad. Sci. USA* **89**, 10252–10256.

Tomb, J.F., Barcak, G.J., Chandler, M.S., Redfield, R.J. and Smith, H.O. (1989). Transposon mutagenesis, characterization, and cloning of transformation genes of *Hemophilus-influenzae* rd. *J. Bacteriol.* **171**, 3796–3802.

Tomb, J.F., Elhajj, H. and Smith, H.O. (1991). Nucleotide-sequence of a cluster of genes involved in the transformation of *Haemophilus influenzae* rd. *Gene* **104**, 1–10.

Toneguzzo, F. and Keating, A. (1986). Stable expression of selectable genes into-duced into human hematopoietic stem cells by electric-field mediated DNA transfer. *Proc. Natl. Acad. Sci. USA.* **83**, 3496–3499.

Tonjum, T., Freitag, N.E., Namork, E. and Koomey, M. (1995). Identification and characterization of *pilG*, a highly conserved pilus-assembly gene in pathogenic *Neisseria. Mol. Microbiol.* **16**, 451–464.

Trautner, T.A. and Spatz, H.C. (1973). Transfection in *B. subtilis. Curr. Top. Microbiol. Immunol.* **62**, 61–88.

Trombe, M.C., Rieux, V. and Baille, F. (1994). Mutations which alter the kinetics of calcium-transport alter the regulation of competence in *Streptococcus pneumoniae. J. Bacteriol.* **176**, 1992–1996.

Tyurin, M.V., Oparina, N.Y. and Livshits, V.A. (1996). Influence of cell preparation conditions and the form of electric pulse on the electrotransformation efficiency in Gram-positive rods and cocci. *Microbiology (Russian)* **65**, 585–590.

Van der Vossen, J.M.B.M., Kok, J., Vanderlelie, D. and Venema, G. (1988). Liposome-enhanced transformation of *Streptococcus lactis* and plasmid transfer by intergeneric protoplast fusion of *Streptococcus lactis* and *Bacillus subtilis. FEMS Microbiol. Lett.* **49**, 323–329.

Van Nieuwenhoven, M.H., Hellingwerf, K.J., Venema, G. and Konings, W.N. (1982). Role of proton motive force in genetic-transformation of *Bacillus subtilis. J. Bacteriol.* **151**, 771–776.

Van Sinderen, D., Luttinger, A., Kong, L.Y., Dubnau, D., Venema, G. and Hamoen, L. (1995). ComK encodes the competence transcription factor, the key regulatory protein for competence development in *Bacillus subtilis. Mol. Microbiol.* **15**, 455–462.

Vijayakumar, M.N. and Morrison, D.A. (1986). Localization of competence-induced proteins in *Streptococcus pneumoniae. J. Bacteriol.* **165**, 689–695.

Vosman, B., Kuiken, G., Kooistra, J. and Venema, G. (1988). Transformation in *Bacillus subtilis*: involvement of the 17-kilodalton DNA-entry nuclease and com-petence specific 18-kilodalton protein. *J. Bacteriol.* **170**, 3703–3710.

Wackernagel, W. (1973). Genetic transformation in *E. coli*: the inhibitory role of the *recBC* DNase. *Biochem. Biophys. Res. Comm.* **51**, 306–311.

Walker, D.C., Aoyama, K. and Klaenhammer, T.R. (1996). Electrotransformation of *Lactobacillus acidophilus* Group A1. *FEMS Microbiol. Lett.* **138**, 233–237.

Wang, Y. and Taylor, D.E. (1990). Natural transformation in *Campylobacter* species. *J. Bacteriol.* **172**, 949–955.

Ward, B.J. and Collins, D.M. (1996). Electrotransformation at elevated tempera-tures substantially improves transformation of slow-growing bacteria. *FEMS Microbiol. Lett.* **145**, 101–105.

Weaver, J.C. (1995) Electroporation theory: concepts and mechanisms In *Methods in Molecular Biology*, Vol. 47. *Electroporation Protocols for Microorganisms* (J.A. Nickoloff, ed.), pp. 1–26. Humana, Towota, New Jersey.

Weber, H. and Weissman, C. (1983). Formation of genes coding for hybrid proteins by recombination between related cloned genes in *Escherichia coli. Nucl. Acids Res.* **11**, 5661–5669.

Weston, A., Humphreys, G.O., Brown, M.G.M. and Saunders, J.R. (1979). Simultaneous transformation of *Escherichia coli* by pairs of compatible and incompatible plasmid DNA molecules. *Mol. Gen. Genet.* **172**, 113–118.

Weston, A., Brown, M.G.M., Perkins, H.R., Saunders, J.R. and Humphreys, G.O. (1981). Transformation of *Escherichia coli* with plasmid deoxyribonucleic-acid – calcium-induced binding of deoxyribonucleic-acid to whole cells and to isolated membrane-fractions. *J. Bacteriol.* **145**, 780–787.

Williams, P.M., Bannister, L.A. and Redfield, R.J. (1994). The *Haemophilus influenzae sxy-1* mutation is in a newly identified gene essential for competence. *J. Bacteriol.* **176**, 6789–6794.

Yother, J., Mcdaniel, L.S. and Briles, D.E. (1986). Transformation of encapsulated *Streptococcus pneumoniae. J. Bacteriol.* **168**, 1463–1465.

Zulty, J. and Barcak, G. J. (1995). Identification of a DNA transformation gene required for *com101A*-plus expression and supertransformer phenotype in *Haemophilus influenzae. Proc. Nat. Acad. Sci. USA* **92**, 3616–3620.

2 The Development of Plasmid Vectors

Lyndsay Radnedge[1] and Hilary Richards[2]

[1]Environmental Molecular Microbiology Laboratory, Lawrence Livermore National Laboratory, Livermore, USA; [2]Department of Biology, University College London, London, UK

◆◆

CONTENTS

◆◆◆◆◆◆ I. INTRODUCTION

The ability to propagate and clone individual segments of DNA by linking them *in vitro* to a replicon indigenous to a particular host cell has become so central a tool in molecular biology that it is sometimes difficult to remember that the first such experiments were done as recently as 1973. The first vectors used were naturally occurring plasmids, and even the sophisticated vectors used today often utilize replication functions of plasmid origin.

Plasmids may be defined as extrachromosomal replicons, made of double-stranded DNA. It was once thought that all plasmids were circular as the methods used for plasmid isolation preferentially separated those molecules. However, since linear plasmids were first isolated (Kinashi *et al.*, 1987) more and more linear plasmids have been isolated from many different bacteria. Bacterial plasmids can now be thought of as mini-chromosomes that can be isolated from bacterial cells and then re-introduced into other cells by conjugation or transformation.

Bacterial plasmids have been studied for many years because many of the special characteristics displayed by bacteria of medical, agricultural, industrial and environmental importance are determined by genes carried on plasmids. Studies of these genes opened up the study of the biology of plasmids, their replication, maintenance and transfer functions as well as their evolution by the acquisition of transposons carrying new genes and which can bring about genome rearrangements.

METHODS IN MICROBIOLOGY, VOLUME 29
0580-9517 $30.00

When plasmids carrying antibiotic resistance genes were first being studied their ability to conjugate to *E. coli* recipients was important for their characterization. Initially, this characterization was limited to antibiotic resistance phenotype and incompatibility and the tests for incompatibility were based on plasmid transfer experiments. However, in Stanley Cohen's laboratory at Stanford University small non-conjugative plasmids were being studied and these were present in multiple copies. Classical genetic methods for studying mutations to single molecules required a method of separating single plasmid molecules and so he developed a method for circular plasmid DNA transformation into *E. coli* using calcium chloride and selecting for transformants by the expression of their antibiotic resistance genes (Cohen *et al.*, 1972). Cells transformed with plasmid DNA grew into a clone of antibiotic-resistant bacteria. Since each cell in the clone contained a DNA species having the same genetic and molecular properties as the plasmid DNA molecule that was taken up by the initial transformant, the procedure made possible the cloning of individual plasmid molecules present in a heterogeneous population.

The cloning of plasmid DNA provided the possibility of cloning other DNA fragments if they were linked to the plasmid's replication function. Stanley Cohen described his plasmid transformation procedure in 1972 at a meeting in Honolulu, Hawaii. At the same meeting Herb Boyer reported experiments indicating that cleavage of DNA by the *Eco*RI endonuclease yielded projecting single-stranded ends. From these two ideas, the concept of using restriction endonucleases to generate DNA fragments which could be linked into plasmid replicons was generated. These first cloning experiments used the large plasmid R6-5 which was cut into 11 fragments using *Eco*RI. Success in these experiments led to the search for a carrier molecule or vector which would contain both the replication machinery and a selectable marker on a single restriction endonuclease fragment. The small plasmid pSC101 proved a more suitable vector: it is 9 kb in size, carries the tetracycline resistance gene, and is cut only once by *Eco*RI. Cohen and his co-workers published the procedure for cloning fragments of *E. coli* plasmids in pSC101 (Cohen *et al.*, 1973), which was then adapted for the cloning of staphylococcal DNA in *E. coli* (Chang and Cohen, 1974).

pSC101 is not an ideal plasmid vector as the site for the restriction endonuclease, *Eco*RI does not inactivate a gene whose loss of function can be assayed. ColE1, another naturally occurring small plasmid, which was also used as a plasmid vector had similar problems. ColE1 produces the small bacteriocin colicin E1 and also codes for immunity to this protein. However, these were not easy functions to assay as colicin-producing colonies had to be replica plated, then killed with chloroform before overlaying a sensitive strain to determine if colicin had been produced. The initial difficulties using these naturally occurring plasmids were soon overcome with the design of improved vectors.

◆◆◆◆◆◆ II. DEVELOPMENT OF PLASMIDS AS CLONING VECTORS

A. Requirements

Bacterial cells often contain DNA molecules in addition to their chromosome. These extrachromosomal elements are known as plasmids, and are found in most types of bacteria. They have been successfully exploited and developed as important tools in molecular biology. Plasmids vary in size from a few thousand to hundreds of thousands of base pairs (comparable in size to bacterial chromosome). Originally, plasmids were believed to be circular DNA molecules, with nucleotides joined by covalent bonds (referred to as covalently closed circular DNA). More recently, linear plasmids have been discovered (e.g. *Streptomyces* sp., *Borrelia burgdorferi*) (See Chapter 7, II.B in this volume).

The rapid spread of antibiotic resistance in hospitals was one of the first indications of such elements subsequently referred to as R-plasmids. Plasmids contain genetic information that confers a variety of phenotypic traits on their host cells. This information is not required for bacterial reproduction, but does impart growth advantages. Plasmids vary greatly in size and may encode a few or several hundred different proteins. The extra information encoded by plasmids increases the ability of the bacterial population as a whole to survive in an increased number of environments. This extra information can be included in the cell without adding to the size of the chromosome; a smaller chromosome which replicates more quickly has a selective advantage.

Plasmids are able to distribute traits to other members of the plasmid-free population, which ensures that a small proportion of the bacterial population will survive in challenging environments (e.g. in the presence of antibiotics). Obviously, plasmids play a crucial role in the dissemination of genetic material necessary for adaptation and evolution. Plasmids have been shown to encode virulence factors (e.g. invasion genes in *Shigella dysenteriae*; Yop proteins of *Yersinia* species), resistance to heavy metals, utilization of unusual nutrient sources (e.g. toluene by Tol plasmid), and resistance to antibiotics (resistance to ampicillin, tetracycline and kanamycin by plasmid RK2). Studies in a wide range of bacteria have shown that they determine a remarkable array of phenotypic traits of medical, agricultural, environmental and commercial importance.

Most plasmids isolated from Gram-negative bacteria have a narrow host range, which means they are stably maintained in closely related species or strains. Some Gram-negative plasmids do have broad host range (e.g. RK2, RP4, pSF1010), although this extended host range does not include Gram-positive bacteria. Generally, most Gram-positive plasmids exhibit an extended host range, but only within Gram-positive bacteria. (See Chapter 6, Table 1 and Chapter 13, VI.E.)

All stable plasmids must contain regions responsible for their autonomous replication (*rep*) including an origin of replication (*ori*). Plasmid replication may be stringently controlled (i.e. low copy number plasmids) or be subject to relaxed control (high copy number plasmids)

(Rownd, 1969). Stringently controlled plasmids encode their own replication proteins, but require protein synthesis by the bacterial cell. Plasmids which replicate under relaxed control plasmids utilize DNA polymerase (PolA) of the host, but do not require protein synthesis. The level of control reflects the copy number and plasmid size. The mode of replication controls the ultimate copy number of the plasmid (average numbers of plasmid molecules per cell). Plasmid copy number is an important consideration for plasmid maintenance. Low copy number plasmids must also encode partition (*par*) functions which ensure that the plasmid is accurately segregated after replication and during cell division. The chances of a low copy number plasmid being lost are high if random segregation were employed. If a plasmid has several copies per cell then chances are each daughter cell will inherit some copies of the plasmid after cell division. The ramifications of plasmid replication and segregation are discussed in more detail in section III below.

Identification and classification of plasmids is ideally based on traits that are ubiquitous and universally inherited such as replication control. Plasmids with the same replication control are incompatible, they cannot coexist stably within a host cell, whereas plasmids with different replication controls are compatible. Consequently incompatibility is a measure of relatedness of plasmid replicons, i.e. those plasmids that share common elements involved in plasmid replication control or equipartitioning. Replication mechanisms of different plasmids are discussed in detail section III below.

A formal scheme of plasmid classification based on incompatibility was developed by Datta and Hedges (1971). Plasmids which are incompatible with each other are assigned to the same incompatibility group. At present about 30 incompatibility groups are represented in enteric bacteria, and 7 in staphylococcal plasmids. Some plasmids have origins that allow them to replicate in many types of bacteria such as RSF1010 (IncQ), RP4 and RK2 (IncP); such plasmids are said to have a broad host range. Plasmids in IncP and IncQ incompatibility groups are often referred to as promiscuous plasmids due to their ability to transfer or be transferred to, and replicate in, different bacteria.

The identification and classification of plasmids are especially important for clinically important traits (e.g. drug resistance, virulence factors) which are frequently plasmid-encoded. Determining the incompatibility group of a resistance plasmid can help trace the source and spread of infection, may establish a diagnosis, and indicate its evolutionary origins.

Plasmids can also be classified according to their ability to spread from one bacterium to another. They are referred to as either conjugative (or self-transmissible) and non-conjugative plasmids. Both types of plasmids are found in Gram-negative and Gram-positive bacteria. A conjugative plasmid differs from a non-conjugative plasmid in the presence of genes encoding plasmid transfer, or *tra* genes. Conjugation is directed transfer of DNA from a donor to recipient bacterial cell (Williams and Skurray, 1980), mediated by *tra* and *mob* (mobilization) genes which encode synthesis of pili and other surface components. As a rule of thumb low copy number, high molecular weight plasmids are usually conjugative, whereas high copy number, low molecular weight plasmids are non-conjugative. The transfer of plasmid DNA by conjugation is discussed in more detail in section V of this chapter.

Autonomously replicating plasmids have proved invaluable in recombinant DNA technology. With the advent of restriction endonucleases it has been possible to splice together these plasmids and inserts of foreign DNA to generate recombinant molecules. Plasmids which are successfully used as cloning vectors must possess the following properties:

- **Multiple copies per cell.** Obviously high copy small plasmids have a pivotal role in recombinant DNA technology since the chief aim is to propagate the insert and allow isolation of large quantities of pure DNA.
- **Replicate autonomously.** Plasmids must be able to replicate and remain stable in the host bacteria during several generations of cell division.
- **Selectable markers.** A means of distinguishing cells that have inherited the plasmid from those which have not. These markers must be absent on chromosome, such as resistance to antibiotics (e.g. β-lactamase, which encodes resistance to ampicillin), so that plasmid-free cells are killed on selective medium (e.g. containing ampicillin). See Table I for a list of antibiotics commonly used in recombinant DNA work.
- **Differentiation.** There must be a means of distinguishing which of the plasmids inherited contain the insert of choice (the recombinant DNA molecule). For example, insertion into β-galactosidase gene (*lacZ*) in pUC plasmids inactivates that non-essential gene by interrupting its open reading frame. β-galactosidase breaks down 5-bromo-4-chloro-3-indolyl-β-D-galactoside (X-gal) to form a blue derivative. Plasmids carrying pUC plasmids will become blue when X-gal is incorporated into the growth medium. Plasmids into which the insert has been successfully inserted (i.e. which interrupt the *lacZ* open reading frame) produce no β-galactosidase, and remain colourless.
- **Unique restriction sites.** Inserting the DNA in the region of choice of the plasmid must not disrupt essential genes and must also leave the selectable marker intact. Multiple cloning sites (polylinkers) have been developed in many plasmids often by the insertion of an oligonucleotide. These polylinkers are designed to contain multiple restriction sites that are unique to the plasmid and which facilitate both insertion and excision of cloned sequences.
- **Control elements for gene expression.** If plasmids are to be used to assay the function of a cloned gene then the vector must also contain the elements necessary to ensure its transcription and translation. These include promoters, which may be inducible or regulatable, and ribosome binding sites in the correct location for expression of the gene of interest.

Since most naturally occurring plasmids do not combine all of the elements above, vectors had to be developed using recombinant DNA technology. One example of a vector that has been progressively improved is pBR322 (Bolivar *et al.*, 1977). *In vitro* recombination techniques were used to construct pBR322, itself derived from pBR313. It replicates under relaxed control and carries resistance genes to ampicillin and tetracycline. pBR322 was constructed with a single PstI site, located in the ampicillin-resistance gene, in addition to four unique restriction sites, EcoRI, HindIII, BamHI and SalI. pBR322 is present in multicopy (10+ copies per chromosome equivalent). Its replicon is ColE1, and the copy number can be

Table I. Antibiotics commonly used to select bacterial plasmids (Ausubel et al., 1987)

Antibiotic	Final Concentration (μg/ml)	Mode of action	Mode of resistance
Ampicillin	50	Bacteriocidal; only kills growing cells; inhibits cell wall synthesis by inhibiting formation of the peptidoglycan cross-link	β-lactamase hydrolyzes ampicillin
Chloramphenicol	20	Bacteriostatic; inhibits protein synthesis, target is 50S ribosomal subunit	Chloramphenicol acetyltransferase inactivates chloramphenicol
Gentamycin	15	Bacteriocidal; inhibits protein synthesis, target is 50S ribosomal subunit	Aminoglycoside acetyltransferase and aminoglycoside nucleotidyltransferase; rplF prevents gentamycin binding
Kanamycin	30	Bacteriocidal; inhibits protein synthesis; inhibits translocation and elicits miscoding	Aminoglycoside phosphotransferase inactivates kanamycin
Nalidixic Acid	15	Bacteriostatic; inhibits DNA gyrase	DNA gyrase mutants prevent nalidixic acid from binding
Rifampicin	150	Bacteriostatic; inhibits RNA synthesis	Mutation in β subunit of RNA polymerase prevents rifampicin from complexing rifampicin; resistance is recessive
Spectinomycin	100	Bacteriostatic; inhibits translocation of peptidyl tRNA from the A site to the P site	Mutations in rpsE prevent spectinomycin from binding; resistance is recessive
Streptomycin	30	Bacteriocidal; inhibits protein synthesis, target is S12 protein of 30S ribosomal subunit	Aminoglycoside phosphotransferase inactivates streptomycin; rpsL (encodes S12) prevents binding; resistance is recessive
Tetracycline	12	Bacteriostatic; inhibits protein synthesis; prevents binding of aminoacyl tRNA to ribosome A site	Active efflux of tetracycline from the cell

amplified to 1500 by treatment with chloramphenicol since its replication is under relaxed control and proceeds in the absence of protein synthesis. One litre of cells can yield 1 mg plasmid DNA. Not all of the unique restriction sites were useful for cloning. Insertion into the EcoRI site of pBR322 does not inactivate a suitable marker. The only way to screen for DNA inserts is to prepare the plasmid DNA, then excise any insert with a restriction endonuclease. A modified version of pBR322, named pBR328, contains a unique EcoRI site which inactivates chloramphenicol acetyl-transferase gene, so recombinant plasmids can be identified in chloramphenicol-sensitive clones.

pBR322 was also the parent of the widely used pUC vectors. pUR2 is 2297 bp PvuII/EcoRI fragment of pBR322. The HaeII *lac* fragment of M13mp2 was inserted into one of the HaeII sites flanking the pBR322 origin in pUR2. This pUR2 derivative resistant to AccI, PstI, HincII was used as the recipient for the multiple cloning site from M13mp7. In these plasmids (designated pUC) the β-lactamase function can be used for selection of the plasmid and the *lacZ* gene used for insertional activation when DNA is inserted into one of the unique restriction sites of the multiple cloning site. One extra benefit of this construction was removal of the copy control locus from pBR322 which results in the higher copy number of pUC plasmids (30–100 per cell). The resulting plasmids pUC8 and pUC9 allow one to clone doubly digested restriction fragments separately with both orientations in respect to the lac promoter (Vieira and Messing, 1982). Later versions of the pUC plasmids also include a polylinker which contains more unique restriction enzyme sites and universal sequencing primer sites to facilitate sequencing (Yanisch-Perron *et al.*, 1985). New strains of *E. coli* were constructed with mutations that improved cloning of unmodified DNA and of repetitive sequences. The complete nucleotide sequences of the M13 and pUC vectors were compiled by Yanisch-Perron *et al.* (1985).

B. Shuttle Vectors

Sometimes there is a requirement to transfer a cloning vector from one organism to another. Even for broad host range plasmids, it is unlikely that a particular plasmid origin will function in two unrelated species. To get around this problem, bifunctional plasmids known as shuttle vectors were created. Shuttle vectors are so named as they are able to "shuttle" genes between two organisms. A shuttle vector has two origins of replication each of which functions in each organism and must also contain selectable genes that can be expressed in both organisms. (See also examples in Chapter 5, IV.A.2 and Chapter 9, III.B and IV.C.)

Since most plasmids have been developed for use in *E. coli* or its close relatives, one of the origins is usually derived from a plasmid capable of replicating in *E. coli*. This facilitates easy manipulation and purification of the plasmid DNA. The shuttle vector can then be transferred to the alternative organism, where it establishes itself using the second replicon. This allows the biochemical or genetic tests to be performed in the other organism.

Some shuttle vectors can replicate in *E. coli* and Gram-positive bacteria, but yet others have been developed for use in shuttling genes between bacteria and eukaryotic cells. For example, the 7.8 kb plasmid pYEP24 has the replication origin of the 2 μm circle (a plasmid from the yeast *Saccharomyces cerevisiae*) and the ColE1 replicon from pBR322 (Botstein *et al.*, 1979). This plasmid is able to replicate in both *S. cerevisiae* and *E. coli*. The selectable markers *URA3* and *bla* are expressed in *S. cerevisiae* and *E. coli* respectively. Transfer of other plasmids has even been possible into mammalian cells using shuttle vectors that utilize the animal simian virus SV40 replicon (Cepko *et al.*, 1984).

In order to develop shuttle vectors for novel bacteria, one first needs to isolate plasmid from species of interest, and then identify its replicon (see section III). In one example, this was successfully carried out for the Gram-negative sulphate-reducing bacterium *Desulphovibrio desulfuricans* (Wall *et al.*, 1993). A 2.3 kb plasmid, designated pBG1, was identified and found to be present in about 20 copies per genome. pBG1 is unable to replicate in *E. coli*. Sequence analysis revealed a small replicon able to support the replication of recombinant plasmids in *Desulphovibrio desulfuricans* and allowed construction of a shuttle vector by creating a composite molecule with the *E. coli* vector pTZ18U. A similar approach generated a shuttle vector for the thermophilic bacterium *Thermus thermophilus* (Koyama *et al.*, 1990; Lasa *et al.*, 1992). Wayne and Xu, (1997) used a pUC19-based vector to select functional thermophilic origins from *Thermus* spp. Kanamycin selection works in *Thermus* and transformants (i.e. clones that can replicate at high temperature) represented newly formed shuttle vectors. The minimal origin, *ori*, was refined to 2.3 kb and enabled characterization of the replicon. It is worth bearing in mind that if a shuttle vector is dependent on a large replicon (or even the entire plasmid) then it may also carry many genes that are non-essential for replication; such large plasmids may not accept insertion of foreign DNA inserts stably.

Rood and co-workers have used different techniques to manipulate plasmids of *Clostridium perfringens* to suit their needs. The native *C. perfringens* plasmid, pIP404 (Garnier and Cole, 1988), was converted into a shuttle vector by ligating the *Clostridium* replicon to pUC18. The resulting construct, pJIR418, had multiple unique cloning sites and two selectable markers, *catP* for chloramphenicol resistance and *ermBP* for erythromycin resistance (Sloan *et al.*, 1992). The laboratory was working on molecular analysis of C. perfringens erythromycin, chloramphenicol and tetracycline resistance determinants and ran into problems because presence of both chloramphenicol and erythromycin resistance on one plasmid was restrictive. These constructs were converted to single antibiotic resistance: pJIR750 was created by cutting pJIR418 with *Ava*II, filling in the ends with Klenow fragment of DNA polymerase I, digesting with *Sca*I and religating the fragment. The resultant plasmid pJIR750 encoded resistance to chloramphenicol but not erythromycin as most of the *ermBP* gene had been deleted. pJIR751 was constructed by cutting pJIR418 with *Ava*II, filling in the ends as above, digesting with *Nae*II and religating. pJIR751 is resistant to erythromycin but not chloramphenicol due to removal of the entire *catP* gene, and also benefited from the removal of a *Sma*I site making *Sma*I site in multiple cloning site unique to the plasmid (Bannam and Rood, 1993).

pJIR750 and pJIR751 were manipulated further by changing them into mobilizable plasmids, which results in more efficient transfer into *C. perfringens* (Lyras and Rood, 1998). These derivatives carry the *oriT* region of plasmid RP4 and were mobilizable by RP4. The 289 bp RP4 *oriT* fragment was excised on EcoRI–HindIII fragment from pVWD2B (Waters *et al.*, 1991), then made blunt ended by treatment with T4 DNA polymerase. This fragment was cloned into XmnI site of pJIR750 or pJIR751 yielding plasmids pJIR1456 and pJIR1457; each plasmid resistant to chloramphenicol and erythromycin respectively. The RP4 *oriT* was shown to be functional as pJIR1456 and pJIR1457 were both mobilized into recipient *E. coli* and *C. perfringens* at high frequency.

◆◆◆◆◆◆ III. PLASMID REPLICATION

Plasmids must control their own replication since they exist separately from the bacterial chromosome. The minimal region which is necessary and sufficient for replication of a particular plasmid is referred to as the "replicon". Plasmids are present in defined copy numbers (average numbers of plasmid molecules per cell) during exponential growth of their hosts. Plasmid copy number is primarily determined by plasmid-encoded functions but is also affected by the host and growth conditions. Replicons have two components: a *cis*-acting site that serves as the origin of replication (*ori*) and *trans*-acting proteins that enable replication at this site; some of these proteins are host encoded. Studies on plasmid replication in both Gram-positive and Gram-negative bacteria have identified important features of these *cis*- and *trans*-acting regions which are involved in replication (and its regulation), partitioning and stability of these extrachromosomal elements. Most of the early studies on plasmid replication were done with *E. coli* plasmids such as ColE1, pSC101, RK2 and F. These were all found to replicate by the theta mechanism. More recently a large number of small multicopy plasmids in different bacteria have been found to replicate by a rolling circle (RC) mechanism originally observed for single-stranded DNA phages of *E. coli*. Strand displacement replication was also described as the mode of replication for IncQ plasmids such as RSF1010.

Linear plasmids and chromosomes were recently identified in both Gram-positive and Gram-negative bacteria and are best characterized in *Streptomyces* species and *Borrelia burgdorferi* (Hinnebusch and Tilly, 1993). Since DNA polymerases require a primer and cannot replicate the extreme 3' ends of linear DNA, plasmids have developed special mechanisms to overcome this problem. Linear plasmids are found in all species of *Borrelia* and range in size from 5 kb to greater than 200 kb. *Borrelia burgdorferi* have telomeric sequences at their ends much like eukaryotic viruses. The two DNA strands are connected forming a perfectly palindromic AT-rich terminal hairpin loop. A conserved 19 bp inverted repeat is present at each end of the plasmid, comprising an inverted terminal repeat. If these plasmids are denatured by strand separation then they will form one large single-stranded circular molecule. After replication the circular intermediates form concatemers which are subsequently resolved (Hinnebusch and Tilly, 1993). The linear plasmids isolated from *Streptomyces* species range in size from 9 kb to more than 600 kb. These plasmids replicate bidirectionally towards their telomeres from a site located near the centre of the molecule (Chang and Cohen, 1994). They contain a terminal protein covalently attached to the 5' end of each strand which serves as a primer for DNA synthesis.

A. Modes of Replication of Circular Plasmids

There are three general replication mechanisms used by circular bacterial plasmids:

1. Theta (θ) type – generally represented by replicons from Gram-negative bacteria.
2. Strand displacement – associated with broad host range plasmids from the IncQ family.
3. Rolling circle (RC) – generally represented by replicons from Gram-positive bacteria.

The initiation of DNA replication generally involves the construction of a complicated physical structure with DNA wrapped around a protein complex. The function of the replication proteins is to open the DNA strands at the origin and use them as templates to synthesize the new DNA strands. Many plasmids recruit some of the proteins used by the host to replicate its own chromosomal origin *oriC* (DnaA – which initiates replication; DnaB – a DNA helicase; DnaC – which loads the DnaB helicase at the replication fork; DnaG – provides the primer for replication; DNA bending proteins such as HU, IHF or FIS). The opening of the DNA strands at *ori* sequences, formation of a replication complex and unwinding of DNA strands by a helicase are features common to all genome replications. Static or intrinsic bends, as well as protein-induced bends, have been shown to contribute to these replicative structures in *oriC* (Shaper and Messer, 1995), lambda (Zahn and Blettner, 1987) P1 (Mukhopadhyay and Chattoraj, 1993) and several others.

1. Theta replication

Theta replication is the predominant mechanism used by Gram-negative bacteria though it has also been reported in plasmids from Gram-positive bacteria. The name stems from the electron microscopic observation of plasmid molecules which revealed that the intermediates of replication appear as bubbles that, as they increase in size, resemble the Greek letter theta (θ). The theta mechanism is the most common form of DNA replication used by several plasmids, and is also used by the bacterial chromosome.

The plasmid begins to replicate when the DNA strands of the plasmid origin open. The plasmid origin is the minimal region of DNA that can support autonomous replication and is where synthesis of the new DNA strand begins.

Features of plasmid origins typically include:

- Binding sites (which may comprise short repeated sequences), for plasmid-encoded replication (Rep) proteins.
- A thermodynamically unstable AT rich region which facilitates strand separation.
- Sequences recognized and bound by the host DnaA initiator protein (*dnaA* boxes).
- Sites for adenine methylation (by *dam* methylase) are important for P1 and pSC101, as well as replication of the *E. coli* chromosome.

In many cases (e.g. P1, F, pSC101, RK2/RP4) the plasmid origin contains short, directly repeated sequences, called iterons, which are the binding sites for the plasmid-encoded replication (Rep) proteins. These iterons are essential for replication. Plasmid R1 is a notable exception which contains no iterons in its origin. The R1 origin contains a RepA-binding region, an AT-rich region and a DnaA box. ColE1, unlike R1 uses the host-encoded DNA polymerase, Pol I, to initiate replication. ColE1 also has sequences that encode the primer of the leading strand in replication (RNAII), sequences that allow that primer to anneal at the origin, sequences that favour processing of this complex by Rnase H to generate the 3′ end to prime leading strand synthesis, a site where the replication complex (primosome) assembles and a site for termination of replication, *terH*.

The initiation of plasmid replication begins with opening of the DNA strands at the origin. Strand opening is mediated by plasmid-encoded Rep and host-encoded DnaA initiator proteins and by RNA polymerase. These proteins are assembled sequentially with the DNA to form a nucleoprotein structure known as the replisome complex. The intrinsic curved nature of the plasmid origin and the presence of factors capable of bending DNA (such as IHF, HU and FIS) serve to maintain the architecture of the replisome.

In theta-replicating plasmids which are dependent on plasmid-encoded initiators the Rep protein binds to specific sequences in the origin, forming a nucleoprotein pre-initiation complex analogous to the one formed by DnaA at *oriC* of the *E. coli* chromosome. The Rep/DNA complex, in combination with DnaA, facilitates the transfer of a helicase complex (DnaB–DnaC) to the origin and the opening of the strands in the thermodynamically unstable AT-rich region. The assembly of the pre-initiation complex and details of the molecular interactions leading to the initiation of replication are well documented for some theta-replicating plasmids (pSC101, P1, R1, RK2, R6K, pAMfl1) and are reviewed by del Solar *et al.* (1998).

One of the best-characterized replicons, ColE1, is independent of plasmid-encoded initiator proteins. Initiation of ColE1 replication involves the consecutive activities of RNA polymerase (RNAP), RNase H, DNA Pol I, and DNA Pol III holoenzyme (Marians, 1992).

Transcription mediated by the host RNAP is required to synthesize the preprimer (RNAII) for leading-strand synthesis. RNAII transcription begins 555 bp upstream of the plasmid origin and is 700 bases long. RNAII hybridizes at the melted plasmid origin, where it is cleaved by RNaseH. The end product of cleavage is a 3' hydroxyl group which provides the start point for leading strand synthesis by Pol I. DNA Pol I extends RNAII by about 400 nucleotides prior to synthesis of the leading strand. A primosome assembly site (*pas*) is then exposed on the displaced strand. The subsequently assembled primosome translocates in the 5' to 3' direction, unwinding the helix as it goes and primes the lagging strand (discontinuous) DNA synthesis. At this point, DNA Pol III holoenzyme takes over for DNA Pol I with a simultaneous increase in processivity. Lagging-strand synthesis, initiated at the *pas* site, extends towards the promoter of RNA II but is arrested at a site (*terH*) 17 bp upstream of the leading-strand initiation site (DasGupta *et al.*, 1987). This arrest of lagging strand synthesis determines the unidirectional pattern of ColE1 replication (del Solar *et al.*, 1998).

Host-encoded DNA helicase catalyses further unwinding of the DNA strands. Generally, RNA primers are synthesized either by RNAP or by bacterial or plasmid primases. DNA synthesis of both strands is coupled and occurs continuously on the leading strand, and discontinuously on the lagging strand and is mediated by DNA Pol III. Unwinding of the DNA strands during replication results in the formation of tightly coiled topoisomers which are resolved by either type I or type II topoisomerases. Genetic analysis revealed the involvement of a specific type II topoisomerase, Topo IV, in the segregation of plasmids and bacterial nucleoids (Kato *et al.*, 1988). A two-stage model for the segregation of the replication products has been proposed (Adams *et al.*, 1992), with DNA gyrase reducing the linking number during elongation of DNA synthesis (stage I) and Topo IV resolving the supercoiled catenates which are the products of replication (stage II).

Theta-type replication terminates by molecular interactions at *ter* sequences (del Solar *et al.*, 1998). In the case of unidirectional replication (such as ColE1) then this site will be very close to the plasmid origin. Sequences arresting lagging-strand synthesis, called *terH*, were found upstream of and close to the *pas* site of ColE1. The arrest of the replication fork seems to be caused by the non-hybridized portion of RNA II (DasGupta *et al.*, 1987). The first *ter* site was identified in plasmid R6K as a block to the unidirectional replication fork. The *ter* region was found to contain a pair of separable 22 bp sites, *terR1* and *terR2*, each of which can block the DNA replication fork (Horiuchi and Hikada, 1988). The description of the essential features of the *ter* site led to the identification of similar sites in plasmids of the IncFII (R1 and R100) and IncFI (repFIC) groups.

2. Strand displacement replication

The best-known examples of plasmids replicating by the strand displacement mechanism are the promiscuous plasmids of the IncQ family, whose prototype is the resistance plasmid RSF1010. IncQ replicons can be propagated in many different hosts and require three plasmid-encoded proteins for initiation of DNA replication. These proteins (RepA, RepB and RepC) promote initiation at a complex origin region, and replication then proceeds bidirectionally by a strand displacement mechanism. Initiation of RSF1010 replication is independent of transcription by host RNA polymerase and is independent of the host-encoded DnaA, DnaB, DnaC, and DnaG proteins, whose roles are played by the combined action of the plasmid-encoded RepA, RepB and RepC proteins (Scherzinger et al., 1991). This independence from the host-encoded replication machinery may account for the broad host range character of the IncQ replicons.

The RSF1010 origin contains three functional loci: three 20 bp iterons, the ssiA region and the ssiB region. The minimal origin includes the iterons plus the 174-bp region that contains a GC-rich and an AT-rich segment. The iterons and the adjacent AT-rich region function as the strand separation region, and the ssiA and ssiB sites constitute the priming region for elongation (Sakai and Komano, 1996). The wild type origin extends further with a non-essential region and two small palindromic sequences (stem loops) containing the ssiA and ssiB sites which are located on opposite strands. A model for initiation of RSF1010 replication was proposed by Scherzinger et al. (1991). Replication initiates at ssiA and ssiB, when these origins are exposed as single-stranded regions. The melting of the DNA strand is dependent on two replication proteins, RepC and RepA, and is facilitated by an AT-rich region that precedes the ssiA and ssiB regions. The plasmid-encoded RepC protein binds the iterons (Haring et al., 1985) where it probably interacts with RepA. RepA is a bifunctional protein with 5' to 3' helicase activity and single-stranded DNA-dependent ATPase activity. It is assumed that the RepA helicase binds to both DNA strands in the AT-rich region, close to the site of interaction of RepC. The melting of the duplex is either dependent on the helicase activity of RepA or the interaction of RepC with the iterons. The RepA/RepC complex exposes the ssi sites in a single-stranded DNA configuration (Scherzinger et al., 1991). The ssiA and ssiB sequences are specifically recognized by the plasmid-encoded RepB primase, which primes continuous replication from both of these sequences (Honda et al., 1989). repB is expressed from two in-frame alternative start codons, resulting in two polypeptides of 36 and 38 kDa, which correspond to two functional forms of the RepB primase: RepB and RepB' (Scholz et al., 1989). The exposure of the stem–loop structure in the ssi sites is probably required for the assembly of the RepB-primase to initiate replication (Miao et al., 1993). Initiation at either ssi site can occur independently, and replication proceeds continuously, with the RepA helicase facilitating displacement of the non-replicated parental strand as a D loop. Continuous replication from each ssi signal in opposite directions would originate a double-stranded DNA theta-shaped structure in the overlapping region and two D loops beyond this region. The ssDNA molecules could correspond to either DNA strand and therefore could contain either the ssiA or ssiB sequences (del Solar et al., 1998).

3. Rolling circle (RC) replication

Most RC plasmids are relatively small, and range in size from 1.3 to approximately 10 kb. Replication by the RC mechanism is widespread among multicopy plasmids from Archaea and Bacteria; examples of RC plasmids include pUB110 and pC194 from *Staphylococcus aureus* and pIJ101 from *Streptomyces lividans*. The current model for RC replications is based predominantly on the results of studies obtained with the plasmids of the pT181 family. Studies on a large number of plasmids that replicate by a rolling circle (RC) mechanism have revealed that they fall into several families based on homology in their initiator proteins and leading-strand origins (Khan, 1997). The term rolling circle replication describes how the template strand can be imagined to roll as it replicates.

The first step in RC replication is an interaction between the plasmid leading-strand origins (DSO) with their initiator proteins. The leading-strand origins contain distinct sequences that are required for binding and nicking by the Rep proteins. DSOs of RC plasmids are usually less than 100 bp, may be embedded within the coding sequence of their Rep proteins and may contain an intrinsic DNA bend (Koepsel and Khan, 1986; Khan, 1997). RC replication begins with the nicking of one strand at the plasmid DSO; the nick site is present in the single-stranded hairpin loop of an extruded cruciform structure (Noirot *et al.*, 1990). The DSO contains two domains: one for binding the Rep protein and another which contains the nick site. In plasmids of the same family the nick site is highly conserved, while the Rep binding sequences are not well conserved (Khan, 1997). Iterons are not generally present in the origins of most RC plasmids.

Rep proteins contain domains that are required for both the initiation and termination of replication. Rep binding and DSO nicking are both essential for initiation of replication (Gennaro *et al.*, 1989). The Rep protein must interact stably with the origin to assemble a replication initiation complex, after which it is closely associated with the replication fork. The 3' OH created by the nick serves as the primer for replication, which proceeds around the circle, displacing the opposite strand. The displaced strand then serves as a template to make another double-stranded DNA molecule starting at the lagging strand origin (SSO). The replication fork proceeds around the circle until the nick site has been regenerated. The fork proceeds approximately 10 nucleotides beyond the Rep nick site and the hairpin structure surrounding the nick site is regenerated and replication is terminated. Replication is complete when the DNA is religated by the initiator protein (Zhao and Khan, 1997). Following termination the Rep protein is inactivated due to the attachment of the 10-mer oligonucleotide to the site in the protein responsible for the nicking closing of the DNA (Rasooly and Novick, 1993). This results in the release of a circular leading-strand ssDNA and a nicked open circular DNA containing the newly replicated leading strand. The nick is then sealed by the host DNA ligase and the DNA is subsequently converted to the supercoiled form with DNA gyrase.

On the displaced strand, replication initiates at the SSO to make another double-stranded replicative form. SSOs usually have extensive secondary structure, and generally are not conserved within families. The SSOs are sequence and orientation specific and are located just upstream of the DSO (Gruss et al., 1987). The SSOs only function as ssDNA and so replication of the lagging strand does not initiate until the leading strand has been almost fully synthesized. RC plasmids generate ssDNA intermediates during replication and are indicative of RC plasmids. The leading- and lagging-strand origins are distinct, and the displaced leading-strand DNA is converted to the double-stranded form by using solely the host proteins. RNA polymerase directed synthesis of a short RNA primer initiates lagging strand replication. If the SSOs do not function (e.g. in a particular host) then single-stranded forms of the plasmid will accumulate, resulting in instability and decreased copy number (Gruss et al., 1987; del Solar et al., 1993). Several types of SSO have been identified based on their secondary structures. Some SSOs are able to function in only their natural hosts, and some in several hosts. The ability of an SSO to function in various hosts may contribute to plasmid promiscuity and horizontal transfer among related bacteria (Khan, 1997).

B. Control of plasmid replication

Another plasmid characteristic controlled by the *ori* region is copy number (average number of plasmid molecules per cell). Finely controlled mechanisms prevent either the over-accumulation or loss of plasmids. High copy number plasmids (e.g. ColE1) need only have a mechanism that inhibits replication when copy number reaches a certain level (relaxed level of control). Low copy number plasmids (e.g. P1) replicate only once or a few times during the cell cycle and have a more stringent mechanism of regulation.

I. Antisense RNA (e.g. ColE I)

Replication of ColE1 plasmids is regulated by the plasmid-encoded antisense RNA called RNAI (Eguchi et al., 1991). RNAI inhibits plasmid replication by interfering with RNAII (the primer for plasmid DNA replication). Ordinarily, RNAII forms DNA–RNA hybrid with the origin, which is then cleaved by RnaseH releasing a 3' hydroxyl group which can be extended by DNA polymerase I. RNAI and RNAII are complementary since they are synthesized from opposite strands of the same sequence. Hybridization of RNAI to RNAII sequesters the RNAII so that initiation of replication is prevented. The pairing of RNAI and RNAII is further enhanced by plasmid-encoded Rop protein (Lacatena and Cesarini, 1981). The corollary of this control mechanism is that the higher the copy number, the more RNAI is synthesized, resulting in more sequestration of RNAII, ultimately followed by less initiation of replication. Rop enhances RNAI/RNAII pairing so that replication is prevented at even lower plasmid copy numbers.

2. Transcriptional repressor and antisense RNA (e.g. R1)

R1 controls its replication by regulating transcription of its replication protein (RepA) so that it is only expressed immediately after cell division and thereafter regulates its copy number using an RNA molecule, CopA.

Plasmid-encoded RepA is required for initiation of plasmid R1. The *repA* gene can be expressed from two promoters p_{copB} and p_{repA}. Expression from p_{copB} yields RepA and CopB. The p_{repA} promoter lies within the *copB* gene and only expresses the RepA protein. p_{repA} is repressed by CopB and so is expressed only immediately after the plasmid enters a cell and before CopB is made. A short burst of RepA expression facilitates replication until the plasmid attains its copy number, at which point CopB represses p_{repA} and *repA* transcription proceeds from the p_{copB} promoter only.

Once the plasmid has attained an appropriate copy number, RepA synthesis (and therefore plasmid replication) is regulated by a small RNA called CopA. CopA affects the stability of the mRNA expressed from the p_{copB} promoter as it is transcribed from the opposite strand of the DNA which encodes the *repA* gene. RNAseIII (which cleaves only double-stranded RNA molecules) cleaves the CopA/RepA mRNA structure, ultimately turning off the synthesis of RepA. Replication of R1 (and other IncFII plasmids) is regulated by CopA concentration, which in turn is dependent on the plasmid copy number.

3. Iterons (e.g. P1)

Many plasmids contain clusters of direct repeats called iterons, which provide binding sites for the plasmid replication proteins. Iterons in the plasmid origin were first demonstrated for F, which has five 22 bp elements, though other plasmids have different arrangements (Tolun and Helinski, 1981). Generally the iterons are spaced so that binding of the replication proteins to the iterons occurs on the same face of the DNA helix. Many iteron-containing plasmids contain a second cluster of direct repeats that are non-essential for replication, but are important for copy number control. The P1 plasmid copy number is controlled by iterons. 19 bp repeat sequences present in the replication origin and in the distal copy control locus *incA*. The P1 iterons are binding sites for the RepA initiator protein. Protein-mediated contacts between the iteron sequence in different copies of the plasmid are thought to limit replication, and thus control copy number. If the iteron sequences are mutated to decrease the binding affinity of RepA then copy number increases (Papp *et al.*, 1994)

Iteron plasmid replication is regulated concurrently in two ways. Firstly, transcriptional autoregulation of the *repA* gene means that as the copy number increases so does RepA concentration. Consequently the higher the RepA concentration, the more it will repress its own synthesis, resulting in a RepA concentration that remains in a narrow range and replication that is strictly regulated. Secondly, the iterons sometimes associate joining the plasmids in a process known as handcuffing. This model is based on experimental evidence that replication is also controlled by the concentration of the iterons. Introduction of more iterons reduces the plasmid copy number.

RepA can link two plasmids by binding iterons and preventing initiation of replication (McEachern *et al.*, 1989; Abeles *et al.*, 1995). In summary, this mechanism of maintaining copy number depends on both the concentration of RepA and the plasmids themselves. Replication proceeds at low copy number and is inhibited at high copy number.

4. Hemimethylation and involvement of the cell membrane

Immediately after replication initiation conditions within the cell are optimal for new rounds of replication to begin. If the origins are made unavailable for replication or sequestered, replication can be prevented until the next cell cycle. Sequestration of both the chromosomal origin and plasmid P1 replication has been demonstrated. The P1 plasmid replication origin P1*oriR* is controlled by methylation of four GATC adenine methylation sites within heptamer repeats. A comparable (13 mer) region is present in the host origin, *oriC*. Immediately after replication the Dam methylation sites of the plasmid origin remain hemimethylated for about 10 min (Campbell and Kleckner, 1990). The sequestration pathway prevents premature reinitiation by blocking the replication of newly initiated DNA which is marked by hemimethylation. Hemimethylated DNA of both the P1 origin and the host chromosome is directly recognized by host protein SeqA. SeqA is thought to bind the hemimethylated region to the cell membrane, thus preventing it from being further methylated (Brendler *et al.*, 1995; Slater *et al.*, 1995). In *seqA* mutants the frequency of replication initiation increases threefold (von Freiesleben *et al.*, 1994).

Evidence for the association of both plasmid and chromosomal DNA with the cell membrane has continued to accumulate (Firshein and Kim, 1997). Jacob *et al.* (1963) proposed a replicon model in which the cell membrane was an essential component. Membrane-associated plasmid DNA replication has been demonstrated for the resistance plasmid RK2 (Mei *et al.*, 1995). Similar results were obtained by Chakraborti *et al.* (1992), who demonstrated that membrane-associated hemimethylated *oriC* from *E. coli* also bound differentially to the two fractions in the same manner as *oriV* from RK2. Host DnaA (up to 50%) has been shown to be associated with the membrane (Sekimizu *et al.*, 1988; Sekimizu and Kornberg, 1988). DnaA is more efficient in a membrane environment (Yung and Kornberg, 1988). Thus it may be that as well as destabilizing AT-rich regions, DnaA could facilitate binding of origins to the membrane (Firshein and Kim, 1997).

C. Replication of Broad Host Range Plasmids

Replication of broad host range plasmids is considerably less well understood than the replication of other prokaryotic replicons. Interestingly, the origins of narrow host range plasmids share features with those of broad host range plasmids (e.g. DnaA boxes located next to AT-rich regions). Many factors influence the ability of plasmids to replicate in different hosts. If initiation of replication is independent of host initiation factors, or the plasmid and host initiation factors are able to act together then the chances of a plasmid being able to replicate in a given host are increased (del Solar *et al.*, 1996). Some plasmids have origins that allow them to replicate in many types of bacteria such as RSF1010 (IncQ), RP4 and RK2 (IncP). Plasmid replication depends on host enzymes and on plasmid-encoded and plasmid controlled *cis* and *trans* determinants. Some plasmids have determinants that are recognized in almost all Gram-negative bacteria and replicate in each host. Other plasmids possess this ability in only some bacteria. Plasmids which replicate in *E. coli*, but which do not replicate in other bacteria, have been widely employed as 'suicide' vectors for the delivery of transposons and mutagenized genes for allelic exchange. This is true of *Clostridia* (Chapter 6, III.A), *Bacteroides* (Chapter 8, Table 2), Archaea (Chapter 10, II, III and IV), *Helicobacter* (Chapter 11, IV.C.1), *Rhodobacter* and *Paracoccus* (Chapter 13, III.B), *Bacillus* (Chapter 14, IV.A).

For example, ColE1 cannot replicate in cell-free extracts of *Pseudomonas*, but partial replication ability is restored when purified *E. coli* DNA gyrase and DNA polymerase are added. Therefore the differences in these enzymes between these bacteria seem to be responsible for the restriction of ColE1 replication to members of the *Enterobacteriaceae* (Diaz and Staudenbauer, 1982). In contrast, broad host range plasmids have gained independence from host enzymes. Consequently, the plasmid-encoded replication functions of broad host range plasmids, such as the IncP and IncQ plasmids, and their control are more complex than those in narrow host range plasmids. They also differ in the fact that the genes that are required for plasmid replication are distributed in several regions of the plasmid (Kues and Stahl, 1989).

The initiation of replication of IncQ plasmids is independent of transcription by host RNA polymerase and is independent of the host-encoded DnaA, DnaB, DnaC and DnaG proteins, whose roles are played by the combined action of the plasmid-encoded RepA, RepB and RepC proteins. This independence may account for the broad host range character of the IncQ replicons.

The replication system of RK2 can be put forward as a model for IncP broad host range plasmids. In addition to an origin and plasmid-encoded replication initiation proteins, RK2 replication also requires the host-encoded proteins DnaA, DnaB, DnaC, DNA gyrase, DnaG primase, DNA polymerase III holoenzyme and SSB (Kittel and Helinski, 1991). Replication in RK2 is in the theta mode, and proceeds unidirectionally from *oriV*. The *oriV* region consists of eight 17 bp iterons, a putative promoter surrounded by DnaA binding sites and an IHF site and a 49 bp AT-rich sequence containing another DnaA box, and a 67 bp GC-rich sequence. The *trfA* gene encodes two replication initiation proteins: TrfA-44 (44 kDa) and TrfA-33 (33 kDa) as a result of an internal translational start in the same open reading frame (Shingler and Thomas, 1984). Either protein alone is sufficient for the replication of RK2 in *E. coli* and a wide range of Gram-negative bacteria, with the exception of *Pseudomonas aeruginosa*, which specifically requires TrfA-44 (Fang and Helinski, 1991). *trfA* mutations that result in loss of binding to the iterons at the RK2 origin are non-functional *in vivo* (Lin and Helinski, 1992). The monomer form of the TrfA protein is active for binding to the iterons at the RK2 replication origin (Toukdarian *et al.*, 1996).

The DnaA protein enhances or stabilizes the formation of the TrfA-mediated open complex (Konieczny *et al.*, 1997), and recruits DnaB to *oriV* (Konieczny and Helinski, 1997). The helical phasing and intrinsic DNA curvature are critical factors for the formation of functional nucleoprotein structure (Doran *et al.*, 1998). The TrfA protein and the *oriV* sequences have been shown to be sufficient for the initiation and control of replication of RK2 in almost all Gram-negative bacteria. The origin was originally described as a 700 bp fragment which contained eight 17 bp repeats (iterons) in two clusters of five and three AT- and GC-rich regions, sequences homologous to DnaA boxes (Stalker *et al.*, 1981). A 393 base pair minimal origin, *oriV*, of plasmid RK2 which contains the five iteron cluster, three of the putative DnaA boxes, and the AT- and GC-rich regions is also a functional origin, though it results in a decreased copy number in *E. coli* (Thomas *et al.*, 1984).

Generally the same origin fragment is used in different bacteria; however, depending on the host, some fine structures of *oriV* are not as important, or can be totally omitted (e.g. number of iterons, location of DnaA boxes; Kues and Stahl, 1989). Mutations in any of the origin features may reduce the host range of the plasmid. Deletion of the upstream three iteron cluster does not inactivate the origin in *E. coli* but does result in an increase in plasmid copy number and some instability (Blasina, 1996).

D. Identification of a Plasmid Replicon

Replicons can be identified by testing smaller fragments of the plasmid to see if they can confer replicative activity. A library of restriction fragments can be ligated to a suitable selectable marker and can then be tested to determine whether or not they are stable. Such an approach was taken to identify the IncFIIA replicon of the 230 kb virulence plasmid pMYSH6000 of *Shigella flexneri* 2a. A pBR322 ligated library of *Sal*I digested fragments of pMYSH6000 was constructed (Sasakawa *et al.*, 1986). Each *Sal*I fragment was assigned a letter according to its size, *Sal*IA being the largest fragment. The third largest fragment, *Sal*IC, was found to contain the essential replication region (Rep) of pMYSH6000. *Sal*IC was ligated to a *Xho*I fragment containing the kanamycin resistance determinant of Tn5. This construct was found to be somewhat unstable unless the 5.7 kb *Sal*IO fragment was also included in the construct (Makino *et al.*, 1988). The *Sal*IO fragment was subsequently identified as a plasmid stability locus which was able to stabilize other replicons (Radnedge *et al.*, 1997).

A similar approach was taken to localize and characterize a minimal 2.3 kb replicon from the thermophilic bacterium *Thermus thermophilus* YS45 (Wayne and Xu, 1997). This fragment contained one significant open reading frame which putatively encodes a 341 amino acid protein of molecular mass of 38.2 kDa. A suitable ribosome binding site and promoter sequences typical of *Thermus* sp. were identified. The open reading frame encoded a novel protein, as determined by a BLAST search against published sequences. The gene was designated *repT*, and its gene product was shown to be necessary for replication by interruption of the open reading frame. Interestingly, no iterons were identified, but sequence analysis revealed two DnaA boxes. The authors suggest that plasmid copy number may be regulated by the relationship between binding of a thermostable DnaA homologue at these sites, and the transcription of RepT. The selection of additional thermophilic origins will help outline the mechanism of plasmid replication in thermophilic bacteria.

◆◆◆◆◆◆ IV. PLASMID STABILITY AND INCOMPATIBILITY

A. Stable Plasmid Maintenance Mechanisms

The copy number control mechanisms of high copy number plasmids maintain the same number of plasmids within the bacterial cell. If one daughter cell contains fewer plasmids after replication and cell division than the other daughter cell, then these mechanisms restore the copy number to the appropriate level. The ramifications of plasmid mis-segregation are much more important for low copy number plasmids, since they will be lost if the daughter plasmids are mis-segregated during cell division. To prevent such instability many low copy plasmids encode genetic systems to ensure their stable retention in the bacterial population. Often the larger low or single copy plasmids will encode more than one stability system to decrease the chances of loss even further. Plasmids ensure hereditary stability either by high copy number or by encoding a specific genetic mechanism such as multimer resolution, plasmid partitioning (par) and postsegregational killing.

I. Multimer resolution

The chances of losing a plasmid during cell division are increased if the plasmids form multimers during replication. Multimers may arise by incomplete termination of replication or by homologous recombination events. The formation of multimers effectively lowers plasmid copy number as a multimer will segregate as a single plasmid. In due course the formation of multimers increases the chance of losing the plasmid and so has serious consequences for a low copy number plasmid. Multimer resolution systems have been identified in a variety of plasmids. They convert multimers to monomers, maximizing the number of independently segregating molecules and minimizing the frequency of plasmid loss.

These systems promote recombination at specific sites on the plasmid when that site occurs more than once (as would occur in a multimer). For example, the cer-Xer site-specific recombination system resolves multimers of ColE1 plasmids (Guhahakurta et al., 1996). The cis-acting cer site is analogous to the dif replication termination site of the E. coli chromosome, and uses the same Xer proteins (XerC and XerD) that interact with dif. Site-specific recombination between directly repeated cer sites in plasmid multimers regenerates monomers and maximizes the number of segregating units. Multimer resolution in *E. coli* also requires chromosomally encoded accessory proteins which promote intramolecular recombination over intermolecular recombination (Colloms *et al.*, 1990; Stirling *et al.*, 1988). The Xer proteins belong to the integrase family of site-specific recombinases which act only at specific sites. Multimer resolution systems have also been described for P1, F, RK2/RP4, R46 and the *Salmonella* virulence plasmid (Helinski *et al.*, 1996).

2. Plasmid partitioning

Some low copy number plasmids encode active *cis*-acting partition systems, e.g. the P1 *par* system, the F *sop* system, which ensures that each daughter cell receives at least one copy of a plasmid molecule after cell division. Most of the partition systems described so far come from Gram-negative bacterial plasmids, though it is likely that they will also be found in plasmids of Gram-positive bacteria.

The best-characterized systems are P1 and F, though similar systems also exist in RK2 (Roberts *et al*, 1990), pTar (Gallie and Kado, 1987), R1/NR1 (Dam and Gerdes, 1994; Jensen *et al*, 1994). All *par* systems encode two *trans*-acting proteins which act at a *cis*-acting site, which is considered to be an analogue of the eukaryotic centromere. These par systems are also capable of stabilizing heterologous replicons. The accurate segregation of P1 during cell division is achieved by an active partition system, a 2.1 kb region of the plasmid with *parA* and *parB* genes and a *cis*-acting site *parS* (Abeles *et al.*, 1985). The P1 *par* system is highly efficient; P1 is lost in only 1 in 10^6 cell divisions (Abeles *et al.*, 1985). Similar *par* systems have not only been identified in a diverse group of plasmids but also in bacterial chromosomes including *Caulobacter crescentus* (Mohl and Gober, 1997) and *Bacillus subtilis* (Webb *et al.*, 1997).

The ParA and ParB proteins of P1 show extensive homology with the SopA and SopB proteins of F. However, the *cis*-acting sites are different. P1 *parS* site contains heptamer repeats which have been shown to bind ParB (Davis and Austin, 1988), while the *sopC* site of F contains twelve 43 bp repeats. In both plasmids the Par proteins are expressed from a promoter upstream of *parA/sopA*. The *parS/sopC* site is necessary and sufficient for partition as long as the two Par/Sop proteins are supplied in *trans*. In P1 the partition complex is macromolecular assembly of these three components (ParA, ParB and *parS*) with the host-encoded DNA-binding protein IHF (Funnell, 1991). ParB binds specifically to *parS* forming a complex that includes IHF (Davis and Austin, 1988; Funnell, 1988). ParA acts to control regulation of the *par* operon by binding to its promoter and also has a direct role in partition (Davis *et al.*, 1996). Autoregulation by ParA is enhanced by P1 ParB protein (Friedman and Austin, 1988). ATPase activity has been demonstrated for ParA (Davis *et al*, 1992) which may provide energy for the partitioning process.

It can be argued that partitioning of daughter plasmids must involve some structure in the cell, and there is some evidence that plasmid partition proteins are membrane associated (Lin and Mallavia, 1998). Williams and Thomas (1992) and Hiraga (1992) have postulated that motor complexes exist in the form of contractile proteins much like the cytoskeletal filaments found in eukaryotic cells. The host-encoded MukBEF proteins identified by Hiraga and co-workers appear to play a role in the positioning and segregation of the *E. coli* chromosome; *muk* mutants generate anucleate cells (Hiraga *et al.*, 1990; Niki *et al.*, 1991; Yamanaka *et al.*, 1996). MukB is not required for accurate segregation of plasmids P1 or F indicating firstly that there are fundamental differences between plasmid partition and chromosome segregation and secondly, that it is unlikely that the plasmid is co-transferred with the chromosome (Ezaki *et al.*, 1991; Funnell and Gagnier, 1995). Interestingly, recent work has suggested that chromosomal segregation of origins in *Caulobacter crescentus* and *Bacillus subtilis* uses proteins that are homologous to the partition proteins of P1 and F (Mohl and Gober, 1997; Webb *et al.*, 1997). Further evidence implicates the cell poles as the target to which the bacterial chromosome origin is aimed by these Par analogues, whereas plasmids P1 and F are localized at the mid-point of the cell (Gordon *et al.*, 1997, Niki and Hiraga, 1997). After association of the chromosomal origin with the cell pole, segregation is completed by condensation of the majority of the chromosomal DNA in a Muk-mediated reaction, whereas the small dimensions of plasmids might explain their independence of the Muk proteins (J. Sawitzke and S. Austin, personal communication).

3. Postsegregational killing systems

A number of plasmid-encoded systems have evolved that kill cells that have lost a plasmid, and whose purpose is to stabilize the plasmid within a bacterial population. Extracellular systems include the bacteriocins and microcins, both of which are polypeptide antibiotics. Cells carrying the plasmids which produce these compounds are immune to their toxic effects. Cells that have lost the plasmid lose their immunity and succumb to the toxin. One type of intracellular system comprises a toxin and antidote both encoded by the plasmid. Since the toxin is more stable than antitoxin, loss of the plasmid will kill the cell since the toxin will linger in the absence of antidote. A second intracellular postsegregational killing system works by preventing the synthesis of a toxin by inhibition of translation by an antisense RNA. Several plasmid based toxin/antitoxin systems that cause the death of plasmid-free segregants have been characterized. The systems comprise two components which are generally a pair of small proteins expressed from a single promoter. If the plasmid is lost then the antidote (usually less stable than the toxin) breaks down leaving the toxin to exert its effect (Jensen and Gerdes, 1995).

The *ccdA/B* operon of F encodes a toxin, CcdB, which alters DNA gyrase (*gyrA*) to cause double-stranded breaks in DNA thereby inducing the SOS response (Jaffe *et al.*, 1985; Bernard and Couturier, 1992). The antidote is CcdA which neutralizes the effect of CcdB. The *ccdA/B* operon is regulated by CcdB, and CcdA enhances the binding of CcdB to the promoter. The toxin is thought to be more stable than the antidote, therefore the toxic activity of CcdB is released in the cells that have lost the plasmid.

P1 harbours a pair of genes that contribute to plasmid stabilization by induction of a lethal response to plasmid loss. The genes that confer *death on curing* (*doc*) and *prevent host death* (*phd*) encode a 126 amino acid toxin and a 73 amino acid antidote. This two-protein system is similar to previously characterized postsegregational killing systems in that the two genes constitute an operon and the synthesis of Doc appears to be translationally coupled to that of Phd. The Phd/Doc system does not induce SOS like the CcdA/CcdB system of F, and the target of Doc remains unclear. Doc and Phd are unrelated to previously described proteins of other plasmids that increase plasmid stability by postsegregational killing (Lenherr *et al.*, 1993).

R1 has two postsegregational killing systems. In the first, the host killer protein, Hok, is a membrane-associated protein which kills by inducing membrane leakiness (Gerdes *et al.*, 1986). Synthesis of Hok is prevented by the antisense Sok RNA. This unstable RNA accelerates the turnover of the otherwise stable Hok messenger and indirectly blocks its translation of the upstream *mok* gene, to which *hok* is translationally coupled (Thisted and Gerdes, 1992). In the second, the *kis/kid* system (and its identical counterpart in R100, PemI/PemK) encodes 9.3 kDa and 12 kDa proteins respectively. The target of the PemK protein is DnaB, and is neutralized by PemI after formation of a tight complex between the two proteins. Again the antidote, PemI, is less stable than the toxin, PemK (Tsuchimoto *et al.*, 1992). The primary effect of the Pem system is to delay cell division, conferring a selective disadvantage on the cured cells, rather than killing plasmid-free segregants. Consequently the *pem* system has a relatively modest stabilization effect (Jensen *et al.*, 1995).

B Stability of the Broad Host Range Plasmid RK2

Some of the large, low copy number plasmids encode more than one stability system as a fail-safe mechanism. The broad host range plasmid RK2 encodes several different mechanisms of ensuring its stability. Two of these mechanisms are encoded in one 3.2 kb region, designated *par*. *par* consists of two divergently arranged operons: *parCBA* (2.3 kb) and *parDE* (0.7 kb) (Roberts and Helinski, 1992). The two operons are divergent and are autoregulated (Davis *et al.*, 1992). The *parCBA* operon encodes a 24 kDa site specific resolvase protein (ParA) and its multimer resolution site (*res*) and two proteins (ParB and ParC) whose functions are unclear. ParB has endonuclease activity, but its contribution to stable maintenance is not known. The *res* site is located in a 100 bp region between the two divergent promoters P_{parCBA} and P_{parDE}. The 2.3 kb *parCBA* fragment alone has been shown to stabilize a mini-RK2 plasmid (Sobecky *et al.*, 1996). Interestingly the multimer resolution activity of *parCBA* does not account for the level of stabilization observed (Roberts and Helinski, 1992; Sobecky *et al.*, 1996). It has been proposed that the *parCBA* operon encodes a plasmid partitioning system (Gerlitz *et al.*, 1990).

The adjacent operon, *parDE*, encodes a toxin/antitoxin system that stabilizes by killing plasmidless cells (Roberts *et al.*, 1993). ParD is known to combine with and neutralize the toxic effects of the ParE protein (Johnson *et al.*, 1996). The contribution to stability of the *parDE* operon depends on the sensitivity of the host in which it is being propagated to the ParE toxin (Easter *et al.*, 1997). The relative importance of *parCBA* and *parDE* varies depending on the host bacterium (Sia *et al.*, 1995).

More stability mechanisms encoded by RK2 have been postulated on the basis of homology to previously identified active partition systems. The four operons of the *kilA*, *kilC* and *kilE* determinants are not essential for replication or conjugation. Since they are conserved within the IncPα family of plasmids and co-regulated with replication it has been suggested that they are involved with plasmid maintenance (Pansegrau *et al.*, 1994; Figurski *et al.*, 1982). The *incC* and *korB* genes of the *korA* operon have been postulated as a plasmid partitioning system on the basis of their sequence identity with those of F and P1. (Motallebi-Veshareh *et al.*, 1990). Gene inactivation studies of *incC* decrease the stability of mini RK2 plasmids (Motallebi-Veshareh *et al.*, 1990; Williams and Thomas, 1992).

In summary, RK2 encodes different mechanisms of ensuring its stability. These mechanisms operate with differing efficiency depending on the situation the plasmid finds itself in. The availability of more than one stability system contributes to the adaptability of RK2, and its effectiveness at being stably maintained in a wide range of bacteria.

C. Problems with Stability in Gram-positive Plasmids

Cloning of long DNA segments (greater than 5 kb) in *Bacillus subtilis* is often unsuccessful when naturally occurring small (less than 10 kb) plasmids are used as vectors. It was generally observed that only short DNA segments can be efficiently cloned in these vectors (Michel *et al.*, 1980), and that longer segments often undergo rearrangements (Ehrlich *et al.*, 1986). Commonly these plasmids accumulate unstable single-stranded DNA molecules as they use the rolling circle mode of replication. Recombination events leading to chromosome rearrangements occur more frequently when DNA is present as a single-stranded molecule, a common intermediate in rolling circle replication. Several studies have indicated that the formation of single-stranded DNA intermediates is an important factor in both structural instability (Bron *et al.*, 1991) and segregational instability (Bron and Luxen, 1985). The formation of high molecular weight plasmid multimers has also been implicated in structural instability and segregational instability (Leonhardt and Alonso, 1991). Illegitimate recombination events, resulting in duplications, deletions, translocations and insertions, have been shown to underlie plasmid instability in *Bacillus subtilis*.

The initial event in establishment of deletions is the introduction of ssDNA nicks or dsDNA breaks by topoisomerases, followed by recombination repair processes. Two classes of deletions have been characterized by Bron and co-workers. Type I deletions occur between non-repeated sequences. The deletions result from Topo I dependent ssDNA nicking, and involve uncoupling of the nicking and closing reactions followed by the joining of unrelated ends producing recombinant molecules. Topo I target sites are seen at deletion end-points, and over-expression of Topo I results in plasmids that are deleted at higher rate (Meima *et al.*, 1998). Type II deletions occur between short direct repeats and are independent of ssDNA replication intermediates, and are therefore independent of the type of replicon used by the plasmid. They are caused by error-prone repair of double strand DNA breaks (DSBs). The model for generating Type II deletions involves exonucleolytic processing of ends followed by annealing of short direct repeats (Meima *et al.*, 1997).

Theta replicating plasmids such as pAMβ1 have been used success-fully in *Bacillus subtilis*. pAMβ1 was originally isolated from and repli-cates via a unidirectional theta mechanism (Bruand *et al.*, 1993). Janniere *et al.*, (1990) showed that vectors derived from the large (26.5 kb) plas-mids pAMβ1 and pTB19 allow efficient cloning and stable maintenance of long DNA segments (up to 33 kb). pAMβ1 replicates use a unidirec-tional theta mode of replication (Bruand *et al.*, 1993). The two large plasmids do not accumulate ss DNA, whereas the rolling-circle replica-tion typical for small plasmids does. In addition, the replication regions of the two large plasmids share no sequence homology with the cor-responding regions of the known small plasmids, which are highly con-served. Recombination occurred much less frequently when carried on large plasmids indicating that large plasmids are structurally much more stable than small ones.

A comparison of the stability of recombinant plasmids based on rolling circle and theta modes of replication in *Lactococcus lactis* was described by Kiewiet *et al.* (1993). Derivatives of the RC plasmid, pWV01, showed size-dependent segregational instability, particularly when large fragments were inserted. In contrast, all the pAMβ1 derivatives were stably main-tained. The pWV01 recombinants generated high molecular weight plasmid multimers that were correlated with plasmid size and inversely correlated with copy numbers of the monomeric forms – phenomena which were absent in pAMβ1 derivatives. Their conclusion was that plasmids employing the theta mode of replication are superior to RC plasmids for cloning in lactococci.

In summary, the inherent instabilities of the single stranded DNA intermediates that arise during rolling circle replication, mean that plasmids which replicate via this method should be avoided whenever possible to clone and maintain long DNA segments in any organism.

D. Incompatibility

Many natural isolates of bacteria have been found to contain more than one type of plasmid. If the plasmids are able to coexist with each other and remain stable, the plasmids are said to be compatible. Not all plas-mids can stably coexist as they interfere with each other's replication and partition systems. If two such plasmids are introduced into a cell then one of them will be lost at a higher then normal rate during subsequent cell divisions. Two plasmids which cannot stably coexist are said to be incompatible, and are members of the same incompatibility group. If two plasmids can stably coexist then they belong to different incompatibility groups. For example RP4 and RK2 both belong to IncP incompatibility group and cannot be maintained in the same cell simultaneously as they have very similar replicons. Plasmids in the same incompatibility group can be incompatible either because of shared replication mechanisms or the same partition mechanisms.

Sharing of any function required for the regulation of plasmid replication by two plasmid elements results in incompatibility (Novick, 1987). Random selection of the daughter plasmids for replication causes imbalance in the copy pools of two co-resident plasmids sharing the same replicon. The disparity cannot be corrected since the copy number control system will select either plasmid for replication (i.e. it does not recognize the two replicons as different) (Helinski *et al.*, 1996). If two plasmids have different replicons (i.e. are compatible) then each plasmid will replicate to reach its own copy number so that at the time of the next cell division the cells will have restored the original number of plasmids, and very few cells will lose the plasmid with each successive generation. If two plasmids have the same replicon (i.e. are incompatible), and plasmid distribution during cell division is asymmetric, then the daughter cells will not receive the same number of plasmids. After cell division the two plasmids replicate until the total number of plasmids reaches the correct copy number (remember each plasmid contributes to the total copy number). The under-representation of one plasmid greatly increases its chance of being lost with each successive round of cell division. A plasmid present in single copy will not be replicated in the presence of another plasmid with the same replicon, and the daughter cell will be cured of one of the two types of plasmid.

ColE1 provides an example in understanding of the mechanistic details of incompatibility. The primary incompatibility determinant of ColE1 is RNAI (Cesarini *et al.*, 1991). Consequently, any other resident plasmid that produces the same RNAI cannot coexist with a ColEI replicon. New incompatibility groups can arise from a single base change in RNA (with the complementary change in the overlapping RNAII primer). The modified RNAI will no longer hybridize with the wild type RNA II, and so each plasmid replicon will be stable since they no longer share replication functions (Helinski *et al.*, 1996).

Plasmids which share partition functions will also exert incompatibility. The plasmid partition mechanism ensures that each daughter cell receives a daughter plasmid. If two different plasmids share the same partition functions it is possible that each will be recognized as the same by the partition functions and will be separated at cell division. This will produce cells cured of one or the other plasmid. The effect of this incompatibility will be the strongest for those plasmids whose copy number is the lowest (Austin and Nordstrom, 1990).

To determine the incompatibility group of an unknown plasmid, it must be introduced into a strain already harbouring a plasmid of a known incompatibility group. If the plasmids are incompatible then loss of the resident plasmid can be determined by monitoring a selectable trait such as antibiotic resistance. Incompatibility testing involves introduction (by conjugation, transduction or transformation) of a plasmid into a strain carrying another plasmid. The two plasmids must have different genetic markers in order to follow their segregation. Selection is usually carried out for the entering plasmid, and the progeny are examined for the continued presence of the resident plasmid. If the resident plasmid is eliminated, the two plasmids are said to be incompatible, and are assigned to the same incompatibility group. In order to perform an incompatibility test both plasmids need suitable selectable markers which are different from each other. It is also worth noting that surface exclusion (inhibition of entry of the donor plasmid) will produce results consistent with incompatibility. The construction of a set of reference mini-plasmids belonging to different incompatibility groups has been created and can be used as recipients in incompatibility tests (Davey *et al.*, 1984). In a recently published example, pSW1201 from *Erwinia stewartii* was shown to be in the same incompatibility group as P1 (IncY). pSW1201 was co-transformed with pGP1 (pGEM into which the *inc* locus of P1 had been cloned) into *E. coli*. Both plasmids were maintained as long as selection by the appropriate antibiotics was provided. In antibiotic-free growth medium pSW1201 was lost at a rapid rate (Fu *et al*, 1997).

E. Method for Determining Plasmid Incompatibility

The general method for testing incompatibility is outlined as follows: the donor (incoming) plasmid is introduced into a recombination-deficient strain carrying the resident plasmid via transduction, conjugation, transformation, with selection for the incoming plasmid. Individual transformants are purified and tested for the presence of the resident plasmid.

If the plasmids are incompatible then the resident plasmid will be displaced and cells bearing only the incoming plasmid will be isolated. If the incompatibility is strong then no transformants will be isolated bearing the resident plasmid, as the incoming plasmid cannot be established in the presence of the resident plasmid. If the plasmids do not share replication functions and the plasmids are compatible, then every colony isolated will contain both plasmids. Each plasmid will have the same copy number that it would have if it were the only replicon in the cells. If every colony contains both plasmids, but the resident plasmid shows reduced copy number (and will eventually be lost after several generations) then the donor exerts weak incompatibility against the resident plasmid.

Below is a protocol for performing an incompatibility test on agar plates using an ampicillin-encoding resident plasmid and an incoming chloramphenicol-encoding plasmid over approximately 25 generations of growth (adapted from Radnedge *et al.*, 1997).

◆◆◆◆◆◆ V. PLASMID CONJUGATION

One of the cornerstones of bacterial genetics is the ability to transfer plasmid DNA from one strain to another. Bacterial conjugation is a specialized process involving unidirectional transfer of DNA from a donor to a recipient cell by a mechanism requiring specific contact. The process utilizes plasmid-encoded functions, *tra*, to transfer DNA between bacteria (see also Chapter 10, IV.B for details of novel conjugation in Archaea). The transfer of DNA by promiscuous plasmids has played a significant role in evolution, constituting the major route for horizontal gene transfer, and many of these promiscuous plasmids were originally isolated as antibiotic resistance factors (e.g. RP4). The transmissibility of R plasmids from one cell to another can quickly transform pathogens into antibiotic-resistant strains making subsequent infections difficult to treat. Plasmids which encode all of the genes necessary for transfer from one cell to another are referred to as self-transmissible, and include F, R1 and RK2. Some plasmids only encode a limited number of mobilization (*mob*) genes for their own DNA processing during conjugation, but require a co-resident self-transmissible plasmid to supply the remaining essential functions. Examples of mobilizable plasmids are RSF1010, ColE1 and pACYC184 (Lanka and Wilkins, 1984). Conjugation is used in many bacteria, see Chapter 5, III.D; Chapter 6, III.D; Chapter 8, II.B and Chapter 13, VI.E.

A. Self-transmissible Plasmids

Briefly summarized, bacterial conjugation involves cell-to-cell contact via the sex pilus, nicking and unwinding of the plasmid DNA at the origin of transfer, transfer of a ssDNA intermediate from the donor to the recipient cell followed by synthesis of the complementary strands to restore the double-stranded molecule. F remains the paradigm of a well-characterized genetic transfer system and was first described by Cavalli et al. (1953). F-like plasmids have been found to encode a range of ecologically important factors such as colicins and metabolic activities, and are found throughout the family Enterobacteriaceae (Jacob et al., 1977). The plasmid transfer (tra) region encodes all of the F loci required for efficient conjugative transfer. The 33.3 kb nucleotide sequence of the entire F tra region was compiled and extensively annotated by Frost and co-workers (1994). This transfer region confers transmissibility on other replicons (Johnson and Willetts, 1980). Some of the tra genes are required to synthesize the sex pilus, while others are more directly involved in DNA transfer and function as endonucleases, helicases, and primases. The functions of the tra genes include: tra operon regulation, pilus assembly, DNA transfer and surface exclusion to prevent the entry of similar plasmids (Firth et al., 1996).

1. Pilus formation

The general model of conjugation begins when two cells are brought together into surface contact by the retraction of the pilus of the donor cell. F pili can be visualized extending 1—2 mm from the surface of the donor cell, though the type of pilus formed depends on species encoding it (Ippen-Ihler and Minkley, 1986). F-like pili are long and flexible and transfer efficiently in liquid cultures, whereas shorter, more rigid pili encoded by other plasmids transfer more efficiently among other surfaces. A donor cell carrying F produces a pilus encoded by the products of the traABCEFGHKLQUVW genes. The pilus contacts a plasmid-free recipient after recognizing a receptor in the recipient cell envelope (Anthony et al., 1994). Retraction of the pilus brings the cells close together where they are stabilized by other specific Tra proteins (Lanka and Wilkins, 1995).

Formation of a stable and functional connection between the two cells is necessary for DNA transfer (Durrenberger et al., 1991). Protein subunits of F pilin (gene product of traA) are arranged helically to form a cylindrical pilus structure 8 nm in diameter and with a 2 mm axial hole (Marvin and Folkard, 1986). The remaining tra genes involved in pilus assembly are involved in transporting the pilin through the cell membrane, assembling the pilus on the cell surface or in DNA transfer to the pilus. F pili are also used as receptors for male-specific phages such as MS2 and M13. It has been shown that conjugation is inhibited by these phages, and provides evidence that the pilus tip is crucial for initial contacts between donor and recipient cells (Novotny et al., 1968).

Stabilization of the mating aggregate represents the conversion of the initial unstable contacts between the donor and recipient cells to a form that is more resistant to shear forces. Mutations were identified in the traG and traN genes which do not interfere with pilus formation, but fail to transfer DNA. These mutants form aggregates inefficiently which led to their characterization as mating aggregate stabilization proteins

(Manning *et al.*, 1981). TraN fractionates with the bacterial outer membrane, and after removal of 18 amino acids, is active as a 63.8 kDa product. A portion of TraN is exposed extracellularly, raising the possibility that it interacts directly with a surface component of the recipient cell envelope (Maneewannakul *et al.*, 1992). TraG appears to be bifunctional, with sequences at the 5' end of the gene being essential to pilus formation, while those at the 3' end are essential for aggregate stabilization only (Manning *et al.*, 1981). TraG appears to be located in the periplasmic space indicating that its interaction with TraN is needed to form a stable and functional connection between conjugating cells. However, the route into the recipient cell through this connection is poorly understood. It may involve a preformed junction between the outer and inner cell membranes, pilus penetration, or DNA import in a two step process involving a periplasmic intermediate (Lanka and Wilkins, 1995).

2. Plasmid DNA transfer

The genes responsible for plasmid DNA transfer include endonucleases which act at the *oriT* sequence to promote transfer, some are helicases that unwind the DNA and some are primases that synthesize RNA to prime the synthesis of complementary DNA. The pilus-mediated cell contact triggers the opening of a specific DNA sequence cleaved at a unique "nick" site (*nic*) in the origin of transfer (*oriT*). *oriT* is the *cis*-acting site, less than 300 bp long, at which plasmid transfer initiates and is separate from the origin of vegetative replication *oriV*. The presence of an *oriT* sequence will convert a non-transmissible to a mobilizable plasmid. *oriT* is also the site at which the plasmid recircularizes after transfer is complete. Plasmids from the same incompatibility group have nucleotide sequence similarities within their *oriT* regions, although differences in the nick regions may vary; *nic* is a short stretch of up to ten nucleotides. Structural features shared by origins of transfer are: (i) they have a higher AT content which facilitates strand opening; (ii) they have extensive secondary structure and intrinsic bends; (iii) they are located so that the *tra* genes enter the recipient last, and include promoters for the expression of *tra* genes. Many of these structural features are similar to those of the origin of replication of rolling-circle plasmids.

The nick at *oriT* is made by plasmid-encoded TraI protein. The *traI* gene product has a nickase function which is an endonuclease specific for the *oriT* sequence. TraI is also an ATP-dependent helicase which is capable of unwinding the two DNA strands in a 5' to 3' direction prior to transfer. The *traI* gene encodes TraI and TraI*. TraI* is derived from a second translational start site within the *traI* open reading frame, and lacks the N-terminal domains essential to nicking but provides additional helicase activity. TraI becomes covalently linked to the 5' terminus of the nicked DNA strand and is implicated as the mediator of transfer termination. An accessory protein, TraY, appears to enhance the nicking activity of TraI (Gao *et al.*, 1994; Luo *et al.*, 1994). TraY also binds the P_{traY} promoter which controls the regulation of the main *tra* operon (Inamoto and Ohtsubo, 1990).

3. Complementary DNA synthesis

The 5' end of the nicked DNA strand is displaced by the helicase activity of TraI and TraI* and it is then transferred to the recipient by other *tra* gene products. Once the entire single-stranded molecule is inside the recipient, the molecule is recircularized to become the template for synthesis of the second strand. The *oriT* termini must be held together, presumably by TraI, to promote religation before synthesis of the complementary strand can begin. The synthesis of double-stranded DNA after transfer requires that primers be synthesized in the recipient cell only (synthesis is primed by the 3' end of the nick at *oriT* in the donor – see below). In a process similar to that of lagging strand replication of rolling-circle plasmids, synthesis of the complementary strand is initiated by the primase, using short RNA primers. The primase must be made in the donor cell as its transcription is dependent on the presence of double-stranded DNA. The primase must therefore be transferred at the same time as the DNA, so that it is available to prime the synthesis of the complementary strand immediately upon entry. Plasmid-encoded primases will increase the range of hosts into which they can be transferred as they will always function in a species-independent manner.

Synthesis of the donor strand is primed by the 3' OH end of the *oriT* nick, again in a process similar to rolling-circle replication. Replacement strand synthesis depends on the *traM*, *traD* and *traI* gene products. TraM is dispensable for pilus assembly, aggregate stabilization and *oriT* nicking. TraM does bind at *oriT* at sites which overlap its own promoter. Binding here is also affected by TraY which is required for maximal *traM* expression (Firth *et al.*, 1996). TraY and TraM binding sequences are required in *oriT* for efficient DNA transfer (Gao *et al.*, 1994). There is evidence that TraM is membrane associated and may anchor *oriT* to the transfer apparatus (Abo *et al.*, 1991; di Laurenzio *et al.*, 1992). *traD* mutants form pili and mating aggregates, and trigger conjugal DNA metabolism, however they fail to transfer DNA (Firth *et al.*, 1996). TraD appears to be required only after aggregate formation so may be involved in transport across the cell envelope. TraD has DNA-dependent ATPase activity which may provide the energy necessary to move DNA during transfer. TraD may tether the TraI helicase to the bacterial inner membrane providing the immobilization needed for strand displacement by DNA unwinding (Dash *et al.*, 1992). The ATP-dependent helicase properties of the C-terminal region of TraI contribute to strand displacement. TraI unwinds DNA at 1200 bp s^{-1} and ATPase activity may energize DNA strand transmission (Willets and Wilkins, 1984). When the mating pair separates, both cells contain double-stranded plasmid, and both are capable of donor activity. Replication of complementary strands does not appear to be essential for transfer as single-stranded plasmids are occasionally found.

Conjugation is still possible for plasmids that are integrated into bacterial chromosomes. If these plasmids start to transfer due to presence of an *oriT* then they will take chromosomal DNA with them; such strains are referred to as Hfr strains (*High frequency of recombination*). If the integrated plasmid has a full complement of *tra* genes then a pilus will be synthesized. Transfer of the integrated plasmid begins from *oriT* and will continue until chromosomal DNA is also transferred, as long as the mating pair is stable. Transfer of the entire chromosome is rare as the DNA is often broken during conjugation. The transferred DNA will be lost unless it is recombined with the homologous sequences in the chromosome of the recipient. This recombination frequently gives rise to recombinants which are different from both the donor and recipient. Strains capable of high frequency of recombination can be exploited to map bacterial chromosomes. The DNA closest to the site of integration will be transferred first, followed by DNA further away, so it is possible to determine the order of genetic markers in the chromosome by measuring the frequency of recombinants.

B. Mobilizable Plasmids

Some plasmids can be transferred by other plasmids and are referred to as mobilizable. Mobilizable plasmids cannot transfer themselves as they lack the genes for pilus synthesis, but can be transferred by certain self-transmissible plasmids present in the same cell. Mobilizable plasmids encode genes for transfer of DNA (*mob* genes) and a unique *oriT*. After mobilization by a self-transmissible plasmid, the recipient ends up with a copy of both the self-transmissible and the mobilizable plasmid (see examples of their use in Chapter 6, III.D). The *mob* genes, which are analogous to the *tra* genes of the self-transmissible plasmid, have similar functions: (i) endonuclease activity for nicking *oriT*; (ii) a helicase activity for strand displacement; (iii) and a recircularization activity for the transferred strand.

Mobilization of the 8.6 kb broad host range plasmid RSF1010 requires a 1.8 kb region of the plasmid. Its *oriT* is 38 bp and is flanked upstream by *mobAB* and downstream by *mobC* which are transcribed divergently. Strand-specific cleavage by purified Mob proteins has been demonstrated with supercoiled DNA, and less efficiently for linear DNA. It is assumed that MobC is brought into position at *oriT* by MobA, where localized destabilization allows MobA protein to interact with its single-stranded recognition sequence that is then cleaved (Lanka and Wilkins, 1995).

Mobilizable plasmids are more useful as cloning vectors as they are smaller than self-transmissible ones. Genes cloned into mobilizable plasmids can be transferred to recipient by the Tra proteins of a self-transmissible plasmid. To ensure only the gene of interest is transferred, the *oriT* of self-transmissible plasmid can be mutated, resulting in the transfer of only the mobilizable plasmid. For example, the *E. coli* donor strain S17-1 carries a chromosomally integrated RP4 molecule that mobilizes appropriate non-conjugative or mobilizable plasmids as efficiently as autonomous RP4 plasmids (Simon *et al.*, 1983). The advantage of this strain is that RP4 cannot be transferred into recipient cells.

C. Transfer Systems in Other Bacteria

Some Gram-positive, self-transmissible plasmids do not encode a pilus, but rather secrete a pheromone-like compound to stimulate mating; consequently the Tra regions are often smaller (Lanka and Wilkins, 1995). These bacteria may not require a pilus since they do not have the elaborate outer membrane of Gram-negative bacteria. For example, some strains of *Enterococcus faecalis* secrete pheromone-like compounds, which are small peptides which may stimulate mating. As a response to the sex pheromones the donor strain synthesizes an adhesin which allows cell-to-cell contact needed for conjugative transfer (Dunny, 1990; Gasson *et al.*, 1992). These pheromones stimulate the expression of *tra* genes in plasmids of neighbouring bacteria. Once the plasmid is received they discontinue synthesis of the pheromone, but synthesis of the transfer proteins is continued to perpetuate mating. This system has the advantage that the *tra* genes are only synthesized under conditions when plasmid-free recipients are available.

Sequence information is also available on the *oriT* regions of Gram-positive bacteria (Lanka and Williams, 1995). This sequence comparison has been instructive in showing that the *oriT* portion of the staphylococcal pGO1 plasmid, and the broad host range streptococcal pIP501 plasmid each contain a nick region very similar or identical to the consensus for the RSF1010 family. These findings are important in showing that the transfer mechanisms of some Gram-positive plasmids are fundamentally similar to Gram-negative counterparts. Some Gram-positive plasma have useful thermosensitive replication properties but these are not always widely applicable (Chapter 6, II).

The Ti plasmid of *Agrobacterium tumofaciens* is the first and only case of a plasmid that is transferable to a different kingdom, i.e. from bacteria to plants. It was also the first known example of the involvement of bacterial plasmids in a plant disease, namely the induction of crown gall tumours by *A. tumofaciens*. The tumour appears on the plant at the crown (where the roots join the stem). Only strains of *A. tumofaciens* which carry the virulence-associated plasmid Ti (Tumor *initiation*) are capable of causing tumour induction. The virulence region of the Ti plasmid provides most of the products that mediate T-DNA movement. The *vir* genes recognize phenolic compound given off by the plant, which induce expression of the genes involved in DNA transfer. A small part of the plasmid, referred to as the T-DNA, is transferred into the plant cell where it integrates into plant DNA. The T-DNA is flanked by two 25 bp sequences, the right and left T-DNA borders. At these *cis*-acting sites, single-stranded scission occurs, initiating the generation of the T-DNA strand to be transferred.

There are many features common to the transfer of DNA by Ti plasmids and the conjugative transfer system of IncP plasmids. Nearly half of RP4 is devoted to conjugative transfer, which is split into two regions: Tra1 and Tra2. Tra1 encodes functions necessary for the initiation of DNA synthesis, and *oriT*, whereas Tra2 encodes the functions needed to establish contact between the donor and recipient, along with TraF from Tra1 (Lessl *et al.*, 1993). The relationship between the two systems suggested by Stachel and Zambryski (1986) was based on the following functional analogies: *oriT* and the T-DNA borders both behave as the *cis*-acting sites where DNA transfer is initiated, the respective DNA processing functions follow a similar succession of events, and both systems encode functions that are required for donor–recipient recognition, and production of a structure through which the DNA is transferred. Evidence was subsequently produced confirming the predicted analogies between the two DNA transfer systems. Notably, the relationships include similarities in amino acid sequence, gene organization, and the physical properties of the gene products (Lessl and Lanka, 1994).

The properties of the Ti plasmid offer a method whereby foreign genes can be introduced into genomes generating transgenic plants. Introduction of a gene into a somatic cell, which can then be cultured to regenerate the whole plant, will create a plant with the inserted foreign gene in all of its cells. Using this approach it has been possible to engineer plants that are resistant to insect pests by introducing genes from *Bacillus thuringiensis* which encode proteins that are specific toxins against these pests. Since their discovery, the Ti plasmids have been engineered for more efficient transfer of genes into plants. The deletion of the genes required for plant tumour formation meant that the transgenic plants would not develop tumours. The incorporation of an expressible kanamycin resistance cassette from Tn5 made transgenic plants resistant to this antibiotic. (See also gene transfer in *Haloferax* sp. Chapter 10, IV.B.)

◆◆◆◆◆◆ VI. ISOLATION AND DETECTION OF PLASMIDS FROM NEW HOSTS

A. Methods of Isolation

Endogenous plasmids have been isolated from several diverse prokaryotes for use as shuttle vectors (see for example Chapter 10, IV.E).

Methods for the isolation of plasmid DNA take advantage of their small size and circular nature. Detailed protocols for these methods are reported in many molecular biology manuals (e.g. Sambrook *et al.*, 1989) and are briefly reviewed here.

Many published protocols are adaptations of the alkaline lysis method described by Birnboim and Doly (1979). The purpose of the procedure is to separate small plasmid molecules from the chromosome. These methods can be subdivided into three discrete steps. First, the cell walls are partially broken down using lysozyme. Secondly, the cells are opened with alkaline SDS which also denatures linear and open circular DNA strands, a process which is also facilitated by the high pH. Supercoiled DNA is unaffected. Finally, addition of neutralizing high salt (0.3 M sodium acetate) causes the chromosome to precipitate as it is – large and bound to proteins. The relatively small plasmids do not precipitate and remain in the supernatant. Circular molecules therefore do not completely denature and rehybridize rapidly once denaturing conditions are removed. The cell walls of Gram-positive bacteria are more resilient to the effects of lysozyme than Gram-negative bacteria. Often the lysis steps for such bacteria will include the addition of lysostaphin to break down the cell walls.

Alkaline lysis on a small scale is often sufficient to isolate DNA for characterization. When larger quantities are required then the protocol can be scaled up and followed by caesium chloride/ethidium bromide centrifugation. Covalently closed circular molecules bind ethidium bromide less effectively than do linear or nicked circular molecules. Intercalation of ethidium bromide is more easily able to unwind linear/nicked DNA resulting in more bound ethidium bromide. The two strands of the covalently closed circles are not free to rotate, allowing less ethidium bromide to bind. Binding of ethidium bromide makes the DNA less dense in salt solution (caesium chloride). Centrifugation of caesium chloride results in the formation of a density gradient. If DNA is added to the centrifugation, the plasmid DNA will band at a lower position than the chromosomal DNA.

Over recent years many commercial kits have been developed which make use of biochemical resins which are able to bind only the smaller molecular weight plasmid DNA (e.g. Wizard Miniprep, Promega; Qiapreps, Qiagen). Most make use of the alkaline lysis procedure described above to prepare cleared lysates for binding to the resin under high salt conditions. The plasmid DNA is then eluted from the column using a low or no salt solution.

B. Methods of Detection

1. Detection by agarose gel electrophoresis

Plasmid DNA can be separated from the chromosome and visualized using agarose gel electrophoresis (Sambrook *et al.*, 1989). Since DNA is negatively charged it will migrate towards the positive electrode upon application of an electric field. The DNA can be visualized with ethidium bromide which fluoresces under UV light. The smaller plasmid DNA will migrate through the gel faster then the chromosomal DNA. Many ultrapure and low melting temperature agaroses are now available which facilitate further manipulation and extraction of the DNA within the gel.

Very large plasmids can be separated using pulsed field gel electrophoresis (PFGE). The electric field applied to these gels periodically changes direction allowing the molecules to reorientate themselves with each shift. Longer molecules reorientate more slowly and migrate more slowly. The lysis protocols described above are fine for high copy number plasmids. However, large, low-copy plasmids are more susceptible to breakage than smaller plasmids; consequently the cell lysis procedures must be gentle. Barton *et al.* (1995) described a method for detecting and estimating the sizes of large bacterial plasmids in the presence of genomic DNA by PFGE in agarose gel plugs. Bacteria were lysed in agarose and the plugs incubated with S1 nuclease which converts the supercoiled plasmid DNA into full length linear molecules. Large plasmids migrate as discrete bands and their sizes can be determined by comparison to standard linear DNA markers. Supercoiled plasmid DNA often migrates anomalously in agarose gels making it difficult to determine their size accurately. Using this technique new plasmids were isolated from *Klebsiella* and *Staphylococcus*.

2. Detection by PCR

It is a laborious process to isolate plasmids after selective cultivation of bacteria and subsequently screen them for the presence of plasmids. As described above, characterization of plasmids is usually performed by incompatibility testing or molecular sizing. Couturier *et al.* (1988) described a collection of *inc/rep* probes which allow replicon typing by DNA hybridization. This approach was extended to use the replicon-specific DNA sequences against which oligonucleotide primers could be designed which would permit detection of plasmids in non-culturable samples (Gotz *et al.*, 1996). In order to be an efficient assay, PCR detection requires that the primers are highly specific for the replicon they were designed against, so that false positive results are not seen. The technique is also dependent on sequence conservation within the replicons of the same Inc group. Prior to this work, little was known about diversity within the replicons (Gotz *et al.*, 1996).

The method was used to determine whether PCR could be used to successfully detect and characterize broad host range plasmids in pure cultures and environmental samples. Sequence-specific primers were designed using published sequences of plasmids of IncP, IncN, IncW, IncQ groups. The IncQ primers were able to generate PCR products with plasmids previously assigned to the IncQ group. Some of the IncP, IncN and IncW primers were also specific, but some yielded no PCR product. The PCR-negative results were confirmed by negative hybridization results for the same replicons. Therefore it was apparent that plasmids assigned to the same Inc groups by traditional methods may differ from their respective reference plasmids at the DNA sequence level.

Recent work to isolate and characterize plasmids from marine bacteria (Dahlberg *et al.*, 1997; Sobecky *et al.*, 1997, 1998) has shown that the extensive characterization of plasmid replicons from clinical isolates, is not necessarily reflected in plasmid populations isolated from alternative environments. About 25% of 297 bacteria isolated from Californian marine sediments typically harboured a single large plasmid of 40–100 kb (Sobecky *et al.*, 1997). Interestingly there was no homology between plasmids isolated from culturable bacteria and the replicon-specific probes developed by Couturier *et al.* (1988). Similarly, of 12 groups of plasmids isolated from marine bacteria in Sweden, none had replication or incompatibility systems similar to the well characterized plasmids (Dahlberg *et al.*, 1997). This would seem to indicate that plasmid replicons isolated from novel environments may not be represented by a library constructed from clinical isolates.

Most of the marine bacteria isolated were non-culturable, so DNA was extracted from the microbial community. PCR probes specific for *oriV* of the broad host range plasmid RK2 produced a specific product indicating a low level of the IncP replicon in the marine microbial community; no product was seen for the narrow host range F (Sobecky *et al.*, 1997). Subsequent analysis of the sequence of the plasmid replicons revealed the presence of features common to the replication origins of well characterized plasmids from clinical isolates, suggesting a similar replication mechanism for plasmids isolated from marine sediment bacteria (Sobecky et al., 1998).

Obviously, DNA amplification techniques work best when the replicons of the plasmids to be identified have close sequence similarities to the reference plasmids from which the PCR primers were designed. More studies on the incompatibility of plasmids isolated from different habitats are needed before the full diversity of natural plasmid replicons can be assessed. This limitation will diminish with the ever increasing numbers of DNA plasmid replicon sequences in the molecular biology databases (such as Genbank and EMBL) which will expand the number of appropriate targets for amplification. As this information becomes available, this PCR-based technique will have the advantage of being much faster than those which involve bacterial culture followed by plasmid DNA isolation, and it can also be applied to non-culturable bacteria present in environmental samples.

References

Abeles, A.L., Friedman, S.A. and Austin, S.J. (1985). Partition of unit-copy mini-plasmids to daughter cells. III. The DNA sequence and functional organization of the P1 partition region. *J. Mol. Biol.* **185**, 261–272.

Abeles, A.L., Reaves, L.D., Youngren-Grimes, B. and Austin, S.J. (1995). Control of P1 plasmid replication by iterons. *Mol. Microbiol.* **18**, 903–912.

Abo, T. and Ohtsubo, E. (1993). Specific DNA binding of the TraM protein to the *oriT* region of plasmid R100. *J. Bacteriol.* **173**, 6347–6354.

Adams, D.E., Shektman, E.M., Zechiedrich, E.L., Schmid, M.B. and Cozzarelli, N.R. (1992). The role of topoisomerase IV in partitioning bacterial replicons and the structure of catenated intermediates in DNA replication. *Cell* **71**, 277–288.

Anthony, K.G., Sherburne, C., Sherburne, R. and Frost, L.S. (1994). The role of the pilus in recipient cell recognition during bacterial conjugation mediated by F-like plasmids. *Mol. Microbiol.* **13**, 939–953.

Austin, S.J. and Nordstrom, K. (1990). Partition mediated incompatibility of bacterial plasmids. *Cell* **60**, 351–354.

Bannam, T.L. and Rood, J.I. (1993). *Clostridium perfringens–Escherichia coli* shuttle vectors that carry single antibiotic resistance determinants. *Plasmid* **29**, 233–235.

Barton, B.M., Harding, G.P. and Zuccarelli, A.J. (1995). A general method for detecting and sizing large plasmids. *Anal. Biochem.* **226**, 235–240.

Bernard, P. and Couturier, M. (1992). Cell killing by the F plasmid CcdB protein involves poisoning of DNA–topoisomerase II complexes. *J. Mol. Biol.* **226**, 735–745.

Birnboim, H.C. and Doly, J. (1979). A rapid alkaline extraction procedure for screening recombinant plasmid DNA. *Nucl. Acids Res.* **7**, 1513–1523.

Blasina, A., Kittell, B.L., Toukdarian, A.E. and Helinski, D.R. (1996). Copy-up mutants of the plasmid RK2 replication initiation protein are defective in coupling RK2 replication origins. *Proc. Natl. Acad. Sci. USA* **93**, 3559–3564.

Bolivar, F., Rodriguez, R.L., Greene, P.J., Betlach, M.C., Heyneker, H.L. and Boyer, H.W. (1977). Construction and characterization of new cloning vehicles. I. Ampicillin-resistant derivatives of the plasmid pMB9. *Gene* **2**, 95–113.

Botstein, D., Falco, S.C., Stewart, S.E., Brennan, M., Scherer, S., Stinchcomb, D.T., Struhl, K. and Davis, R.W. (1979). Sterile host yeasts (SHY): a eukaryotic system of biological containment for recombinant DNA experiments *Gene* **8**, 17–24.

Brendler, T., Abeles, A.L. and Austin, S.J. (1995). A protein that binds to the P1 origin core and the *oriC* 13mer region in a methylation-specific fashion is the product of the host *seqA* gene. *EMBO J.* **14**, 4083–4089.

Bron, S. and Luxen, E. (1985). Segregational instability of pUB110-derived recombinant plasmids in *Bacillus subtilis*. *Plasmid* **14**, 95–244.

Bron, S., Holsappel, S., Venema, G. and Peeters, B.P.H. (1991). Plasmid deletion formation between short direct repeats in *Bacillus subtilis* is stimulated by single stranded rolling circle replication intermediates. *Mol. Gen. Genet.* **226**, 88–96.

Bruand, C., Le Chatelier, E., Ehrlich, S.D. and Janniere, L. (1993). A fourth class of β-theta-replicating plasmids: The pAMfl1 family from Gram-positive bacteria. *Proc. Natl. Acad. Sci. USA* **90**, 11668–11672.

Campbell, J.L. and Kleckner, N. (1990). *E. coli oriC* and *dnaA* gene promoters are sequestered from dam methyltransferase following the passage of the chromosomal replication fork. *Cell* **62**, 967–979.

Cavalli, L.L., Lederberg, E. and Lederberg, J.M. (1953). An infective factor controlling sex compatibility in *Bacterium coli*. *J. Gen. Microbiol.* **8**, 89–103.

Cepko, C.L., Roberts, B.E. and Mulligan, R.C. (1984). Construction and applications of a highly transmissible murine retrovirus shuttle vector. *Cell* **37**, 1053–1062.

Cesarini, G., Helmer, C. M. and Castagnoli, L. (1991). Control of ColE1 plasmid replication by antisense RNA. *Trends Genet.* **7**, 230–235.

Chakraborti, A., Gunji, S., Shakibai, N., Cubeddu, J. and Rothfield, L.I. (1992). Characterization of the *Escherichia coli* membrane domain responsible for binding *oriC* DNA. *J. Bacteriol.* **174**, 7202–7206.

Chang, A.C.Y. and Cohen, S.N. (1974) Genome construction between bacterial species in vitro: replication and expression of *Staphylococcus* plasmid genes in *Escherichia coli*. *Proc. Natl. Acad. Sci. USA* **71**, 1030.

Chang, P.C. and Cohen, S.N. (1994). Bidirectional replication from an internal origin in a linear *Streptomyces* plasmid. *Science* **265**, 952–954.

Cohen, S.N., Chang, A.C.Y. and Hsu, L. (1972). Nonchromosomal antibiotic resistance in bacteria: genetic transformation of *Escherichia coli* by R-factor DNA. *Proc. Natl. Acad. Sci. USA* **69**, 2110.

Cohen, S.N., Chang, A.C.Y., Boyer, H.W. and Helling, R.B. (1973). Construction of biologically functional bacterial plasmids in vitro. *Proc. Natl. Acad. Sci. USA* **70**, 3240.

Colloms , S.D., Sykora, P., Szatmari, G. and Sherratt, D.J. (1990). Recombination at ColE1 *cer* requires the *Escherichia coli xerC* gene product, a member of the lambda integrase family of site specific recombinases. *J. Bacteriol.* **172**, 6973–6980.

Couturier, M., Bex, F., Bergquist, P.L. and Maas, W.K. (1988). Identification and classification of bacterial plasmids. *Microbiol. Rev.* **52**, 375–395.

Dahlberg, C., Linberg, C., Torsvik, V.L. and Hermansson, M. (1997). Conjugative plasmids isolated from bacteria in marine environments show various degrees of homology to each other and are not closely related to well-characterized plasmids. *Appl. Environ. Microbiol.* **63**, 4692–4697.

Dam, M. and Gerdes, K. (1994). Partitioning of plasmid R1. Ten direct repeats flanking the *parA* promoter constitute a centromere-like partition site *parC*, that expresses incompatibility. *J. Mol. Biol.* **236**, 1289–1298.

DasGupta, S., Masukata, H. and Tomizawa, J.I. (1987). Multiple mechanisms for initiation of ColE1 DNA replication: DNA synthesis in the presence and absence of ribonuclease H. *Cell* **51**, 1113–1122.

Dash, P.K., Traxler, B.A., Panicker, M.M., Hackney, D.D. and Minkley, E.G. Jr (1992). Biochemical characterization of *Escherichia coli* DNA helicase I. *Mol. Microbiol.* **6**, 1163–1172.

Datta, N. and Hedges, R.W. (1971). Compatibility groups among fi – R factors. *Nature (London)* **234**, 222–223.

Davey, R.B., Bird, P.I., Nikoletti, S.M., Praskier, J. and Pittard, J. (1984). The use of mini-Gal plasmids for rapid incompatibility grouping of conjugative R plasmids. *Plasmid* **11**, 234–242.

Davis, M.A. and Austin, S.J. (1988). Recognition of the P1 plasmid centromere analog involves binding of the ParB protein and is modified by a specific host factor. *EMBO J.* **7**, 1881–1888.

Davis, M.A., Martin, K.A. and Austin, S.J. (1992). Biochemical activities of the ParA partition protein of the P1 plasmid. *Mol. Microbiol.* **6**, 1141–1147.

Davis, M.A., Radnedge, L., Martin, K.A., Hayes, F., Youngren, B. and Austin, S.J. (1996). The P1 ParA protein and its ATPase activity play a direct role in the segregation of plasmid copies to daughter cells. *Mol. Microbiol.* **21**, 1029–1036.

Davis, T.L., Helinski, D.R. and Roberts, R.C. (1992). Transcription and autoregulation of the stabilizing functions of broad-host-range plasmid RK2 in *Escherichia coli*, *Agrobacterium tumefaciens* and *Pseudomonas aeruginosa*. *Mol. Microbiol.* **6**, 1981–1994.

del Solar, G., Kramer, G., Ballester, S. and Espinosa, M. (1993). Replication of the promiscuous plasmid pLS1, a region encompassing the minus origin of replication is associated with stable plasmid inheritance. *Mol. Gen. Genet.* **241**, 97–105.

del Solar, G., Alonso, J.C., Espinosa, M. and Diaz-Orejas, R. (1996). Broad host range plasmid replication: an open question. *Mol. Microbiol.* **21**, 661–666.

del Solar, G., Giraldo, R., Ruiz-Echevarria, M.J., Espinosa, M. and Diaz-Orejas, R. (1998). Replication and control of circular bacterial plasmids. *Microbiol. Mol. Biol. Revs.* **62**, 434–464.

Diaz, R. and Staudenbauer, W.L. (1982). Replication of the broad host range plasmid RSF1010 in cell-free extracts of *Escherichia coli* and *Pseudomonas aeruginosa*. *Nucl. Acids Res.* **10**, 4687–4702.

Di Laurenzio, L., Frost, L.S. and Paranchych, W. (1992). The TraM protein of the conjugative plasmid F binds to the origin of transfer of the F and ColE1 plasmids. *Mol. Microbiol.* **6**, 2951–2959.

Doran, K.S., Konieczny, I. and Helinski, D.R. (1998). Replication origin of the broad host range plasmid RK2. *J. Biol. Chem.* **273**, 8447–8453.

Dunny, G.M. (1990). Genetic functions and cell–cell interactions in the pheromone-inducible plasmid transfer system of *Enterococcus faecalis*. *Mol. Microbiol.* **4**, 689–696.

Durrenberger, M.B., Villiger, W. and Bachi, T. (1991). Conjugational junctions: morphology of specific contacts in conjugating *Escherichia coli* bacteria. *J. Struct. Biol.* **107**, 146–156.

Easter, C.L., Sobecky, P.A. and Helinski, D.R. (1997). Contribution of different segments of the par region to stable maintenance of the broad host range plasmid RK2. *J. Bacteriol.* **179**, 6472–6479.

Eguchi, Y., Itoh, T. and Tomizawa, J. (1991). Antisense RNA. *Annu. Rev. Biochem.* **60**, 631–652.

Ehrlich, S.D., Niorot, P., Petit, M.A., Janniere, L., Michel, B. and te Riele, H. (1986). In *Genetic Engineering* (J.K. Setlow and A. Hollaender, eds), Vol. 8, pp. 71–83. Plenum, New York.

Ezaki, B., Ogura, T., Niki, H. and Hiraga, S. (1991). Partitioning of a mini-F plasmid into anucleate cells of the *mukB* null mutant. *J. Bacteriol.* **173**, 6643–6646.

Fang, F. and Helinski, D.R. (1991). Broad host range properties of plasmid RK2, importance of overlapping genes encoding the plasmid replication initiation protein, TrfA. *J. Bacteriol.* **173**, 5861–5868.

Farrand, S.K. (1993). In *Bacterial Conjugation* (D.B. Clewell, ed.), pp. 255–291. Plenum, New York.

Figurski, D.H., Pohlman, R.F., Bechhofer, D.H. Prince, A.S. and Kelton, C.A. (1982). Broad host range plasmid encodes multiple *kil* genes potentially lethal to host *E. coli* host cells. *Proc. Natl. Acad. Sci. USA* **79**, 1935–1939.

Firshein, W. and Kim, P. (1997). Plasmid replication and partition in *Escherichia coli*: is the cell membrane the key? *Mol. Microbiol.* **23**, 1–10.

Firth, N., Ippen-Ihler, K. and Skurray, R.A. (1996). In *Escherichia coli* and *Salmonella typhimurium*: Cellular and Molecular Biology (F.C. Niedhardt, J.L. Ingraham, K.B. Low, B. Magasanik, M. Schaechter and H.E. Umbarger, eds), pp. 2377–2401. ASM Press.

Friedman, S.A. and Austin, S.J. (1988). The P1 plasmid-partition system synthesizes two essential proteins from an autoregulated operon. *Plasmid* **19**, 103–112.

Frost L.S., Ippen-Ihler, K. and Skurray, R.A. (1994). Analysis of the sequence and gene products of the transfer region of the F sex factor. *Microbiol. Rev.* **58**, 162–210.

Fu, J-F., Ying, S-W. and Liu, S-T. (1997). Cloning and characterization of the *ori* region of pSW1200 of *Erwinia stewartii*: similarity with plasmid P1. *Plasmid* **38**, 141–147.

Funnell, B.E. (1988). Mini-P1 plasmid partitioning: excess ParB protein destabilizes plasmids containing the centromere *parS*. *J. Bacteriol.* **170**, 954–969.

Funnell, B.E. (1991). The P1 plasmid partition complex at *parS*. *J. Biol. Chem.* **266**, 14328–14337.

Funnell, B.E. and Gagnier, L. (1995). Partition of P1 plasmids in *Escherichia coli mukB* chromosomal partition mutants. *J. Bacteriol.* **177**, 2381–2386.

Gallie, D.R. and Kado, C.I. (1987). *Agrobacterium tumefaciens* pTAR *parA* promoter region involved in autoregulation, incompatibility and plasmid partitioning. *J. Mol. Biol.* **193**, 465–478.

Gao, Q., Luo, Y. and Deonier, R.C. (1994). Initiation and termination of DNA transfer at F plasmid *oriT*. *Mol. Microbiol.* **11**, 449–458.

Garnier, T. and Cole. S.T. (1988). Complete nucleotide sequence and genetic organization of the bacteriocinogenic plasmid, pIP404, from *Clostridium perfringens*. *Plasmid* **19**, 134–150.

Gasson, M.J., Swindell, S., Maeda, S. and Dodd, M.E. (1992). Molecular rearrangement of lactose plasmid DNA associated with high-frequency transfer and cell aggregation in *Lactococcus lactis* 712. *Mol. Microbiol.* **6**, 3213–3223.

Gennaro, M.L., Iordanescu, S., Novick, R.P., Murray, R.W., Stech, T.R. and Khan, S.A. (1989). Functional organization of the plasmid pT181 replication origin. *J. Mol. Biol.* **205**, 355–362.

Gerdes, K., Rasmussen, P.B. and Molin, S. (1986). Unique type of plasmid maintenance function: post segregational killing of plasmid free cells. *Proc. Natl. Acad. Sci. USA* **83**, 3116–3120.

Gerlitz, M.O., Hrabak, O. and Schwab, H. (1990). Partitioning of broad-host-range plasmid RP4 is a complex system involving site-specific recombination. *J. Bacteriol.* **172**, 6194–6203.

Gordon, G.S., Sitnikov, D., Webb, C.D., Teleman, A., Straight, A., Losick, R., Murray, A.W. and Wright, A. (1997). Chromosome and low copy plasmid segregation in *E. coli*: visual evidence for distinct mechanisms. *Cell* **90**, 1113–1121.

Gotz, A., Pukall, R., Smit, E., Tietze, E., Prager, R., Tschape, H., van Elsas, J.D. and Smalla, K. (1996). Detection and characterization of broad host range plasmids in environmental bacteria by PCR. *Appl. Env. Microbiol.* **62**, 2621–2628.

Gruss, A., Ross, H.F. and Novick, R.P. (1987). Functional analysis of a palindromic sequences required for normal replication of several staphylococcal plasmids. *Proc. Natl. Acad. Sci. USA* **84**, 2165–2169.

Guhathakurta, A., Viney, I. and Summers, D. (1996). Accessory proteins impose site selectivity during ColE1 dimer resolution. *Mol. Microbiol.* **20**, 613–620.

Haring, V., Scholz, P., Scherzinger, E., Frey, J., Derbyshire, K., Hatfull, G., Willetts, N.S. and Bagdasarian, M. (1985). Protein RepC is involved in copy number control of the broad host range plasmid RSF1010. *Proc. Natl. Acad. Sci. USA* **82**, 6090–6094.

Helinski, D.R., Toukdarian, A.E. and Novick, R.P. (1996). In *Escherichia coli and Salmonella typhimurium: Cellular and Molecular Biology* (F.C. Niedhardt, J.L. Ingraham, K.B. Low, B. Magasanik, M. Schaechter and H.E. Umbarger, eds), pp. 2377–2401. ASM Press.

Hinnebusch, J. and Tilly, K. (1993). Linear plasmids and chromosomes in bacteria. *Mol. Microbiol.* **10**, 917–922.

Hiraga, S. (1992). Chromosome and plasmid partition in *Escherichia coli. Annu. Rev. Biochem.* **61**, 283–306.

Hiraga, S., Niki, H., Ogura, T., Ichinose, C. and Mori, H. (1990). Positioning of replicated chromosomes in *Escherichia coli. J. Bacteriol.* **172**, 31–39.

Honda, Y., Sakai, H., Komano, T. and Bagdasarian, M. (1989). RepB' is required in trans for the two single-strand DNA initiation signals in *oriV* of plasmid RSF1010. *Gene* **80**, 155–159.

Horiuchi, T. and Hidaka, M. (1988). Core sequence of two separable terminus sites is a 20 bp inverted repeat. *Cell* **54**, 515–523.

Inamoto, S. and Ohtsubo, E. (1990). Specific binding of the TraY protein to *oriT* and the promoter region for the *traY* gene of plasmid R100. *J. Biol. Chem.* **265**, 6461–6466.

Ippen-Ihler, K.A. and Minkley, Jr. E.G. (1986). The conjugation system of F, the fertility factor of F. *Annu. Rev. Genet.* **20**, 593–624.

Jacob, A.E., Shapiro, J.A., Yamamoto, L., Smith, D.L., Cohen, S.N. and Berg, D. (1977). In *DNA Insertion Elements, Plasmids and Episomes* (A.I. Bukhari, J.A. Shapiro and S.L. Adhya, eds), pp. 607–638. Cold Spring Harbor Press.

Jacob, F., Brenner, S. and Cuzin, F. (1963). On the regulation of DNA replication in bacteria. Cold Spring Harbor Symp. *Quant. Biol.* **28**, 329–348.

Jaffe, A., Ogura, T. and Hiraga, S. (1985). Effects of the *ccd* function of the F plasmid on bacterial growth. *J. Bacteriol.* **163**, 841–849.

Janniere L., Bruand C. and Ehrlich S.D. (1990). Structurally stable *Bacillus subtilis* cloning vectors. *Gene* **87**, 53–61.

Jensen, R.B. and Gerdes, K. (1995). Programmed cell death in bacteria: proteic plasmid stabilization systems. *Mol. Microbiol.* **17**, 205–210.

Jensen, R.B., Dam, M. and Gerdes, K. (1994). Partitioning of plasmid R1. The *parA* operon is autoregulated by ParR and its transcription is highly stimulated by a downstream activating element. *J. Mol. Biol.* **236**, 1299–1309.

Jensen, R.B., Grohmann, E., Schwab, H., Diaz-Orejas, R. and Gerdes, K. (1995). Comparison of *ccd* of F, *parDE* of RP4, and *parD* of R1 using a novel conditional replication control system of plasmid R1. *Mol. Microbiol.* **17**, 211–220.

Johnson, D.A. and Willetts, N.S. (1980). Construction and characterization of multicopy plasmids containing the entire F transfer region. *Plasmid* **4**, 292–304.

Johnson, E.P., Strom, A.R. and Helinski, D.R. (1996). Plasmid RK2 toxin protein ParE: purification and interaction with the ParD antitoxin protein. *J. Bacteriol.* **178**, 1420–1429.

Kato, J., Nishimura, Y., Imamura, R., Niki, H., Hiraga, S. and Suzuki, H. (1988). Gene organization in a region containing a new gene involved in chromosome partitioning in *Escherichia coli. J. Bacteriol.* **170**, 3967–3977.

Khan, S.A. (1997). Rolling-circle replication of bacterial plasmids. *Microbiol. Mol. Biol. Revs.* **61**, 442–455.

Kiewiet, R., Bron, S., de Jonge, K., Venema, G. and Seegers, J.F. (1993). Theta replication of the lactococcal plasmid pWVO2. *Mol. Microbiol.* **10**, 319–327.

Kinashi, H., Shimaji, M. and Sakai, A. (1987). Giant plasmids in *Streptomyces* which code for antibiotic biosynthesis genes. *Nature* **328**, 454–456.

Kittell, B.L. and Helinski, D.R. (1991). Iteron inhibition of plasmid RK2 replication in vitro: evidence for intermolecular coupling of replication origins as a mechanism for RK2 replication control. *Proc. Natl. Acad. Sci. USA* **88**, 1389–1393.

Koepsel, R.R. and Khan, S.A. (1986). Static and initiator protein enhanced bending of DNA at a replication origin. *Science* **233**, 1316–1318.

Konieczny, I. and Helinski, D.R. (1997). Helicase delivery and activation by DnaA and TrfA proteins during the initiation of replication of the broad host range plasmid RK2. *J. Biol. Chem.* **272**, 33312–33318.

Konieczny, I., Doran, K.S., Helinski, D.R. and Blasina, A. (1997). Role of TrfA and DnaA proteins in origin opening during initiation of DNA replication of the broad host range plasmid RK2. *J. Biol. Chem.* **272**, 20173–20178.

Koyama, Y., Arikawa, Y. and Furukawa, K. (1990). A plasmid vector for an extreme thermophile, *Thermus thermophilus*. *FEMS Microbiol. Lett.* **60**, 97–101.

Kues, U. and Stahl, U. (1989). Replication of plasmids in Gram-negative bacteria. *Microbiol. Revs.* **53**, 491–516.

Lacatena, R.M. and Cesarini, G. (1981). Base pairing of RNAI with its complementary sequence in the primer inhibits ColE1 replication. *Nature (London)* **294**, 623–626.

Lanka, E. and Wilkins, B.M. (1995). DNA processing reactions during bacterial conjugation. *Annu. Rev. Biochem.* **64**, 141–169.

Lasa, I., de Grado, M., de Pedro, M.A. and Berenguer, J. (1992). Development of *Thermus-Escherichia* shuttle vectors and their use for expression of the *Clostridium thermocellum celA* gene in *Thermus thermophilis*. *J. Bacteriol.* **74**, 6424–6431.

Lehnherr, H., Maguin, E., Jafri, S. and Yarmolinsky, M.B. (1993). Plasmid addiction genes of bacteriophage P1: *doc* which causes cell death on curing of prophage and *phd* which prevents host death when prophage is retained. *J. Mol. Biol.* **233**, 414–428.

Leonhardt, H. and Alonso, J.C. (1991). Parameters affecting plasmid stability in *Bacillus subtilis*. *Gene* **103**, 107–111.

Lessl, M. and Lanka, E. (1994). Common mechanisms in bacterial conjugation and Ti mediated T-DNA transfer to plant cells. *Cell* **77**, 321–324.

Lessl, M., Balzer, D., Weyrauch, K. and Lanka, E. (1993). The mating pair formation system of plasmid RP4 defined by RSF1010 mobilization and donor-specific phage propagation. *J. Bacteriol.* **175**, 6415–6425.

Lin, J. and Helinski, D.R. (1992). Analysis of mutations in *trfA*, the replication initiation gene of the broad host range plasmid RK2. *J. Bacteriol.* **174**, 4110–4119.

Lin, Z. and Mallavia, L.P. (1998). Membrane association of active plasmid partitioning protein A in *Escherichia coli*. *J. Biol. Chem.* **273**, 11302–12.

Luo, Y., Gao, Q. and Deonier, R.C. (1994). Mutational and physical analysis of F plasmid TraY protein binding to *oriT*. *Mol. Microbiol.* **11**, 459–469.

Lyras, D. and Rood, J.I. (1998). Conjugative transfer of RP4-*oriT* shuttle vectors from *Escherichia coli* to *Clostridium perfringens*. *Plasmid* **39**, 160–164.

Makino, S., Sasakawa, C. and Yoshikawa, M. (1988). Genetic relatedness of the basic replicon of the virulence plasmid of *Shigellae* and enteroinvasive *Escherichia coli*. *Microb. Path.* **5**, 267–274.

Maneewannakul, S., Kathir, P. and Ippen-Ihler, K. (1992). Characterization of the F plasmid mating aggregation gene *traN* and of a new F transfer region locus *trbE*. *J. Mol. Biol.* **225**, 299–311.

Manning, P.A., Morelli, G. and Achtman, M. (1981). traG protein of the F sex factor of Escherichia coli K-12 and its role in conjugation. *Proc. Natl. Acad. Sci. USA* **78**, 7487–7491.

Marians, K. (1992). Prokaryotic DNA replication. *Annu. Rev. Biochem.* **61**, 673–719.

Marvin, D.A. and Folkhard, W. (1986). Structure of F-pili: reassessment of the symmetry. *J. Mol. Biol.* **191**, 299–300.

McEachern, M.J., Bott, M.A., Tooker, P.A. and Helinski, D.R. (1989). Negative control of plasmid R6K replication: possible role of intermolecular coupling of replication origins. *Proc. Natl. Acad. Sci. USA* **86**, 7942–7946.

Mei, J., Benashki, S. and Firshein, W. (1995). Interactions of the origin of replication (*oriV*) and initiation proteins (TrfA) of plasmid RK2 with submembrane domains of *Escherichia coli*. *J. Bacteriol.* **177**, 6766–6772.

Meima, R., Haijema, B.J., Dijkstra, H., Haan, G.-J., Venema, G. and Bron, S. (1997). Role of enzymes of homologous recombination in illegitimate plasmid recombination in *Bacillus subtilis*. *J. Bacteriol.* **179**, 1219–1229.

Meima, R., Haan, G.-J., Venema, G., Bron, S. and de Jong, S. (1998). Sequence specificity of illegitimate plasmid recombination in *Bacillus subtilis*: possible recognition sites for DNA topoisomerase I. *Nucl. Acids Res.* **26**, 2366–2373.

Miao, D.-M., Honda, Y., Tanaka, K., Higashi, A., Nakamura, T., Taguchi, Y., Sakai, H., Komano, T. and Bagdasarian, M. (1993). A base-paired hairpin structure essential for the functional priming signal for DNA replication of the broad host-range plasmid RSF1010. *Nucl. Acids Res.* **21**, 4900–4903.

Michel, B., Palla, E., Niaudet, B. and Ehrlich, S.D. (1980). DNA cloning in *Bacillus subtilis*. III. Efficiency of random-segment cloning and insertional inactivation vectors. *Gene* **12**, 147–154.

Mohl, D.A. and Gober, J.W. (1997). Cell cycle-dependent polar localization of chromosome partitioning proteins in *Caulobacter crescentus*. *Cell* **88**, 675–684.

Motallebi-Veshareh, M., Rouch, D.A. and Thomas, C.M. (1990). A family of ATPases involved in active partitioning of diverse bacterial plasmids. *Mol. Microbiol.* **4**, 1455–1463.

Mukhopadhyay, G. and Chattoraj, D.K. (1993). Conformation of the origin of P1 plasmid replication. Initiator protein induced wrapping and intrinsic unstacking. *J. Mol. Biol.* **231**, 19–28.

Niki, H. and Hiraga, S. (1997). Subcellular distribution of actively partitioning F plasmid during the cell division cycle in *E. coli*. *Cell* **90**, 951–957.

Niki, H., Jaffe, A., Imamura, R., Ogura, T. and Hiraga, S. (1991). The new *mukB* codes for a 177 kd protein with coiled-coil domains involved in chromosome partitioning of *E. coli*. *EMBO J.* **10**, 183–193.

Noirot, P., Bargonetti, J. and Novick, R.P. (1990). Initiation of rolling circle replication in pT181 plasmid: initiator protein enhances cruciform extrusion at the origin. *Proc. Natl. Acad. Sci. USA* **87**, 8560–8564.

Novick, R.P. (1987). Plasmid incompatibility. *Microbiol. Revs.* **51**, 381–395.

Novotny, C.P., Knight, W.S. and Brinton, Jr. C.C. (1968). Inhibition of bacterial conjugation by ribonucleic acid and deoxyribonucleic acid male-specific bacteriophages. *J. Bacteriol.* **95**, 314–326.

Pansegrau, W., Lanka, E., Barth, P.T., Figurski, D.H., Guiney, D.G., Haas, D., Helinski, D.R., Schwab, H., Stanisch, V.A. and Thomas, C.M. (1994). Complete nucleotide sequence of Birmingham IncPα plasmids: compilation and comparative analysis. *J. Biol. Chem.* **266**, 12536–12543.

Papp, P.P., Mukhopadhyay, G. and Chattoraj, D.K. (1994). Negative control of plasmid replication by iterons. *J. Biol. Chem.* **269**, 23563–23568.

Radnedge, L., Davis, M.A., Youngren, B. and Austin, S.J. (1997). Plasmid maintenance functions of the large virulence plasmid of *Shigella flexneri*. *J. Bacteriol.* **179**, 3670–3675.

Rasooly, A. and Novick, R.P. (1993). Replication-specific inactivation of the pT181 plasmid initiator protein. *Science* **262**, 1048–1050.

Roberts, R.C., Burioni, R. and Helinski, D.R. (1990). Genetic characterization of the stabilizing functions of a region of broad-host-range plasmid RK2. *J. Bacteriol.* **172**, 6204–6216.

Roberts, R.C. and Helinski, D.R. (1992). Definition of a minimal plasmid stabilization system from the broad-host-range plasmid RK2. *J. Bacteriol.* **174**, 8119–8132.

Roberts, R.C., Spangler, C. and Helinski, D.R. (1993). Characteristics and significance of DNA binding activity of plasmid stabilization protein ParD from the broad-host-range plasmid RK2. *J. Biol. Chem.* **268**, 27109–27117.

Rownd, R. (1969). Replication of a bacterial episome under relaxed control. *J. Mol. Biol.* **44**, 387–402.

Sakai, H. and Komano, T. (1996). DNA replication of the IncQ broad-host-range plasmids in gram-negative bacteria. *Biosci. Biotechnol. Biochem.* **60**, 377–382.

Sambrook, J., Fritsch, E.F. and Maniatis, T. (1989). *Molecular Cloning: A Laboratory Manual.* 2nd edn. Cold Spring Harbor Laboratory Press, Cold Spring Harbor, New York.

Sasakawa, C., Kamata, K., Sakai, T., Murayama, S.Y., Makino, S. and Yoshikawa, M. (1986). Molecular alteration of the 140-megadalton plasmid associated with loss of virulence and Congo red binding activity in *Shigella flexneri*. *Infect. Immun.* **51**, 470–475.

Scherzinger, E., Haring, V., Lurz, R. and Otto, S. (1991). Plasmid RSF1010 replication *in vitro* promoted by purified RSF1010 RepA, RepB and RepC proteins. *Nucl. Acids Res.* **19**, 1203–1211.

Scholz, P., Haring, V., Wittman-Liebod, B., Ashman, K., Bagdasarian, M. and Scherzinger, E. (1989). Compete nucleotide sequence and gene organization of the broad host range plasmid RSF1010. *Gene* **75**, 271–288.

Sekimizu, K. and Kornberg, A. (1988). Cardiolipin activation of *dnaA* protein, the initiation protein of replication in *Escherichia coli*. *J. Biol. Chem.* **263**, 7131–7135.

Sekimizu, K., Yung, B.Y.M. and Kornberg, A. (1988). The *dnaA* protein of *Escherichia coli*. Abundance, improved purification, and membrane binding. *J. Biol. Chem.* **263**, 7136–7140.

Shaper, S. and Messer, W. (1995). Interaction of the initiator protein DnaA of *Escherichia coli* with its DNA target. *J. Biol. Chem.* **270**, 17622–17626.

Shingler, V. and Thomas, C.M. (1984). Analysis of the *trfA* region of the broad host range plasmid RK2 by transposon mutagenesis and identification of polypeptide products. *J. Mol. Biol.* **17**, 229–249.

Sia, E.A., Roberts, R.C., Easter, C., Helinski, D.R. and Figurski, D.H. (1995). Different relative importances of the *par* operons and the effect of conjugal transfer on the maintenance of intact promiscuous plasmid RK2. *J. Bacteriol.* **177**, 2789–2797.

Simon, R., Priefer, U. and Puhler, A. (1983). A broad host range mobilization system for *in vivo* genetic engineering; transposon mutagenesis. *BioTechnology* **1**, 37–45.

Slater, S., Wold, S., Lu, M., Boye, E., Skarstad, K. and Kleckner N. (1995). *E. coli* SeqA protein binds *oriC* in two different methyl-modulated reactions appropriate to its roles in DNA replication initiation and origin sequestration. *Cell* **82**, 927–936.

Sloan, J., Warner, T.A., Scott, P.T., Bannam, T.L., Berryman, D.I. and Rood, J.I. (1992). Construction of a sequenced *Clostridium perfringens–Escherichia coli* shuttle plasmid. *Plasmid* **27**, 207–219.

Sobecky, P.A., Easter, C.L., Bear, P.D. and Helinski, D.R. (1996). Characterization of the stable maintenance properties of the *par* region of broad host range plasmid RK2. *J. Bacteriol.* **178**, 2086–2093.

Sobecky, P.A., Mincer, T.J., Chang, M.C. and Helinski D.R. (1997). Plasmids isolated from marine sediment microbial communities contain replication and incompatibility regions unrelated to those of known plasmid groups. *Appl. Environ. Microbiol.* **63**, 888–895.

Sobecky, P.A., Mincer, T.J., Chang, M.C., Toukdarian, A. and Helinski, D.R. (1998). Isolation of broad-host-range replicons from marine sediment bacteria. *Appl. Environ. Microbiol.* **64**, 2822–2830.

Stachel, S.E. and Zambryski, P.C. (1986). *Agrobacterium tumefaciens* and the susceptible plant cell: a novel adaptation of extracellular recognition and DNA conjugation. *Cell* **47**, 155–157.

Stalker, D.M., Thomas, C.M. and Helinski, D.R. (1981). Nucleotide sequence of the region of the origin of replication of the antibiotic resistance plasmid RK2. *Mol. Gen. Genet.* **181**, 8–12.

Sterling, C.J., Szatmari, G. and Sherratt, D.J. (1988). The arginine repressor is essential for plasmid stabilizing site specific recombination at the ColE1 *cer* locus. *EMBO J.* **7**, 4389–4395.

Thisted, T. and Gerdes, D. (1992). Mechanism of postsegregational killing by the *hok/sok* system of plasmid R1. Sok antisense RNA regulates *hok* gene expression indirectly through the overlapping *mok* gene. *J. Mol. Biol.* **223**, 41–54.

Thomas, C.M., Cross, M.A., Hussain, A.A.K. and Smith, C.A. (1984). Analysis of copy number control elements in the region of the vegetative replication origin of the broad host range plasmid RK2. *EMBO J.* **3**, 1513–1519.

Tolun, A. and Helinski, D. R. (1981). Direct repeats of the plasmid *incC* region express F incompatibility. *Cell* **24**, 687–694.

Toukdarian, A.E., Perri, S. and Helinski, D.R. (1996). The plasmid RK2 initiation protein binds to the origin of replication as a monomer. *J. Biol. Chem.* **271**, 7072–7078.

Tsuchimoto, S., Nishimura, Y. and Ohtsubo, E. (1992). The stable maintenance system pem of plasmid R100: degradation of PemI protein may allow PemK protein to inhibit cell growth. *J. Bacteriol.* **174**, 4205–4211.

Vieira, J. and Messing, J. (1982). The pUC plasmids, an M13mp7-derived system for insertion mutagenesis and sequencing with synthetic universal primers. *Gene* **19**, 259–268.

von Freiesleben, U., Rasmussen, K.V. and Schaechter, M. (1994). SeqA limits DnaA activity in replication from *oriC* in *E. coli. Mol. Microbiol.* **14**, 763–772.

Wall, J.D., Rapp-Giles, B.J. and Rousset, M. (1993). Characterization of a small plasmid from *Desulfovibrio desulfuricans* and its use for shuttle vector construction. *J. Bacteriol.* **175**, 4121–4128.

Waters, V.L., Hirata, K.H., Pansegrau, W., Lanka, E. and Guiney, D.G. (1991). Sequence identity in the nick regions of IncP plasmid transfer origins and sequence identity in the nick regions of IncP plasmid transfer origins and T-DNA borders of *Agrobacterium* Ti plasmids–DNA borders of *Agrobacterium* Ti plasmids. *Proc. Natl. Acad. Sci. USA* **88**, 1456–1460.

Wayne, J. and Xu, S.Y. (1997). Identification of a thermophilic plasmid origin and its cloning within a new *Thermus-E. coli* shuttle vector. *Gene* **195**, 321–328.

Webb, C.D., Teleman, A., Gordon, S., Straight, A., Belmont, A., Lin, C.-H., Grossman, A., Wright, A. and Losick, R. (1997). Bipolar localization of the replication origin regions of chromosomes in vegetative and sporulating cells of *Bacillus subtilis. Cell* **88**, 667–674.

Willetts, N. and Wilkins, B. (1984). Processing of plasmid DNA during bacterial conjugation. *Microbiol. Rev.* **48**, 24–41.

Williams, N. and Skurray, R. (1980). The conjugation system of F-like plasmids. *Ann. Rev. Genet.* **14**, 41–76.

Williams, D.R. and Thomas, C.M. (1992). Active partitioning of bacterial plasmids. *J. Gen. Microbiol.* **138**, 1–16.

Yamanaka, K., Ogura, T., Niki, H. and Hiraga, S. (1996). Identification of two new genes, *mukE* and *mukF*, involved in chromosome partitioning in *Escherichia coli*. *Mol. Gen. Genet.* **250**, 241–251.

Yanisch-Perron C., Vieira J. and Messing J. (1985). Improved M13 phage cloning vectors and host strains: nucleotide sequences of the M13mp18 and pUC19 vectors. *Gene* **33**, 103–119.

Yung, B.Y. and Kornberg, A. (1988). Membrane attachment activates *dnaA* protein, the initiation protein of chromosome replication in *Escherichia coli*. *Proc. Natl. Acad. Sci. USA* **85**, 7202–7205.

Zahn, K. and Blattner, F.R. (1987). Direct evidence for DNA bending at the lambda replication origin. *Science* **236**, 416–422.

Zhao, A.C. and Khan, S.A. (1997). Sequence requirements for the termination of rolling circle replication of plasmid pT181. *Mol. Microbiol.* **24**, 535–544

3 Exploitation of Bacteriophages and their Components

Margaret CM Smith[1] and Catherine ED Rees[2]

[1]Institute of Genetics, Queens Medical Centre, University of Nottingham, Nottingham, UK;
[2]Division of Food Sciences, University of Nottingham, Loughborough, Leicestershire, UK

◆◆◆

CONTENTS

◆◆◆◆◆◆ I. INTRODUCTION

Studies on the growth and replication of phages have provided many precedents for understanding fundamental biological principles at the molecular level. Classical examples include the operation of a genetic switch, control of a simple developmental pathway, mechanisms of DNA replication and recombination, and macro-molecular assembly (Casjens and Hendrix, 1988; Geiduschek and Kassavetis, 1988; Kornberg and Baker,1992; Ptashne, 1992; Greenblatt *et al.*, 1993; Leach, 1996). However, as well as being simple model organisms, phage derivatives and phage components are essential tools to the molecular geneticist. From the classical experiments on transduction to the development of phages as cloning vectors, there has been no lack of imagination on how phages can be exploited for use in genetic analysis. The amazing success of the phage λ cloning systems in *E. coli* has had a deeply profound effect on the analysis of complex genomes. In this chapter we are going to discuss the exploitation of bacteriophages for genetic analysis and *in vitro* recombinant DNA techniques in diverse bacterial species. Bacteriophages are obligate intracellular parasites and are uniquely adapted to growth in their hosts. They will inevitably, therefore, encode genetic elements that are easily accessible as tools for the genetic analysis of their hosts.

◆◆◆◆◆◆ II. TYPES OF BACTERIOPHAGE

Bacteriophages cannot be classified using the species concept. Instead, a new set of classification criteria must be used to group phages and these are still under consideration by the International Committee for Taxonomy of Viruses (ICTV). Two very obvious characteristics of phages are the nature of their nucleic acid (ds or ssDNA, ds or ssRNA) and their morphology, but these alone cannot be used to classify phages as there is still too much diversity within each group. For example, most phages (95% of known bacterial viruses) are dsDNA tailed phages with icosahedral (sometimes elongated) heads (Ackerman and DuBow, 1987; Casjens *et al.*, 1992). These include coliphages λ, P1, P2, T-even, T-odd and most of the phages that are described in this chapter. In spite of their similar morphologies, these phages show great diversity in their DNA sequences, lifestyles and genome organization. Similarly, φX174 and MS2 are both very tiny icosahedral non-tailed phages but the former is a ssDNA phage and the latter a ssRNA phage. Other morphologies include the filamentous phages such as the ssDNA fd family and at least one enveloped phage. Further classification of phages makes reference to the host cell type (e.g. coliphage, actinophage, myxophage, mycobacteriophage, etc). Within phages that infect a host, there are families of phages, sometimes termed "quasi-species", defined by a similar life cycle, genome and transcriptional organization and opportunities for DNA exchange (Campbell, 1994; Casjens *et al.*, 1992). The best studied example of a quasi-species is the family of phages where all the members are related to phage λ, i.e. the lambdoid phage. Analysis of the genomes of the lambdoid phages showed that there is much exchange of DNA between family members leading to mosaics of relatedness (Casjens *et al.*, 1992). Recently the sequences of many dsDNA phage genomes have been completed and their analysis has shown that phages that infect different hosts and belong to different families contain related genes (Hartley *et al.*,1994; Esposito *et al.*, 1996; van Sinderen *et al.*, 1996; Stanley *et al.*, 1997; Ford *et al.*, 1998). Data of this type strongly suggest that all dsDNA phage have access to a large pool of genes and that they are undergoing profuse genetic exchange (Hendrix *et al.*, 1998).

◆◆◆◆◆◆ III. BACTERIOPHAGE LIFE CYCLES

Bacteriophages can grow lytically during which time the host cell is completely reprogrammed and becomes dedicated to replicating the infecting phage. This ultimately ends in lysis of the host cell and the release of many progeny phages. Exceptions are the filamentous phages, which are extruded into the surrounding media without causing cell lysis; in a bacterial lawn a plaque is visible due to retardation of growth of the infected cells. Alternatively, some phages enter into the lysogenic cycle and become dormant. Lysogeny describes the state when a host cell contains a prophage in which the lytic genes are repressed by the phage-encoded repressor. It is the activity of the phage repressor that confers superinfection immunity to homoimmune phages, i.e. phages which have the same repressor-operator specificity. However, a culture of lysogenic cells contains free phages due to spontaneous induction of the prophage into the lytic cycle in some cells.

The prophage persists in the lysogen as a plasmid (e.g. phage P1) or integrated into the chromosome either at a specific site (e.g. λ) or at a random site (e.g. Mu). Phages that insert site-specifically encode a recombinase which catalyses recombination between the phage attachment site (*attP*) and the bacterial attachment site (*attB*), resulting in integration of the phage genome into the bacterial chromosome. The integrated prophage is flanked by the hybrid recombination sites, *attL* and *attR*. In a reaction resembling the reverse of the integration reaction, *attL* and *attR* recombine to excise the phage DNA and regenerate the *attP* and *attB* sites. For phage λ the chromosomal *attB* site is a specific sequence which lies outside a transcriptional unit of the *E. coli* chromosome, but for many phages the integration target is a tRNA gene (for a review see Campbell, 1992; see also Table 1). Usually integration of the phage genome into a gene containing an *attB* site does not result in an insertion mutation due to homologous sequences contained within the phage *attP* site which allow regeneration of an intact tRNA following the recombination event (e.g. Dupont *et al.*, 1995).

While the phage lytic genes themselves are repressed during the establishment of lysogeny, expressions of other non-essential genes carried on the phage genome is not always tightly controlled. In this way strains may be phenotypically transformed following establishment of lysogeny (lysogenic conversion). A classical example of this is the introduction of toxin genes encoded by phage such as *Vibrio cholerae* CTXφ and those for *Corynebacterium diphtheriae* (Groman, 1984; Waldor and Mekalanos, 1996) giving a pathogenic phenotype, but more recently other phenotypes such as loss of O surface antigen expression in *Acetobacter methanolicus* have been reported (Mamat *et al.*, 1995). This becomes a problem when classical typing of bacteria recovered from natural environments is carried out using an array of biochemical tests and suggests that lysogenic conversion may account in part for the appearance of non-classical variants of bacterial species. The prophage can be induced into lytic growth as a response to changes in the host physiology; the classic example and one that induces many prophages is the presence of a mutagen such as mitomycin C or UV (Ptashne, 1992; van de Guchte *et al.*, 1994a; McVeigh and Yasbin, 1996; Walker *et al.*, 1998). Not all phages can enter into the lysogenic state; those that can undertake both lifestyles are termed temperate phages and those that can only enter into the lytic pathway are termed virulent.

Table I. Site-specific integrases and integration vectors for diverse prokaryotes

Bacterial hosts	Source of integrase[a]	Vector	attB site	Reference
Amylcolatopsis methanolica	pMEA300			Vrijbloed *et al.*, 1994
Arthrobacter aureus, Brevibacterium lactofermentum, Corynebacterium glutamicum	φAAU2	p5510		LeMarrec *et al.*, 1994, 1996
Enterococcus faecalis	φFC1			Kim *et al.*, 1996
Haemophilus influenzae	φHP1		tRNA(Leu)	Goodman and Scocca, 1989; Hauser and Scocca, 1992
Klebsiella pneumoniae	φP4	P4D0104 P4D0105		Ow and Ausubel, 1983
Lactobacillus delbrueckii subsp. *bulgaricus, L. planitarum, L. casei, Lactococcus lactis* subsp. *cremoris, Enterococcus faecalis, Streptococcus pneumoniae*	φmv4	pMC1	tRNA(ser)	Auvray *et al.*, 1997; Dupont *et al.*, 1995
Lactobacillus gasseri	φadh	pTRK182		Raya *et al.*, 1992
Lactococcus lactis subsp. *cremoris*	φTuc2009	pIN1		Van de Guchte *et al.*, 1994b
Lactococcus lactis subsp. *cremoris*	φTP901-1	pBC170		Christiansen *et al.*, 1994, 1996
Lactococcus lactis subsp. *lactis*	φLC3	pINT2		Lillehaug *et al.*, 1997
Leuconostoc oenos	φ10MC		tRNA(leu)	Gindreau *et al.*, 1997
Mycobacteria smegmatis, M. tuberculosis, BCG	φL5	pMH5 pMH94	tRNA(gly)	Lee *et al.*, 1991a
Mycobacteria smegmatis	φD29	pRM64		Ribeiro *et al.*, 1997
Mycobacteria smegmatis, M bovis BCG	φFRAT1	pJRD184		Haeseleer *et al.*, 1992, 1993
Myxococcus xanthus	φMx8			Tojo *et al.*, 1996; Salmi *et al.*, 1998
Rhizobium meliloti	φ16-3		tRNA(pro)	Papp *et al.*, 1993
Prevotella ruminicola	φAR29			Gregg *et al.*, 1994
Pseudomonas aeruginosa	φCTX	pINT pINTS		Wang *et al.*, 1995
Saccharopolyspora erythreae, Streptomyces lividans	pSE101		tRNA(thr)	Brown *et al.*, 1994
Saccharopolyspora erythreae	pSE211	pTKB270	tRNA(phe)	Brown *et al.*, 1990; Katz *et al.*, 1991
Staphylococcus aureus	φL54a	pCL83 pCL84		Lee *et al.*, 1991
Streptococcus pyogenes	φT12	pWM139	tRNA(ser)	McShan *et al.*, 1997

Streptococcus thermophilus	φSfi21		tRNA(arg)	Bruttin *et al.*, 1997
Streptomyces spp.	φC31	pSET152, pKC796		Bierman *et al.*, 1992; Kuhstoss *et al.*, 1991
Streptomyces ambofaciens, S. coelicolor, S. lividans, Mycobacteria smegmatis	pSAM2	pKC824 pPM927 pTSN39	tRNA(thr)	Boccard *et al.*, 1989; Smokvina *et al.*, 1990; Kuhstoss *et al.*, 1989, 1991; Martin *et al.*, 1991
Streptomyces lividans	SLP1			Brasch *et al.*, 1993
Streptomyces griseus	pSG1		tRNA(ser)	Bar-Nir *et al.*, 1992
Streptomyces parvulus	φR4	pAT98		Matsuura *et al.*, 1996
Streptomyces rimosus	φRP3		tRNA(arg)	Gabriel *et al.*, 1995
Sulfolobus shibatae	SSV1			Muskhelishvili *et al.*, 1993

[a] "φ" indicates the source of the integrase and *attP* site is a bacteriophage. Other systems are from integrating plasmids.

[b] Vectors ready-for-use are indicated.

Bacteriophages are model systems for understanding temporal gene control. During the lytic cycle there are some events that must occur early, for example switching off host RNA synthesis and initiating phage DNA replication. The synthesis of the phage coat and tail proteins occurs later. Phage heads assemble from oligomers of the major capsid proteins, the DNA is packaged inside the heads formed by this oligomerization and then tails and tail fibres are attached. The final stage of the lytic process is release of the newly synthesized progeny phages into the surrounding environment by lysis of the infected cell. In most phages the lytic developmental pathway is controlled by a cascade of transcription factors. The cascade begins immediately after injection of the DNA, with the transcription of phage promoters that are recognized by the host RNA polymerase. The products of these "immediately-early" genes encode proteins that subvert the host into expressing preferentially the next wave of phage genes, i.e. the "early" genes. One of the early gene products will be an activator required for expression of the late genes. All kinds of mechanisms imaginable are utilized for controlling gene expression in bacteriophages (Geiduschek and Kassavetis, 1988; Ptashne, 1992). As the transcriptional circuitry of phages other than those infecting the coliforms are deduced, one might expect even greater diversity to be revealed (e.g. Brown *et al.*, 1997; Wilson and Smith, 1998).

◆◆◆◆◆◆ IV. ISOLATION OF BACTERIOPHAGES

Bacteriophages are generally found wherever their bacterial host cells are found. Estimates of the numbers of phage in any given environment suggest that there are generally tenfold more phages than bacteria (Bergh *et al.*, 1989; Whitman *et al.*, 1998). Procedures for the isolation of new phages simply entail filtration of an aqueous sample derived from an environment which contains the host bacteria and plating on a suitable indicator lawn. In some cases the phages may be bound to particulate matter and so are removed by filtration. This problem sometimes occurs with phages which infect soil bacteria such as *Streptomyces*. If the soil sample is inoculated with a spore suspension of indicator bacteria, incubated overnight, filtered and then plated, this enriches for the phages infecting that particular host and helps to achieve release of the phages into the supernatant (Diaz *et al.*, 1989; Dowding, 1973). Although genetically identical bacteriophages from a single environmental sample can be isolated, from geographically different environments it is far more usual to isolate similar, but not identical, phages. For example, there are many relatives of phage λ that have been isolated and make up the lambdoid family, but λ itself has only been isolated once from nature (Casjens *et al.*, 1992). This diversity probably reflects an amazingly rapid evolution of bacteriophages.

◆◆◆◆◆◆ V. TRANSDUCTION

Transduction is a form of gene transfer from one bacterium to another, mediated by bacteriophage particles. The process of transduction is typically resistant to DNase I treatment of samples, as the DNA to be transferred is protected by the phage coat proteins. Phages can mediate two types of transduction: specialized and generalized. Classically, specialized transduction refers to the ability of some temperate phages to transfer a small number of genetic markers (i.e. one or two). The specialized transducing phages usually arise by aberrant excision of the dormant prophage from the host chromosome during which DNA adjacent to their specific integration sites is incorporated into the phage heads. In many cases the phages are still capable of replication, i.e. in addition to the chromosomal DNA they still have essential phage genes for lytic growth, but some inessential phage genes have often been deleted to accommodate the incorporation of the additional host cell DNA. Since the advent of recombinant DNA techniques, specialized transduction now includes the use of phages as cloning vectors into which specific DNA fragments have been inserted *in vitro*. The development of phages as cloning vectors is considered further below (section VII).

Recently, specialized transduction of a pathogenicity island encoding the toxic shock syndrome toxin-1 (TSST-1) from *Staphylococcus aureus* by staphylococcal phage 80α was reported (Lindsay *et al.*, 1998). TSST-1 is located as part of a 15.2 kbp genetic element, SaPI1, which also encodes a site-specific recombinase. During infection of *S. aureus* carrying SaPI1 with 80α, SaPI1 is excised, replicated and packaged to form SaPI1 transducing particles at a frequency close to that of the plaque-forming titre. In this novel system, elements of phage 80α, normally a generalized transducing phage for *S. aureus*, are apparently being exploited by SaPI1 to aid its spread.

A. Generalized Transduction

In generalized transduction the phage can transfer any part of the bacterial chromosome. This occurs due to occasional mispackaging of chromosomal DNA into phage particles. Upon infection of a recipient cell by the transducing particle, the chromosomal fragment can recombine with the recipient chromosome and the recombinants can be detected by selection for genetic markers transferred from the original lysate. Transduction was originally observed in *Salmonella typhimurium* by Zinder and Lederberg (1952) using the temperate phage P22 and later using *E. coli* phage P1 (Lennox, 1955). Transduction quickly became an essential tool for the fine-structure mapping of genes and for transferring chromosomal mutations into different strains. An especially important application is to transfer a newly induced mutation into a clean genetic background.

B. Isolation of General Transducing Phage

The original observation of transduction (Zinder and Lederburg, 1952) was performed simply by mixing two strains of *S. typhimurium*, LT-2 and LT-22, and selecting for appearance of recombinant marker gene pairs. After it was realized that the genetic material was transferred via a filterable agent it soon became clear that phage particles were the mediators. Screening of a phage collection for transducing phages can be performed by testing for transfer of easily selectable markers from a donor (e.g. *trp*[+], resistance to streptomycin) to a recipient lacking those growth characteristics (e.g. *trp*[−], sensitivity to streptomycin). Ideally a demonstration of co-transduction of two markers is desirable.

The development of generalized transducing phages for genetic analysis is considered to be an extremely desirable, if not essential, genetic tool. Mapping by transduction is still useful for analysis of poorly characterized genomes and often is the only available method for the movement of genetic markers to different genetic backgrounds. Consequently considerable effort has gone into the isolation and use of phages for transduction in many bacterial species. Transducing phages as genetic tools are reviewed in *Methods in Enzymology* volume 204 for *E. coli* and *Salmonella* (Sternberg and Maurer, 1991), *Bacillus subtilis* (Hoch, 1991), *Vibrio* spp. (Silverman *et al.*, 1991), *Myxobacteria* (Kaiser, 1991), *Staphylococcus* (Novick, 1991) *Caulobacter* (Ely, 1991), *Rhizobium* (Glazebrook and Walker, 1991), *Streptococcus* (Caparon and Scott, 1991) and *Pseudomonas* (Rothmel *et al.*, 1991). In addition, generalized transducing phages have been isolated for *Streptomyces venezualae* (Stuttard, 1979) for *Erwinia carotovora* subsp. *carotovora* (Toth *et al.*, 1993), *E. carotovora* subsp. *atroseptica* (Toth *et al.*, 1997) the spirochete, *Serpulina hyodysenteriae* (Humphrey *et al.*, 1997; see Chapter 7, IV.C.2) and for Methano bacterium thermoauto trophicum (Chapter 10, III.C). While there has been lack of success in the isolation of a generalized transducing phage for *Rhodobacter sphaeroides*, *R. capsulatus* spontaneously releases a "gene transfer agent" (GTA) capable of generalized transfer of about 4 to 5 kbp of DNA (Donohue and Kaplan, 1991). There has been little success in isolating a generalized transducing phage for *Mycobacteria* (Jacobs *et al.*, 1991; but see Chapter 9, III.C) or for *Neisseria* (Seifert and So, 1991), a genus for which no phage has yet been described.

C. Properties of General Transducing Phage

Generalized transduction depends on the random mispackaging of chromosomal DNA into phage heads. The length of the mispackaged DNA reflects the packaging constraints of the phage and these can vary widely. The *B. subtilis* phage PBS1 will package between 200 and 300 kbp of DNA (5–10% of the host genome) whereas the *Salmonella* phage P22 only packages about 44 kbp. Obviously it is important that the phage infection process itself does not induce wholesale degradation of the chromosomal DNA, otherwise fragments of sufficient length to fill the phage heads will not be available. The frequency of mispackaging, and therefore the number of transducing particles, is determined by the mechanism of DNA packaging. (For reviews on phage DNA packaging see Black, 1988; Catalano *et al.*, 1995; Hendrix, 1998). Phages that package DNA via the "headful" mechanisms are more likely to be capable of generalized transduction than those that package via other strategies. Packaging via the "headful" mechanism relies on a locus called the *pac* site, which is recognized by the packaging apparatus to initiate the process. Phage DNA is incorporated into the prohead until the head is filled. The phage DNA substrate is actually a concatemer of phage genomes and the filled head contains just over one genome length of phage DNA; the DNA is cleaved from the concatameric form and the packaging apparatus then uses this cleaved end to initiate filling another empty prohead. The mechanism is therefore processive and results in a population of phages that contain slightly more than one genome length, i.e. the DNA is terminally redundant and cyclically permuted. Phages which use the *pac*/headful mechanism occasionally package chromosomal DNA instead of phage DNA, initiating the packaging process from a random site in the chromosome which does not appear to have any homology with the *pac* sequence. It is these phage particles containing mispackaged chromosomal DNA which form the transducing particles. The frequency of mispackaging can be enhanced by inserting a *pac* site on the non-phage DNA to be transferred; for example P22 transduction frequency increases 100–1000-fold if the donor DNA contains a *pac* site (Schmidt and Schmieger, 1984; Vogel and Schmieger, 1986). Phage P1 and P22 are typical examples of transducing phage for *E. coli*, *Salmonella* and their relatives giving frequencies of 3×10^{-4} to 1×10^{-2} and 5×10^{-7} to

5×10^{-4} transductants per infected donor cell, respectively (Sternberg and Maurer, 1991).

An alternative mechanism for packaging of DNA into phage heads involves cleaving the DNA concatamers at a specific site (the *cos* site) located at the end of a single phage genome unit to generate 5′ or 3′ protruding complementary (cohesive) ends. Packaging in these phages starts by recognition of a *cos* site, packaging of the phage genome until the apparatus reaches a second *cos* site on the same linear molecule where it cleaves again, thus filling the phage head with one complete genome. Even though these phages recognize a specific site in the genome, they are still reliant on packaging of a defined length of DNA to fill the prohead. Phages with cohesive ends are usually less useful than those using the *pac*/headful mechanism as generalized transducers because the chances of two *cos* sites occurring at random in the bacterial genome to give the correct intervening size of DNA to complete the packaging process are virtually nil. However, it has been possible to use at least one *cos*-containing phage, phage λ, as a general transducer using a specially devised procedure (Sternberg, and Weisberg, 1975; Sternberg and Maurer, 1991). Like the *pac*/headful packaging phage, λ occasionally initiates packaging at a random sequence in the chromosome and fills a prohead. However, without a nearby *cos* site, the remaining protruding DNA cannot be cleaved and assembly is blocked. If the lysate containing these blocked particles is treated with DNase I, the remaining maturation steps, i.e. attachment of tails, etc., can occur. This procedure yields between 7×10^{-7} to 1.5×10^{-5} transductants per infected donor.

Phages that package via the headful mechanism (and which are therefore candidate generalized transducing phage) are generally less sensitive to chelating agents, such as tetrasodium pyrophosphate, than those that package by a *cos* mechanism. Stuttard (1989) has used the sensitivity to sodium pyrophosphate as a mechanism to enrich for potential transducing phages for genetic analysis in *Streptomyces* and transducing phages were isolated for *S. venezualae*. Phages that package with a *cos* mechanism and survive the treatment frequently contain deletions and this is a useful property for defining those regions of the phage that are dispensable for growth and has been exploited in the construction of phage-cloning vectors (Parkinson and Huskey, 1971; Chater *et al.*, 1981; Stuttard, 1989; see section VII).

A practical requirement for a transducing phages is that the transductants must be recoverable above the background of infectious phage particles. Thus temperate phages are preferred to lytic phages unless the frequency of transduction is sufficiently high in the latter that the phages can be diluted or inactivated. Methods to detect transductants are discussed below.

D. Strategies to Increase the Yield of Transductants

The basic transduction procedure involves preparation of a lysate on the donor strain containing the genetic marker, then infecting a recipient and screening/selection for the desired phenotype. Two problems must be overcome in screening the recipients. Firstly, the recipients are in danger of being killed by the infectious phage in the lysate, especially if the frequency of transducing particles is low compared to the phage titre. Secondly, abortive transductants may vastly outnumber the stable transductants as is frequently observed with P22 and P1 transduction in *Salmonella* and *E. coli*. Abortive transductants occur when the DNA has not been stably integrated into the recipient chromosome and instead at cell division the DNA is inherited by one daughter cell only. The mechanism resulting in abortive transductants is that the free DNA ends, which would normally be recombinogenic, are held together by phage proteins and become resistant to involvement in strand invasion required for homologous recombination (Ikeda and Tomizawa, 1965). UV treatment of a P1 lysate increases the numbers of transductants because the UV damage of the incoming phage induces recombination in the recipient cell (Newman and Masters, 1980; Sternberg and Maurer, 1991).

The UV treatment of phage P1 has the added benefit of decreasing the numbers of infectious phages about tenfold and so decreases the chances of superinfection. This approach has also been used to kill the virulent *Erwinia carotovora* subsp. *atroseptica* transducing phage ϕM1. In this case the phage titre reduces about 100-fold but the numbers of transductants increase about threefold (Toth *et al.*, 1997). Similarly UV treatment of the virulent *Caulobacter crescentus* transducing phage ϕCR30 reduces the phage titre from about 10^{10} to about 10^5 pfu ml^{-1} but still results in efficient levels of transduction (Ely and Johnson, 1977; Ely, 1991). Moreover ϕCR30 is inhibited when plated on minimal media and this further reduces the killing of transductants. Indeed various growth media for propagation of the host strain have been used to selectively inhibit the growth of the phage; for example the *Rhizobium* virulent phage ϕM12 adsorbs poorly to the bacterial cell surface in media containing less than 0.25 M Ca^{2+} ions and *E. coli* phage P1 and *S. aureus* phage ϕ11 are inhibited by sodium citrate (Finan *et al.*, 1984; Novick, 1991; Sternberg and Maurer, 1991). Another strategy is to use temperature-sensitive mutations to prevent killing of transductants; myxophage Mx4 is a virulent phage which has had a temperature-sensitive mutation incorporated along with a second mutation to increase host range to form an efficient transducing phages (Geisselsoder, *et al.*, 1978). Other means of inactivating or removing phages in the donor lysate which might compromise the growth of transductants are the use of centrifugation (e.g. the use of streptococcal phage A25; Caparon and Scott, 1987, 1991), antibody, or dilution. However, dilution can only realistically be relied upon if the frequency of transduction is sufficiently high such as occurs in the marine *Vibrio* phage AS-3, which even without UV inactivation can transduce markers at a frequency of 10^{-5} and 10^{-6} per donor cell (Ichige *et al.*, 1989). Of course if the transducing phage is temperate, use of a recipient which is lysogenic for the same or a similarly regulated phage will permit detection of the transductants following plating on selective media (e.g. myxophage Mx8, staphylococcal phage, ϕ11; Kaiser, 1991; Novick 1991).

Another strategy to increase the yield of transductants is to place a copy of the phage *pac* site close to markers to be transferred. The strategy has been described utilizing the *pac* site from phage CP-T1, a generalized transducing, temperate phage for *V. cholerae* El Tor and classic biotypes. The genome of this phage is approximately 43 kbp and is thought to package via a *pac*/headful mechanism. Thus the CP-T1 *pac* site has been inserted into a Tn5 derivative to form Tn*pac* with the aim of obtaining very high frequency of transduction of markers close to the transposon after random insertion in the chromosome (Manning, 1988; Silverman *et al.*, 1991).

E. Use of the Specialized Transducing *Bacillus* Phage, SPβ, as a General Transducer

B. subtilis is well served for generalized transducing phage (Cutting and Vander Horn, 1990; Hoch, 1991). PBS1 packages approximately 200 kbp of DNA and before the *Bacillus* genome was sequenced, transduction was the method of choice for initiating a genetic mapping project. Once linkage is established between markers by PBS1-mediated transduction, co-transformation of chromosomal DNA can be used to locate the marker to within 30 kbp. The temperate phage SPβ has been developed to transduce DNA for complementation analysis (Cutting and Vander Horn, 1990; Zahler, 1982). SPβ normally integrates into a specific site in the *B. subtilis* chromosome. SPβ *c2 int-5* mutants are defective in the phage repressor (allowing heat induction of the prophage) and in the recombinase required for integration at the specific site. A derivative of this carrying Tn*917*Cm (Tn*917* encoding chloramphenicol resistance) has been constructed. The strategy employed entails using this SPβ *c2 int-5*::Tn*917*Cm phage to infect a non-lysogen containing Tn*917*Em (Tn*917* encoding erythromycin resistance) already inserted into the gene of interest. Selection for chloramphenicol and erythromycin-resistant colonies recovers cells in which the circular SPβ *c2 int-5*::Tn*917*Cm has inserted into the chromosome via homologous recombination with Tn*917*Em. SPβ released from this lysogen will include some transducing particles containing flanking markers; infection of a recipient lysogenic for SPβ at its normal integration site and selection for chloramphenicol resistance will yield clones that can be used for complementation analysis.

F. Plasmid Transduction

Plasmid transduction is a useful method to transfer plasmids from one strain to another, thereby eliminating any constraint imposed by a requirement either to induce competence or to make and regenerate protoplasts. *B. subtilis* phage, SPP1, *Lactobacillus gasseri* phage, φadh and *S. aureus* phage, φ11 have been used to transduce plasmids at high frequency between strains. The efficiency of transfer of plasmid molecules can be dramatically increased by cloning random pieces of the transducing phage genome into the plasmid vector (Deichelbohrer *et al.*, 1985; Novick *et al.*, 1986; Raya and Klaenhammer, 1992). In *Streptomyces* species, phage FP43 has been used to transfer plasmids containing a fragment of FP43 DNA into 24 strains of *Streptomyces*, including many to which the phage is refractory for plaque formation, and between genera (McHenney and Baltz, 1988). Cosmids have been constructed by insertion of *cos* sites from the *Streptomyces* temperate phage R4, φC31 and RP3 into autonomously replicating plasmids and these can be transduced between strains at frequencies as high as 2×10^{-2} transductants per infectious phage particle (Morino *et al.*, 1985; Kobler *et al.*, 1991; Kinner, *et al.*, 1994).

G. Intergeneric Transduction

Host range may sometimes be a consideration in developing a phage for transduction. For example a broad host range phage could be used for transfer of plasmids and markers with ease between different strains, or once a single phage has been developed for use, that phage will serve as a transducer for many strains. Mutation of a transducing phage to increase host range has been reported (e.g. myxophage Mx4: Kaiser, 1991; phage P1; Yarmolinsky and Sternberg, 1988). Phage P1 naturally has an extremely broad host range spanning across many genera (Yarmolinsky and Sternberg, 1988). P1 contains two sets of genes encoding tail fibre proteins – the major determinants of host range in bacteriophages. An invertible segment, the C segment, of DNA switches expression of the tail fibre proteins between the two forms. Depending on the orientation of the tail fibre genes during replication, the progeny phage either can or cannot plaque on *E. coli* K12. P1 can multiply in *Klebsiella*, *Erwinia* and *Pseudomonas* and deliver DNA (but not replicate) in *Myxobacteria*, *Yersinia pestis*, *Flavobacterium*, *Agrobacterium*, some *Vibrio* strains and *Alcaligenes*. The ability of P1 to inject its DNA but not to replicate has facilitated delivery of DNA into, for example, *Myxobacteria* and *Vibrio*. Thus P1 will transduce plasmids from *E. coli* into these bacteria, but of most use are P1 phage that deliver Tn5 or mini-Mu derivatives (Kuner and Kaiser, 1981; Belas *et al.*, 1982, 1984, 1985; Martin *et al.*, 1989; see Chapter 4).

Under special circumstances intergeneric transduction can be obtained with phage λ. In the genera *Erwinia* and *Yersinia enterocolitica*, strains engineered to contain the λ receptor (LamB) are now sensitive to λ infection and plaques will form, suggesting that the phage promoters, regulatory mechanisms and necessary host factors are all functional in these heterologous hosts (Salmond *et al.*, 1986; Brzostek *et al.*, 1995; Chapter 12). In other bacteria, e.g. *Myxobacteria* and *Vibrio*, provision of the LamB protein is only sufficient to obtain infection and the phages cannot form plaques. In both situations λ can be used as a suicide vector for the delivery of Tn5 derivatives (Way *et al.*, 1984; Salmond *et al.*, 1986; Silverman *et al.*, 1991).

◆◆◆◆◆◆ VI. MU-LIKE BACTERIOPHAGES

Bacteriophage Mu has had an enormous impact on genetic analysis in *E. coli* and close relatives (Howe, 1987; van Gijsegem *et al.*, 1987; Groisman, 1991; Chapter 4, III; Chapter 12, II.2(a)). Mu is a "mutator" phage, i.e. Mu lysogens have a high frequency of mutation detectable initially by screening for auxotrophy. Subsequent studies demonstrated that this phage replicates its DNA via a transposition process. Among other applications, Mu and its derivatives are used as generalized transducing phage, for *in vivo* cloning and for the isolation of gene fusions. Mu derivatives, in particular those which contain the P1 C region (which determines the host range of P1, see section V.G above) to form a Mu-P1 hybrid, can be used in a wide range of closely related genera including *Shigella*, several *Salmonella* species, *Serratia*, *Citrobacter*, *Enterobacter*, *Klebsiella* and *Erwinia*. Mu derivatives have also been used in *Rhizobium* and *Agrobacterium*. Mu-like phage have been isolated from *P. aeruginosa*, *Agrobacterium tumefaciens* and *V. cholerae* (Gerdes and Romig, 1975; Expert and Tourneur, 1982; Rothmel *et al.*, 1991).

A. Use of Phage D3112 and its Derivatives in *Pseudomonas aeruginosa*

One particular Mu-like phage, D3112, isolated from *P. aeruginosa* has been developed as a versatile tool for studies of *Pseudomonas* genetics (Rothmel *et al.*, 1991). D3112 encapsulates its DNA via a headful packaging mechanism and can incorporate approximately 40 kbp of DNA. It is a generalized transducing phage and can transduce chromosome markers at frequencies of approximately 10^{-9}. However, it is the mini-D3112 (mini-D) derivatives which are most useful in genetic analyses. In the mini-D derivatives, sequences from the phage termini flank various selectable markers and broad host range plasmid replicons. These mini-D constructs are perfectly stable in *P. aeruginosa* but can be induced to transpose or replicate in *P. aeruginosa*::D3112*cts* lysogens which provide the helper functions deleted from the D3112 genome. Mini-D elements have been used for insertional mutagenesis and analysis of the insertion mutations generated shows that approximately 80–90% are single insertions and their distribution in the chromosome is apparently random. Mini-D derivatives can transduce more than 35 kbp of chromosomal DNA and are about 700-fold more efficient than D3112*cts* alone and are more efficient than other *P. aeruginosa* transducing phages F116L and G101. If the mini-D contains an origin of replication and a selectable marker a form of *in vivo* cloning is a useful available technique. The mini-D is introduced into a thermoinducible D3112cts lysogen so that transposition can be induced by heat treatment. Multiple rounds of transposition result in mini-D elements flanking a gene of interest "x" in the chromosome. Some of the phage particles resulting from this induction may contain "x" and the two flanking mini-D elements which, after injection into a sensitive *P. aeruginosa* host, can recombine to generate a recombinant plasmid.

◆◆◆◆◆◆ VII. BACTERIOPHAGES AS CLONING VECTORS

When the tools became available to insert foreign DNA into heterologous hosts, *E. coli* phage λ was poised as an ideal phage system to develop as a cloning vector. The availability of well-characterized deletion mutants meant that the molecular biologist knew which regions of the λ genome could be discarded without affecting replication. Also derivatives that lacked restriction targets could be selected by plating on restriction+ strains, thus obtaining phage variants with unique cloning sites. Moreover, an understanding of the transcriptional circuitry, replication, recombination and other salient characteristics of the biology of λ led to rational design of vectors (Murray, 1983, 1991). A browse through the catalogues for companies selling molecular biology reagents provides an idea of just how sophisticated these λ vectors have become. Vectors are available which can accept between 9 and 23 kbp of DNA using versatile restriction enzyme sites for cloning and analysis of inserts, T7 and T3 promoters are used to generate end-specific transcripts for the generation probes (e.g. for chromosome walking), an *in vivo* excision system can be used to excise the insert along with a ColE1 origin of replication and promoters are arranged to express the cloned genes either in bacteria or in eukaryotic cells. Another milestone for phage λ as a cloning vector was in the development of *in vitro* DNA packaging systems which greatly increased the efficiency of introduction of DNA into the host cell (Rosenberg, 1987; Sambrook *et al.*, 1989). Later, cosmids were constructed, i.e. plasmids carrying the *cos* site of phage λ (Sambrook *et al.*, 1989 and references therein). Large DNA fragments inserted into these cosmid vectors are packaged efficiently providing the *cos* sites are separated by 37–52 kbp, i.e. a unit length of the λ genome. As the cosmids are only about 10 kbp in size, 30–40 kbp can be inserted into the vectors. The efficiency of cloning in λ and the ease of introduction of large DNA sequences via the *in vitro* packaging system greatly favoured λ as a vector of choice for genomic library construction. Furthermore, vectors which are capable of forming lysogens are frequently chosen for insertion of sequences in single copy into the *E. coli* chromosome and expression vectors based on λ are advantageous when screening libraries with antibody-linked probes, since each individual plaque represents a small-scale induced protein lysate of the cloned genes.

No other bacteriophage has been developed as extensively as λ for cloning in any other bacteria. However, phage vectors for cloning in several bacteria, including *Bacillus*, using φ105, and *Streptomyces*, using φC31, have been developed (see also Chapter 9, IV.C for a mycobacteriophage vector and Chapter 10, V.D for a *Sulfolobus* phage-derived replicon). As in the development of phage λ as a cloning vector, it was important that both φ105 and φC31 had regions of the genome that could be deleted and still yield viable phage.

A. Development of the Streptomyces Temperate Phage φC31 as a Cloning Vector

The development of the temperate *Streptomyces* phage φC31 as a cloning vector followed very much the same kind of steps as those used for phage λ and this subject has been reviewed by Chater (1986) and Hopwood *et al.* (1987). An important consideration in the development of φC31 as a cloning vector is the host range; a broad host range is desirable as many different species of *Streptomyces*, producing different secondary metabolites, are studied at any time. In fact φC31 infects about half of the 137 strains that have been tested. The phage vectors for cloning in *Streptomyces* complement a comprehensive range of other genetic tools (Hopwood *et al.*, 1985). However, some *Streptomyces* strains have proved very difficult to introduce DNA by any means other than by use of φC31 (e.g. *S. hygroscopicus*; Lomovskaya *et al.*, 1997). The repertoire of φC31 vectors now available provides capacity to clone up to 9 kbp of DNA using a variety of restriction enzymes and includes vectors designed to perform specific functions (e.g. promoter fusions, gene replacement.)

The first cloning experiment with φC31 entailed partial digestion of a φC31 deletion derivative (obtained by selection for resistance to a chelating agent; section V.C) with a restriction enzyme, ligation to linearized pBR322 and introduction of the ligation mixture into *S. lividans* protoplasts. The resulting chimera could replicate in *E. coli* as a multicopy vector and in *Streptomyces* as a temperate phage (Suarez and Chater, 1980). The consequence of introducing pBR322 into the phage was the insertion of several unique restriction sites. Deletions of this recombinant phage were isolated and these were found to have removed over 4 kbp from a region inessential for lytic growth. An antibiotic resistance marker suitable for selection in *Streptomyces* was inserted (Chater *et al.*, 1982). The first vector used to generate a library of *S. coelicolor* A3(2) chromosomal DNA was KC401 (Piret and Chater, 1985). This vector contains a 6.7 kbp *Bam*HI fragment which can be excised to leave only 33.8 kbp of phage DNA – too small to package efficiently. Insertion of DNA fragments between 2 and 9 kbp favours packaging of recombinant phage vectors. As KC401 is *attP⁻* (i.e. lacks the phage attachment site) and defective in the repressor gene, *c* (required for maintenance of lysogeny), the phage can only grow lytically. To use the library constructed in KC401 for complementation analysis, the library was used to superinfect a φC31 lysogen selecting for the resistance marker present on the vector. The superinfecting genome could integrate either by homologous recombination between the phage sequence or into the chromosome via the cloned DNA carried in the vector. This strategy was successfully used to identify DNA fragments from the library which complemented mutants in a sporulation gene, *bldA* (Piret and Chater, 1985).

B. Mutational cloning with φC31

An ingenious alternative strategy for identification of cloned DNA in φC31 vectors is to use "mutational cloning" (Chater *et al.*, 1985; Chater,

1986). This technique permits simultaneous isolation of DNA encoding a gene of interest and generation of a mutant. The strategy requires the use of a phage which can establish lysogeny (i.e. c^+) but is deleted for the *attP* site and therefore cannot integrate at the normal bacterial attachment site. Thus the phage can only form lysogens via homologous recombination between the inserted DNA and chromosomal DNA. If the inserted DNA fragment is entirely within a transcriptional unit then the integration will generate a mutation (see Chapter 10, II, III and IV). One copy of the gene is deleted at the upstream end and the other deleted at the downstream end. Phages released from the clone containing the mutant phenotype (by homologous recombination between duplicated sequences) contains DNA originating from the desired gene or operon. This approach was used to clone the methylenomycin biosynthesis genes (Chater and Bruton, 1983, 1985). The technique can also be used to map the transcriptional organization of gene clusters because if the DNA inserted into the ϕC31 vector contains either end of the transcription unit a complete copy will be regenerated and no mutant phenotype will be observed (Malpartida and Hopwood, 1986; see Chapter 10, II, III and IV). The ability to generate mutants by insertion of the phage has been used to analyse gene function by directed gene replacement; this allows the recovery of mutants even without the concomitant generation of a new phenotype (e.g. Buttner *et al.*, 1990).

Other ϕC31-derived vectors have been constructed which incorporate reporter genes for the analysis of promoter function. A particular aim was to use the mutational cloning technique to generate transcriptional fusions in the chromosome. A promoterless viomycin resistance gene was placed downstream of a fragment of DNA derived from the *gyl* (glycerol utilization) operon in a ϕC31 vector. The recombinant phage was integrated into the chromosome via the homology with *gyl* and viomycin resistance was then inducible by growth in the presence of glycerol (Rodicio *et al.*, 1985). ϕC31 derivatives containing a promoterless *lacZ* gene were also tested but were not pursued, due to (among other problems) poor levels of expression in *Streptomyces* (King and Chater, 1986). Later derivatives containing a promoterless *xylE* reporter gene were constructed (Bruton *et al.*, 1991).

C. Phage Vectors for Use in *Bacillus*

Another phage that has been extensively developed as a cloning vector is phage ϕ105, a temperate phage of *B. subtilis* (Errington, 1990). In this case cloning with phage vectors overcomes specific problems of structural instability, frequently encountered when using high copy number plasmids in *Bacillus*. The single copy nature of the lysogens greatly reduces the risk of rearrangements due to recombination between plasmid inserts and chromosomal homologues. Furthermore, problems have been encountered in preparing libraries of *Bacillus* DNA in *E. coli*, often due to the presence of strong promoters. Consequently the ϕ105 phage cloning system became the method of choice for the generation of gene libraries in *Bacillus* and for complementation analysis. Although the use of integrating plasmids has become popular for many applications, phage vectors offer a major advantage in that the cloned DNA is easier to re-isolate. Vectors based on φ105 have a limited number of restriction sites but when used in conjunction with a partial digest, end-fill procedures are fairly versatile. Moreover partial in-fill procedures prevent the vector from self-religation and so increases the proportion of chimeric phage. Phage φ105J119 can accommodate up to 4 kbp of DNA and has a thermoinducible repressor so that heat induction can be used to prepare large amounts of phage vector DNA. At low temperature the phage can establish lysogeny and can be used for complementation analysis. φ105 vectors permitting controllable expression of an inserted gene have been constructed (East and Errington, 1989; Gibson and Errington, 1992; Thornewell *et al.*, 1993). These phages are defective for lysis and carry a temperature-sensitive repressor gene. Genes inserted into these vectors in the same orientation as phage transcription are expressed, without comcomitant cell lysis, at a high level after a temperature shift.

D. Cloning DNA by "Prophage Transformation" in *Bacillus* phage SPβ

A novel cloning procedure, prophage transformation (Figure 1), takes advantage of the natural ability of *B. subtilis* to be transformed with linear DNA fragments, SPβ or φ105 phage cloning vectors and the versatile transposon Tn917 derivatives (see Chapter 4, III). The major advantage of prophage transformation with SPβ, as described by Poth and Youngman (1988), is that larger DNA fragments can be isolated than is currently possible with the φ105 vectors. The rationale behind prophage transformation is as follows: a plasmid, pCV1, contains part of the SPβ phage genome. Inserted into a non-essential region is a Tn917 derivative containing a chloramphenicol-resistance gene, Cm^r, a pBR322 origin of replication, and ampicillin resistance gene, Ap^r, for selection in *E. coli* and a cloning site (e.g. *Bam*HI). Insertion of DNA into the cloning site to form concatemers of SPβ DNA arms flanking the cloned DNA are transformed into SPβ::Tn917 lysogens selecting for the chloramphenicol-resistance marker. Double crossovers between the SPβ arms and the resident prophage integrate the cloned DNA into the prophage. Screening for the correct clone is achieved by complementation. SPβ is a large phage and analysis of the cloned insert is difficult. Systems, such as "recombinational subcloning" (Poth and Youngman, 1988; Figure 1), have therefore been devised to rescue the insert from the prophage. Transformation of a *B. subtilis* strain carrying a Tn917Em-containing plasmid, pTV5, with DNA from the complementing phage and selection for chloramphenicol resistance results in a shuttle plasmid that contains an *in vivo* recombinant of the cloned insert from the Tn917Cm^r strain and the resident plasmid. Another way to rescue the recombinant is to prepare chromosomal DNA from the

A Prophage transformation

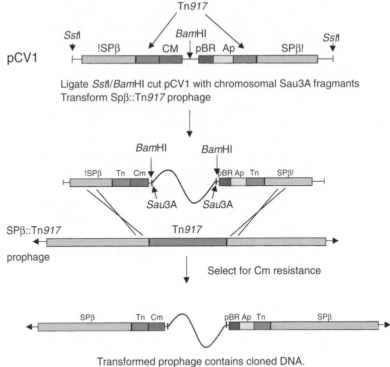

Figure 1. SPβ prophage transformation. (a) Chromosomal fragments are inserted into a *Bam*HI cloning site in the plasmid pCV1, which contains SPβ arms and a Tn*917* derivative containing a chloramphenicol resistance marker, Cm, a pBR322 origin of replication and an ampicillin-resistance gene, Ap. The ligation mixture is introduced into an SPβ::Tn*917* prophage selecting for Cm resistance. Screening for the correct clone is by complementation. The SPβ derivative so created may be induced to produce specialized transducing phage. (*continued*)

transformed prophage, restrict with an appropriate enzyme that does not cut between the *E. coli* plasmid origin of replication, the ampicillin-resistance marker and the cloned insert, self-ligate the DNA and recover clones containing the plasmid replicon (and associated cloned chromosomal DNA fragments) in an *E. coli* host.

◆◆◆◆◆◆ **VIII. REPORTER PHAGES**

The development of reporter phages grew from studies of phage genome organization and gene transduction. In essence these represent a very spe-

B "Recombinational subcloning" of the DNA insert from SPβ specialized transducting phage

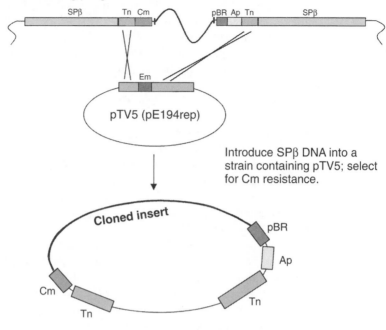

Introduce SPβ DNA into a strain containing pTV5; select for Cm resistance.

Shuttle plasmid (replicates in both Bacillus and *E. coli*) contains cloned DNA

Figure 1. (*continued*) (b) "Recombinational subcloning" of inserts from the specialized transducing phage created by prophage transformation. Phage DNA from the correct clone is introduced into a strain carrying pTV5, which contains a Tn*917* derivative. Recombination between the Tn*917* arms in the SPβ derivative and the Tn*917* arms in pTV5 creates a shuttle plasmid that can replicate in *E. coli*.

cialized form of transduction where the gene to be transduced (in this case a marker gene) is specifically introduced into the phage genome rather than being incorporated by chance packaging. In this way the phages are often acting more as vectors to transfer the gene into a new host cell rather than as a transducing particle. Another essential difference here is that the phages are usually designed not to become lysogenic or to become established in the cell by integration into the host cell chromosome. Indeed entry into the lytic pathway is often essential for the success of the reporter phages as a detection system since expression of the marker gene is often coupled to a phage late gene promoter. This provides the high level of gene expression required for amplification of the signal generated from low numbers of cells and the host–phage interaction allows the detection of specific bacteria in a mixed population without the need to purify to homogeneity.

Reporter phages based on the bacterial luciferase genes (*luxAB*) have been successfully generated for *E. coli* and *Salmonella* (Kodikara *et al.*, 1991; Chen and Griffiths, 1996) and the detection of *E. coli* has also been achieved using an assay based on the expression of the bacterial ice-nucleation gene (*ina*; Kozloff *et al.*, 1992). Early attempts to generate a *lux*-based reporter phage specific for *Listeria monocytogenes* were unsuccessful since the 2.1 kbp size of the *luxAB* cassette appeared to be too large to be accommodated within the packaging constraints of the temperate phage used (genome size estimated to be ~40 kbp). However, Loessner *et al.* (1996) have successfully introduced the *luxAB* genes into a larger virulent listeriaphage and have demonstrated its usefulness in rapid detection of *Listeria* in food samples (Loessner *et al.*, 1997).

The strategy used to construct these *Listeria lux* phages was a directed approach which requires a minimum amount of genome analysis to allow successful insertion of the reporter genes while retaining all phage essential functions. The strategy is generically applicable to all phages which can accommodate the insertion of the marker gene without exceeding the packaging constraints of the phages. To begin the construction, proteins which constitute a major component of the phage structure (and are therefore encoded by highly expressed genes) were purified and subjected to *N*-terminal amino acid analysis. The amino acid sequence was then used to deduce the corresponding DNA sequence and degenerate oligonucleotide probes complementary to this sequence used to identify the region of the phage genome encoding the protein by DNA hybridization. DNA fragments carrying the gene were then cloned and sequenced and the *lux* genes introduced into the cloned phage DNA as operon fusions with the identified structural genes. Using an appropriate shuttle vector, these constructs were then introduced into a propagating strain for the phage. Lysates recovered following infection of the plasmid-containing strains with wild-type phage were screened for the presence of light plaques. In the case of the generation of the virulent *Listeria lux* phages, double recombinants were found at an unexpectedly high frequency of 5×10^{-4} (Loessner *et al.*, 1996). This type of strategy is only effective if the packaging constraints of the phages are sufficient to allow incorporation of the marker genes into the phage genome. If this is not the case, functional phages can be generated if compensating deletions of inessential material are made, however this requires further analysis of the phage genome so that appropriate regions can be identified.

One of the best described and evaluated set of reporter phages developed to date are the mycobacteriophage reporter derivatives (Sarkis *et al.*, 1995; Pearson *et al.*, 1996; Carriere *et al.*, 1997). In this case both temperate phages such as L5 and virulent phage such as TM4 have been used to express the firefly *luc* genes. In the case of the temperate L5 *luc* phage, a constitutive promoter is fortuitously generated when the phage becomes integrated into the chromosome. Thus cells can be detected by production of light either by expression of *luc* from a phage-encoded promoter during the lytic cycle or from this newly generated hybrid promoter during lysogeny (Sarkis *et al.*, 1995) which overcomes the dormancy problem associated with using a temperate phage. An additional advantage here is that the firefly luciferase–luciferin reaction is dependent on a supply of ATP provided by the living bacterial cell. These workers have exploited this to develop not only a rapid detection system for mycobacteria in pathological samples but also a method to determine antibiotic sensitivity of strains (Jacobs *et al.*, 1993; Carriere *et al.*, 1997). When cultured in the absence of antibiotics, all strains susceptible to the phage produce light when infected with the *luc* phage. However if an antibiotic is introduced into the medium and the strain is senstitve to it, cell viability (and therefore the level of intracellular ATP) is reduced. Hence the levels of light resulting from a *luc* phage infection are also reduced, rapidly indicating the effect of the antibiotic on the strain tested avoiding the extended periods of time normally required for microbiological culture of these bacteria.

◆◆◆◆◆◆ IX. VECTORS BASED ON SITE-SPECIFIC RECOMBINATION SYSTEMS

Plasmids that integrate site-specifically into the bacterial chromosome to give a stable single copy insertion are relatively easy to construct and are used widely (Table 1). These vectors are different from integrative systems requiring homologous recombination between a fragment on the vector and its homologue in the chromosome. In the site-specific integration vectors, the integration is dependent on the recombination elements of temperate phages or integrative plasmids (see review by Nash, 1996). They integrate site-specifically into a site in the bacterial chromosome, the bacterial attachment site or *attB*. The reaction is normally catalysed by a phage- (or plasmid-) encoded site-specific integrase and requires the phage (or plasmid) attachment site, *attP*. As *attP* is on a circular molecule, recombination between *attP* and *attB* results in integration of whole circle flanked by the hybrid sites *attL* and *attR*. For excision to occur another phage or plasmid-encoded protein, Xis, is normally required and this acts in association with the integrase on the *attL* and *attR* sites in a reaction resembling the reverse of the integration reaction.

A. Construction of a Site-Specific Integration Vector

The ease with which these vectors can be constructed has meant that characterization of the integrase has become a primary aim in the analysis of any new phage or integrating plasmid. The basic components of an integration vector are the integrase gene, the *attP* site and a selectable marker. The integration vector does not usually contain an origin of replication functional in the bacteria in which it is to be used; normally the plasmids are constructed in *E. coli* using the ColE1 origin of replication. Integration vectors which carry *oriT* for intergeneric conjugal transfer from *E. coli* have proved to be extremely popular for facilitating introduction of the recombinant vectors into, for example, *Streptomyces* (Bierman *et al.*, 1992). As the integrating plasmid cannot replicate after introduction into the host bacteria, selection for the marker yields a clone carrying the whole plasmid integrated at the *attB* site. Presence of the excision (*xis*) gene(s) on integration vectors usually leads to instability of the inserted DNA. *xis* genes are generally located close to the *int/attP* region (although there are exceptions; e.g. Esposito and Scocca, 1994), but are not easily identified by sequence analysis as they tend to be small and highly divergent. For this reason most integration vectors are designed to avoid inclusion of complete ORFs present on either side of the *attP/int* region. If the plasmid is constructed avoiding the phage or plasmid-encoded excision genes, the plasmids are generally extremely stable and selection is not normally necessary to maintain the integrated plasmid. The disadvantage to this is that recovery of the plasmid is not as easy as using a phage or autonomously replicating vector.

Integrating vectors have been constructed from several temperate phages and plasmids (Table 1). Frequently the host range for the vector is very broad; for example vectors derived from the *Streptomyces ambofaciens* plasmid, pSAM2, integrate into the genomes of many *Streptomyces* strains. This is because the *attB* site recognized by pSAM2 integrase lies within the tRNA(thr) gene. As the *attB* site is also conserved in mycobacteria, pSAM2-derived integrating plasmids have also been used in this heterologous system. The host range for the integrating vector is determined partly by the presence or absence of the *attB* site and partly by the host-encoded accessory factors (e.g. integration host factors), which may be required. As many *attB* sites are within tRNA genes (Table 1), which are generally highly conserved, existing integration systems may be functional in quite a broad range of bacteria. Plasmids that integrate at different *attB* sites can be used together, in the same strain so long as they carry different antibiotic resistance markers (e.g. pSAM2- and φC31-derived vectors can be use together in *Streptomyces*).

What is the most direct way to identify the components for integration from a newly characterized temperate phage? The most straightforward way is to use Southern blot analysis to identify a restriction fragment from the phage which contains the *attP* site. This restriction fragment will be altered in the lysogen when compared to the pattern of fragments generated from the free phage due to the recombination event. Most *attP* sites lie adjacent to the integrase gene. Sometimes, as in the Mx8 mycobacterial phage, the *attP* site lies within the *int* gene and recombination causes a structural change to *int* (Tojo *et al.*, 1996). All phage integrases described to date fall into either of the two groups of site-specific recombinases, i.e. those resembling the λ integrase family, which is by far the largest group, or those related to the resolvase/invertases family. Both families of integrases are fairly divergent in sequence but both contain invariant amino acid residues which can be recognized by searches using the programs PROSITE (http://www2.ebi.ac.uk/ppsearch) and PFAM (http://www.sanger.ac.uk/Software/Pfam/), both available at the

Sanger Centre (http://www2.ebi.ac.uk/services.html). To identify the nature of the *attB* site the usual approach is to generate a library from the lysogen, probing it with the *attP* site to pick out clones containing *attL* and *attR* and then designing PCR primers to amplify the *attB* site directly from the chromosome.

◆◆◆◆◆◆ X. EXPLOITATION OF PHAGE PROMOTERS AND REPRESSORS

It is often necessary to express genes in their homologous host for the purposes of complementation, to study the effects of over-expression or for purification. A strong regulatable promoter is a necessary requirement for these kinds of approaches. In *E. coli* phage promoters have been exploited for these purposes; the λ PR promoter, regulated by the temperature-sensitive repressor produced from the λcI^{857} allele has been widely used as simple heat shock is sufficient to de-repress the promoter. Another, even more popular, *E. coli* system used especially for high levels of protein expression, is based on phage T7 promoters which are recognized specifically by T7 RNA polymerase (Studier *et al.*, 1990). The *E. coli* strains possess T7 RNA polymerase expressed from the *lac* promoter; induction of T7 RNA polymerase with IPTG is required in order to induce expression from the T7 promoter located upstream of the gene to be over-expressed. The success of T7 in *E. coli* has been an incentive for people to use this system in diverse organisms including *Lactococcus lactis* (Wells *et al.*, 1993) and *Ps. aeruginosa* (Brunschwig and Darzins, 1992). The *L. lactis* system utilizes the *lac* promoter from *L. lactis lac* genes to express T7 RNA polymerase in media containing lactose and therefore induce expression of genes inserted downstream of its cognate promoter on the expression vector. Another regulatable system for use in *L. lactis* has been developed without the use of heterologous DNA, a desirable strategy for food grade (GRAS) bacteria. This control system utilizes the repressor, Rro, which binds to three operators, O1, O2 and O3 that control the promoter, P2, from the temperate lactococcal phage r1t. Addition of mitomycin C to cultures containing this system induces P2 by about 70-fold. Thermoinducible derivatives of Rro have now been obtained which could be used instead of mitomycin C (Nauta *et al.*, 1997).

The repressor–operator elements from the *B. subtilis* temperate phage φ105 have also been exploited for thermoregulated expression (Dhaese *et al.*, 1984; Osburne *et al.*, 1985). This system, however, is not used as widely in *B. subtilis* as the more popular *spac* promoter; a hybrid expression system containing a strong phage promoter from phage SPO-1 regulated by the *lac* operator and repressor from *E. coli* (Yansura and Henner, 1984; Henner, 1990). The *spac* promoter consists of a *lac* operator inserted downstream of a promoter from phage SPO-1 and a *lacI* gene placed under the control of a *B. licheniformis* penicillinase promoter to ensure constitutive expression of the repressor. IPTG is used to induce *spac*. The *spac* promoter has been applied in the analysis of genes of unknown function arising from the sequencing of the genome. Briefly, a vector, pMUTIN2 (V. Vagner, Genbank accession AF072806), contains the *spac* promoter upstream of a multiple cloning site on a plasmid unable to replicate in *B. subtilis*. A DNA fragment containing the 5′ end of the gene to be analysed (without its promoter region) is inserted into the multiple cloning site and the plasmid is introduced into *B. subtilis*. Integration occurs by homologous recombination to form a disrupted copy of the gene and an intact copy, the latter being read from the inducible *spac* promoter. Thus if the gene is essential, the strain will be dependent on IPTG for growth; if the gene is non-essential, the phenotype of the null mutant can be investigated.

Other uses for phage repressors arise due to their ability to confer immunity to superinfection by homoimmune phage. Thus the repressor from mycobacteriophage L5, gp71, has been used as a marker, effective in the maintenance of plasmids as an alternative to the use of an antibiotic resistance determinant (Donnelly-Wu *et al.*, 1993). In *Vibrio cholerae* live-attenuated vaccine strains, a repressor conferring superinfection immunity to the filamentous, toxin-encoding phage, CTXφ, has been use to protect these strains against possible reversion to toxigenicity (Kimsey and Waldor, 1998).

◆◆◆◆◆◆ XI. EXPLOITATION OF PHAGE LYSINS

Bacteriophage lysins are synthesized late during the phage replication process and are designed to facilitate the release of the mature virion particles by inducing lysis of the host cell (reviewed by Young, 1992). There are two families of lysins, one of which is exemplified by the *E* gene of φX174. This in itself does not have murein-degrading activity but, although their exact mechanism of action remains unknown, they are believed to induce activation of host-encoded autolysins. Accordingly, at least one locus (*slyD*) has been shown to be essential for phage-induced lytic activity (Maratea *et al.*, 1985). A second family is that exemplified by the phage λ *R* gene which are efficient and specific peptidoglycan-degrading enzymes (murein hydrolase) that attack either glyscosidic or peptide bonds which are integral parts of the polymer structure. During phage infection, the lysins access the cell wall structure following penetration of the cytoplasmic membrane. Since these lysin genes do not possess secretion signals, passage across the membrane is often facilitated by a co-expressed holin gene which is believed to form oligomeric structures in the membrane which allow the lysins to access the peptidoglycan layer. The intrinsic lytic activity of this second family of lysin genes has been recognized as useful and has resulted in their exploitation as a biochemical tool.

For cloning purposes, the lytic activity itself can be used to screen a library of random clones prepared from phage genomic DNA (this is only true for the holin–lysin family of proteins). The DNA fragments are cloned into an expression vector and transformed into *E. coli*. Since this family of lysin genes require the holin to access the cell wall, even if the *E. coli* peptidoglycan is a suitable substrate for the cloned lysin gene the recombinant plasmids should not be lethal unless the holin gene is present on the same gene fragment. However, since the holin and lysin genes are often arranged into a so-called lysis cassette, it is advisable to generate a library of smaller phage fragments to try and circumvent this problem. Cells containing individual clones from the library are grown to colonies on media to induce expression of the cloned genes and are then lysed using chloroform treatment to release expressed proteins. Cells of the target organism (i.e. the host bacteria for the phage) are incorporated into a soft agar overlay and poured over the lysed colonies and lysin-containing clones identified by the appearance of zones of clearing surrounding the positive clone (see Loessner *et al.*, 1995b). Alternatively, either heat-killed cells or preparations of bacterial host cell walls can be incorporated into the overlay which will allow increased incubation times to allow zones of clearing to develop but avoids the problem of the zones being obscured due to cell growth. Once cloned and the gene sequence identified, the lysin genes can be readily adapted for affinity column purification by *N*-terminal modification (either His-tag or GST-fusion vectors have been successfully used) and pure samples of relatively high activity can be simply prepared.

Due to the evolution of the specific phage–host cell interaction, lysins isolated from phages specific for a particular host tend to be most active against the cell wall structure of these bacteria, or other bacteria possessing a similar peptidoglycan structure. Hence various applications of phage lysins have been proposed, from surface sanitization of food products to the generation of lysin-expressing recombinant starter cultures for the specific eradication of pathogens from dairy products (see Stewart *et al.*, 1996). However, for molecular biology purposes the main applications have been achieving rapid lysis of cells which are normally refractive to enzymatic degradation by enzymes such as lysozyme or muramidase (such as *Listeria*). When used to lyse cells for DNA, RNA or protein purification result in a simpler and more efficient procedure than is often required, especially for Gram-positive bacteria (Loessner *et al.*, 1995a).

A second application is in studies of cell invasion by pathogenic bacteria. The ability of intracellular pathogens to invade eukaryotic cells is often measured by enumerating bacteria recovered following lysis of infected cell lines. The assay relies on the addition of an antibiotic (commonly gentamycin) to kill extracellular bacteria which do not undergo invasion. However, for long incubation times the level of antibiotic used is crucial since it may penetrate the host cells and kill intracellular bacteria thus invalidating the assay. The use of the antibiotic can be replaced with a lytic enzyme, e.g. the use of lysostaphin for *S. aureus* (Yao *et al.*, 1995) or phage lysin for *Listeria*, thus avoiding the problems associated with antibiotic leakage. However, this is only effective if the efficiency of lysis is sufficiently high under the conditions used for incubation of the eukaryotic cells to ensure complete eradication of all external bacterial cells.

◆◆◆◆◆◆ XII. FUTURE PROSPECTS

Research on phage biology fuels the use of phages and their components as tools for genetic analysis. In *E. coli* and *Salmonella*, an understanding of the phage developmental gene control systems has led to the generation of sophisticated molecular tools such as the Challenge Phage (reviewed by Numrych and Gardner, 1995) for the investigation of DNA–protein interactions. It can be predicted that as such detailed understanding of bacteriophages specific for more diverse bacterial genera is gained, similar tools will in turn be developed to aid the elucidation of genetic control circuits in their host bacteria. The numbers of papers relating to phage research would indicate that the interest in phage has not peaked. Indeed the numbers of papers reporting new integrases are increasing at an apparently exponential rate. As microbiology moves into detailed analysis of diverse species, examining their evolution, ecology and physiology, phages are both integral to these processes (e.g. the movement of pathogenicity islands in the evolution of harmful bacteria) and easily exploitable. Phages are the most numerous and most diverged entities on this planet. This must be a good place to look for new genetic tools.

Acknowledgements

Research in the authors' laboratories is funded by the Wellcome Trust, MRC, BBSRC and Amersham International.

References

Ackerman, H. and DuBow, M. (1987). *Viruses of Prokayotes*, Vol I and II. CRC Press, Boca Raton.

Auvray, F., Coddeville, M., Ritzenthaler, P. and Dupont, L. (1997). Plasmid integration in a wide range of bacteria mediated by the integrase of *Lactobacillus delbrueckii* bacteriophage mv4. *J. Bacteriol.* **179**, 1837–1845.

Bar-Nir, D., Cohen, A. and Goedeke, (1992). tDNA[ser] sequences are involved in the excision of *Streptomyces griseus* plasmid pSG1. *Gene* **122**, 71–76.

Belas, R., Mileham, A., Cohn, D., Hilman, M., Simon, M. and Silverman, M. (1982). Bacterial bioluminescence – isolation and expression of the luciferase genes from *Vibrio harveyi*. *Science* **218**, 791–793.

Belas, R., Mileham, A., Simon, M. and Silverman, M. (1984). Transposon mutagenesis of marine I spp. *J. Bacteriol.* **158**, 890–896.

Belas, R., Simon, M. and Silverman, M. (1986). Regulation of lateral flagella gene transcription in *Vibrio parahaemolyticus*. *J. Bacteriol.* **167**, 210–218.

Bergh, O., Borsheim, K.Y., Bratbak, G. and Heldal, M. (1989). High abundance of viruses found in aquatic environments. *Nature* **340**, 467–468.

Bierman, M., Logan, R., OíBrien, K., Seno, E.T., Rao, R.N. and Schoner, B.E. (1992). Plasmid cloning vectors for the conjugal transfer of DNA from *Escherichia coli* to *Streptomyces* spp. *Gene* **116**, 43–49.

Black, L.W. (1988). DNA packaging in dsDNA bacteriophages. In *The Bacteriophages*, Vol. II (R. Calender, ed.), pp. 321–373. Plenum Press, New York.

Boccard, F., Smokvina, T., Pernodet, J.L., Friedmann, A. and Guerinaeu, M. (1989). The integrated conjugative plasmid pSAM2 of *Streptomyces ambofaciens* is related to temperate bacteriophages. *EMBO J.* **8**, 973–980.

Brasch, M.A., Pettis, G.S., Lee, S.C. and Cohen, S.N. (1993). Localisation and nucleotide sequences of genes mediating site-specific recombination of the SLP1 element in *Streptomyces lividans*. *J. Bacteriol.* **175**, 3067–3074.

Brown, D.P., Idler, K.B. and Katz, L. (1990). Characterisation of the genetic elements required for site-specific integration of plasmid pSE211 in *Saccharopolyspora erythraea*. *J. Bacteriol.* **172**, 1877–1888.

Brown, D.P., Idler, K.B., Backer, D.M., Donadio, S. and Katz, L. (1994). Characterisation of the genes and attachment sites for site-specific integration of plasmid pSE101 in *Saccharopolyspora erythraea* and *Streptomyces lividans*. *Mol. Gen. Genet.* **242**, 185–193.

Brown, K.L., Sarkis, G.J., Wadsworth, C. and Hatfull, G.F. (1997). Transcriptional silencing by the mycobacteriophage L5 repressor. *EMBO J.* **16**, 5914–5921.

Brunschwig, E. and Darzins, A. (1992). A 2-component T7-system for the overexpression of genes in *Pseudomonas aeruginosa*. *Gene* **111**, 35–41.

Bruton, C.J., Guthrie, E.P. and Chater, K.F. (1991). Phage vectors that allow monitoring of transcription of secondary metabolism genes in *Streptomyces*. *Bio/technology* **9**, 652–656.

Bruttin, A., Foley, S. and Brussow, H. (1997). The site-specific integration system of the temperate *Streptococcus thermophilus* bacteriophage φSfi21. *Virology* **237**, 148–158.

Brzostek, K., Heleszko, H. and Hrebenda, J. (1995). Production of *Escherichia coli* LamB protein in *Yersinia enterocolitica*. *FEMS Microbiol. Lett.* **127**, 17–21.

Buttner, M.J., Chater, K.F. and Bibb, M.J. (1990). Cloning, disruption and transcriptional analysis of three RNA polymerase sigma factor genes of *Streptomyces coelicolor* A3(2). *J. Bacteriol.* **172**, 3367–3378.

Campbell, A.M. (1992). Chromosomal insertion sites for phages and plasmids. *J. Bacteriol.* **174**, 7495–7499.

Campbell, A. (1994). A comparative molecular biology of the lambdoid phages. *Ann. Rev. Microbiol.* **48**, 193–222.

Caparon, M.G. and Scott, J.R. (1987). Identification of a gene that regulates expression of M protein, the major virulence determinant of group-a Streptococci. *Proc. Natl. Acad. Sci. USA* **84**, 8677–8681.

Caparon, M.G. and Scott, J.R. (1991). Genetic manipulation of pathogenic streptococci. *Meth. Enzymol.* **204**, 556–586.

Carriere, C., Riska, P.F., Zimhony, O., Kriakov, J., Bardarov, S., Burns, J., Chan, J. and Jacobs, W.R. (1997). Conditionally replicating luciferase reporter phages: Improved sensitivity for rapid detection and assessment of drug susceptibility of *Mycobacterium tuberculosis*. *J. Clin. Microbiol.* **35**, 3232–3239.

Casjens, S. and Hendrix, R. (1988) Control mechanisms in dsDNA bacteriophage assembly. In *The Bacteriophages*, Vol. I (R. Calender, ed.), pp. 15–91. Plenum Press, New York.

Casjens, S., Hatfull, G. and Hendrix, R. (1992). Evolution of dsDNA tailed-bacteriophage genomes. *Semin. Virol.* **3**, 383–397.

Catalano, C.E., Cue, D. and Feiss, M. (1995). Virus DNA packaging: the strategy used by phage λ. *Mol. Microbiol.* **16**, 1075–1086.

Chater, K.F. (1986). *Streptomyces* phages and their applications to *Streptomyces* genetics. In *The Bacteria*, Vol. 9 (S.W. Queener and L.E. Day, eds), pp. 119–157.

Chater, K.F. and Bruton, C.J. (1983). Mutational cloning in *Streptomyces* and the isolation of antibiotic production genes. *Gene* **26**, 67–78.

Chater, K.F. and Bruton, C.J. (1985). Resistance, regulatory and production genes for the antibiotic methylenomycin are clustered. *EMBO J.* **4**, 1893–1897.

Chater, K.F., Bruton, C.F., Springer, W. and Suarez, J.E. (1981). Dispensable sequence and packaging contraints of DNA from the *Streptomyces* temperate phage φC31. *Gene* **15**, 249–256.

Chater, K.F., Bruton, C.J., King. A.A. and Suarez, J.E. (1982). The expression of *Streptomyces* and *Escherichia coli* drug-resistance determinants cloned into the *Streptomyces* phage φC31. *Gene* **19**, 21–32.

Chater, K.F., King, A.A., Rodicio, R., Bruton, C.J., Fisher, S.H., Piret, J.M., Smith, C.P. and Foster, S.G. (1985). Cloning and analysis of Streptomyces DNA in φC31-derived vectors. In *Microbiology-1985* (G. Hegemon and C. Hershberger, eds), pp. 421–426. *American Society of Microbiology* Washington DC.

Chen, J. and Griffiths M.W. (1996). *Salmonella* detection in eggs using *lux⁺* bacteriophages. *J. Food Protect.* **59**, 908–914.

Christiansen, B., Johnsen, M.G., Stenby, E., Vogensen, F.K. and Hammer, K. (1994). Characterisation of the lactococcal temperate phage TP901-1 and its site-specific integration. *J. Bacteriol.* **176**, 1069–1076.

Christiansen, B., Brondsted, L., Vogensen, F.K. and Hammer, K. (1996). A resolvase-like protein is required for the site-specific integration of the temperate lactococcal bacteriophage TP901-1. *J. Bacteriol.* **178**, 5164–5173.

Cutting, S.M. and Vander Horn, P.B. (1990). Genetic analysis. In *Molecular Biological Methods in* Bacillus (C.R. Harwood and S.M. Cutting, eds), pp. 27–74. John Wiley, Chichester.

Deichelbohrer, I., Alonso, J.C., Luder, G. and Trautner, T.A. (1985). Plasmid transduction by *Bacillus subtilis* bacteriophage SPP1: Effects of DNA homology between plasmid and bacteriophage. *J. Bacteriol.* **162**, 1238–1243.

Dhaese, P., Hussey, C. and VanMontagu, M. (1984). Thermo-inducible gene expressison in *Bacillus subtilis* using transcriptional regulatory elements from temperate phage φ105. *Gene* **32**, 181–194.

Diaz, L.A., Hardisson, C. and Rodicio, M.R. (1989). Isolation and characterisation of actinophages infecting *Streptomyces* species and their interaction with host restriction-modification systems. *J. Gen. Microbiol.* **135**, 1847–1856.

Donnelly-Wu, M.K., Jacobs, W.R. and Hatfull, G.F. (1993). Superinfection immunity of mycobacteriophage L5 – applications for genetic transformation of *Mycobacteria*. *Mol. Microbiol.* **76**, 163–176.

Donohue, T.J. and Kaplan, S. (1991). Genetic techniques in *Rhodospirillaceae*. *Meth. Enzymol.* **204**, 459–485.

Dowding, J.E. (1973). Characterisation of a bacteriophage virulent for *Streptomyces coelicolor* A3(2). *J. Gen. Microbiol.* **76**, 163–176.

Dupont, L., Boisetbonhoure, B., Coddeville, M., Auvray, F. and Ritzenthaler, P.

(1995). Characterisation of genetic elements required for site-specific integration of *Lactobacillus delbrueckii* subsp *bulgaricus* bacteriophage mv4 and construction of an integration proficient vector for *Lactobacillus plantarum*. *J. Bacteriol.* **177**, 586–595.

East, A.K. and Errington, J. (1989). A new bacteriophage vector for cloning in *Bacillus subtilis* and the use of φ105 for protein synthesis. *Gene* **81**, 35–43.

Ely, B. (1991) Genetics of *Caulobacter crescentus*. *Meth. Enzymol.* **204**, 372–384.

Ely, B. and Johnson, R.C. (1977). Generalised transduction in Caulobacter crescentus. *Genetics* **87**, 391–399.

Errington,J. (1990). Gene cloning techniques. In *Molecular Biological Methods for Bacillus*. (C.R. Harwood, and S.M. Cutting, eds), pp. 175–220. John Wiley, Chichester.

Esposito, D. and Scocca, J.J. (1994). Identification of an HP1 phage protein required for site-specific excision. *Mol. Microbiol.* **13**, 685–695.

Esposito, D., Fitzmaurice, W.P., Benjamin, R.C., Goodman, S.D., Waldman, A.S. and Scocca, J.J. (1996). The complete nucleotide sequence of bacteriophage HP1 DNA. *Nucl. Acids Res.* **24**, 2360–2368.

Expert, D. and Tourneur (1982). Psi, a temperate phage of *Agrobacterium tumefaciens*, is mutagenic. *J. Virol.* **42**, 283–291.

Finan, T.M., Hartweig, K., Lemieux, K., Bergman, K., Walker, G.C. and Signer, E.R. (1984). General transduction in *Rhizobium meliloti*. *J. Bacteriol.* **159**, 120–124.

Ford, M.E., Sarkis, G.J., Belanger, A.E., Hendrix, R.W. and Hatfull, G.F. (1998). Genome structure of mycobacteriophage D29: Implications for phage evolution. *J. Mol. Biol.* **279**, 143–164.

Gabriel, K., Schmid, H. Schmidt, U. and Rausch, H. (1995). The actinophage RP3 DNA integrates site-specifically into the putative tRNA (arg)(AGG) gene of *Streptomyces rimosus*. *Nucl. Acids Res.* **23**, 58–63.

Geiduschek, E.P. and Kassavetis, G.A. (1988). Changes in RNA polymerase. In *The Bacteriophages*, Vol I (R. Calender, ed.) pp. 93–115. Plenum Press, New York.

Geisselsoder, J., Campos, J.M. and Zusman, D.R. (1978). Physical characterisation of bacteriophage mx4, a generalized transducing phage for *Myxococcus xanthus*. *J. Mol. Biol.* **119**, 179–189.

Gerdes, J.C. and Romig, W.R. (1975). Complete and defective bacteriophage of classical *Vibrio cholerae*: Relationship to the kappa type bacteriophage *J. Virol.* **15**, 1231–1238.

Gibson, R.M. and Errington, J. (1992). A novel *Bacillus subtilis* expression vector based on bacteriophage φ105. *Gene* **121**, 137–142.

Gindreau, E., Torlois, S. and Lonvaud-Funel, A. (1997). Identification and sequence analysis of the region encoding the site-specific integration system from *Leuconostoc oenos* (*Oenococcus oeni*) temperate bacteriophage φ10MC. *FEMS Mirobiol. Lett.* **147**, 279–285.

Glazebrook, J. and Walker, G.C. (1991). Genetic techniques in *Rhizobium meliloti*. *Meth. Enzymol.* **204**, 398–418.

Goodman, S.D. and Scocca, J.J. (1989). Nucleotide sequence and expression of the gene for the site-specific integration protein from bacteriophage HP1 of *Haemophilus influenzae*. *J. Bacteriol.* **171**, 4232–4240.

Greenblatt, J., Nodwell, J.R. and Mason, S.W. (1993). Transcriptional antitermination. *Nature* **364**, 401–406.

Gregg, K., Kennedy, B.G. and Klieve A.V. (1994). Cloning and DNA sequence analysis of the region containing attP of the temperate phage φAR29 of *Prevotella ruminicola* AR29. *Microbiology UK* **140**, 2109–2114.

Groisman, E.A. (1991). *In vivo* genetic engineering with bacteriophage Mu. *Meth. Enzymol.* **204**, 180–212.

Groman, N.B. (1984). Conversion by corynephages and its role in the natural history of diphtheria. *J. Hyg. Camb.* **93**, 405–417.

Haeseleer, F., Pollet, J.F., Bollen, A. and Jacobs, P. (1992). Molecular cloning and sequencing of the attachment site and integrase of the temperate mycobacteriophage FRAT1. *Nucl. Acids Res.* **20**, 1420.

Haeseleer, F., Pollet, J.F., Haumont, M., Bollen, A. and Jacobs, P. (1993). Stable integration and expression of the *Plasmodium falciparum* circumsporozoite protein coding sequence in *Mycobacteria. Mol. Biochem. Parasitol.* **57**, 117–126.

Hartley, N.M., Murphy, G.J.P., Bruton, C.J. and Chater, K.F. (1994). Nucleotide sequence of the essential early region of φC31, a temperate phage of *Streptomyces* spp; with unusual features in its lytic development. *Gene* **47**, 29–40.

Hauser, M.A. and Scocca, J.J. (1992). Site-specific integration of the *Haemophilus influenzae* bacteriophage HP1: Location of the boundaries of the phage attachment site. *J. Bacteriol.* **174**, 6674–6677 .

Hendrix, R., Smith, M.C.M., Burns, R.N., Ford, M.E. and Hatfull, G.F. (1998). Evolutionary relationships among diverse bacteriophages and prophages: All the world's a phage *Proc. Natl. Acad. Sci. USA* **96**, 2191–2197.

Hendrix. R.W. (1998). Bacteriophage DNA packaging: RNA gears in a DNA transport machine. *Cell* **94**, 147–150.

Henner, D.J. (1990). Inducible expression of regulatory genes in *Bacillus subtilis. Meth. Enzymol.* **185**, 223–228.

Hoch, J.A. (1991). Genetic analysis in *Bacillus subtilis. Meth. Enzymol.* **204**, 305–320.

Hopwood, D.A., Bibb, M.J., Chater, K.F., Kieser, T., Bruton, C.J., Kieser, H.M., Lydiate, D.J., Smith, C.P., Ward, J.M. and Schrempf, H. (1985). *Genetic Manipulation of* Streptomyces; *A Laboratory Manual*. John Innes Institute, Norwich.

Hopwood, D.A., Bibb, M.J., Chater, K.F. and Kieser, T. (1987). Plasmid and phage vectors for gene cloning and analysis in *Streptomyces. Meth. Enzymol.* **153**, 116–167.

Howe, M. (1987) Phage Mu: an overview. In *Phage Mu* (N. Symonds, A. Toussaint, P. van de Putte and M.M. Howe, eds), pp. 25–39. Cold Spring Harbor Laboratory Press, New York.

Humphrey, S.B., Stanton, T.B., Jensen, N.S. and Zuerner, R.L. (1997). Purification and characterisation of VSH-1, a generalized transducing bacteriophage of *Serpulina hyodysenteriae J. Bacteriol.* **179**, 323–329.

Ichige, A., Matsutani, S., Oishi, K., and Mizushima, S. (1989). Establishment of gene transfer systems for and construction of the genetic map of a marine *Vibrio* strain. *J. Bacteriol.* **171**, 1825–1834.

Ikeda, H. and Tomizawa, J.I. (1965). Transducing fragments in generalized transduction by phage P1. I. Molecular origin of fragments. *J. Mol. Biol.* **14**, 85–109.

Jacobs W.R., Jr, Kalpana, G.V., Cirillo, J.D., Pascopella, L., Snapper, C.B., Udani. R.A., Jones W., Barletta, R.G. and Bloom, B.R. (1991). Genetic systems for *Mycobacteria. Meth. Enzymol.* **204**, 537–555.

Kaiser, D. (1991). Genetic systems in *Myxobacteria. Meth. Enzymol.* **204**, 357–372.

Katz, L., Brown, D.P. and Donadio, S. (1991). Site-specific recombination in *Escherichia coli* between the *att* sites of plasmid pSE211 from *Saccharopolyspora erythraea. Mol. Gen. Genet.* **227**, 155–159.

Kim, M., Lee, J.Y., Kim, Y.W., Sung, H.C. and Chang, H.I. (1996). Molecular characterisation of the region encoding integative functions from enterococcal bacteriophage φFC1. *J. Biochem. Mol. Biol.* **29**, 448–454.

Kimsey, H.H. and Waldor, M.K. (1998). CTX phi immunity: Application in the development of cholera vaccines. *Proc. Natl. Acad. Sci. USA* **95**, 7035–7039.

King, A.A. and Chater, K.F. (1986). The expression of the *Escherichia coli lacZ* gene in *Streptomyces. J. Gen. Microbiol.* **132**, 1739–1752.

Kinner, E., Pocta, D., Ströer, S. and Schmieger, H. (1994). Sequence analysis of cohesive ends of the actinophage RP3 genome and construction of a transducible vector. *FEMS Microbiology Lett.* **118**, 283–290.

Kobler, L., Schwertfirm, G., Schmieger, H., Bolotin, A. and Sladkova, I. (1991). Construction and transduction of a shuttle vector bearing the cos site of Streptomyces phage φC31 and determination of its cohesive ends. *FEMS Microbiol. Lett.* **78**, 347–354.

Kodikara, C.P., Crew, H.H and Stewart, G.S.A.B. (1991). Near on-line detection of entric bacteria using lux recombinant bacteriophage. *FEMS Microbiol. Lett.* **83**, 261–266.

Kornberg, A. and Baker, T. (1992). *DNA replication*, 2nd edn. Freeman and Co., New York.

Kozloff L.M., Lute M., Arellano F. and Turner M.A. (1992). Bacterial ice nucleation activity after T4 bacteriophage infection. *J. Gen. Microbiol.* **138**, 941–944.

Kuhstoss, S. Richardson, M.A. and Rao, R.N. (1989). Site-specific integration in *Streptomyces ambofaciens*: localisation of integration functions in *S. ambofaciens* plasmid pSAM2. *J. Bacteriol.* **171**, 16–23.

Kuhstoss, S., Richardson, M.A. and Rao, R.N. (1991). Plasmid cloning vectors that integrate site-specifically in *Streptomyces* spp. *Gene* **97**, 143–146.

Kuner, J.M. and Kaiser, D. (1981). Introduction of transposon Tn5 into *Myxococcus* for analysis of developmental and other non-selectable mutants. *Proc. Natl. Acad. Sci. USA* **78**, 425–429.

Leach, D.R.F. (1996). *Genetic Recombination.* Blackwell Science, Oxford.

Lee, M.H., Pascopella, L., Jacobs, W.R. and Hatfull, G.F. (1991a). Site-specific integration of mycobacteriophage L5 integration proficient vectors for *Mycobacterium smegmatis*, *Mycobacterium tuberculosis* and bacille Calmette-Guerin. *Proc. Natl. Acad. Sci USA* **88**, 3111–3115.

Lee, C.Y., Buranen, S.L. and Ye, Z.H. (1991b). Construction of single-copy integration vectors for *Staphylococcus aureus. Gene* **103**, 101–105.

LeMarrec, C., Michotey, V., Blanco, C., and Trautwetter, A. (1994). φAAU2, a temperate bacteriophage, specific for *Arthrobacter aureus*, whose integrative functions work in other *Corynebacteria. Microbiology UK*, **140**, 3071–3077.

LeMarrec, C., Moreau, S., Loury, S., Blanco, C. and Trautwetter, A. (1996). Genetic characterisation of site-specific integration functions of φAAU2 infecting *Arthrobacter aureus* C70. *J. Bacteriol.* **178**, 1996–2004.

Lennox, E.S. (1955). Transduction of linked genetic characters of the host by bacteriophage P1. *Virology* **1**, 190–206.

Lillehaug, D., Nes, I.F. and Birkeland, N.K. (1997). A highly efficient and stable system for site-specific integration of genes and plasmids into the phage φLC3 attachment site (*attB*) of the *Lactococcus lactis* chromosome. *Gene* **188**, 129–136.

Lindsay, J.A., Ruzin, A., Ross, H.F., Kurepina, N. and Novick, R.P. (1998). The gene for toxic shock toxin is carried by a family of mobile pathogenicity islands in *Staphylococcus aureus. Mol. Microbiol.* **29**, 527–543.

Loessner, M.J., Schneider, A., Scherer, S. (1995a) A new procedure for efficient recovery of DNA, RNA and proteins from *Listeria* cells by rapid lysis with a recombinant bacteriophage endolysin. *Appl. Environ. Microbiol.* **61**, 1150–1152.

Loessner, M.J., Wendlinger, G. and Scherer, S. (1995b). Heterogeneous endolysins in *Listeria monocytogenes* bacteriophages – a new class of enzymes and evidence for conserved holin genes within the siphoviral lysis cassettes. *Mol. Microbiol.* **16**, 1231–1241.

Loessner, M.J., Rees, C.E.D., Stewart, G.S.A.B. and Scherer, S. (1996). Construction of luciferase reporter bacteriophage A511::*luxAB* for rapid and sensitive detection of viable *Listeria* cells *Appl. Environ. Microbiol.* **62**, 1133–1140.

Loessner, M.J., Rudolf, M. and Scherer, S. (1997). Evaluation of luciferase reporter bacteriophage A511::*luxAB* for detection of *Listeria monocytogenes* in contaminated foods. *Appl. Environ. Microbiol.* **63**, 2961–2965.

Lomovskaya, N., Fonstein, L., Ruan, X., Stassi, D., Katz,L. and Hutchinson, C.R. (1997). Gene disruption and replacement in the rapamycin-producing *Streptomyces hygroscopicus* strain ATCC 29253. *Microbiology* **143**, 875–883.

Malpartida, F. and Hopwood, D.A. (1986). Physical and genetic characterisation of the gene cluster for the antibiotic actinorhodin in *Streptomyces coelicolor* A3(2). *Mol. Gen. Genet.* 66–73.

Mamat, U., Rietschel, E.T. and Schmidt, G. (1995). Repression of lipopolysaccharide biosynthesis in *Escherichia coli* by an antisense RNA of *Acetobacter methanolicus* phage ACM1. *Mol. Microbiol.* **15**, 1115–1125.

Manning, P.A. (1988) Molecular genetic approaches to the study of *Vibrio cholerae*. *Microbiol. Sci.* **5**, 196–201.

Maratea, D., Young, K. and Young, R. (1985). Deletion and fusion analysis of the φX174 lysis gene *E. Gene* **40**, 39–46.

Martin, C., Mazodier, P., Mediola, M.V., Gicquel, B., Smokvina, T., Thompson, C.J. and Davies, J. (1991). Site-specific integration of the *Streptomyces* plasmid pSAM2 in *Mycobacterium smegmatis*. *Mol. Microbiol.* **5**, 2499–2502.

Martin, M., Showalter, R. and Silverman, M. (1989) Identification of a locus controlling expression of luminescence genes in *Vibrio harveyi*. *J. Bacteriol.* **171**, 2406–2414.

Matsuura, T., Noguchi, T., Yamaguchi, D., Aida, T., Asayama, M., Takahashi, H. and Shirai, M. (1996). The sre gene (ORF469) encodes a site-specific recombinase responsible for integration of the R4 phage genome. *J. Bacteriol.* **178**, 3374–3376.

McHenney, M.A. and Baltz, R.H. (1988). Transduction of plasmid DNA in *Streptomyces* spp. and related genera by bacteriophage FP43. *J. Bacteriol.* **170**, 2276–2282.

McShan, W.M., Tang, Y.F. and Ferretti, J.J. (1997). Bacteriophage T12 of *Streptococcus pyogenes* integrates into the gene encoding a serine tRNA. *Mol. Microbiol.* **23**, 719–728.

McVeigh, R.R. and Yasbin, R.E. (1996). Phenotypic differentiation of "smart" versus "naïve" bacteriophages of *Bacillus subtilis*. *J. Bacteriol.* **178**, 3399–3401.

Morino, T., Takahashi, H. and Saito, H. (1985). Construction and characterisation of a cosmid of *Streptomyces lividans*. *Mol. Gen. Genet.* **198**, 228–233.

Murray, N.E. (1983) Phage lambda and molecular cloning. In *Lambda II* (R.W. Hendrix, J.W. Roberts, F.W. Stahl, and R.A. Weisberg, eds), pp. 395–432. Cold Spring Harbor Laboratory, New York.

Murray, N.E. (1991). Special uses of lambda phage for molecular cloning. *Meth. Enzymol.* **204**, 280–301.

Muskhelishvili, G., Palm, P. and Zillig, W. (1993). SSV1-encoded site-specific recombination system in *Sulfolobus shibatae*. *Mol. Gen. Genet.* **237**, 334–342.

Nash, H.A. (1996). Site–specific recombination: Integration, excision, resolution, and inversion of defined DNA segments. In Escherichia coli *and* Salmonella, *Cellular and Molecular Biology*, Vol. 1 2nd edn (F.C. Neidhardt, ed.). ASM Press, Washington.

Nauta, A., van Sinderen, D., Karsens, H., Smit, E., Venema, G. and Kok, J. (1996). Inducible gene expression mediated by a repressor-operator system isolated from *Lactococcus lactis* bacteriophage r1t. *Mol. Microbiol.* **19**, 1331–1341.

Newman, B.J. and Masters, M. (1980). The variation in frequency with which markers are transduced by P1 is primarily a result of discrimination during recombination. *Mol. Gen. Genet.* **180**, 585–589.

Novick, R.P. (1991). Genetic systems in staphylococci. *Meth. Enzymol.* **204**, 587–636.

Novick, R.P., Edelman, I. and Lofdahl, S. (1986). Small *Staphylococcus aureus* plasmids are transduced as linear multimers. *J. Mol. Biol.* **192**, 209–220.

Numrych, T.E. and Gardner, J.F. (1995). Characterising protein–nucleic acid interactions with challenge phages. *Semin. Virol.* **6**, 5–13.

Osburne, M.S., Craig, R.J. and Rothstein, D.M. (1985) Thermo-inducible transcription system from *Bacillus subtilis* that utilises control elements from temperate phage ϕ105. *J. Bacteriol.* 1101–1108.

Ow, D.W. and Ausubel, F.M. (1983). Conditionally replicating plasmid vectors that can integrate into the *Klebsiella pneumoniae* chromosome via bacteriophage P4 site-specific recombination. *J. Bacteriol.* **155**, 704–713.

Papp, I., Dorgai, L., Papp, P., Jonas, E., Olasz, F. and Orosz, L. (1993). The bacterial attachment site of the temperate *Rhizobium* phage 16–3 overlaps the 3′ ends of a putative proline tRNA gene. *Mol. Gen. Genet.* **240**, 258–264.

Parkinson, J.S. and Huskey, R.J. (1971). Deletion mutants of bacteirophage lambda. I. Isolation and initial characterisation. *J. Mol. Biol.* **56**, 369–384.

Pearson, R.E., Jurgensen, S., Sarkis, G.J., Hatfull, G.F., and Jacobs, W.R. (1996). Construction of D29 shuttle phasmids and luciferase reporter phages for detection of *Mycobacteria*. *Gene* **183**, 129–136.

Piret, J.M. and Chater, K.F. (1985). Phage-mediated cloning of bldA, a region involved in *Streptomyces* coelicolor morphological development, and its analysis by genetic complementation. *J. Bacteriol.* **163**, 965–972.

Poth, H. and Youngman, P. (1988). A new cloning system for *Bacillus subtilis* comprising elements of phage, plasmid and transposon vectors. *Gene* **73**, 215–226.

Ptashne, M. (1992). *A Genetic Switch*, 2nd edn. Cell Press and Blackwell Scientific Publications, Cambridge, USA.

Raya, R.R. and Klaenhammer, T.R. (1992). High frequency plasmid transduction by *Lactobacillus gasseri* bacteriophage ϕadh. *Appl. Environ. Microbiol.* **58**, 187–193.

Raya, R.R., Fremaux, C., Deantoni, G.L. and Klaenhammer, T.R. (1992). Site-specific integration of the temperate bacteriophage ϕadh into the *Lactobacillus gasseri* chromosome and molecular characterisation of the phage (*attP*) and the bacterial (*attB*) attachment sites. *J. Bacteriol.* **174**, 5584–5592.

Ribeiro, G., Viveiros, M., David, H.L. and Costa, J.V. (1997). Mycobacteriophage D29 contains an integration system similar to that of the temperate mycobacteriophage L5. *Microbiology UK* **143**, 2701–2708.

Rodicio, M.R., Bruton, C.J. and Chater, K.F. (1985). New derivatives of the

Streptomyces temperate phage φC31 useful for the cloning and functional analysis of *Streptomyces* DNA. *Gene* **34**, 283–292.

Rosenberg, S.M. (1987). Improved *in vitro* packaging of lambda DNA. *Meth. Enzymol.* **153**, 95–103.

Rothmel, R.K., Chakrabarty, A.M., Berry, A. and Darzins, A. (1991). Genetic systems in *Pseudomonas*. *Meth. Enzymol.* **204**, 485–514.

Salmi, D., Magrini, V., Hartzell, P.L. and Youderian, P. (1998). Genetic determinants of immunity and integration of *Myxococcus xanthus* phage Mx8. *J. Bacteriol.* **180**, 614–621.

Salmond, G.P.C., Hinton, J.C.D., Gill, D.R. and Perombelon, M.C.M (1986). Transposon mutagenesis of *Erwinia* using phage λ vectors. *Mol. Gen. Genet.* **203**, 524–528.

Sambrook, J., Fritsch, E.F. and Maniatis, T. (1989). *Molecular Cloning; A Laboratory Manual*, 2nd edn. Cold Spring Harbor Laboratory, Cold Spring Harbor, New York.

Sarkis, G.J., Jacobs, W.R. and Hatfull, G.F. (1995). L5 Luciferase reporter mycobacteriophages – a sensitive tool for the detection and assay of live *Mycobacteria*. *Mol. Microbiol.* **15**, 1055–1067.

Schmidt, C. and Schmieger, H. (1984). Selective transduction of recombinant plasmids with cloned *pac* sites by *Salmonella* phage P22. *Mol. Gen. Genet.* **196**, 123–128.

Seifert, H.S. and So, M. (1991). Genetic systems in pathogenic Neisseriae. *Meth. Enzymol.* **204**, 342–357.

Silverman, M., Showalter, R. and McCarter, L. (1991). Genetic analysis in *Vibrio*. *Meth. Enzymol.* **204**, 515–536.

Smokvina, T., Mazodier, P., Boccard, F., Thompson, C.J. and Guerinaeu, M. (1990). Construction of a series of pSAM2-based integrative vectors for use in actinomycetes. *Gene* **94**, 53–59.

Stanley, E., Fitzgerald, G.F., Le Marrec, C., Fayard, B. and van Sinderen, D. (1997). Sequence analysis and characterisation of phiO1205, a temperate bacteriophage infecting *Streptococcus thermophilus* CNRZ1205. *Microbiology* **143**, 3417–3429.

Sternberg, N. and Maurer, R. (1991). Bacteriophage-mediated generalized transduction in *Escherichia coli* and *Salmnella typhimurium*. *Meth. Enzymol.* **204**, 18–43.

Sternberg, N. and Weisberg, R. (1975). Packaging of prophage and host DNA by coliphage λ. *Nature* **256**, 97–103.

Stewart, G.S.A.B., Loessner, M.J. and Scherer, S. (1996). The bacterial *lux* gene bioluminescent biosensor revisited *ASM News* **62**, 297–301.

Studier, F.W., Rosenberg, A.H., Dunn, J.J. and Dubendorff, J.W. (1990). Use of T7 RNA polymerase to direct expression of cloned genes. *Meth. Enzymol.* **185**, 60–89.

Stuttard, C. (1979). Transduction of auxotrophic markers in a chloramphenicol-producing strain of *Streptomyces*. *J. Gen Microbiol.* **110**, 479–482.

Stuttard, C. (1989). Generalized transduction in *Streptomyces* species. In *Genetics and Molecular Biology of Industrial Micro-organisms* (C.L. Hershberger, S.W. Queener and G. Hegeman, eds), pp. 157–162. American Society for Microbiology, Washington, DC.

Suarez, J.E. and Chater K.F. (1980). DNA cloning in *Streptomyces*; a bifunctional replicon comprising pBR322 inserted into a *Streptomyces* phage. *Nature* **276**, 527–529.

Thornewell, S.J., East, A.K. and Errington, J. (1993). An efficient expression and secretion system based on *Bacillus subtilis* phage φ105 and its use for the production of *Bacillus cereus* β-lactamase I. *Gene* 133, 47–53.

Tojo, N., Sanmiya, K., Sugawara, H., Inouye, S. and Komano, T. (1996). Integration of bacteriophage Mx8 into the *Myxococcus xanthus* chromosome causes a structural alteration at the C-terminal region of the IntP protein. *J. Bacteriol.* **178**, 4004–4011.

Toth, I.K., Perombelon, M.C.M. and Salmond, G.P.C. (1993). Bacteriophage φKP mediated generalised transduction in *Erwinia carotovora* subsp. *carotovora. J. Gen. Microbiol.* **139**, 2705–2709.

Toth, I.K., Mulholland, V., Cooper, V., Bentley, S., Shih, Y.-L., Perombelon, M.C.M. and Salmond, G.P.C. (1997). Generalised transduction in the potato blackleg pathogen *Erwinia carotovora* subsp. *atroseptica* by bacteriophage φM1. *Microbiology* **143**, 2433–2438.

Van de Guchte, M., Daly, C., Fitzgerald, G.F. and Arendt, E.K. (1994a). Identification of the putative repressor-encoding gene cI of the temperate lactococcal bacteriophage Tuc2009. *Gene* **144**, 93–95.

Van de Guchte, M., Daly, C., Fitzgerald, G.F. and Arendt, E.K. (1994b). Bacteriophage Tuc2009 and their use for site-specific plasmid integration in the chromosome of Tuc2009 resistant *Lactococcus lactis* MG1363. *Appl. Environ. Microbiol.* **60**, 2324–2329.

Van Gijsegem, F., Toussaint, A. and Casadaban, M. (1987). Mu as a genetic tool. In *Phage Mu* (N. Symonds, A. Toussaint, P. van de Putte and M.M. Howe, eds), pp. 215–250. Cold Spring Harbor Laboratory Press, New York.

Van Sinderen, D., Karsens, H., Kok, J., Terpstra, P., Ruiters, M.H.J., Venema, G. and Nauta, A. (1996). Sequence analysis and molecular characterisation of the temperate lactococcal bacteriophage r1t. *Mol. Microbiol.* **19**, 1343–1355.

Vogel, W. and Schmieger, H. (1986). Selection of bacterial *pac* sites recognised by *Salmonella* phage P22. *Mol. Gen. Genet* **205**, 563–567.

Vrijbloed, J.W., Madon, J. and Dijkhuizen, L. (1994). A plasmid from the methylotrophic actinomycete *Amycolatopsis methanolica* capable of site-specific integration. *J. Bacteriol.* **176**, 7087–7090 .

Waldor, M.K. and Mekalanos, J.J. (1996). Lysogenic conversion by a filamentous phage encoding cholera toxin. *Science* **272**, 1910–1914.

Walker, S.A., Dombroski, C.S. and Klaenhammer, T.R. (1998). Common elements regulating gene expression in temperate and lytic bacteriophages of *Lactococcus lactis. Appl. Environ, Microbiol.* **64**, 1147–1152.

Wang, Z., Xiong, G. and Lutz, F. (1995). Site-specific integration of the phage φCTX genome into the *Pseudomonas aeruginosa* chromosome – Characterisation of the functional integrase gene located close to and upstream of *attP. Mol. Gen. Genet.* **246**, 72–79.

Way, J.C., Davis, M.A., Morisato, D., Roberts, D.E. and Kleckner, N. (1984). New Tn10 derivatives for transposon mutagenesis and for construction of *lacZ* operon fusions by transposition. *Gene* **32**, 369–379.

Wells, J.M., Wilson, P.W., Norton, P.M., Gasson, M.J. and Lepage, R.W.F. (1993). *Lactococcus lactis* – high level expression of tetanus toxin fragment-C and protection against lethal challenge. *Mol. Microbiol.* **8**, 1155–1162.

Whitman, W.B., Coleman, D.C. and Wiebe, W.J. (1998). Prokaryotes: the unseen majority. *Proc. Natl. Acad. Sci. USA* **95**, 6578–6583.

Wilson, S.E. and Smith, M.C.M. (1998). Oligomeric properties and DNA binding specificities of repressor isoforms from the Streptomyces bacteriophage φC31. *Nucl. Acids Res.* **26**, 2457–2463.

Yansura, D. and Henner, D.J. (1984). Use of the *Escherichia coli lac* repressor and operator to control gene expression in *Bacillus subtilis Proc. Natl, Acad. Sci. USA* **81**, 439–443.

Yao, L., Bengualid, V., Lowy, F.D., Gibbons, J.J., Hatcher, V.B. and Berman, J.W.

(1995). Internalization of *Staphylococcus aureus* by endothelial cells induces cytokine gene expression. *Infect. Immun.* **63**, 1835–1839.

Yarmolinsky, M.B. and Sternberg, N. (1988). Bacteriophage P1. In *The Bacteriophages* (R. Calender, ed.), pp. 291–438. Plenum Press, New York and London.

Young, R.Y. (1992). Bacteriophage lysis – Mechanism and regulation. *Microbiol. Revs.* **56**, 430–481.

Zahler, S.A. (1982). Specialised transduction in *Bacillus subtilis*. In *The Molecular Biology of the Bacilli* (D. Dubnau, ed.), pp. 269–305. Academic Press, New York. .

Zinder, N.D. and Lederberg, J. (1952). Genetic exchange in *Salmonella*. *J. Bacteriol.* **64**, 679–699.

4 Isolation and Development of Transposons

Paul J Dyson

School of Biological Sciences, University of Wales Swansea, Swansea, UK

◆◆◆

CONTENTS

◆◆◆◆◆◆ I. INTRODUCTION

Mutant isolation is central to the genetic analysis of any organism and, in the last twenty years, transposable elements have become indispensable tools for this purpose, particularly in the context of where there is little or no information about the nature of the genes under investigation. Transposition is a DNA recombination reaction resulting in translocation of a discrete DNA segment termed the transposable element, insertion sequence or transposon from a donor site to one of many non-homologous target sites. By virtue of this insertion into coding sequences, these elements provide a means of potential random mutagenesis, with the inherent advantage to the investigator that the mutated gene is "tagged" by the element. This in turn greatly facilitates isolation of the gene – an

METHODS IN MICROBIOLOGY, VOLUME 29
0580-9517 $30.00

aspect not afforded by conventional mutagenesis protocols. Typically, diverse bacterial species represent "black boxes" with regard to their genetics, so that random mutagenesis offers the only realistic approach to their analysis. In exceptional cases, this is not necessarily the case: the sequence information of candidate genes identified by genome projects can change the strategy of mutagenesis so that a non-random, directed method can be employed. However, even in this context, the elaboration and application of specialist transposable elements as, for example, promoter probes can short-circuit a laborious step-by-step analysis of regulatory networks affecting many dispersed genes. This chapter reviews the general principles of identifying appropriate tranposable elements for use in diverse prokaryotes, how these elements can be optimized for specific purposes, and the development of suitable delivery systems. Where possible, specific examples are described of how transposons have been applied in diverse species, together with some other less transparent potential uses of relevance to organisms for which a substantial sequence database exists. Inevitably, it is often necessary to cite examples concerning the biology and application of transposons in *E.coli*. This is justifiable given the wealth of knowledge which has accrued from many years of experimentation in this species; this information is invaluable for the successful use of transposons in other species. Transposon insertions and transposon-promoted deletions can also be harnessed in a strategy to provide mobile priming sites for DNA sequence analysis of cloned genes. However, as this can normally be achieved in a standard *E.coli* cloning host, it will not be addressed further here, and the reader is referred to Berg *et al.* (1994) and Berg and Berg (1996) for description of appropriate systems and procedures.

◆◆◆◆◆◆ II. HETEROLOGOUS OR HOMOLOGOUS TRANSPOSONS?

The overall structure of many transposons, independent of source, is very well conserved, reflecting similar mechanisms by which these elements move about. There are one or two exceptions, for example the conjugative transposon Tn916 which is discussed later. Typically, transposons possess terminal inverted repeats of 18–35 bp and at least one gene encoding a protein, transposase, whose role is to create DNA strand breaks at the boundary of the inverted repeats and the flanking sequence of the original "donor" location. Strand breaks are also introduced in the target sequence. Double-strand breaks in the target sequence are typically staggered and the length of the stagger is a hallmark of the individual transposase (or "transpososome", as other accessory proteins are also implicated). The 3′ OH ends of the transposon and the 5′ ends of the target sequence are then joined, essentially to recombine the transposon into its new location. DNA replication, primed from the exposed 3′ ends of the target DNA, is required as a minimum to "fill-in" the gap resulting from staggered breakage at the target site, although for many elements DNA replication then proceeds through the length of the transposon resulting in replicative transposition; the contrasting mechanisms of "replicative" and "cut and paste" transposition are illustrated in Figure 1. Many elements yield only conservative cut and paste products (typical for many insertion sequences and compound transposons), whereas for others, like the IS6 and Tn3 families, a replication pathway is preferred. The Tn3 family also encode a site-specific recombination function to resolve the co-integrate intermediate formed by replicative transposition. A detailed description of the various steps involved in replicative and cut and paste transposition is given by Craig (1996). However, as is clear from studies on enteric transposons in which transposition has been assayed both in different genetic backgrounds of the same species and *in vitro*, successful transposition requires more than just expression of active transposase. The selection of target DNA and the building of a stable synaptic complex involves the participation of accessory proteins, some of which may be transposon-encoded while others are host-specific. A list of 15 host factors involved in transposition in *E.coli* has been compiled by Craig (1996). It should be said that even for the well-studied enteric transposons, the detail on how some of these host proteins function in the transposition

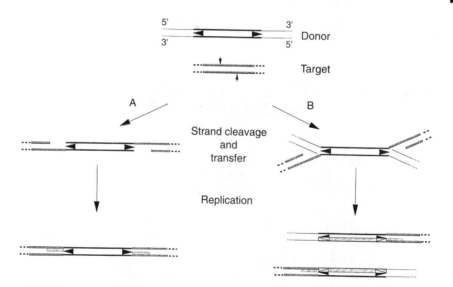

Figure 1. Mechanism of "cut and paste" non-replicative and replicative transposition. The transposon in the donor molecule is represented by the double-stranded bold segment delineated by the inverted arrow heads representing the terminal inverted repeats. Transposase introduces staggered double-strand cuts in the target molecule, as indicated by arrows, and, in pathway (A) for cut-and paste non-replicative transposition, at both 5' and 3' ends of the transposon on both strands in the donor molecule. For replicative transposition, pathway (B), transposase cuts are at only the 5' ends of the transposon. The free 3' ends of the transposon are then joined to the free 5' ends of the target molecule. For pathway (A), replication repairs the gaps flanking the 5' ends of the transposon, creating a characteristic target site duplication indicated by the hatched boxes. For pathway (B), replication proceeds through the entire transposon, not only duplicating the target site, but also the transposon, as indicated by the hatched boxes.

process is not clearly understood. It is believed that some interact with specific DNA-binding sites on the transposon, but it is also conceivable that interactions with transposase itself and/or other transposon-encoded proteins are involved. These protein–DNA and protein–protein interactions may be the molecular basis which limits the host-range of a given transposon. In addition, expression of transposition functions may limit host range as transcription and to a lesser extent translation signals can be quite divergent in different species. The consequence is that well-documented transposons, typically the enteric elements such as Mu, Tn5 or Tn10, which have been exploited in the development of a broad range of useful genetic tools, cannot simply be taken "off the shelf" and used in all prokaryotes. Many diverse species require their own specific elements, or at least transposons from closely related species in which host factors are well conserved. Moreover, there is an argument for choosing a heterologous transposon from a related organism rather than an homologous element. If hybridization studies show that the heterologous element is not present in the genome of the host in which the transposon is required to function, subsequent transposition assays will not be confused by homologous recombination events which can integrate the new element independent of transposition. Subsequent screening of cloned tagged sequences will also be facilitated if the transposon is to be employed as a probe in colony hybridization.

◆◆◆◆◆◆ III. BROAD HOST-RANGE AND CONJUGATIVE TRANSPOSONS WORTH TRYING

Developing new transposon tools can take a considerable time and so, as a short cut to transposon mutagenesis in an untried host species, it is well worth considering elements with a proven track record, especially if there is an indication that they can function in a variety of species. Table 1 gives an indication of the host range of a number of the commonly used bacterial transposons, from which an informed decision can be made about likely candidates for use in other species. For some of these elements, several derivatives have been developed which incorporate features which offer potential for use in a range of genetic analyses. If the progenitor transposon can be shown to function in the organism under investigation, these other applications become immediately possible, without recourse to first finding a new element and then customizing it for special uses.

The enteric compound transposon Tn5 has been widely used in a range of Gram-negative species, in cyanobacteria and, more recently, in Gram-positive *Streptomyces* (Table 1). It transposes via a conservative cut and paste mechanism and exhibits little target specificity – less so than its counterpart Tn10, although mutants of this latter element show reduced specificity for insertion (see section IV below). Tn10 has also been used in the Gram-positive host *Bacillus subtilis* (Petit *et al.*, 1990). In some of the proteobacteria where Tn5 transposes poorly, bacteriophage Mu has been employed with greater success. A list of useful engineered versions of these transposons has been compiled by Berg and Berg (1996), and some of these genetic tools are described in greater detail in subsequent sections. More recently, the composite *Bacteroides* transposon Tn*4351* (see Chapter 8, III) has been shown to function in a number of Gram-negative species, in particular those belonging to the *Cytophaga-Flavobacterium–Bacteroides* subgroup of eubacteria in which enteric transposons apparently do not (Table 3; McBride and Baker, 1996).

Tn*917*, a member of the Tn*3* family, transposes via a replicative pathway and has been employed for mutagenesis of a number of Gram-positive species (Table 1). It was originally described in group D streptococci (Tomich and Clewell, 1980). Unlike many enteric Tn*3*-like transposons, Tn*917* transposes at reasonable frequency into the chromosome. It is believed to exhibit little target site-specificity, although in *Bacillus subtilis* (see Chapter 14) the insertions tend to cluster at the terminus of the chromosome (Hoch, 1991). A number of Tn*917* derivatives and delivery systems have been developed (Youngman *et al.*, 1985). Transposition of Tn*917* into plasmid targets has been demonstrated in *E.coli* (Kuramitsu and Casadaban, 1986).

Conjugative transposons are a special class of elements which can cross the Gram-negative–Gram-positive divide. They encode not only the ability to transpose, but also the genetic systems required for transfer between organisms by conjugation (reviewed by Scott and Churchward, 1995; Salyers *et al.*, 1995; see also Chapter 6, II). Moreover, the mechanism by which they transpose is quite distinct from the typical replicative or conservative cut and paste mechanisms described above. Much of the detail on the mechanism has come from studies with Tn*916*, an 18 kb element which, as a paradigm for conjugative transposons, has been the most widely exploited as a genetic tool. This transposon was originally found in *Enterococcus faecalis* (Franke and Clewell, 1981), and related elements have subsequently been discovered in a number of other Gram-positive species. Another family of these elements has been found in gram-negative *Bacteroides* (Shoemaker *et al.*, 1989; see also Chapter 8, III.D in this volume). The mechanism of transposition involves excision from the donor molecule, circularization of the transposon and then insertion into a target site in the recipient molecule. Moreover, insertion is not accompanied by duplication of target sequences. The circular transposition intermediate is generated by an excision reaction dependent on transposon-encoded Xis and Int proteins which produce 6 bp staggered cuts at each end of the element. One strand is nicked at the exact boundary of the 26 bp transposon inverted repeat and the donor sequence, whereas the other strand is nicked within the donor DNA. The excised molecule then circularizes, involving heteroduplex formation between the two single-stranded nonidentical donor sequences; this region is termed the coupling sequence. At the same time, the donor site is repaired either to restore the original DNA sequence prior to insertion, or to create a 6 bp substitution. Re-integration of the circular intermediate requires the Int protein which, in contrast to the Xis protein, shares homology with the equivalent bacteriophage λ protein. A 6 bp staggered break is introduced at the target site. Different target sites share little or no sequence homology, although insertion occurs

Table I. Host range of commonly used transposons

Transposon	Group and species	Reference
	Alpha proteobacteria	
Tn5 and derivatives	*Acetobacter methanolicus*	Dobrowolski and Grundig, 1990
	Agrobacterium tumefaciens	Cangelosi *et al.*, 1991
	Azospirillum brasilense	Vanstockem *et al.*, 1987
	Brucella abortus	Smith and Heffron, 1987
	Caulobacter crescentus	Ely and Croft, 1982
	Gluconobacter oxydans	Gupta *et al.*, 1997
	Rhizobium meliloti	Beringer *et al.*, 1978
	Rhodobacter sphaeroides	Morenovivian *et al.*, 1994
	Rhodopseudomonas palustris	Elder *et al.*, 1993
	Rhodospirillum rubrum	Ghosh *et al.*, 1994
	Beta proteobacteria	
	Alcaligenes eutrophus	Srivastava *et al.*, 1982
	Methylobacterium extorquens	Lee *et al.*, 1991
	Gamma proteobacteria	
	Actinobacillus actinomycetemcomitans	Kolodrubetz and Kraig, 1994
	Azotobacter vinelandii	Contreras *et al.*, 1991
	Erwinia carotovora	Zink *et al.*, 1984
	Halomonas elongata	Kunte and Galinski, 1995
	Lysobacter brunescens	Lin and McBride, 1996
	Proteus mirabilis	Belas *et al.*, 1991
	Pseudomonas putida	Herrero *et al.*, 1990
	Yersinia enterocolitica	Zhang and Skurnik, 1994
	Xenorhabdus nematophilus	Xu *et al.*, 1991
	Delta proteobacteria	
	Desulfovibrio desulfuricans	Wall *et al.*, 1996
	Myxococcus xanthus	Kuner and Kaiser, 1981
	Stigmatella aurantiaca	Silakowski *et al.*, 1996
	Cyanobacteria	
	Anabaena sp.	Borthakur and Haselkorn, 1989
	Nostoc sp.	Cohen *et al.*, 1994
	Gram +ve actinomycetes	
	Streptomyces lividans	Volff and Altenbuchner, 1997
	Gamma proteobacteria	
Tn10 and derivatives	*Acinetobacter calcoaceticus*	Leahy *et al.*, 1993
	Actinobacillus pleuropneomoniae	Tascon *et al.*, 1993
	Legionella pneumophila	Pope *et al.*, 1994
	Pasteurella multocida	Lee and Henk, 1996
	Xenorhabdus bovienii	Francis *et al.*, 1993
	Yersinia enterocolitica	Yamamoto *et al.*, 1994
	Gram +ve endospore-forming rods and cocci	
	Bacillus subtilis	Petit *et al.*, 1990

Bacteriophage Mu and derivatives	**Alpha proteobacteria**	
	Zymomonas mobilis	Pappas *et al.*, 1997
	Gamma proteobacteria	
	Erwinia carotovora	Jayaswal *et al.*, 1984
	Klebsiella aerogenes	Groisman, 1991
	Legionella pneumophila	Mintz and Shuman, 1987
	Vibrio fischeri	Graf *et al.*, 1994
	Yersinia pestis	Goguen *et al.*, 1984
Tn916	**Beta proteobacteria**	
	Neisseria meningitidis	Kathariou *et al.*, 1990
	Gamma proteobacteria	
	Actinobacillus actinomycetemcomitans	Sato *et al.*, 1992
	Haemophilus ducreyi	Gibson *et al.*, 1997
	Haemophilus influenzae	Holland *et al.*, 1992
	Gram +ve cocci	
	Staphylococcus aureus	Jones *et al.*, 1987
	Streptococcus mutans	Caufield *et al.*, 1990
	Streptococcus pyogenes	Caperon and Scott, 1991
	Gram +ve endospore-forming rods and cocci	
	Bacillus anthracis	Ivins *et al.*, 1988
	Clostridium botulinum	Lin and Johnson, 1991
	Clostridium acetobutylicum	Mattson and Rogers, 1994
	Gram +ve non-sporulating rods	
	Listeria monocytogenes	Kathariou *et al.*, 1987
	Mycoplasmas	
	Acholeplasma laidlawii and *Mycoplasma pulmonis*	Dybvig and Cassell, 1987
Tn917 and derivatives	**Gram +ve cocci**	
	Enterococcus faecalis	Ike *et al.*, 1990
	Lactococcus lactis	Israelsen *et al.*, 1995
	Staphylococcus epidermidis	Heilman *et al.*, 1996
	Streptococcus mutans	Gutierrez *et al.*, 1996
	Streptococcus pyogenes	Eichenbaum and Scott, 1997
	Gram +ve endospore-forming rods and cocci	
	Bacillus licheniformis	Pragai *et al.*, 1994
	Bacillus subtilis	Youngman *et al.*, 1984
	Clostridium perfringens	Babb *et al.*, 1993
	Gram +ve non-sporulating rods	
	Lactobacillus plantarum	Cosby *et al.*, 1989
	Listeria monocytogenes	Camilli *et al.*, 1990

preferentially at regions of bent DNA. The circular intermediate is also subject to a 6 bp staggered cut, nicks occurring on either strand at the boundary of the inverted repeat and the coupling sequence. The ligation of this molecule to the target site creates two flanking 6 bp regions of heteroduplex which are resolved either by mismatch repair or by replication through each region.

Transposition can occur either intra- or intercellularly, the latter involving conjugal transfer of a single strand of the circular transposition intermediate. Replication in both donor and recipient cells regenerates double-stranded intermediates proficient for integration. The transposons are remarkably promiscuous and their conjugative functions permit their transfer between a large number of species belonging to different genera. Generally, transfer into another Gram-positive species is more efficient than into Gram-negative species, although transfer can be efficient between Gram-negative species. Both excision and integration functions appear to function well in either group of bacteria. Tn*916* has now been employed as a tool in genetic analysis of a number of species (Table 1), with the advantage that issues relating to the development of appropriate delivery systems are taken care of to a large extent by the transposon itself (although to mutagenize *Haemophilus paragallinarum*, Tn*916* was introduced on a suicide vector by electroporation (see Chapter III, D.5); Gonzales *et al.*, 1996).

◆◆◆◆◆◆ IV. FISHING FOR NEW TRANSPOSONS

Transposons are ubiquitous, so a search for a novel element that can be exploited in a new species can be initiated with a large degree of confidence. Small-scale sequencing projects, especially of genes encoding "exotic" functions not immediately related to primary metabolism, can often turn up a transposable element; many otherwise cryptic insertion sequences have been identified this way as they flank these regions and have probably been involved in the evolution and/or acquisition of these novel metabolic functions. Many of these genes are plasmid-encoded and indeed several labs have stumbled on transposable elements in the characterization of endogenous plasmids which have had potential for development as cloning vectors. Another related approach, which can often pay dividends, is to clone out antibiotic or heavy metal resistance genes. These are often components of compound transposons which consist of two largely identical copies of an insertion sequence flanking the resistance marker, an example being Tn*610* from *Mycobacterium fortuitum* which carries a sulphonamide resistance determinant (Martin *et al.*, 1990). Alternatively, as in the case of the Tn*3*-family, the resistance gene is an integral part of a discrete transposon. However, as a cautionary note, this approach can be unfruitful; antibiotic-producing *Streptomyces* contain a plethora of immobile resistance genes.

Southern hybridization using the transposase gene of a heterologous transposon as a probe can light-up resident transposable elements. However, as a strategy this has limited application in finding new and useful elements: if the homology is detectable by hybridization it is reasonable to expect that the probe transposon could function in the new host. This really highlights that although overall genetic organization of transposons is typically conserved, transposase genes themselves are quite divergent in sequence. Comparison of transposase amino acid sequences has permitted their grouping into several families (Craig, 1996). One of these families share homology with retroviral integrases in the region presumed to be the active site. However, even this conservation is limited and predominantly concerns a D, D (35) E motif (D = asparagine, E = glutamine), the asparagine residues being separated by between 50 and 60 amino acids (Polard and Chandler, 1995). An implication is that identification of entirely novel transposons using a PCR strategy to amplify DNA between conserved domains in the transposase gene is generally not feasible.

In addition to these approaches, a transposon trap can be employed to fish out useful elements. These traps are plasmid or phage vectors carrying selectable markers which can be passaged through the bacterium of interest. The vectors can then be screened for transposable elements disrupting a marker gene. In practice this is facilitated if there is a positive selection for the disrupted gene; normally transposition events are rare so that large numbers of clones need to be screened. An example of a plasmid-based transposon trap, pCZA126, was employed to isolate the insertion sequence IS493 from *Streptomyces lividans* (Solenberg and Burgett, 1989). This shuttle vector can replicate in both *E.coli* and *Streptomyces* spp. and contains a gene coding for the lambda *c*I857 repressor (see Chapter 3, X) and a gene, under repressor control, coding for apramycin resistance (Am^r). Mutation of the *c*I857 gene results in an Am^r phenotype. The plasmid was passaged through *S.lividans*, and then screened for transposable elements disrupting the *c*I857 gene by transforming *E.coli* to Am^r. The procedure is subject to "noise" as other types of mutation can result in the same phenotype, but genuine insertions can be discriminated on the basis of an easily detectable increase in size of the plasmid. Of the three Am^r transformants obtained, one resulted from insertion of the transposable element IS493 which has subsequently been exploited as a genetic tool in these important bacteria. Ideally, the base composition of the target gene in the transposon trap should be similar to that of the DNA in which the element will be required to function. This will favour isolation of transposons with low target specificity which is clearly an asset for subsequent random mutagenesis; as the authors acknowledged, the *c*I857 gene, with a G+C content of 45% was not optimal for trapping transposons for use in *Streptomyces* which have a G+C content of over 70%.

Before a commitment can be made to optimizing a transposon as a genetic tool, it is important to learn a little about its biology. For example, if the goal is to provide a random mutator, it needs to be demonstrated that the element has low target specificity in the organism to be studied. Transposons can display a wide spectrum of target site specificity. For example, γδ and Tn5 display little insertion selectivity in *E.coli* (although γδ, like other members of the Tn3 family, rarely targets chromosomal sequences and inserts preferentially into plasmids), whereas Tn7 inserts into a single site called *att*Tn7, an attribute determined by recognition of this site by the Tn7-encoded protein TnsD which, in turn, interacts with the other Tn7 transposition proteins (Bainton *et al.*, 1993). In order to determine its target site specificity, the transposon needs to be introduced into the host and a system for selection of transposants devised (see section VI below). A limited number of independent transposants should be isolated and the locations of the insertion in each clone established. This can be achieved by hybridization, although conclusive data on site specificity can only be gained by cloning and sequencing the junctions of the transposon and the host DNA. Alignment of these junction sequences can indicate the nature and extent of any target site preference. Another, often more lengthy approach is to construct a library (say 2000 clones) of independent transposants and screen these for auxotrophies. A library containing a range of different auxotrophic mutations is a good indication of random transposition.

A transposon exhibiting high target specificity can be mutated to reduce the degree of insertion specificity. For example, in the absence of TnsD, and in the presence of another Tn7-encoded targeting protein, TnsE, Tn7 inserts into many different sites which are unrelated to *att*Tn7 in nucleotide sequence (Bainton *et al.*, 1993). Mutation of the transposase gene itself can also alter target selection as has been illustrated for Tn*10*. The Tn*10* transposase recognizes a 9 bp target site which, from alignment of different insertion sites, has the consensus sequence 5'NGCTNAGCN3' (Kleckner *et al.*, 1991). The transposon inserts into sites with a close match to this sequence at higher frequency than into other related but more divergent sequences. The context of the 9 bp target sequence also influences the efficiency of insertion. However, two mutants have been isolated which exhibit only slightly reduced transposition activity but possess altered target specificity (ATS) transposases (Bender and Kleckner, 1992). Examining the distribution of insertions into the *lacZ* gene demonstrated that the ATS transposases utilize a much larger number of sites than the wild type.

◆◆◆◆◆◆ V. FASHIONING A HIGH-HOPPING MINI-TRANSPOSON

To mutagenize a genome, the transposon is introduced into the bacterium on a delivery vector. Subsequent intermolecular transposition results in insertion of the element within the genome. Transposons can also mediate a variety of intramolecular rearrangements, including deletions and inversions, which affect DNA adjacent to the transposon. Some of these rearrangements are simply the intramolecular equivalents of complete intermolecular events, whereas others may reflect aborted transposition reactions. The concern to the investigator lies in the fact that a desired phenotype may not be a consequence of simple insertion, but may arise due to deletion of several genes as a result of intramolecular transposition *subsequent* to the initial delivery of the transposon into the genome.

Moreover, in relation to secondary transposition events, there is also a worry that a given mutant may acquire additional insertions at new sites over prolonged storage. However, in the majority of cases the consequences of secondary transposition can be countered by engineering a transposon so that the transposase gene itself is divorced from the inverted repeat boundaries. The transposing segment, or mini-transposon, which does not itself encode transposase, can then be supplied with the enzyme transiently to catalyse intermolecular transposition. If the source of transposase is subsequently lost or silenced, the element is then crippled for further movement. How this can be achieved is discussed later in section VI.

Another benefit to constructing a mini-transposon is that in the course of manipulating the transposase gene, it can be fused to a strong, inducible promoter and good ribosome binding site (see also Chapter 3, VI.A for the use of mini-D constructs in *Pseudomonas* aeruginosa). This can permit inducible overproduction *in vivo* of this protein, which can in turn increase the frequency of transposition and hence the number of potential mutants that can be obtained in any one experiment. Transposition is generally infrequent, typically at a frequency of between 10^{-4} and 10^{-6} per element per generation. This is a consequence of regulatory systems evolved to limit transposition and the potentially deleterious effects on the host due to, for example, insertion into and inactivation of an essential gene. A variety of regulatory mechanisms have been reviewed by Kleckner (1990). The frequency of transposition thus reflects a balance between maintaining the transposon and safeguarding the viability of the host. For many elements, the frequency of transposition is directly determined by the level of transposase. Transposase expression is therefore critical and can be modulated at the transcriptional, translational or post-translational level. Transcriptional regulation can involve a weak promoter or may be influenced by the binding of regulatory proteins. In IS1 (Machida and Machida, 1989; Zerbib *et al.*, 1990) and Tn7 (Craig, 1996), transposase expression appears to be autoregulated, whereas for the Tn3 family, transposase transcription is modulated by the specific DNA-binding activity of another element-encoded recombinase, the resolvase (Sherratt, 1989). Host proteins like IHF are believed to regulate transcription of bacteriophage Mu transposase (Alazard *et al.*, 1992; Gamas *et al.*, 1992). The deleterious effects of cut and paste transposition, which leaves a double-strand break at the donor site, can be countered by ensuring that transposition occurs preferentially just after the replication fork has passed through the element in the donor site. As a consequence, there is an intact copy of the site which can ensure cell survival and also serve as a template for DNA repair of the break. The link between transposition and replication can be mediated by the action of a methylase enzyme. For example, the transposase promoters of Tn10 and Tn5 both possess overlapping GATC sequences which are recognized by the *E.coli* Dam methylase (Yin *et al.*, 1988; Kleckner, 1990). If both strands are methylated at these sequences, the promoters are silent. DNA replication results in a transient period when the DNA is methylated on just one strand (the "old" strand), or is hemimethylated, and during this period the promoter is active.

The efficiency of transposase translation can also play a key role in controlling the level of transposase. Indeed, for many elements, inefficient translation, often because of suboptimal initiation signals, appears to be the primary method of ensuring low levels of transposase. For example, IS10 mRNA is one of the most poorly translated mRNAs in *E.coli* (Raleigh and Kleckner, 1986). Alternatively, translational regulation can occur during the elongation phase. Expression of functional IS1 and IS3 transposases involves a rare frameshifting event (Chandler and Fayet, 1993). At least for IS1, the non-frameshifted polypeptide acts as negative regulator of transposition, probably by excluding intact transposase from the ends of the element by competitive binding (Machida and Machida, 1989; Zerbib *et al.*, 1990; Escoubas *et al.*, 1991). This example of post-translational modulation of transposase activity is not unique: an alternate promoter and ribosome binding site allows synthesis of a truncated form of Tn5 transposase which acts as an inhibitor of the full length protein (Johnson *et al.*, 1982). In this case, the transposase and transposase inhibitor interact to form inactive mixed multimers, thereby reducing transposase activity (de la Cruz *et al.*, 1993). It also appears that Tn5 transposase has a suboptimal DNA-binding motif: hypertransposing transposase mutants have been isolated which recognize and bind the ends of the transposon with greater efficiency (Zhou and Reznikoff, 1997). This may be the case for transposases encoded by other elements, and reflect evolutionary pressures to maintain a balance between transposition proficiency and host viability.

Examination of the DNA sequence of the transposase gene to be manipulated can give a good indication of the types of control mechanism in play, ideas on cloning strategies which can be employed to fuse the gene to a strong promoter, and if the reading frame encodes an overlapping inhibitor which could be disarmed by site-directed mutagenesis. With the aim of driving increased transposition of mini-Tn10 transposons, the altered target site *ats1 ats2* and wild-type transposase genes have been fused to the strong isopropyl-β-D-thiogalactopyranoside (IPTG)-inducible P^{tac} (Kleckner *et al.*, 1991): this system is now widely used for mutagenesis in enteric bacteria. Likewise, over-expression in *Streptomyces* of the transposase gene of the mycobacterial insertion sequence IS6100 driven by the thiostrepton-inducible P^{tipA} results in a 100-fold increase in transposition frequencies (Smith and Dyson, 1995). To gain maximum benefit from manipulating the transposase gene for over-expression, it should be expressed *in cis* relative to the mini-transposon: the instability of transposase appears to be another mechanism in play to limit transposition frequencies as has been shown for IS903 (Derbyshire *et al.*, 1990). There is proof that transposase is preferentially *cis*-acting for many other elements, which may be due to it being physically sequestered close to its point of synthesis by non-specific DNA binding. A sensible strategy, therefore, is to engineer the recombinant transposase gene into the delivery vector for the mini-transposon.

Another interesting observation in relation to increasing transposition frequencies comes from an engineered derivative, Tn5099, of the *Streptomyces* insertion sequence IS493. Redesigning this transposable element included inserting a promoterless copy of the *xylE* reporter gene close to the left end (Hahn *et al.*, 1991). In fact the cloning step removed the inner, less well-conserved half (from base pair 56) of the left inverted repeat (IR-L), which had no observable effect on transposition frequency. A spontaneous deletion mutant of Tn*5099* was then isolated, in which IR-L sequences from base pair 49 and the entire *xylE* gene were lost (Solenberg and Baltz, 1994). This derivative, Tn*5099-10*, was shown to transpose at 10- to 1000-fold higher frequency than the parental transposon, suggesting that sequences from IR-L, from base pair 50 to 56, may be involved in regulation. This hypertransposing derivative showed no altered target site specificity.

Designing a mini-transposon from scratch permits the incorporation of many useful features between the inverted repeats. A basic provision of major importance is a suitable selectable marker, so that transposants can be readily identified. Native insertion sequences are cryptic, so converting them to "transposons" carrying a selectable antibiotic resistance gene is vital. Moreover, the selectable marker carried by a wild-type transposon may not be utilizable in the host to be mutagenized necessitating replacement by a more appropriate gene. The choice of marker should be compatible with those present in commonly used cloning vectors for the particular bacterial host, as it may be necessary to complement a given mutation *in trans* in the course of analysing gene function.

There are several other features which can be built in to provide customized mini-transposons: these are considered individually in relation to different applications described below.

◆◆◆◆◆◆ VI. DELIVERY SYSTEMS AND THE SELECTION OF TRANSPOSANTS

Delivery of the transposable element to a range of different target sites depends on intermolecular transposition from a vector which must be introduced into the organism under analysis. The vector should be unable to integrate via any other type of recombination mechanism, and it must be possible to discriminate between cells in which a transposition event has occurred and those simply retaining the vector. The latter is conventionally achieved by using a "suicide" vector – one which cannot replicate so that the resistance marker carried by the transposable element can only be inherited if it has undergone transposition into the genome of the host. These considerations have a bearing on the choice of vector, and another factor is whether or not the vector can be introduced with high efficiency. Consequently, the choice may lie with utilizing a vector based on a temperate phage or a plasmid. Phage suicide vectors are designed to have defects in integration or replication: for example, vectors lack the phage attachment site or are unable to replicate because of synthesis of repressor. These vectors offer the most efficient means by which to introduce the transposon, although the hurdles to their development for use in diverse species are: (i) the need to identify and genetically characterize an appropriate temperate phage (see Chapter 3); and (ii) the manipulation of these vectors to introduce an engineered transposon may not be as straightforward as for plasmids. In the absence of a suitable temperate phage for a given host, it is possible to extend the host range of a characterized phage of another species. For example, λ suicide vectors carrying Tn5- and Tn10-derived transposons have been used in strains of *Erwinia* (Chapter 12) *Salmonella typhimurium* (de Vries *et al.*, 1984) and *Xenorhabdus bovienii* (Francis *et al.*, 1993) expressing a constitutive allele of the *E.coli* λ receptor gene *lamB* (see Chapter 3, V.G). After phage absorption and infection, colonies carrying the antibiotic resistance marker of the transposon are simply selected. In the case of either Tn5- or Tn10-derived transposons, these arise due to cut and paste insertion into the host genome and all phage DNA sequences are lost. A large library of independent insertions can be obtained very easily in this manner. The same outcome can be expected if generalized transducing particles are used to deliver a transposon. In this case, there is no requirement for engineering the phage genome and the transposon can be carried by a plasmid in the donor strain. P22-mediated transduction is used to deliver mini-Tn10 *ats* elements to *Salmonella typhimurium* (Altman *et al.*, 1996).

A number of different strategies can be employed to ensure loss of a plasmid-based transposon vector after transposition has occurred. Broadly speaking, these can be categorized into strategies in which either the vector is first replicated and maintained in the host, prior to manipulating its loss, or, after introduction, the plasmid is unable to replicate and so is lost at cell division. Intuitively, the first category offers a larger "window of opportunity" for transposition to occur which might be important if transposition functions have not been deregulated beforehand. One way to maintain the plasmid, and then subsequently lose it, is to use a plasmid with temperature-sensitive (ts) replication functions. This may be the phenotype of a naturally occurring plasmid, for example all vectors based on the *Streptomyces ghanaensis* plasmid pSG5 are temperature sensitive (Muth *et al.*, 1988). Alternatively, a routinely used plasmid vector can be mutagenized in order to derive a ts version. A ts derivative of the mycobacterial replicon pAL5000 has been used to deliver Tn611 into the chromosome of *Mycobacterium smegmatis* (Gavigan *et al.*, 1995). Transposants can then be identified by virtue of their antibiotic resistance phenotype when or after colonies are grown at the non-permissive temperature for plasmid replication. Use of ts plasmid delivery vectors is now a tried and tested approach to transposon mutagenesis particularly in Gram-positive bacteria; various examples are listed in Table 2. However, with a large window of opportunity for transposition to occur, there is a concern that a single early transposition event, a so-called "jackpot", can subsequently be over-represented in a population of transposants if this approach is adopted. Not only can this skew the distribution of insertions in a transposon library, but it can also lead to an overestimation of the transposition frequency. This is particularly a problem with transposition assays carried out in liquid culture; at least on solid media, clones are physically separated and can be picked off individually, if necessary. For example, independent insertions of streptomycete transposons delivered

Table 2. Examples of the use of temperature-sensitive transposon delivery vectors

Species	Transposon	ts plasmid	Reference
Lactococcus lactis *Enterococcus faecalis* *Streptococcus thermophilus*	ISS1	pG(+)host	Maguin *et al.*, 1996
Streptococcus mutans	Tn917	*repA*(ts) pWV05	Gutierrez *et al.*, 1996
Staphylococcus aureus	Tn551	*repA*36 pI258	Patee, 1981
Bacillus licheniformis	Tn917PFI	pTnPFI	Pragai *et al.*, 1994
Bacillus subtilis	Tn10	*repA*(ts) pWVO5 (*Lactococcus lactis*)	Maguin *et al.*, 1992
Bacillus subtilis	Tn917ac1	pD917 (ts pE194 replication functions)	Chang *et al.*, 1994
Mycobacterium smegmatis	Tn611	ts conjugative derivative of pAL500	Gavigan *et al.*, 1995
Streptomyces lividans	IS6100	pUCS30 (pSG5 replication functions)	Smith and Dyson, 1995
Streptomyces griseofuscus *S. fradiae* *S. roseosporus*	Tn5099	pSG5-derived vectors	Baltz *et al.*, 1997

from pSG5-derived vectors can be generated by plating the host at low colony density on selective medium and growing initially at the permissive temperature before transfer to the non-permissive temperature. Growth of individual colonies then ceases unless a transposition event has integrated the relevant antibiotic resistance markers within the host genome. This is apparent from colony morphology, with the formation of actively dividing outgrowths from the moribund master colony (Solenberg and Baltz, 1991; Smith and Dyson, 1995). Another caveat with the method concerns growth at high temperature which in itself may be mutagenic to some organisms. When a pSG5-derived vector was used to deliver Tn5096 in *Streptomyces coelicolor* in order to obtain developmental "bald" mutants defective in aerial mycelium formation, only 4% of the isolated bald mutations (1 from 25 analysed) were linked to the antibiotic resistance marker of the transposon (Nodwell *et al.*, 1996). The one mutant showing linkage in fact arose due to recombinational integration of almost the entire vector, and was not due to the expected transposition event. Despite this, the same system has proven a reliable means of obtaining non-developmental mutants in several other *Streptomyces* species (Baltz *et al.*, 1997), suggesting that mutation by growth at high temperature only affects certain phenotypic traits in some species.

Other ways of evicting a resident plasmid include engineering a counterselectable marker on the plasmid. The *Bacillus subtilis* levansucrase (*sacB*) gene, which confers sensitivity to sucrose, was used as a counterselectable marker on a plasmid used for Tn5 mutagenesis of *Xenorhabdus nematophilus* (Xu *et al.*, 1991; see also Chapter 9, V.B and Chapter 13, VI.B.2). Alternatively, a second plasmid may be introduced which is incompatible with the transposon vector. The second plasmid must carry an appropriate marker gene to permit its selection at the expense of the donor plasmid. This approach was used to generate insertions of the Tn5-derived transposon Tn*phoA* (see section XI below) into virulence genes of *Vibrio cholerae* (Taylor *et al.*, 1989). The tetracycline-resistant donor IncP plasmid was mated into *V. cholerae* from *E.coli*. Kanamycin-resistant transposants were then identified after mating in a second IncP plasmid carrying a gentamicin resistance gene to displace the donor plasmid. This delivery system was compared to mating in a transposon vector which could not itself replicate in *V.cholerae* (see below). However, whereas the use of plasmid incompatibility (see Chapter 2, IV.D) generated "clean" insertions expressing active *phoA* fusions, for the non-replicating plasmid system about 20% of fusion strains carried an integrated copy of the donor plasmid which could have arisen via a non-transposition-dependent recombination event which may have been non-random. This would suggest that the extra steps involved in mating and selection with the incompatible plasmid are possibly worth while.

Other plasmid-based delivery systems are based on introduction of a non-replicating vector. After the vector has entered the cell, and before it is subsequently lost due to its inability to replicate and segregate at cell division, there is the possibility for the transposon to jump on to a resident replicon. Hence, selection can only be for the transposon marker if the element transposes either by a cut and paste mechanism or by replicative transposition followed by co-integrate resolution. Given that a sensible host for all transposon and vector construction is *E.coli*, and that typical *E.coli* plasmids do not replicate in many other prokaryote species, this is a very attractive strategy if, as is often the case, there is a limited choice of alternative replicating vectors for the organism to be analysed. Perhaps the major hurdle to overcome is exactly how the DNA can be introduced to generate a sufficiently large number of individual transposants. The possibilities include manual introduction of the DNA by a transformation or electroporation protocol, or, as is being increasingly applied, via intergeneric conjugation. Whatever the choice, for some species this can place a limit on the numbers of individual transposants that can easily be generated, probably due to the inefficiency of how the DNA is taken up (see Chapter 1) and a low frequency of transposition. Intergeneric conjugation (see Chapter 2, II.B), when it has been shown to work, is probably the most reliable means to derive large numbers of transposon mutants, and there have now been many succesful applications in many diverse Gram-negative species (Table 3). There are fewer examples involving uptake of naked DNA. *Brevibacterium flavum* has been mutagenized by transformation with an *E.coli* vector carrying a mini-Tn*31831*: 10^4 transposants per microgram of DNA were obtained (Vertes *et al.*, 1994). Similar numbers were obtained by electroporation of *Haemophilus ducreyi* with a vector carrying Tn*1545*-Delta 3 (Stevens *et al.*, 1995). However, electroporation of *Spiroplasma citri* with a plasmid carrying the *Staphylococcus aureus* transposon Tn*4001* resulted in only 10 transposants per 100 µg DNA (Foissac *et al.*, 1997b). Despite this, it has been possible to exploit this system to isolate mutants affected in plant pathogenicity (Foissac *et al.*, 1997a).

Table 3. Examples of the use of conjugative or mobilizable transposon delivery vectors

Species	Transposon	Conjugative/mobilizable plasmid	Reference
Rhodopseudomonas palustris	Tn5	pSUP2021	Elder et al., 1993
Rhodospirillum rubrum	Tn5	pSUPEG11	Ghosh et al., 1994
Cytophaga johnsonae C. hutchinsonii C. succinicans Flavobacterium meningosepticum Flexibacter canadensis Sporocytophaga myxococcoides	Tn4351	R751	McBride and Kempf, 1996 McBride and Baker, 1996
Vibrio sp. S141 Pseudomonas sp. S91	Tn10	pLBT (based on RP4)	Albertson et al., 1996
Pseudomonas putida Klebsiella pneumoniae	Tn10 & Tn5	various based on RP4	Herrero et al., 1990
Yersinia enterocolitica	Tn10	R6K	Yamamoto et al., 1994
Halomonas elongata	Tn5 & Tn1732	pSUP101 & pSUP102Gm	Kunte and Galinski, 1995
Rhizobium meliloti	Tn5	pRK600	Glazebrook and Walker, 1991

◆◆◆◆◆◆ VII. PROVING TRANSPOSITION HAS TAKEN PLACE

Confirming that the transposon has jumped into a resident replicon is usually straightforward. For elements transposing via cut and paste or by replicative transposition followed by site-specific resolution, the outcome will be inheritance of the resistance marker of the transposon and loss of any markers carried by the donor vector. These criteria can be checked rapidly using selective plating of individual putative transposants. To gain confidence in the system adopted, it is worth while to isolate genomic DNA from a random selection of clones and probe the DNA by Southern hybridization to check for a distribution of independent insertions. This may be necessary for each experiment if "jackpots" are suspected (see section VI). Replicative transposition of elements which do encode co-integrate resolution functions can also be verified by hybridization: in this case vector resistance markers will be integrated and it is important to check for genuine co-integrates which are flanked by two copies of the transposon.

For many investigators, the ultimate goal is to create a library of independent insertion mutants and screen this for representatives displaying a certain phenotype. These clones will then form the basis for detailed molecular analysis, for example for cloning out the disrupted gene (see section IX below). With a reliable system of transposon mutagenesis, it may well be possible to go ahead to this next stage of genetic analysis with some confidence. However, when applying a new system in a not well-characterized host organism, it can pay to err on the side of caution. Cloning out the disrupted sequence and using this subsequently to fish out the wild-type copy for full characterization can be a thankless task if it subsequently turns out that the mutation of interest arose independently of the insertion – that is, they are in fact unlinked. Proving linkage in some species can be quite straightforward: if there is generalized transducing phage available (see Chapter 3), it can be quickly ascertained if the phenotype of interest is co-transduced with the antibiotic-resistance marker of the transposon.

Other systems of natural gene transfer can also be utilized, for example by proving linkage in mating experiments. For some systems one can also exploit reversion of transposon insertions: recombination across the short direct repeats of target DNA flanking an insertion can precisely excise the element, although the frequency of such events is rare, generally occurring at less than 10^{-8}. If loss of the transposon resistance marker and reversion of phenotype of interest are coincident, it is safe to assume linkage. Gene replacement techniques can also be harnessed: a library of wild-type DNA segments (or a subset of an ordered library if the approximate location of the insertion is known from, for example, physical analysis by pulsed-field gel electrophoresis) can be introduced into the mutant and restoration of the wild-type phenotype selected. Linkage can be assumed if the double crossover also results in loss of the antibiotic resistance phenotype of the transposon. Should none of these tests be feasible, it is worth considering a suitable control experiment to be run in parallel with the original transposition assay. For instance, a version of the transposon vector lacking the element but with the same resistance marker, or with a copy with a defective transposase gene, could be introduced. This host–vector combination should be treated under similar conditions in parallel with the genuine transposition assay, in order to ascertain the frequencies of events not related to transposition causing either inheritance of the resistance marker or acquisition of the relevant phenotype. For the transposition system to be reliable, these frequencies in the control experiment should be several orders of magnitude less than in the genuine transposition assay. In an attempt to utilize Tn5 in *Neisseria gonorrhoeae* (see Chapter 5), apparent random integration of transposon sequences via a RecA-dependent, transposase-independent, mechanism was observed when control experiments were followed (Thomas *et al.*, 1996). There are other anecdotal reports of pseudo-transposition of various transposons in other species, so it is worth while being rigorous in applying the various criteria to prove genuine transposition.

◆◆◆◆◆◆ VIII. IF ALL ELSE FAILS: SHUTTLE MUTAGENESIS

The problems with developing a transposon mutagenesis protocol in *N. gonorrhoeae* led to elaboration of a procedure to mutagenize *N. gonorrhoeae* genes in *E.coli* prior to inactivation of the corresponding chromosomal genes in *N. gonorrhoeae* by transformation and allelic exchange (Seifert *et al.*, 1986; Seifert *et al.*, 1990; see also Chapter 5, IV.A). In an elegant example of this procedure of shuttle mutagenesis, mutants defective in invasion of epithelial cells were isolated (Kahrs *et al.*, 1994). A plasmid library of 20 000 independent clones carrying 3–4 kb *N. gonorrhoeae* DNA inserts was prepared in *E.coli*. The library was divided into separate pools and mutagenized with a mini-transposon Tn*Max4*, derived from Tn*1721*, after induction of expression of the transposase and resolvase genes which are under control of the P^{trc} promoter. Like other members of the Tn3-family, and in contrast to typical cut and paste transposable elements, Tn*Max4* has a strong preference for transposing into plasmids rather than the chromosome, and so is optimal for mutagenesis of plasmid clones. The transposon carries the *fd* origin of replication, so to identify transposants, plasmid DNA was isolated and transformed into a *polA1 E.coli* strain. Only mutated plasmids carried both the resistance marker of the target plasmid together with the *fd* origin which functions in this host. Tn*Max4* also carries a promoterless *phoA* reporter gene (see section X below), which can be used for targeting secretory determinants. Plasmid DNA was extracted from PhoA⁺ *E.coli* clones and used to transform *N. gonorrhoeae*. These bacteria are naturally competent for DNA uptake, and Tn*Max4* carries its own DNA-uptake signal. The plasmids cannot replicate in *N. gonorrhoeae*, so selection of the antibiotic resistance marker of the transposon permitted identification of *N. gonorrhoeae* clones in which a double crossing-over event had led to allelic replacement of the wild-type gene with the mutagenized copy. Of 232 mutants constructed this way, 51 proved to be defective in epithelial cell invasion, leading to the identification of about 30 distinct genetic loci determining invasiveness. These disrupted genes could be subsequently recovered by restriction of chromosomal DNA, self-ligation and transformation of the *polA1 E.coli* host by virtue of the selectable marker and *fd* origin of replication carried by the transposon (see section IX).

Thus the problems in getting a system of transposon mutagenesis up and running in a new species can be circumvented by a procedure like shuttle mutagenesis. The same methodology has been employed in other gonococci, *Helicobacter pylori* (Haas *et al.*, 1993; Chapter 11, IV.C) and mycobacteria (Kalpana *et al.*, 1991).

◆◆◆◆◆◆◆ **IX. CLONING TAGGED GENES**

The outstanding advantage in adopting transposons for random mutagenesis lies in the fact that the mutated gene is tagged by the element and can therefore be cloned with ease, without resort to cloning by complementation. The cloning strategy to be adopted depends largely on the nature of the transposon, but can be facilitated by engineering useful features into mini-transposons. As a starting point, the antibiotic resistance gene of the transposon provides a means for positive selection for cloning a linked sequence.

Using appropriate known restriction sites in the element, one or other flanking sequences may be cloned as part of a DNA fragment consisting of the resistance gene, one inverted repeat and the extent of flanking DNA up until the next restriction site. From this clone, it is then possible to isolate the corresponding wild-type DNA sequence by screening a wild-type gene library by hybridization. By choosing a restriction enzyme that does not cut within the transposon to digest the DNA of the mutant, the resistance gene linked to both flanking sequences may be cloned. In this case, it may be possible to recover the wild-type allele as a result of homologous recombination with the cloned, disrupted copy. For this, the original cloning should be into an appropriate shuttle vector that can be moved from the *E.coli* cloning host into the wild-type species under analysis. A double crossover will permit *in vivo* cloning of the wild-type allele; detection of such an event is facilitated by incorporating a counterselectable marker within the transposon as in the case of Tn5-*rpsL* (Stojiljkovic *et al.*, 1991). Other *in vivo* cloning technologies are described in Chapter 3, VI.A and VII.D.

If a plasmid replication origin is incorporated within a transposon, the genomic DNA from a mutant can be restricted with an enzyme which does not cleave within the transposon, self-ligated and used to transform an appropriate host. In practice, such a strategy should employ the replication origin of an *E.coli* plasmid which does not function in the species of interest, as insertion of a functional plasmid origin of replication into the chromosome can be deleterious. An example is Tn5-V, containing the plasmid pSC101 replication origin, which can be used to mutagenize *Myxococcus xanthus* in which the origin of replication is non-functional (Furuichi *et al.*, 1985). The DNA of mutated genes has been cloned directly into *E.coli* by appropriate restriction of genomic DNA, self-ligation and transformation. The same principle was applied in *B.subtilis*, using a derivative of Tn*917* carrying pBR322 (Youngman *et al.*, 1984).

PCR technology can also be used to clone out tagged sequences whereby the transposon contributes one or both PCR priming sites. For the former, a second priming site may be provided by ligation of genomic restriction fragments with a linker of known sequence (vectorette or cassette ligation; Riley *et al.*, 1990; Isegawa *et al.*, 1992). PCR will then amplify the intervening region consisting of transposon and flanking genomic sequences bracketed by the priming sites; the product can be directly sequenced or cloned. Alternatively, the second priming site may be provided by a dispersed repetitive (REP) sequence (Subramanian *et al.*, 1992), if these have been identified in the species in question, or a linked gene of known sequence. These alternatives may soon become more feasible as procedures for long-distance PCR are improved (Barnes 1994; Cheng *et al.*, 1994). The method of inverse PCR can also be used (Rich and Willis, 1990). This employs two transposon priming sites and involves restriction of mutant DNA with an enzyme not recognizing transposon sequences, circularization at low DNA concentration to avoid intermolecular ligation, followed by PCR priming from either end of the element. The product will be a chimera of both flanking sequences; it can be sequenced directly or separated into its constituent halves by restriction with the same enzyme used to cleave the original genomic DNA of the mutant.

X. REFINING A TRANSPOSON FOR SPECIALIST USE: MOBILE REPORTER ELEMENTS FOR ENGINEERING TRANSCRIPTIONAL FUSIONS

The elaboration of specialist transposons to extend their utility for genetic analysis includes designing elements to incorporate promoterless reporter genes. The power of these transposons is that their insertion within a transcription unit can be monitored by expression of the reporter gene. Screening a transposon library can identify mutants in which the reporter gene is conditionally expressed and thereby lead to the identification of sets of genes which are under co-ordinate regulation. The paradigm of such an element is a derivative of phage Mu, Mud1, which was engineered to contain an inwardly facing promoterless *lacZ* gene near one end (Casadaban and Cohen, 1979). Synthesis of β-galactosidase encoded by the gene can be detected by growing transposon mutants on media containing the chromogenic compound X-gal, which, when metabolized, releases a blue pigment which stains the colony (see Chapter 9). Employing this element in *E.coli* led to the discovery of several genes co-ordinately regulated so that they are expressed in response to DNA damage (Keynon and Walker, 1980). Subsequently, similar mobile reporter transposons have been utilized to identify other conditionally or developmentally activated genes in all manner of bacterial species. For instance, Tn*917*-LTV1, again containing a promoterless copy of *lacZ*, has been used to make a transposon library in *Lactococcus lactis* (Israelsen *et al.*, 1995). Screening revealed mutants in which β-galactosidase expression was regulated either by temperature, pH, growth phase or the concentration of arginine in the media. Examples of these specifically regulated promoters could then be cloned, using standard procedures for cloning out transposon-tagged sequences, to permit further characterization of their regulation and the genes and their products whose expression is regulated.

When constructing these transposons, an appropriate reporter gene must be chosen whose activity can be monitored in the host to be mutagenized. There is a reasonable choice now available, although for some hosts it may be necessary to optimize the codon usage of the gene if it is to be sensitive enough. Typical requisites are that the gene product should be easily detected by non-invasive means and that there should be little or no endogenous activity which would mask expression of the reporter. For some purposes it may be beneficial if the amount of product can be quantified, also non-invasively. A range of β-galactosidase activity can be visualized crudely in certain bacteria from the intensity of the blueness of the colony, but a more reliable estimate can only be gained by enzyme assay. The *lux* cassette from *Vibrio fischeri* and the green flourescent protein (GFP) gene from the jellyfish *Aequorea victoria* offer perhaps the greatest possibilities for non-invasive quantification and are increasingly being employed as reporter genes in a variety of biological systems.

As the reporter gene is required to be as sensitive as possible to the activity of transcription reading through the transposon, it should be engineered immediately adjacent to one inverted repeat. Translation should not be limiting, so the reading frame should be preceded by an efficient ribosome binding site. Translation of the fusion transcript emanating from without the transposon is likely to produce an out-of-frame polypeptide and may well interfere with efficient translation initiation at the start codon of the reporter gene. To counter this, a sensible step is to incorporate a sequence with stop signals in all three reading frames between the inverted repeat and reporter gene. The sequence of the inverted repeat and immediate internal region should also be carefully scrutinized for features which may interfere with either readthrough transcription or efficient translation of the reporter gene. This is important as certain elements have evolved these features to prevent external activation of the transposase promoter. For example, the regions adjacent to the transposase genes of several insertion sequences (IS1, IS2, IS3, IS5 and IS30) contain sequences which act as transcription terminators in the corresponding mRNAs (Galas and Chandler, 1989). The sequences immediately preceding the transposase gene of Tn10 also promote secondary structure formation in the corresponding mRNA, but in this case in a manner to occlude the translation initiation signals of the transposase gene (Kleckner, 1990).

◆◆◆◆◆◆ XI. MOBILE REPORTER ELEMENTS FOR GENERATING TRANSLATIONAL FUSIONS

Translational fusions can be generated by transpositional insertion within a target gene if the reporter gene lacks not only a promoter but also a translational initiation site. Assuming completely random transposition, only one-sixth of insertions within a gene will be in the appropriate orientation and correct reading frame for this to occur. Hence, in practice, a considerably larger number of independent insertions within a target gene need to be screened to find such a translational fusion, compared to insertions either causing inactivation of the target (six times more probable) or permitting expression of a reporter gene by transcriptional fusion (three times more probable). Moreover, a number of genuine in-frame translational fusions will not be detected if the biological activity of the reporter protein is dysfunctional due to incorrect folding of the fusion polypeptide. The importance of generating active translational fusions with certain reporter genes, however rare they may be, is that they can be used to determine the cellular compartment in which the target protein is localized.

To date, the most widely used reporter genes for this purpose are phoA, encoding alkaline phosphatase, and lacZ for β-galactosidase (Slauch and Silhavy, 1991). Alkaline phosphatase is active only if it has traversed the cytoplasmic membrane, whereas β-galactosidase is only functional as a cytoplasmic protein. Hence, translational fusions with the former, which confer a PhoA⁺ phenotype on the mutant, imply that the target gene encodes a transmembrane (and that the insertion is within an extracellular domain) or secreted protein. Likewise, cells expressing active β-galactosidase result from translational fusions with cytoplasmic proteins. If there is no endogenous activity in the bacterial species to be analysed, either phenotype can be readily scored by colony colour on medium supplemented with chromogenic substrates, or, more quantitatively, by direct enzyme assay.

The paradigm for a mobile reporter to detect translational fusions is Tn*phoA*, a derivative of Tn5, which was developed for use in *E.coli* to identify genes encoding proteins possessing secretion signal sequences, as well as being a tool to analyse membrane protein topology (Manoil and Beckwith, 1985, 1986). This element has subsequently been exploited in several pathogenic Gram-negative species to detect genes involved in colonization or virulence. The protein products which are important for interaction with the host are usually localized either on the cell surface or extracellularly, making the genes excellent targets for generating active PhoA fusions. One-tenth of *Salmonella typhimurium* mutants selected for a PhoA⁺ phenotype after random Tn*phoA* mutagenesis were avirulent in mice, even though the technique only selects for genes expressed on surface cultures grown on agar medium (Miller *et al.*, 1989). Similarly, virulence determinants have been targeted with the same element in *Vibrio cholerae* (Taylor *et al.*, 1989). Other pathogenic species have been analyzed using different *phoA* transposons: transposition of phage Mu is more efficient than Tn5 in *Legionella pneumophila*, and consequently a Mud*phoA* element has been utilized in this species (Albano *et al.*, 1992). Virulence genes of *Neisseria gonorrhoeae* have been identified using Tn*Max4* carrying *phoA* (see section VII above and Chapter 5, IV) and a Tn3 derivative, mTn*CmphoA*, after shuttle mutagenesis (Kahrs *et al.*, 1994; Boylevavra and Seifert, 1995).

In a further refinement of analysing membrane protein topology by generating transposon-induced translational fusions, the complementary properties of β-galactosidase and alkaline phosphatase have been exploited (Wilmes-Riesenberg and Wanner, 1992). Sets of Tn*phoA* and Tn5*lacZ* (the latter represented by elements with potential for generating both transcriptional and translational fusions) transposons were constructed containing different antibiotic resistance genes and regions of DNA homology flanking the reporter and resistance genes. By virtue of this homology, the reporter and resistance genes present in any one mutant can be switched by homologous recombination. This then permits testing of transcription, translation, or cell surface localization determinants at the same site within a gene. This system was utilized to study the phosphate regulon in *E.coli*, but could be utilized in other bacterial species, although it might be necessary to increase the extent of DNA homology (*ca.* 200 bp) to permit efficient homologous recombination in certain bacteria.

◆◆◆◆◆◆ XII. INDUCIBLE MOBILE PROMOTER ELEMENTS

Many naturally occurring transposable elements can switch on silent genes, or deregulate others, by virtue of insertion within upstream sequences and transcriptional readthrough from an internal promoter within the element into the gene itself. Transposons can also be engineered to contain strong inducible promoters to provide mobile switches to regulate adjacent genes. Insertion within an upstream region can disrupt normal promoter activity and consequently have a polar effect on the gene which is "repaired" if the internal promoter is subsequently induced. An insertion in the opposite orientation and downstream from the gene may well be phenotypically silent until inducer is applied: the activity of the promoter can produce an antisense RNA which can interfere with normal expression of the gene. These elements are therefore useful for generating conditional mutations important for the analysis of essential genes. This is an important feature in relation to the data generated by genome sequencing projects: to ascribe function to unknown ORFs requires construction of "knock-out" mutations which can only be generated if the gene is not essential. Transposons for generating conditional mutations of essential genes are an important addition in the armoury of tools to address this particular problem. Theoretically, insertions adjacent to an essential ORF, which could provide potentially conditional mutants, could be identified by a technique of "site-selected mutagenesis" previously developed to isolate P-element insertions within genes of unknown function but known sequence in *Drosophila melanogaster* (Ballinger and Benzer, 1989; Kaiser and Godwin, 1990). The procedure depends on use of PCR to amplify between a transposon primer and a gene-specific primer. A library of insertion mutants is subdivided into pools from which DNA is isolated, and each pooled DNA preparation is the template for the PCR reaction. Given knowledge of the gene sequence, the length of an amplified PCR product is a reliable indicator of an insertion at a given distance either upstream or downstream from the gene. If a PCR product of desired size is detected in a pool, this can then be subdivided and the process reiterated until the relevant mutant is isolated.

In practice, published reports indicate that mobile promoter elements based on Tn5 have been used to generate conditional mutants in *E.coli* (Chow and Berg, 1988), *Bordetella pertussis* (Cookson *et al.*, 1990) and *Pseudomonas* (de Lorenzo *et al.*, 1993). In the former species, an outward-facing isopropyl-β-D-thiogalactopyranoside (IPTG)-inducible P_{tac} promoter was employed. Both types of conditional mutation were isolated: those repaired by addition of IPTG (Chow and Berg, 1988), and those induced by addition of inducer (Neuwald *et al.*, 1993). A derivative of Tn917 carrying a constitutive outward-facing strong promoter was used to identify a cryptic β-galactosidase gene in *B.subtilis* (Zagorec and Steinmetz, 1991).

◆◆◆◆◆◆ XIII. GENOME MAPPING WITH THE HELP OF TRANSPOSONS

Mutations caused by insertion of a transposon lend themselves to straightforward mapping by conjugational or transductional crosses by virtue of the associated antibiotic resistance marker. In refining genetic mapping in Gram-negative bacteria with the aid of transposons, versions of Tn5 (Yakobson and Guiney, 1984) and mini-Mu (Ratet *et al.*, 1988) were developed containing the origin of transfer (*oriT*) of broad host range conjugative plasmids. Not only can any mutations caused by insertion of these elements be readily mapped by conjugal transfer, but a panel of strains containing insertions of these elements distributed around the chromosome can act as donors to map genes throughout the genome. Moreover, versions of phage Mu have been developed that allow for rapid mapping of *Salmonella typhimurium* genes by transductional crosses (Benson and Goldman, 1992). These transposons contain a defective "locked-in" version of the generalized transducing P22 prophage lacking an attachment site (*att⁻*), together with an antibiotic resistance marker (Youderian *et al.*, 1988). After insertion via Mu-sponsored transposition, multiple rounds of P22 replication are induced by addition of mitomycin C. The replication forks progress through the genes surrounding the site of insertion, generating an "onion-skin" amplification of the immediate area of the chromosome. The induced packaging functions of the phage then assemble headfuls of DNA, commencing at the *pac* site within the transposon and progressing unidirectionally into the surrounding over-replicated DNA, so that the mutated locus and immediate closely linked genes to one side are abundantly represented in transducing particles, greatly facilitating fine mapping of the region. Further non-transposon-induced mutations can also be targeted to the specific region of the genome transduced this way. For this, the transducing particles are treated with a chemical or physical mutagen before infecting a recipient.

Increasingly, physical mapping is assuming more importance, especially in genetic analysis of diverse species for which no efficient systems of natural gene transfer are available. Physical mapping is dependent on establishing a restriction map of rare-cutting enzyme sites for the chromosome as determined from using pulsed-field gel electrophoresis (PFGE). The locations of specific cloned genes can be determined by Southern hybridization against a profile of restriction fragments separated by PFGE. In the same way, the position of a transposon insertion and hence the disrupted gene can be ascertained by hybridization of a transposon-specific probe against DNA fragments from the mutant separated by PFGE (Smith and Kolodner, 1988). This can be further refined if the transposon itself is engineered to carry a site for the rare-cutting enzyme (Wong and McClelland, 1992). Thus, as a result of insertion, the parental large fragment will be replaced by two smaller fragments. By estimating the size of these smaller fragments, an approximate position of the point of insertion within the parental fragment can be determined. A physical map of the circular *Listeria monocytogenes* chromosome has been established with the help of a Tn*917* derivative carrying portable *Not*I and *Sma*I restriction sites (He and Luchansky, 1997).

◆◆◆◆◆◆ XIV. SIGNATURE-TAGGED MUTAGENESIS

A conventional route to establish comprehensively both the number and identity of non-essential genes involved in a particular biological process is to establish a library of mutants and screen them individually under appropriate conditions. However, given the large numbers of mutants to test in a truly random library, certainly for some test conditions the exercise beomes impracticable. A case in point concerns bacterial virulence genes: it would be neither feasible nor ethically recommendable to test, one by one, tens of thousands of individual mutants of a pathogenic organism for their ability to infect an animal. To circumvent this problem, a transposon-mutagenesis system, termed signature-tagged mutagenesis, has been developed (Hensel *et al.*, 1995). In this process, each transposon mutant is tagged with a different DNA sequence, permitting the identification of bacteria recovered from an animal previously infected with a mixed population of mutants. By subtraction, the attenuated mutants not represented in the recovered population, but present in the original library, can be recognized. Using this approach to identify virulence genes of *Salmonella typhimurium* in a murine model of typhoid fever, 40 mutants were identified after infection of just 24 mice (each of the 12 pools of mutants was injected into two mice). Further characterization showed that 13 of these mutations were in previously known virulence genes, while 15 represented mutations in genes not previously implicated in pathogenicity.

In practice, the same methodology could be applied to investigate genes in other bacterial species involved in any number of diverse biochemical, developmental, regulatory and signalling pathways, as well as the ability to adapt to certain environmental conditions. To generate individually tagged mutants, a synthetic linker of variable sequence is inserted within a transposon prior to random transposon mutagenesis. For generating a random library of sequence-tagged *S.typhimurium* mutants, the sequence tag employed ensured that the same sequence could only occur once in 2×10^{17} molecules. This sequence tag was engineered into a mini-Tn5, and the transposon library generated after conjugating in the plasmid transposon delivery vector from *E.coli* into a *S.typhimurium* strain in which it could not replicate. A library of mutants was arrayed in microtitre dishes, and the 96 mutants from each dish pooled before injecting a mouse. Three days after injection, bacteria were recovered, DNA extracted and the population of sequence tags amplified by PCR for use as a hybridization probe against a colony blot of the original microtitre dish. Non-hybridizing mutants represented mutants with attenuated virulence.

A similar approach has been adopted using variably tagged Ty*1* retrotransposons in *Saccharomyces cerevisiae* to identify genes required under specific growth conditions (Smith *et al.*, 1995): this is a powerful technology which should see application in a wide range of microbial species in the future.

◆◆◆◆◆◆ XV. CONCLUDING REMARKS

Applied transposon mutagenesis has come a long way since the pioneering work of Barbara McClintock in identifying mobile genetic elements (a history collated in McClintock, 1987). Transposon-based genetic tools have been developed and exploited to the greatest extent in bacteria, and are now an essential part of the armoury to genetically dissect these organisms. As we diversify and investigate a broader spectrum of microbial species, so it will become necessary to continue these developments. Likewise, new applications can be expected to follow from technical advances and the wealth of information generated by genome-sequencing projects. These will all impinge on our understanding of the biology of the prokaryotes, how bacteria can be exploited in biotechnology, and the approaches we adopt to combat bacterial diseases.

REFERENCES

Alazard, R., Betermier, M. and Chandler, M. (1992). *Escherichia coli* integration host factor stabilizes bacteriophage Mu repressor interactions with operator DNA *in vitro*. *Mol. Microbiol.* **5**, 1701–1714.

Albano, M.A., Arroyo, J., Eisenstein, B.I. and Engleberg, N.C. (1992). PhoA gene fusions in *Legionella pneumophila* generated in vivo using a new transposon, Mud*phoA*. *Mol. Microbiol.* **6**, 1829–1839.

Albertson, N.H., Stretton, S., Pongpattanakitshote, S., Ostling, J., Marshall, K.C., Goodman, A.E. and Kjelleberg, S. (1996). Construction and use of a new vector/transposon, pLBT::miniTn*10 lac* Kan, to identify environmentally responsive genes in a marine bacterium. *FEMS Microbiol. Lett.* **140**, 287–294.

Altman, E., Roth, J.R., Hessel, A. and Sanderson, K.E. (1996). Transposons currently in use in genetic analysis of *Salmonella* species. In Escherichia coli *and* Salmonella: *Cellular and Molecular Biology*, Vol. 2 (F.C. Neidhardt, R. Curtiss III, J.L. Ingraham, E.C.C. Lin, K.B.Low, B. Magasanik, W.S. Reznikoff, M.Riley, M. Schaechter and H.E. Umbarger, eds), pp. 2613–2626, American Society for Microbiology, Washington, DC.

Babb, B.L., Collett, H.J., Reid, S.J. and Woods, D.R. (1993). Transposon mutagenesis of *Clostridium acetobutylicum* P262 – isolation and characterization of solvent deficient and metronidazole-resistant mutants. *FEMS Microbiol. Lett.* **114**, 343–348.

Bainton, R., Kubo, K.M., Feng, J. and Craig, N.L. (1993). Tn7 transposition: target DNA recognition is mediated by multiple Tn7-encoded proteins in a purified *in vitro* system. *Cell* **72**, 931–943.

Ballinger, D.G. and Benzer, S. (1989). Targeted gene mutations in *Drosophila. Proc. Natl. Acad. Sci. USA* **86**, 9402–9406.

Baltz, R.H., McHenney, M.A., Cantwell, C.A., Queener, S.W. and Solenberg, P.J. (1997). Applications of transposition mutagenesis in antibiotic producing streptomycetes. *Anton. Leeuwenhoek Int. J. Gen. Microbiol.* **71**, 179–187.

Barnes, W.M. (1994). PCR amplification of up to 35-kb with high fidelity and high yield from λ bacteriophage templates. *Proc. Natl. Acad. Sci. USA* **91**, 2216–2220.

Belas, R., Erskine, D. and Flaherty, D. (1991). Transposon mutagenesis in *Proteus mirabilis. J. Bacteriol.* **173**, 6289–6293.

Bender, J. and Kleckner, N. (1992). IS*10* transposase mutations that specifically alter target site recognition. *EMBO J.* **11**, 741–750.

Benson, N.R. and Goldman, B.S. (1992). Rapid mapping in *Salmonella typhimurium* with Mu*d*-P22 prophages. *J. Bacteriol.* **174**, 1673–1681.

Berg, C.M. and Berg, D.E. (1996). Transposable element tools for microbial genetics. In Escherichia coli *and* Salmonella: *Cellular and Molecular Biology*, Vol. 2 (F.C. Neidhardt, R. Curtiss III, J.L. Ingraham, E.C.C. Lin, K.B. Low, B. Magasanik, W.S. Reznikoff, M.Riley, M. Schaechter, and H.E. Umbarger, eds), pp. 2588–2612. American Society for Microbiology, Washington, DC.

Berg, C.M., Isono, W.G., Kasai, H. and Berg, D.E. (1994). Transposon-facilitated large-scale DNA sequencing. In *Automated DNA Sequencing and Analysis*. (C. Fields, M.D. Adams, and J.C. Venter, eds), pp. 51–59, Academic Press, London.

Beringer, J.E., Benyon, J.L., Buchanan-Wollaston, A.V. and Johnston, A.W.B. (1978). Transfer of the drug-resistance transposon Tn5 to *Rhizobium*. *Nature* **276**, 633–634.

Borthakur, D. and Haselkorn, R. (1989). Tn5 mutagenesis of *Anabaena* sp. strain PCC 7120: isolation of a new mutant unable to grow without combined nitrogen. *J. Bacteriol.* **171**, 5759–5761.

Boylevavra, S. and Seifert, H.S. (1995). Shuttle mutagenesis – a mini-transposon for producing PhoA fusions with exported proteins in *Neisseria gonorrhoeae*. *Gene* **155**, 101–106.

Camilli, A., Portnoy, D.A. and Youngman, P. (1990). Insertional mutagenesis of *Listeria monocytogenes* with a novel Tn917 derivative that allows direct cloning of DNA flanking transposon insertions. *J. Bacteriol.* **172**, 3738–3744.

Cangelosi, G.A., Best, E.A., Martinetti, G. and Nester, E.W. (1991). Genetic analysis of *Agrobacterium*. *Meth. Enzymol.* **204**, 384–397.

Caparon, M.G. and Scott, J.R. (1991). Genetic manipulation of pathogenic *Streptococci*. *Meth. Enzymol.* **204**, 556–586.

Casadaban, M.J. and Cohen, S.N. (1979). Lactose genes fused to exogenous promoters in one step using a Mu-*lac* bacteriophage: *in vivo* probe for transcriptional control sequences. *Proc. Natl. Acad. Sci. USA* **76**, 4530–4533.

Caufield, P.W., Shah, G.R. and Hollingshead, S.K. (1990). Use of transposon Tn916 to inactivate and isolate a mutacin-associated gene from *Streptococcus mutans*. *Infec. Immunity* **58**, 4126–4135.

Chandler, M. and Fayet, O. (1993). Translational frameshifting in the control of transposition in bacteria. *Mol. Microbiol.* **7**, 497–503.

Chang, L.K., Chen, C.L., Chang, Y.S., Tschen, J.S.M., Chen, Y.M. and Liu, S.T. (1994). Construction of Tn917acI, a transposon useful for mutagenesis and cloning of *Bacillus subtilis* genes. *Gene* **150**, 129–134.

Cheng, S., Fockler, C., Barnes, W.M. and Higuchi, R. (1994). Effective amplification of long targets from cloned inserts and human genomic DNA. *Proc. Natl. Acad. Sci. USA* **91**, 5695–5699.

Chow, W.-Y. and Berg, D.E. (1988). Tn5tac1, a derivative of Tn5 that generates conditional mutations. *Proc. Natl. Acad. Sci. USA* **85**, 6468–6472.

Cohen, M.F., Wallis, J.G., Campbell, E.L. and Meeks, J.C. (1994). Transposon mutagenesis of *Nostoc* sp. strain ATCC-29133, a filamentous cyanobacterium with multiple cellular differentiation alternatives. *Microbiology-UK* **140**, 3233–3240.

Contreras, A., Maldonado, R. and Casadesus, J. (1991). Tn5 mutagenesis and insertion replacement in *Azotobacter vinelandii*. *Plasmid* **25**, 76–80.

Cookson, B.T., Berg, D.E. and Goldman, W.E. (1990). Mutagenesis of *Bordatella pertussis* with transposon Tn5*tac1*: conditional expression of virulence-associated genes. *J. Bacteriol.* **172**, 1681–1687.

Cosby, W.M., Axelsson, L.T. and Dobrogosz, W.J. (1989). Tn*917* transposition in *Lactobacillus plantarum* using the highly temperature-sensitive plasmid pTV1ts as a vector. *Plasmid* **22**, 236–243.

Craig, N.L. (1996). Transposition. In Escherichia coli *and* Salmonella: *Cellular and Molecular Biology*, Vol. 2 (F.C. Neidhardt, R. Curtiss III, J.L. Ingraham, E.C.C. Lin, K.B.Low, B. Magasanik, W.S. Reznikoff, M.Riley, M. Schaechter, and H.E. Umbarger, eds), pp. 2339–2362. American Society for Microbiology, Washington, DC.

de la Cruz, N.B., Weinreich, M.D., Wiegand, T.W., Krebs, M.P. and Reznikoff, W.S. (1993). Characterization of the Tn*5* transposase and inhibitor proteins: a model for the inhibition of transposition. *J. Bacteriol.* **17**, 6932–6938.

de Lorenzo, V., Eltis, L., Kessler, B. and Timmis, K.N. (1993). Analysis of *Pseudomonas* gene products using *lacI*ᵠ/*Ptrp-lac* plasmids and transposons that confer conditional phenotypes. *Gene* **123**, 17–24.

Derbyshire, K.M., Kramer, N. and Grindley, N.D. (1990). Role of instability in the *cis* action of the insertion sequence IS*903*. *Proc. Natl. Acad. Sci. USA* **87**, 4048–4052.

de Vries, G.E., Raymond, C.K. and Ludwig, R.A. (1984). Extension of bacteriophage λ host range: selection, cloning, and characterization of a constitutive λ receptor gene. *Proc. Natl. Acad. Sci. USA* 81, 6080–6084.

Dobrowolski, P. and Grundig, M.W. (1990). Mutagenesis of *Acetobacter methanolicus* MB58 with the transposon Tn*5*. *J. Basic Microb.* **30**, 295–296.

Dybvig, K. and Cassell, G.H. (1987). Transposition of gram-positive transposon Tn*916* in *Acholoplasma laidlawii* and *Mycoplasma pulmonis*. *Science* **235**, 1392–1394.

Eichenbaum, Z. and Scott, J.R. (1997). Use of Tn*917* to generate insertion mutations in the group A *Streptococcus*. *Gene* **186**, 213–217.

Elder, D.J.E., Morgan, P. and Kelly, D.J. (1993). Transposon Tn*5* mutagenesis in *Rhodopseudomonas palustris*. *FEMS Microbiol. Lett.* **111**, 23–30.

Ely, B. and Croft, R.H. (1982). Transposon mutagenesis in *Caulobacter crescentus*. *J. Bacteriol.* **149**, 620–625.

Escoubas, J.M., Prere, M.F., Fayet, O., Salvignol, I., Galas, D., Zerbib, D. and Chandler, M. (1991). Translational control of transposition activity of the bacterial insertion sequence IS*1*. *EMBO J.* **10**, 705–712.

Foissac, X., Danet, J.L., Saillard, C., Gaurivaud, P., Laigret, F., Pare, C. and Bove, J.M. (1997a). Mutagenesis by insertion of Tn*4001* into the genome of *Spiroplasma citri*: characterization of mutants affected in plant pathogenicity and transmission to the plant by the leafhopper vector *Circulifer haematoceps*. *Mol. Plant Microbe Interaction.* **10**, 454–461.

Foissac, X., Saillard, C. and Bove, J.M. (1997b). Random insertion of transposon Tn*4001* in the genome of *Spiroplasma citri* strain GII3. *Plasmid* **37**, 80–86.

Francis, M.S., Parker, A.F., Morona, R. and Thomas, C.J. (1993). Bacteriophage lambda as a delivery vector for Tn*10*-derived transposons in *Xenorhabdus bovienii*. *Appl. Environ. Microbiol.* **59**, 3050–3055.

Franke, A.E. and Clewell, D.B. (1981). Evidence for a chromosome-borne resistance transposon (Tn*916*) in *Streptococcus faecalis* that is capable of conjugal transfer in the absence of a conjugative plasmid. *J. Bacteriol.* **145**, 494–502.

Furuichi, T., Inouye, M. and Inouye, S. (1985). Novel one-step cloning vector with a transposable element: application in *Myxococccus xanthus* genome. *J. Bacteriol.* **164**, 270–275.

Galas, D.J. and Chandler, M. (1989). Bacterial insertion sequences. In *Mobile DNA*

(D.E. Berg and M.M. Howe, eds), pp. 109–162. American Society for Microbiology, Washington, DC.

Gamas, M.J., Toussaint, A. and Higgins, N.P. (1992). Stabilization of bacteriophage Mu repressor–operator complexes by the *Escherichia coli* integration host factor protein. *Mol. Microbiol.* **6**, 1715–1722.

Gavigan, J.A., Guihot, C., Gicquel, B. and Martin, C. (1995). Use of conjugative and thermosensitive cloning vectors for transposon delivery to *Mycobacterium smegmatis*. *FEMS Microbiol. Lett.* **1127**, 35–39.

Ghosh, R., Elder, D.J.E., Saegesser, R., Kelly, D.J. and Bachofen, R. (1994). An improved procedure and new vectors for transposon Tn5 mutagenesis of the phototrophic bacterium *Rhodospirillum rubrum*. *Gene* **150**, 97–100.

Gibson, B.W., Campagnari, A.A., Melaugh, W., Phillips, N.J., Apicella, M.A., Grass, S., Wang, J., Palmer, K.L. and Munson, R.S. (1997). Characterization of a transposon Tn*916*-generated mutant of *Haemophilus ducreyi* 35000 defective in lipooligosaccharide biosynthesis. *J. Bacteriol.* **179**, 5062–5071.

Glazebrook, J. and Walker, G.C. (1991). Genetic techniques in *Rhizobium meliloti*. *Meth. Enzymol.* **204**, 398–418.

Goguen, J.D., Yother, J. and Straley, S.C. (1984). Genetic analysis of the low calcium response in *Yersinia pestis* Mudi (Ap*lac*) insertion mutants. *J. Bacteriol.* **160**, 842–848.

Gonzales, A.E., Glisson, J.R. and Jackwood, M.W. (1996). Transposon mutagenesis of *Haemophilus paragallinarum* with Tn*916*. *Vet. Microbiol.* **48**, 283–291.

Graf, J., Dunlap, P.V. and Ruby, E.G. (1994). Effect of transposon-induced motility mutations on colonization of the host light organ by *Vibrio fischeri*. *J. Bacteriol.* **176**, 6986–6991.

Groisman, E.A. (1991). *In vivo* genetic engineering with bacteriophage Mu. *Meth. Enzymol.* **204**, 180–212.

Gupta, A., Verma, V. and Qazi, G.N. (1997). Transposon induced mutation in *Gluconobacter oxydans* with special reference to its direct glucose oxidation metabolism. *FEMS Microbiol. Lett.* **147**, 181–188.

Gutierrez, J.A., Crowley, P.J., Brown, D.P., Hillman, J.D., Youngman, P. and Bleiweis, A.S. (1996). Insertional mutagenesis and recovery of interrupted genes of *Streptococcus mutans* by using Tn*917* – preliminary characterization of mutants displaying acid sensitivity and nutritional requirements. *J. Bacteriol.* **178**, 4166–4175.

Haas, R., Meyer, T.F. and van Putten, J.P.M. (1993). Aflagellated mutants of *Helicobacter pylori* generated by genetic transformation of naturally competent strains using transposon shuttle mutagenesis. *Mol. Microbiol.* **8**, 753–760.

Hahn, D.R., Solenberg, P.J. and Baltz, R.H. (1991). Tn*5099*, a *xylE* promoter probe transposon for *Streptomyces* spp. *J. Bacteriol.* **173**, 5573–5577.

He, W. and Luchansky, J.B. (1997). Construction of the temperature-sensitive vectors pLUCH80 and pLUCH88 for delivery of Tn*917*::Not I/*Sma*I and use of these vectors to derive a circular map of *Listeria monocytogenes* Scott A, a serotype 4b isolate. *Appl. Environ. Microbiol.* **63**, 3480–3487.

Heilmann, C., Gerke, C., Perdreauremington, F. and Gotz, F. (1996). Characterization of Tn*917* insertion mutants of *Staphylococcus epidermidis* affected in biofilm formation. *Infec. Immun.* **64**, 277–282.

Hensel, M., Shea, J.E., Gleeson, C., Jones, M.D., Dalton, E. and Holden, D.W. (1995). Simultaneous identification of bacterial virulence genes by negative selection. *Science* **269**, 400–403.

Herrero, M., de Lorenzo, V. and Timmis, K.N. (1990). Transposon vectors containing non-antibiotic resistance selection markers for cloning and stable chromosomal insertion of foreign genes in gram-negative bacteria. *J. Bacteriol.* **172**, 6557–6567.

Hoch, J.A. (1991). Genetic analysis in *Bacillus subtilis. Meth. Enzymol.* **204**, 305–320.

Holland, J., Towner, K.J. and Williams, P. (1992). Tn*916* insertion mutagenesis in *Escherichia coli* and *Haemophilus influenzae* type-B following conjugative transfer. *J. Gen. Microbiol.* **138**, 509–515.

Ike, Y., Clewell, D.B., Segarra, R.A. and Gilmore, M.S. (1990). Genetic analysis of the *pad1* hemolysin bacteriocin determinant in *Enterococcus faecalis* – Tn*917* insertional mutagenesis and cloning. *J. Bacteriol.* **172**, 155–163.

Isegawa, Y., Sheng, J., Sokawa, Y., Yamanishi, K., Nakagomi, O. and Ueda, S. (1992). Selective amplification of cDNA sequence from total RNA by cassette-ligation mediated polymerase chain reaction (PCR): application to sequencing 6.5 kb genome segment of hantavirus strain B-1. *Mol. Cell. Probes* **6**, 467–475.

Israelsen, H., Madsen, S.M., Vrang, A., Hansen, E.B. and Johansen, E. (1995). Cloning and partial characterization of regulated promoters from *Lactococcus lactis* Tn*917-lacZ* integrants with the new promoter probe vector, pAK80. *Appl. Environ. Microbiol.* **61**, 2540–2547.

Ivins, B.E., Welkos, S.L., Knudson, G.B. and Leblanc, D.J. (1988). Transposon Tn*916* mutagenesis in *Bacillus anthracis. Infec. Immun.* **56**, 176–181.

Jayaswal, R.K., Bressan, R.A. and Handa, A.K. (1984). Mutagenesis of *Erwinia carotovora* subsp. *carotovora* with bacteriophage Mu D1 (*apr lac cts62*) – construction of *his–lac* gene fusions. *J. Bacteriol.* **158**, 764–766.

Johnson, R.C., Yin, J.C. and Reznikoff, W.S. (1982). Control of Tn*5* transposition in *Escherichia coli* is mediated by protein from the right repeat. *Cell* **30**, 873–882.

Jones, J.M., Yost, S.C. and Pattee, P.A. (1987). Transfer of the conjugal tetracycline resistance transposon Tn*916* from *Streptococcus faecalis* to *Staphylococcus aureus* and identification of some insertion sites in the staphylococcal chromosome. *J. Bacteriol.* **169**, 2121–2131.

Kahrs, A.F., Bihlmaier, A., Facius, D. and Meyer, T.F. (1994). Gneralized transposon shuttle mutagenesis in *Neisseria gonorrhoeae*: a method for isolating epithelial cell invasion-defective mutants. *Mol. Microbiol.* **12**, 819–831.

Kaiser, K. and Goodwin, S.F. (1990). "Site-selected" transposon mutagenesis of *Drosophila. Proc. Natl. Acad. Sci. USA* **87**, 1686–1690.

Kalpana, G.V., Bloom, B.R. and Jacobs, W.R. (1991). Insertional mutagenesis and illegitimate recombination in mycobacteria. *Proc. Natl. Acad. Sci. USA* **88**, 5433–5437.

Kathariou, S., Metz, P., Hof, H. and Goebel, W. (1987). Tn*916*-induced mutations in the hemolysin determinant affecting virulence of *Listeria monocytogenes. J. Bacteriol.* **169**, 1291–1297.

Kathariou, S., Stephens, D.S., Spellman, P. and Morse, S.A. (1990). Transposition of Tn*916* to different sites in the chromosome of *Neisseria meningitidis* – a genetic tool for meningococcal mutagenesis. *Mol. Microbiol.* **4**, 729–735.

Kaufman, M.R. and Taylor, R.K. (1994). Identification of bacterial cell-surface virulence determinants with Tn*phoA. Meth. Enzymol.* **235**, 426–448.

Keynon, C.J. and Walker, G.C. (1980). DNA-damaging agents stimulate gene expression at specific loci in *Escherichia coli. Proc. Natl. Acad. Sci. USA* **77**, 2819–2823.

Kleckner, N. (1990). Regulation of transposition in bacteria. *Ann. Rev. Cell Biol.* **6**, 297–327.

Kleckner, N., Bender, J. and Gottesman, S. (1991). Uses of transposons with emphasis on Tn*10. Meth. Enzymol.* **204**, 139–180.

Kolodrubetz, D. and Kraig, E. (1994). Transposon Tn*5* mutagenesis of *Actinobacillus actinomycetemcomitans* via conjugation. *Oral Microbiol. Immunol.* **9**, 290–296.

Kuner, J.M. and Kaiser, D. (1981). Introduction of transposon Tn*5* into *Myxococcus* for analysis of developmental and other nonselectable mutants. *Proc. Natl. Acad. Sci. USA* **78**, 425–429.

Kunte, H.J. and Galinski, E.A. (1995). Transposon mutagenesis in halophilic eubacteria – conjugal transfer and insertion of transposon Tn5 and Tn1732 in *Halomonas elongata*. *FEMS Microbiol. Lett.* **128**, 293–299.

Kuramitsu, H.K. and Casadaban, M.J. (1986). Transposition of the gram-positive transposon Tn917 in *Escherichia coli*. *J. Bacteriol.* **167**, 711–712.

Leahy, J.G., Jonesmeehan, J.M., Pullias, E.L. and Colwell, R.R. (1993). Transposon mutagenesis in *Acinetobacter calcoaceticus* Rag-1. *J. Bacteriol.* **175**, 1838–1840.

Lee, K.E., Stone, S., Goodwin, P.M. and Holloway, B.W. (1991). Characterization of transposon insertion mutants of *Methylobacterium extorquens* AM1 (*Methylobacterium* strain AM1) which are defective in methanol oxidation. *J. Gen. Microbiol.* **137**, 895–904.

Lee, M.D. and Henk, A.D. (1996). Tn10 insertional mutagenesis in *Pasteurella multocida*. *Vet. Microbiol.* **50**, 143–148.

Lin, D.L. and McBride, M.J. (1996). Development of techniques for the genetic manipulation of the gliding bacteria *Lysobacter enzymogenes* and *Lysobacter brunescens*. *Can. J. Microbiol.* **42**, 896–902.

Lin, W.J. and Johnson, E.A. (1991). Transposon Tn916 mutagenesis in *Clostridium botulinum*. *Appl. Environ. Microbiol.* **57**, 2946–2950.

Machida, C. and Machida, Y. (1989). Regulation of IS1 transposition by the insAgene product. *J. Mol. Biol.* **208**, 567–574.

Maguin, E., Duwat, P., Hege, T., Ehrlich, D. and Gruss, A. (1992). New thermosensitive plasmid for gram-positive bacteria. *J. Bacteriol.* **174**, 5633–5638.

Maguin, E., Prevost, H., Ehrlich, S.D. and Gruss, A. (1996). Efficient insertional mutagenesis in *Lactococci* and other gram-positive bacteria. *J. Bacteriol.* **178**, 931–935.

Manoil, C. and Beckwith, J. (1985). TnphoA: a transposon probe for protein export signals. *Proc. Natl. Acad. Sci. USA* **82**, 8129–8133.

Manoil, C. and Beckwith, J. (1986). A genetic approach to analyzing membrane protein topology. *Science* **233**, 1403–1408.

Martin, C., Timm, J., Rauzier, J., Gomez-Lus, R., Davies, J. and Gicquel, B. (1990). Transposition of an antibiotic resistance element in Mycobacteria. *Nature* **345**, 739–743.

Mattsson, D.M. and Rogers, P. (1994). Analysis of Tn916-induced mutants of *Clostridium acetobutylicum* altered in solventogenesis and sporulation. *J. Ind. Microbiol.* **13**, 258–268.

McBride, M.J. and Baker, S.A. (1996). Development of techniques to genetically manipulate members of the genera *Cytophaga, Flavobacterium, Flexibacter*, and *Sporocytophaga*. *Appl. Environ. Microbiol.* **62**, 3017–3022.

McBride, M.J. and Kempf, M.J. (1996). Development of techniques for the genetic manipulation of the gliding bacterium *Cytophaga johnsonae*. *J. Bacteriol.* **178**, 583–590.

McClintock, B. (1987). The discovery and characterization of transposable elements: the collected papers of Barbara McClintock. Garland, New York.

Miller, I., Maskell, D., Hormaeche, C., Johnson, K., Pickard, D. and Dougan, G. (1989). Isolation of orally attenuated *Salmonella typhimurium* following TnphoA mutagenesis. *Infect. Immun.* **57**, 2758–2763.

Mintz, C.S. and Shuman, H.A. (1987). Transposition of bacteriophage Mu in the legionnaires disease bacterium. *Proc. Natl. Acad. Sci. USA* **84**, 4645–4649.

Morenovivian, C., Roldan, M.D., Reyes, F. and Castillo, F. (1994). Isolation and characterization of transposon Tn5 mutants of *Rhodobacter sphaeroides* deficient in both nitrate and chlorate reduction. *FEMS Microbiol. Lett.* **115**, 279–284.

Muth, G., Wohlleben, W. and Pühler, A. (1988). The minimal replicon of the *Streptomyces ghanaensis* plasmid pSG5 identified by subcloning and Tn5 mutagenesis. *Mol. Gen. Genet.* **211**, 424–429.

Neuwald, A.F., Krishnan, B.R., Ahrweiler, P.M., Frieden, C. and Berg, D.E. (1993). Conditional dihydrofolate reductase deficiency due to transposon Tn5tac1 insertion downstream from the *folA* gene in *Escherichia coli. Gene* **125**, 69–73.

Nodwell, J.R., McGovern, K. and Losick, R. (1996). An oligopeptide permease responsible for the import of an extracellular signal governing aerial mycelium formation in *Streptomyces coelicolor. Mol. Microbiol.* **22**, 881–893.

Pappas, K.M., Galani, I. and Typas, M.A. (1997). Transposon mutagenesis and strain construction in *Zymomonas mobilis. J. Appl. Microbiol.* **82**, 379–388.

Patee, P.A. (1981). Distribution of Tn551 insertion sites responsible for auxotrophy on the *Staphylococcus aureus* chromosome. *J. Bacteriol.* **145**, 479–488.

Petit, M.A., Bruand, C., Jannière, L. and Ehrlich, S.D. (1990). Tn10 derived transposons active in *Bacillus subtilis. J. Bacteriol.* **172**, 6736–6740.

Polard, P. and Chandler, M. (1995). Bacterial transposases and retroviral integrases. *Mol. Microbiol.* **15**, 13–23.

Pope, C.D., Dhand, L. and Cianciotto, N.P. (1994). Random mutagenesis of *Legionella pneumophila* with mini-Tn10. *FEMS Microbiol. Lett.* **124**, 107–111.

Pragai, Z., Tran, S.L.P., Nagy, T., Fulop, L., Holczinger, A. and Sik, T. (1994). Transposon Tn917PFI mutagenesis in *Bacillus licheniformis. Microbiology-UK* **140**, 3091–3097.

Raleigh, E.A. and Kleckner, N. (1986). Quantitation of insertion sequence IS10 transposase gene expression by a method generally applicable to any rarely expressed gene. *Proc. Natl. Acad. Sci. USA* **83**, 1787–1791.

Ratet, P., Schell, J. and de Bruijn, F.J. (1988). Mini-Mu-*lac* transposons with wide host-range origins of conjugal transfer and replication for gene transfer studies in *Rhizobiaceae. Gene* **63**, 41–52.

Rich, J.J. and Willis, D.K. (1990). A single oligonucleotide can be used to rapidly isolate DNA sequences flanking a transposon Tn5 insertion by the polymerase chain reaction. *Nucl. Acids Res.* **18**, 66–76.

Riley, J., Butler, R., Ogilvie, D., Finniear, R., Jenner, D., Powell, S., Anand, R., Smith, J.C. and Markham, J.F. (1990). A novel, rapid method for the isolation of terminal sequences from yeast artificial chromosome (YAC) clones. *Nucl. Acids Res.* **18**, 2887–2890.

Salyers, A.A., Shoemaker, N.B., Stevens, A.M. and Li, L.-Y. (1995). Conjugative transposons: an unusual and diverse set of integrated gene transfer elements. *Microbiol. Revs.* **59**, 579–590.

Sato, S., Takamatsu, N., Okahashi, N., Matsunoshita, N. and Inoue, M. (1992). Construction of mutants of *Actinobacillus actinomycetemcomitans* defective in serotype-B-specific polysaccharide antigen by insertion of transposon Tn916. *J. Gen. Microbiol.* **138**, 1203–1209.

Scott, J.R. and Churchward, G.G. (1995). Conjugative transposition. *Ann. Rev. Microbiol.* **49**, 367–397.

Seifert, H.S., Chen, E.Y., So, M. and Heffron, F. (1986). Shuttle mutagenesis: a method of transposon mutagenesis for *Saccharomyces cerevisiae. Proc. Natl. Acad. Sci. USA* **83**, 735–739.

Seifert, H.S., Ajioka, R.S., Paruchuri, D., Heffron, F. and So, M. (1990). Shuttle mutagenesis of *Neisseria gonorrhoeae*: pilin null mutations lower DNA transformation competence. *J. Bacteriol.* **172**, 40–46.

Sherratt, D. (1989). Tn3 and related tranposable elements: site-specific recombination and transposition. In *Mobile DNA* (D.E. Berg and M.M. Howe, eds), pp. 163–184. American Society for Microbiology, Washington, DC.

Shoemaker, N.B., Barber, R. and Salyers, A.A. (1989). Cloning and characterization of a *Bacteroides* conjugal tetracycline resistance element using a shuttle cosmid vector. *J. Bacteriol.* **171**, 1294–1302.

Silakowski, B., Pospiech, A., Neumann, B. and Schairer, H.U. (1996). *Stigmatella aurantiaca* fruiting body formation is dependent on the *fbfA* gene encoding a polypeptide homologous to chitin synthases. *J. Bacteriol.* **178**, 6706–6713.

Slauch, J.M. and Silhavy, T.J. (1991). Genetic fusions as experimental tools. *Meth. Enzymol.* **204**, 213–248.

Smith, B. and Dyson, P. (1995). Inducible transposition in *Streptomyces lividans* of insertion sequence IS6100 from *Mycobacterium fortuitum*. *Mol. Microbiol.* **18**, 933–941.

Smith, C.L. and Kolodner, R.D. (1988). Mapping of *Escherichia coli* chromosomal Tn5 and F insertions by pulsed field gel electrophoresis. *Genetics* **119**, 227–236.

Smith, L.D. and Heffron, F. (1987). Transposon Tn5 mutagenesis of *Brucella abortus*. *Infec. Immun.* **55**, 2774–2776.

Smith, V., Botstein, D. and Brown, P.O. (1995). Genetic footprinting: a strategy for determining a gene's function given its sequence. *Proc. Natl. Acad. Sci. USA* **92**, 6479–6483.

Solenberg, P.J. and Baltz, R.H. (1991). Transposition of Tn5096 and other IS493 derivatives in *Streptomyces griseofuscus*. *J. Bacteriol.* **173**, 1096–1104.

Solenberg, P.J. and Baltz, R.H. (1994). Hypertransposing derivatives of the streptomycete insertion sequence IS493. *Gene* **147**, 47–54.

Solenberg, P.J. and Burgett, S.G. (1989). Method for selection of transposable DNA and characterization of a new insertion sequence, IS493, from *Streptomyces lividans*. *J. Bacteriol.* **171**, 4807–4813.

Srivastava, S., Urban, M. and Friedrich, B. (1982). Mutagenesis of *Alcaligenes eutrophus* by insertion of the drug resistance transposon Tn5. *Arch. Microbiol.* **131**, 203–207.

Stevens, M.K., Cope, L.D., Radolf, J.D. and Hansen, E.J. (1995). A system for generalized mutagenesis of *Haemophilus ducreyi*. *Infect. Immun.* **63**, 2976–2982.

Stojiljkovic, I., Trgovcevic, Z. and Salaj-Smic, E. (1991). Tn5-rpsL: a new derivative of transposon Tn5 useful in plasmid curing. *Gene* **99**, 101–104.

Subramanian, P., Versalovic, J., McCabe, E.R.B. and Lupski, J.R. (1992). Rapid mapping of *E. coli*::Tn5 insertion mutations by REP-Tn5 PCR. *PCR Meth. Appl.* **1**, 187–194.

Tascon, R.I., Rodriguezferri, E.F. and Gutierrezmartin, C.B. (1993). Transposon mutagenesis in *Actinobacillus pleuropneumoniae* with a Tn10 derivative. *J. Bacteriol.* **175**, 5717–5722.

Taylor, R.H., Manoil, C. and Mekalanos, J.L. (1989). Broad-host-range vectors for delivery of TnphoA: use in genetic analysis of secreted virulence determinants of *Vibrio cholerae*. *J. Bacteriol.* **171**, 1870–1878.

Thomas, C.E., Carbonetti, N.H. and Sparling, P.F. (1996). Pseudo-transposition of a Tn5 derivative in *Neisseria gonorrhoeae*. *FEMS Microbiol. Lett.* **145**, 371–376.

Tomich, P.K. and Clewell, D.B. (1980). Properties of erythromycin-inducible transposon Tn917 in *Streptococcus faecalis*. *J. Bacteriol.* **141**, 1366–1374.

Vanstockem, M., Michiels, K., Vanderleyden, J. and Vangool, A.P. (1987). Transposon mutagenesis of *Azospirillum brasilense* and *Azospirillum lipoferum* – physical analysis of Tn5 and Tn5-mob insertion mutants. *Appl. Environ. Microbiol.* **53**, 410–415.

Vertes, A.A., Asai, Y., Inui, M., Kobayashi, M., Kurusu, Y. and Yukawa, H. (1994). Transposon mutagenesis of coryneform bacteria. *Mol. Gen. Genet.* **245**, 397–405.

Volff, J.N. and Altenbuchner, J. (1997). High frequency transposition of the Tn5 derivative Tn5493 in *Streptomyces lividans*. *Gene* **194**, 81–86.

Wall, J.D., Murnan, T., Argyle, J., English, R.S. and Rappgiles, B.J. (1996). Transposon mutagenesis in *Desulfovibrio desulfuricans* – development of a random mutagenesis tool from Tn7. *Appl. Environ. Microbiol.* **62**, 3762–3767.

Wilmes-Riesenberg, M.R. and Wanner, B.L. (1992). Tn*phoA* and Tn*phoA'* elements for making and switching fusions for study of transcription, translation, and cell surface localization. *J. Bacteriol.* **174**, 4558–4575.

Wong, K.K. and McClelland, M. (1992). Dissection of the *Salmonella typhimurium* genome by use of a Tn5 derivative carrying rare restriction sites. *J. Bacteriol.* **174**, 3807–3811.

Xu, J.M., Olson, M.E., Kahn, M.L. and Hurlbert, R.E. (1991). Characterization of Tn5-induced mutants of *Xenorhabdus nematophilus* ATCC 19061. *Appl. Environ. Microbiol.* **57**, 1173–1180.

Yakobson, E.A. and Guiney, D.G., Jr. (1984). Conjugal transfer of bacterial chromosomes mediated by the RK2 plasmid transfer origin cloned into transposon Tn5. *J. Bacteriol.* **160**, 451–453.

Yamamoto, T., Hanawa, T., Murayama, S.Y. and Ogata, S. (1994). Isolation of thermosensitive mutants of *Yersinia enterocolitica* by transposon insertion. *Plasmid* **32**, 238–243.

Yin, J.C., Krebs, M.P. and Reznikoff, W.S. (1988). Effect of *dam* methylation on Tn5 transposition. *J. Mol. Biol.* **199**, 34–45.

Youderian, P., Sugiono, P., Brewer, N.P. Higgins, N.P. and Elliot, T. (1988). Packaging specific segments of the *Salmonella* chromosome with locked-in Mud-P22 prophages. *Genetics* **118**, 581–592.

Youngman, P., Perkins, J.B. and Losick, R. (1984). A novel method for the rapid cloning in *Escherichia coli* of *Bacillus subtilis* chromosomal DNA adjacent to Tn*917* insertions. *Mol. Gen. Genet.* **195**, 424–433.

Youngman, P., Zuber, P., Perkins, J.B., Sandman, K., Igo, M. and Losick, R. (1985). New ways to study developmental genes in spore-forming bacteria. *Science* **228**, 285–291.

Zagorec, M. and Steinmetz, M. (1991). Construction of a derivative of Tn*917* containing an outward-directed promoter and its use in *Bacillus subtilis*. *J. Gen. Microbiol.* **137**, 107–112.

Zerbib, D., Polard, P., Escoubas, J.M., Galas, D. and Chandler, M. (1990). The regulatory role of the IS1-encoded InsA protein in transposition. *Mol. Microbiol.* **4**, 471–477.

Zhang, L.J. and Skurnik, M. (1994). Isolation of an R$^-$M$^+$ mutant of *Yersinia enterocolitica* serotype 0–8 and its application in construction of rough mutants utilizing mini-Tn5 derivatives and lipopolysaccharide-specific phage. *J. Bacteriol.* **176**, 1756–1760.

Zhou, M. and Reznikoff, W.S. (1997). Tn5 transposase mutants that alter DNA binding specificity. *J. Mol. Biol.* **271**, 362–373.

Zink, R.T., Kemble, R.J. and Chatterjee, A.K. (1984). Transposon Tn5 mutagenesis in *Erwinia carotovora* subsp. *carotovora* and *Erwinia carotovora* subsp. *atroseptica*. *J. Bacteriol.* **157**, 809–814.

PART II
Case Studies

◆◆

5 Genetic Methods in *Neisseria*

T Schwan[1]**, T Rudel**[2] **and TF Meyer**[3]

[1]*Byk Gulden Pharmazeutika, FB3, Byk Gulden Str. 2, 78467 Konstanz, Germany,* [2]*Max Planck Institute für Biologie, Abteilung Infektionsbiologie, Spemannstrasse 34, Tübingen, Germany and Max Planck Institut für Infectionsbiologie, Abteilung Molekulare Biologie, Monbijoustr. 2, 10117 Berlin, Germany*

◆◆

CONTENTS

◆◆◆◆◆◆ **I. GENETIC INSTABILITY – A PERPETUAL THEME IN HANDLING THE PATHOGENIC *NEISSERIA* SPECIES**

A prominent feature of the pathogenic *Neisseria* species is the phenotypic variation of surface structures implicated in the molecular interplay between pathogen and host. This phenotypic variation is often referred to as phase and antigenic variation or, addressing the mechanistic basis, as genetic variation. Phase variation can be defined as the successive loss and reappearance of a detectable structure (domain, molecule, organelle), while antigenic variation is an immunologically detectable change of composition of a given structure. Depending on the variation mechanism, the frequency of variation in *Neisseria gonorrhoeae* is in the range of 10^{-2} to 10^{-3} per gene (Schneider *et al.*, 1988). As the frequency of spontaneous mutations is approximately 10^{-7}, this difference and the fact that genetic variation is a reversible process implies the existence of distinctive mechanisms at distinct genetic sites (Robertson and Meyer, 1992).

Thoroughly described examples of phase-variable structures of both pathogenic neisserial species include type 4 pili, proteinaceous appendages protruding from the cell surface. Type 4 pilus biogenesis in *Neisseria* constitutes an important requisite for the ability to adhere to host cells, and for competence of DNA transformation. Pili are composed of one main subunit, pilin, which is encoded by the chromosomal pilin expression locus *pilE*. The silent, non-expressed *pilS* loci contain partial gene copies encoding variant sequences of pilin. Intragenic recombination events between *pilE* and *pilS* cause both the on and off switching of the piliation phenotype and the structural variation of pilin. At least three pathways exist by which *pil* genes can recombine, one of which involves transformation with species-specific DNA, while other pathways involve intracellular events (Meyer *et al.*, 1994; Seifert *et al.*, 1996). Both, intragenic recombination and homologous recombination using exogenous DNA are RecA-dependent processes. Prolonged cultivation of *Neisseria* (> 1 day) results in increased autolysis of bacteria and subsequent liberation of chromosomal and plasmid DNA. The natural transformation competence of *Neisseria* depends on the recognition of a specific DNA uptake signal (5'GCCGTCTGAA3'; Goodman and Socca, 1988) and the formation of pili (Fussenegger *et al.*, 1997).

Another mechanism exploited to create genetic diversity in *N. gonorrhoeae* is the use of short repeated sequence motifs at the 5' end of genes and within promoter sequences. The repeated motif can be as small as a single base pair, resembling a homopolymeric repeat. Loss or gain of single repeat units confers phase variation by a process termed slipped-strand mispairing. Slipped-strand mispairing can induce frameshift mutations which lead to premature transcriptional termination. Examples for homopolymeric repeats are given by *pilC* (Jonsson *et al.*, 1991), a gene encoding the pilus-associated adhesin (Rudel *et al.*, 1995b), and certain *lgt* genes involved in the biosynthesis of lipopolysaccharide (Gotschlich, 1994). In both cases, poly-G tracts at the 5' coding region of genes account for reversible on and off switching of functional translation. A more complex repetitive motif is present in *opa* genes coding for opacity (Opa) proteins. The Opa proteins constitute a family of outer membrane proteins which determine the cellular tropisms of *N. gonorrhoeae* to various host tissues. Opa proteins are encoded by a constitutively transcribed multi-gene family of 11 alleles scattered around the chromosome of *N. gonorrhoeae*, and 3–4 alleles in *N. meningitidis*. The number of pentanucleotide coding repeat units of the sequence 5'CTTCT3' encoding the hydrophobic part of the leader peptide is subject to phase variation and determines the expression state (Stern *et al.*, 1986). Other examples of variable repeats include the outer membrane proteins PorA (van der Ende *et al.*, 1995) and Opc (Sarkari *et al.*, 1994) of *Neisseria meningitidis*. In both cases variation in the number of bases within a poly-C tract between the −35 and −10 region regulates promoter activity. Common to these alterations in the number of repeat units is the independence of RecA function.

Phase variation can be conveniently monitored by viewing the colony morphology with a stereo microscope. Plating serial dilutions of *Neisseria* gives rise to distinct single colonies varying in shape and transparency. Variable oblique substage lighting reveals colony phenotypes ranging from clear, fully translucent colonies (O⁻) to strongly opaque (O⁺) ones. This opacity phenotype largely correlates with the absence and presence of Opa proteins in the outer membrane of *N. gonorrhoeae*. In *N. meningitidis* the opacity phenotype is hidden by the expression of a polysaccharide capsule. Another colony phenotype correlates with the pilus formation. Colonies consisting of piliated bacteria (P⁺) differ in shape from non-piliated (P⁻) colonies. Due to enhanced autoagglutination piliated bacteria produce bulkier colonies covering smaller agar surface areas, thus producing a typical shade at the colony border. Non-piliated bacteria, in contrast, give rise to flat colonies covering larger surface areas lacking the shade typical of P^+ bacteria. Since only P^+ variants are naturally competent for DNA uptake, monitoring this aspect of colony phenotype is crucial for transformation. P^+ bacteria have to be selected for prior to transformation experiments using *Neisseria* as recipients for the uptake of extraneous DNA. Generally, all possible combinations of variable phenotypes can be selected for. Selection of specific phenotypes requires visual differentiation of variants, which is dependent upon reproducible handling and conditions, including the choice of appropriate media for cultivating bacteria. Regarding piliation and opacity, P^-O^-, P^-O^+, P^+O^-, and P^+O^+ colonies can be identified and variant bacteria be easily enriched by streaking single colonies on agar plates. Since the number of variable genes is constantly growing, understanding underlying genetic mechanisms leading to phase and antigenic variation is essential.

◆◆◆◆◆◆ II. CULTIVATING *NEISSERIA* – FIRST STEPS IN HANDLING A SENSITIVE BUG

Cultivating *Neisseria* is generally not connected with major problems. However, sporadic difficulties may arise resulting from subtle changes of laboratory and management routine, initially not noticed by practitioners. To circumvent such situations it is recommended to take certain common practice precautions. This may imply control of water quality, testing charges of growth media and checking for residual detergents in glassware used for preparing media or growing bacteria. In the case of doubt it may be useful to rinse flasks successively with 0.1% HCl and deionized water or with liquid growth medium prior to inoculation of liquid cultures. Ultrapure water used for preparing media is produced by a Milli-Q PF Plus system (Millipore), the performance of which is controlled by electric resistance measurement. In addition, such high quality water provides the basis of buffers used for storage or in restriction/modification reactions with nucleic acids as substrates. All other buffers used for analytical purposes such as gel electrophoresis are prepared with deionized water. Instead of the water drawn from a Millipore system, double-distilled water can be used as well. Media for cultivating and storing bacteria generally include liquid media, solid media, storage media, and supplements. Solid media are mostly liquid media solidified by the addition of agar.

A. Liquid Media

Culture preparation and handling of bacteria is carried out in a Biohazard hood to minimize the risk of contamination and for personal safety. Liquid medium is inoculated with a single colony picked using a sterile glass rod or a disposable plastic pipette and dipped into the medium. When multiple colonies are used for inoculation, respective colonies are harvested with a sterile cotton swab. Adherent bacteria are collected in a small volume of medium in a sterile 1.5 ml centrifuge tube, a small flask or any other vessel, allowing convenient and sterile handling of the resulting suspension. The latter can then be used for inoculating the actual growth medium within an Erlenmeyer flask. The total volume of this Erlenmeyer flask should be at least 5 times the volume of the culture. The culture is grown to the desired density at 37°C with agitation not exceeding 150 rpm. The concentration of bacteria in culture can be determined retrospectively by streaking serial dilutions on agar plates and counting resulting colonies after growth or actually by using a spectrophotometer and measuring OD_{550nm} of the undiluted or diluted suspension. Obtained values should be smaller than 1. Spontaneous lysis of the culture may occure at $OD_{550nm} > 0.8$.

I. PPM

PPM (proteose peptone medium) is a commonly used liquid growth medium. It provides the basis for liquid media used for culturing and storing *Neisseria*, also for use in transformation and conjugation. The components outlined below are made up with distilled H_2O to 1000 ml. If necessary pH is adjusted to 7.5. After autoclaving, liquid media are stored at 4°C. Prior to use 1% vitamin mix and 5 mM sterile filtered $NaHCO_3$ are added.

Proteose peptone no.3 (Difco Laboratories)	15 g
NaCl	5 g
Starch (soluble)	1 g
KH_2PO_4	1 g
K_2HPO_4	4 g

2. Transformation medium

PPM, 1% vitamin mix, 5 mM $NaHCO_3$, 10 mM $MgCl_2$, pH 7.5.

3. Conjugation medium

PPM, 1% vitamin mix, 5 mM $NaHCO_3$, 100 µg ml^{-1} DNase I, pH 7.5.

4. Storage medium

Make up 15–20% glycerol in PPM. Optionally, 10 mM $MgCl_2$ can be added in order to stabilize the bacterial outer membrane.

B. Solid Media

The method of choice for maintainence of *Neisseria* in live culture during experimentation is daily streaking of the bacteria on fresh agar plates using a heat-sterilized inoculation loop or a glass rod. Since autolysis is a pronounced feature, it is essential to transfer bacteria on fresh, pre-warmed (>20°C) agar plates daily. Generally, individual cultures should not be grown for more than 16 to 18 h in order to prevent spontaneous lysis and excessive recombination by horizontal exchange. Streaking on solid media allows isolation of single colonies thus enabling selection of specific phenotypes. Single colonies can be obtained by virtue of sector-ized streaking or, more efficiently, by streaking serial dilutions, followed by overnight growth. However, selecting and streaking single colonies will never ensure obtaining 100% homogenous cultures for any variable trait. Phase variation implies that every selected culture contains 0.1–1% heterogeneity for a given variable trait.

1. GC agar plates

For agar plates, 36 g GC agar base (Creatogen Biosciences) and 15 g pro-teose peptone No 3 (Difco Laboratories) is made up to 1000 ml with dis-tilled water and autoclaved separately from additional supplements. After the agar has cooled down to 50°C, 1% vitamin mix and appropriate volumes of additionally required stock solutions (e.g. antibiotics) are added. A magnetic stirring bug can be added to the medium before auto-claving to facilitate easier mixing.

2. PPM agar plates

Solid media based on GC agar base tend to be somewhat turbid. This may occlude colony morphology especially if subtle differences have to be judged. To obtain more translucent plates especially enhancing identifica-tion of different opacity phenotypes, liquid PPM medium can be solidi-fied by adding 1.5% Bacto agar (Difco Laboratories). As for GC agar plates, 1% vitamin mix is added shortly before pouring the plates.

C. Supplements

Supplements are prepared separately to avoid chemical decomposition of thermolabile substances during autoclaving and because they represent additives only required for specific strains or applications. Prepared as concentrated stock solutions, they should be sterilized by filtration through a 0.2 μm filter. Non-aqueous solutions based on ethanol do not need to be sterilized.

1. CO$_2$ source

Aqueous solutions of NaHCO$_3$ deliver an appropriate CO$_2$ source for proper growth of *Neisseria* in liquid media. 200 × concentrated stock is prepared as an 8.4% solution. Just before inoculating liquid media with bacteria, add 5 ml^{-1} of sterile filtered stock.

2. Vitamin mix

Vitamin mix is an essential supplement to liquid and solid media used for growing *Neisseria*. It is composed of an intricate mixture of different compounds and due to different solubility it is prepared initially as two separate solutions. Subsequently both solutions are mixed, the volume adjusted to 1000 ml, the mix sterilized by filtration through a 0.2 μm filter and aliquoted into 50 ml aliquots for storage at -20°C. Vitamin mix is sensitive to light.

Ingredients of solution 1 are dissolved in approximately 500 ml distilled water:

D(+)-Glucose monohydrate	100 g
L-Glutamine	10 g
L-Cysteine hydrochloride monohydrate	26 g
Cocarboxylase (chloride)	100 mg
Fe(NO$_3$)$_3$ x 9H$_2$O	20 mg
Vitamin B$_1$ hydrochloride	3 mg
β-Nicotinamide adenine dinucleotide (NAD)	250 mg
Vitamin B$_{12}$	10 g

Solid components of solution 2 are stirred into a mixture of 300 ml distilled water and 15 ml 32% HCl:

L-Cystine	1.1 g
Adenine	1.0 g
Guanine hydrochloride	30 mg
Uracil	500 mg
L-Arginine monohydrochloride	150 mg
4-aminobenzoic acid	13 mg

3. Antibiotic concentrations used for recombinant *N. gonorrhoeae* strain MS11

Depending on individual strain properties and characteristics of the genetic construct, antibiotic concentrations may be subject to change. For specific constructs please refer to concentrations cited in literature.

Chloramphenicol	10 μg ml^{-1}
Erythromycin	7 μg ml^{-1}
Kanamycin	40 μg ml^{-1}
Tetracycline	10 μg ml^{-1}

◆◆◆◆◆◆ III. GENETIC METHODS – POWERFUL TOOLS FOR MANIPULATING PATHOGENIC *NEISSERIA* SPECIES

A. Plasmid preparation

Because of the lack of easy to handle, autonomously replicating plasmids, extraction of recombinant plasmids from *Neisseria* is generally not a routine laboratory technique. However, if neisserial plasmids such as the cryptic plasmid need to be isolated, this can easily be achieved by the method outlined below. The boiling method described below is a modification of the original protocol of Holmes and Quigley (1981). Other techniques have also successfully been applied to purify large conjugative plasmids, e.g. p*tet*M25.2 (Kupsch *et al.*, 1996), from *N. gonorrhoeae*.

1. Suspend pelleted bacteria in 600 μl STET (8% sucrose, 5% Triton X-100, 50 mM EDTA pH 8.0, 50 mM Tris-HCl pH 8.0).
2. Prepare a fresh solution of lysozyme in STET (10 mg ml⁻¹). Add 30 μl solution to bacterial suspension and incubate 5 to 10 min on ice.
3. Lyse bacteria by boiling 2 min at 100°C.
4. Sediment cell debris and chromosomal DNA by centrifugation at 12 000 × g for 10 min at room temperature in a bench-top microfuge.
5. Remove pellet with a sterile toothpick.
6. Precipitate nucleic acids by adding 400 μl 2-propanol and place 10 min at −20°C.
7. Collect plasmids by centrifugation at 12 000 × g for 15 min at room temperature.
8. Dry pellet and dissolve in 100 μl water.
9. Remove co-precipitated proteins by standard phenol extraction.
10. Collect the upper aqueous phase (100 μl). Add 10 ml 3 M sodium acetate (pH 5.2) and 2.5 volumes absolute ethanol (275 μl). Vortex and incubate 5 to 10 min on ice.
11. Collect precipitated DNA by centrifugation at 12 000 × g for 15 min at room temperature. Wash the pellet with 1 ml 70% ethanol and recentrifuge at 12 000 × g for 5 min at room temperature in a microfuge.
12. Dry pellet at 37°C and dissolve nucleic acids in 30 to 50 μl of water.

B. Preparation of Chromosomal DNA

Isolation of whole genome DNA provides the basis for a variety of experiments. Dissecting gene function by the construction of genomic mutants or complementation studies may critically depend on the availability of high-quality preparations of chromosomal DNA. If the locus to be manipulated is not accessible by the polymerase chain reaction, protocols relying on the subsequent lysis of bacteria and purification of DNA have to be followed. The following section will outline three methods for the isolation of chromosomal DNA which might be slightly modified to obtain best results for the specific strain used.

1. Total lysis (Laible *et al.*, 1989)

1. Harvest bacteria with a cotton swab from three agar plates and resuspend in 6 ml PBS.
2. Add 6 ml 2× cholate buffer (0.1 M EDTA pH 8.0, 0.005% Na-deoxycholate, 0.05% SDS). Incubate 5 min at room temperature until a clear solution forms.
3. Recover DNA by gently mixing with three volumes absolute ethanol (36 ml) and precipitate 5 min at room temperature.
4. Centrifuge 3000 rpm for 5 min at room temperature in a Hermle centrifuge.
5. Recentrifuge pellet with 10 ml 70% ethanol.
6. Discard supernatant and dissolve nucleic acids in 10 ml sodium citrate buffer (0.03 M sodium citrate pH 9.5).
7. Add 80 µg ml^{-1} proteinase K and incubate at 55°C for 2 h or overnight.
8. Remove residual contaminating proteins by phenol–chloroform extraction.
9. Precipitate nucleic acids with ethanol and collect by centrifugation.
10. Recentrifuge with 10 ml 70% ethanol for 5 min at 3000 rpm.
11. Dissolve pellet in an appropriate volume of water.
12. Digest with RNase A prior to further use.

2. Simple and rapid method (Chen and Kuo, 1993)

1. Pelleted bacteria from approximately 1.5 ml stationary phase culture are resuspended in 200 µl lysis buffer (40 mM Tris-acetate pH 7.8, 20 mM sodium acetate, 1 mM EDTA, 1% SDS) by vigorous pipetting.
2. RNA is degraded by incubating with RNase A for 30 min at 37°C.
3. Mix with 66 µl 5 M NaCl and centrifuge 10 min 14 000 rpm at 4°C.
4. Add equal volume chloroform to supernatant and mix by inverting at least 50 times.
5. Centrifuge 3 min 14 000 rpm.
6. Precipitate supernatant with ethanol.
7. Wash twice with 70% ethanol and dry pellet.
8. Dissolve DNA pellet in an appropriate volume of water.

3. CsCl preparation

1. Harvest overnight grown bacteria from half an agar plate and suspend in 1 ml Tris/EDTA (100 mM Tris-HCl pH 8.5, 100 mM EDTA).
2. Gently mix with 100 µl 10% SDS until a clear solution forms.
3. Add 1.5 ml Tris/EDTA and 80 µg ml^{-1} proteinase K. Incubate 2 h at 55°C.
4. Extract clear solution twice with one volume phenol and once with chloroform.
5. Dissolve 2.5 g CsCl in bacterial lysate and add 40 µl 10 mg ml^{-1} ethidium bromide.
6. Using a disposable syringe and a hypodermic 19G needle place the sample into a quickseal tube (Beckman). The remainder of the tube is filled with 1 g ml^{-1} CsCl solution in Tris/EDTA. Seal the tube.
7. Centrifuge 4 h at 100 000 rpm and 20°C in a TLA 100.3 angle rotor using a Beckman Optima TLX ultracentrifuge.

8. Collect DNA with a 19G needle and remove ethidium bromide by extraction with water-saturated 1-butanol.
9. CsCl can be removed by overnight dialysis against repeatedly changed 10 mM Tris-HCl, 1 mM EDTA. However, a more time-saving method is differential ethanol precipitation and subsequent collection of DNA by centrifugation.
10. Precipitate DNA with 100% ethanol and wash pellet twice with 70% ethanol.
11. Air-dry pellet and redissolve in water.

C. Transformation of *Neisseria*

Transformation of bacteria with extraneous DNA is an essential method in order to modify and assess genetic properties. It further provides the basis for amplification and cloning of those DNA molecules used in genetic engineering and transformation procedures. Two broadly different methods for transformation of Gram-negative bacteria with DNA have been elaborated. The classical procedure, also termed chemical transformation, relies on a preliminary treatment of bacteria at low temperature in a solution of divalent cations. Subsequent incubation of bacteria with nucleic acids and application of a certain heat regimen leads to DNA uptake. Electroporation delivers a second technique to temporarily induce transformation competence. Highly concentrated bacteria take up target DNA from solution if they are subjected to a high-voltage electric field (see Chapter 1, III.D.5). Preceding treatment of cells is limited to the preparation of a bacterial suspension of very low ionic strength. To date no protocol for the electroporation of *Neisseria* species has been published.

Pathogenic *Neisseria* are naturally competent for the uptake and chromosomal integration of DNA (see Chapter 1, II.C.2.(b)). No special treatment of cells is required to induce competence. One prerequisite to be fullfilled is the expression of pilin since transformation competence and type 4 pilus biogenesis at a certain stage seem to be functionally linked processes. As outlined earlier, DNA uptake is governed by certain aspects of sequence specificity (Goodman and Socca, 1988) with extensions to the range of both donor and recipient species (Frosch and Meyer, 1992). Horizontal exchange of genetic material and selection in the host contributes to adaptive radiation of strains leading to the evolution of neisserial species. Furthermore, naturally occurring transformation-mediated recombination is thought to have pronounced influence on intraspecies variation. Interspecies transfer of genetic markers and certain intraspecies variation like pilin variation demonstrate strong DNase I sensitivity, an observation which contributes to the significance of both processes.

Functional transformation can lead to the establishment of replicating episomes or chromosomal recombinants. Different to *E. coli*, construction of neisserial integrative recombinants does not require special genotypes to be used. Homologous recombination is a matter of a simple transformation experiment which resembles cultivating bacteria in liquid media in the presence of donor DNA. This procedure readily improves studying gene function by facilitating the construction of isogenic strains differing

179

only in the nucleic acid fragment previously cloned and manipulated. With two exceptions, wild-type *Neisseria* contain a multiplicity of different plasmids but no generally used shuttle vectors or cloning vectors that stably replicate in *Neisseria* have been published. Exceptions are the vectors pLES2 and pLEE10 introduced by D. Stein, the latter of which proved useful in the preparation of a mini-gene bank in *N. gonorrhoeae* and subsequent phenotypic selection (Stein *et al.*, 1983; Sandlin *et al.*, 1993). The second system provides a functional combination between integrative recombination and freely replicating episomes (Kupsch *et al.*, 1996). A series of shuttle vectors, termed Hermes, can be used for maintenance and manipulation of *Neisseria* genes within *E. coli*. Subsequent transformation into a neisserial donor containing the conjugative plasmid p*tet*M25.2 leads to the formation of stable co-integrates between both plasmids. These can then be transferred conjugation between different *Neisseria* strains. Conjugation proves to be a very efficient process possibly because DNA is transferred in a single-stranded state not providing a target for strain-specific restriction-modification systems. During transformation the majority of DNA molecules are taken up into the cytoplasm in a double-stranded DNase-resistant form, also in the case of circular plasmid DNA (Biswas *et al.*, 1986). These randomly cleaved linear molecules can then be religated to form circular plasmids or integrated into the chromosome. Transformation of *Neisseria* is essentially dependent on functional RecA activity.

1. Transformation by transient cultivation in liquid culture

1. Harvest piliated *Neisseria* from a GC plate with sterile cotton swab and resuspend in transformation medium (PPM, 1% vitamin mix, 10 mM $MgCl_2$).
2. Dilute suspension 1 : 20 and measure OD_{550nm}.
3. Dilute to OD_{550nm} = 0.2.
4. Pierce holes in the lid of a 1.5 ml micro-centrifuge tube with a hot hypodermic needle to prepare the transformation vessel. Incubate in it 250 µl bacteria and <1 µg DNA for 4 h at 37°C in 5% CO_2. Alternatively, transformation can be performed within a small Erlenmeyer flask which is gently shaken for 4 h at 37°C.
5. Plate 100 µl and 100 µl 1:10 dilution (10^{-1}) of the culture on GC plates containing appropriate antibiotic. For calculation of transformation efficiency plate 100 µl aliquots of 10^{-5} and 10^{-6} dilutions on GC plates without antibiotic.
6. Incubate plates overnight at 37°C in 5% CO_2.

2. Spot dilution method (Drazek *et al.*, 1995)

1. Plate serial dilutions of transformation proficient *Neisseria* on GC plates and allow suspensions to be soaked into agar.
2. Spot prewarmed solution containing 1 µg donor DNA on dried bacterial suspension and incubate overnight.
3. Transfer colonies on new GC plates from an area where DNA was spotted.
4. Examine transformants.

D. Conjugation of *Neisseria*

Besides transformation, many bacterial species have evolved additional systems to efficiently exchange genetic material. The importance for *in vivo* and *in vitro* conjugation of pathogenic *Neisseria* is a clearly established fact. However, transduction, the phage-mediated transfer of non-viral genetic information between bacterial cells, could not be demonstrated. To date no bacteriophages that infect *Neisseria* have been identified. Thus conjugation provides the second important means for DNA exchange exploited by *Neisseria* which could be successfully adapted for *in vitro* manipulation.

Conjugation is a process of DNA transfer from a donor to a recipient cell that is dependent on close physical contact between the exchanging cells (see Chapter 2, V.A and B). Conjugal proficiency of the donor is determined through the presence of a self-transmissible plasmid that is able to disseminate to other bacteria. A multiplicity of conjugative plasmids were isolated from pathogenic *Neisseria*, some of them conferring resistance to antibiotics. In addition to self-transmissible plasmids, systems were developed for the mobilization of plasmids lacking transfer functions. Biswas *et al.* (1980) reported conjugation of such a plasmid, pFA3, encoding β-lactamase activity. By virtue of the presence of a 36 kb conjugative plasmid, pFA2, high-frequency mobilization of pFA3 could be achieved. Interestingly, mobilization of chromosomal genes could not be detected. This might imply a very low frequency of recombination with the chromosome. On the other hand, detectable chromosomal recombination frequency might be achievable under conditions which have not yet been identified. Availability of conjugative plasmids led to the elaboration of suicide systems avoiding the apparent insufficiencies of the neisserial transformation system (Kupsch *et al.*, 1996). Formation of co-integrates between two plasmids permits the exploitation of neisserial conjugation for the conduction of functional studies and for the propagation of DNA sequences in a wild-type environment. Changing the cellular environment of a DNA molecule by transformation may have profound influence on its physical integrity although the host species may not be changed. Stein *et al.* (1988) have shown that interstrain barriers for the acquisition of foreign DNA by transformation exist. Differences in restriction/modification systems may partially prevent stable perpetuation of incoming DNA. However, if novel DNA is acquired by conjugation, no such difficulties occur. Thus transfer of DNA by conjugation provides an attractive technique because of avoiding problems caused by strain-specific restriction/modification systems. This may be explained by the failure of single-stranded DNA to serve as a substrate for restriction endonucleases.

I. *In vitro* conjugation by filter mating

1. Donor and recipient are harvested with a sterile cotton swab from agar plates and are separately resuspended in 1 ml conjugation medium.
2. Dilute to an OD_{550nm} of 0.15 and mix same volumes of suspensions.

3. Collect bateria by vacuum filtration onto nitrocellulose filters with 0.2 mm pores.
4. Incubate filters with bacteria on GC agar plates for 3 h at 37°C, 5% CO_2 in humidified atmosphere.
5. Using a cotton swab, resuspend bacteria in 2 ml liquid medium and prepare two successive 1:10 dilutions. Plate 100 µl each on appropriate selective agar plates and incubate overnight.

◆◆◆◆◆◆ IV. GENETIC ENGINEERING IN *NEISSERIA* – IDENTIFYING GENES BY FUNCTIONAL GENOMICS

A. Site-directed vs. Random Mutagenesis

The ability of *Neisseria* to import species-related DNA and the subsequent integration of incoming DNA into the chromosome via homologous recombination has greatly facilitated the generation of gene knock-out mutants. (A more detailed discussion of gene knockouts and integration via homologous recombination in to the chromosome is in Chapter 10, II, III and IV.) As a general approach, genes conferring antibiotic resistance are inserted into a cloned neisserial gene, amplified in *E. coli* host strains and then returned into *Neisseria*. Via the mechanism of integration, the wild-type gene is replaced by the mutated allele and the recombinant bacteria can be selected for by antibiotic selection. This allows the generation of isogenic strains differing only in the functional integrity of the gene or region of interest. Resistance genes which have successfully been used with this approach include the β-lactamase (*bla*) gene (Koomey *et al.*, 1982), chloramphenicol-acetyl-transferase (*cat*) gene (Haas *et al.*, 1993), the *ermC* gene that confers erythromycin resistance (Projan *et al.*, 1987) and the *apha-3* gene from Tn*1545* for kanamycin resistance (Trieu-Cuot *et al.*, 1985). The resistance markers differ in their effectiveness; while *cat* and *ermC* are excellent markers with only very low background growth of untransformed bacteria, selection under kanamycin results sometimes in the isolation of spontaneously resistant colonies. This requires careful verification of the mutation not only during the isolation of a mutant clone but also during all the subsequent passages. Although this phenomenon is not fully understood, increased kanamycin resistance seems to be connected to the switch to an uptake-resistant phenotype involving loss of pili formation (Gibbs *et al.* 1989). The *bla* gene marker confers only very low resistance to ampicillin or penicillin and has thus been replaced in most laboratories.

Single gene knock-outs are not always helpful to study pathogenic factors in *Neisseria*. Many genes are present in multiple copies and probably fulfil redundant functions, so that mutagenesis of several genes in the same strain has become an important goal. Still, the knock-out of a reasonable number of genes can be achieved by inserting different resistance markers consecutively into the loci of interest. This has been done for the *pilC* genes where both copies have been deleted (Jonsson *et al.*, 1991;

Nassif et al., 1994; Rudel et al., 1995b). Also, three gene knock-outs have been created by consecutively transforming with cat genes conferring low, then high levels of chloramphenicol resistance (Cam) and the ermC' gene (Rudel et al., 1995a). In theory, the generation of a mutant with defects in up to four genes is possible by using the whole panel of resistance genes available.

However, the limitation of this approach becomes dramatically apparent in light of the fact that there are, for example, 11 opa genes and 7 pil gene loci containing 17 partial pil copies in gonococcal strain MS11. The ideal solution would be a "recyclable" knock-out system and this has been solved in two totally different ways. The first system depends on a "jumping" resistance gene which can be mobilized after insertion into the chromosome of Neisseria. So-called resolution sites (res) from transposon Tn1721 flank the resistance gene and, under the control of an inducible promoter, the gene for the enzyme resolvase. Upon induction, the resolvase mobilizes the res-flanked marker–resolvase cassette and a single res site remains left over from the whole process. The system is now ready for the next round of insertional/deletional mutagenesis. One problem with the "res system" might be that several inserted res sites render the chromosome assessable for rearrangements induced by the powerful resolvase.

The second system, first described by Stibitz et al. (1986) for mutagenesis in Bordetella and later adapted for Neisseria (Johnston et al., 1996), depends on the use of a two-gene cassette for the knock-out containing a selectable and a counter-selectable marker. The first step involves selection for the integration of the cassette into the chromosome by Erm resistance. The counter-selectable marker codes for an allele of the ribosomal S12 protein which renders the knock-out mutant streptomycin (SM) sensitive. Now, a second transformation is performed using a deletion derivative or linker insertion of the mutated gene and the replacement of the Erm^r/SM^s cassette by selection on SM. The resulting strain exhibits the same resistance phenotype as the parental strain but now harbours a deletion or insertion in the gene of interest.

A very common way to identify unknown factors includes generalized mutagenesis to disrupt the function of a factor followed by selection for the defect of interest. Chemical mutagenesis has been used but is very inefficient (Campbell and Yasbin, 1984). In addition, most of the transposons from enterobacteria do not function well in Neisseria. Tn916 (Kathariou et al., 1990) and Tn1543-Δ3 (Nassif et al., 1991) are mobile in Neisseria, however, these transposons are relatively large and show only low frequencies of transposition in the range of 10^{-7} in gonococci which complicates the generation of complete mutant libraries. Shuttle transposon mutagenesis has thus become an up-to-date technique in Neisseria. The term shuttle mutagenesis stands for a procedure where a cloned DNA fragment is mutagenized in E. coli and then reintroduced into Neisseria by natural transformation (see also Chapter 4). A variety of minitransposons have been developed for shuttle mutagenesis in Neisseria (Boyle and Seifert, 1993,1995; Haas et al., 1993; Kahrs et al., 1995; Mehr and Seifert, 1997). The following procedure describes how a complete mutated plasmid library can be generated by this technique. First, a genomic plasmid library of *Neisseria*, e.g. in the mobile minimal vector pMin1 (ori_{ColE1}; *oriT*; Tet^r) is transformed into an *E. coli* strain which expresses the fd-phage protein II and already harbours the cloned mini-transposon of the Tn*Max* series. The mobile unit of Tn*Max* transposons contains a resistance marker and the origin of replication of phage fd (ori_{fd})which allows replication in the presence of the phage protein II, flanked by inverted repeats. The transposase, the enzyme which mobilizes Tn*Max*, is cloned in the same plasmid under the control of the isopropyl-β-thio-galactoside (IPTG)-inducible P_{trc}-promoter. *E. coli* transformants now harbouring two replicating plasmids are plated on Amp/Cam and IPTG to induce transposition. The mutated target plasmids are now mobilized under appropriate selection into a *polA1* mutant *E. coli* strain which allows only replication by the Tn*Max*-borne ori_{fd}. This procedure results in a very strong selection for target plasmids which contain a copy of Tn*Max*, since the original Tn*Max* plasmid is not mobile and the target plasmid needs Tn*Max* for replication. Now the plasmids are isolated from individual *E. coli* clones, pooled and transformed into a *Neisseria* recipient strain to obtain a mutant library. *Neisseria* mutant libraries generated by this or a modified method (Mehr and Seifert, 1997) have successfully been used to identify factors involved in transformation competence (Fussenegger *et al.*, 1996), invasion into epithelial cells (Kahrs *et al.*, 1994) and pilin antigenic variation (Mehr and Seifert, 1997). The advantages of the transposon shuttle mutagenesis approach are simple techniques and a relatively high frequency of mutagenesis. Disadvantages are the sequence preference for insertion of some transposons and the possible loss of clones which exhibit a toxic effect if expressed in *E. coli*.

The random transposon shuttle mutagenesis approach described above and the *in vitro* random marker insertion approach explained below largely depend on the mechanism of DNA uptake by *Neisseria* which involves linearization and stable integration into the chromosome by homologous recombination. The same mechanism of DNA uptake is found in *Haemophilus influenzae* where the random marker insertion technique has been described first (Sharetzky *et al.*, 1991). Chromosomal DNA is partially digested and religated under conditions where the formation of circles is favoured. A second partial digest with a different enzyme opens the chromosomal DNA circles at different sides. Then, a resistance marker is inserted and the hybrid circles are used to directly transform *Neisseria* (Kupsch *et al.*, 1996). Dependent on the second digest the marker is embedded in a more (insertion) or less (deletion) continuous DNA fragment and thus can recombine into the homologous region of the chromosome.

1. Preparation of circular DNA molecules for mutagenesis

1. Partially digest 400 μg of chromosomal DNA with *Hpa*II.
2. Fractionate DNA on a sucrose gradient (15–30% sucrose, 1 M NaCl, 20 mM Tris-HCl, 5 mM EDTA, pH 8.0) for 24 h at 27 000 rpm in a SW41 rotor at 20°C.

3. Pool and precipitate the 10–20 kb fragments by addition of 2 volumes of ethanol.
4. Ligate the DNA at a concentration of 2 µg ml^{-1} for 48 h at 4°C.
5. Partially digest the circles with a second enzyme, e.g. *Mbol*.
6. Collect and purify the DNA over a column (e.g. Qiagen tip 20).
7. Ligate a resistance marker flanked by compatible restriction sites (e.g. *Bam*HI) and directly transform into *Neisseria*.

An advantage of the *in vitro* mutagenesis approach is its absolute randomness of insertion and the lack of any amplification step which might result in the loss of clones. However, perhaps due to the advanced technical skills required to create the mutated chromosomal DNA, this technique is used rarely compared to transposon shuttle mutagenesis.

An important issue in conjunction with mutagenesis is genetic complementation. A more commonly used procedure depends on the insertion of the gene of interest into a chromosomal region which is dispensable for the experiment. Researchers have successfully introduced genes into the *iga* locus and the *recA* locus under the control of a constitutive or an inducible promoter. A general design for a complementation unit includes a resistance marker for the selection adjacent to the gene of interest both flanked by targeting sequences of the locus where integration should occur. A disadvantage of the gene replacement approach for genetic complementation is the need for transformation competence. Many mutations, however, affect the natural transformation competence, such as mutations in pilin and pilus assembly factors, factors involved in DNA uptake and recombination. Therefore, genetic complementation by conjugal transfer of plasmids is a more applicable method. The so-called Hermes shuttle system has been successfully used to functionally complement transformation competent and non-competent gonococcal mutants (Kupsch *et al.*, 1996). The system works as follows. A Hermes vector consists of a plasmid backbone for replication in *E. coli* and a multiple cloning site (MCS) close to a resistance marker for selection. Both the marker and the MCS are framed by targeting sequences for the gonococcal conjugative plasmid p*tet*M25.2 mediating stable integration of any cloned gene into a permissive site of p*tet*M25.2.

2. Introduction of cloned genes into *Neisseria* by the Hermes shuttle system

1. Transformation of the recombinant Hermes vector into a competent gonococcal strain which already harbours the p*tet*M25.2 vector.
2. Antibiotic selection for recombinant p*tet*M25.2 carrying the transferable unit of Hermes.
3. Conjugative transfer of the recombinant p*tet*M25.2 into any gonococcal strain (Kupsch *et al.*, 1996). For selection of transconjugants, the recipient strain has to contain a selection marker which is not present in the donor strain. If necessary, recipient strains can be selected for spontaneous resistance for rifampicin.

B. Stabilization of Gene Expression

Neisseria display an extremely high frequency of variation in the expression of surface components. These phenotypic alterations result primarily from the antigenic and phase variation of pathogenic factors like pilin, Opa proteins and LPS. Controlled expression of individual variable factors is, however, essential for the study of their function in the context of the other factors. Two major mechanisms account for most of the variation events in *Neisseria*: homologous recombination controls predominantly the antigenic variation of pilin and slipped-strand mispairing the antigenic variation of LPS and the ON/OFF switches of Opa protein expression. Since homologous recombination depends on functional RecA protein, mutations in *recA* consequently prevent antigenic variation of pilin (Koomey *et al.*, 1987). The advantage of a *recA* knock-out mutation which prevents variation of the pilin protein is disadvantageous in that these mutants are not accessible to further mutagenesis by gene replacement. One solution to this problem offers inducible *recA* alleles where the original *recA* promoter has been replaced by a IPTG-inducible promoter controlled by the LacI repressor (Seifert, 1997; Carol P. Gibbs and TFM, unpublished). The *recA*-inducible strains constructed so far exhibit, under permissive conditions, a similar but not identical phenotype to the *recA* knock-out mutants. In contrast to *recA* knock-outs, the natural transformation competence and pilin antigenic variation is drastically reduced but not totally prevented. The decision whether to use the more secure knock-out or the more flexible inducible system depends on the specific requirements of the experiment.

The variation of gene expression by alterations in repetitive sequence has been recognized as a commonly occurring mechanism. The mechanism is believed to involve slipped-strand mispairing, a slippage of the DNA-dependent DNA polymerase leading to either small deletions or insertions in the repetitive sequence independent of RecA. As a consequence the reading frames of genes are changed or promoter elements are modified. In order to stabilize this type of variation each gene has to be modified separately. Since the main reason for the slippage of the polymerase is the repetitive sequence, introduction of silent mutations which destroy the repetitive character of the sequence stabilize the expression significantly. Using this approach, the expression of Opa proteins (Kupsch *et al.*, 1993), PilC proteins (Rudel *et al.*, 1995a, 1995b) and the Lgt proteins (Yang and Gotschlich, 1996) was successfully stabilized.

◆◆◆◆◆◆ V. PROSPECTS OF GENOME SEQUENCING – *NEISSERIA IN SILICO*

Analysing whole genome sequences provides a systematic approach to study gene function. Initially, bioinformatics makes use of homology searches to identify genes and thereby potentially encoded functions through the identification of significant sequence identities. Such searches performed on an intra- and intergenomic scale have clearly emphasized the anticipation of a highly flexible genome. Within the neisserial genomes a whole wealth of past and present time genome rearrangements have been identified. Paralogous genes originating from local or large-scale genomic duplications and individual mutation tell of intragenomic flexibility. Mosaicism resulting from the exchange of protein-coding fragments reveals the modular organization of individual open reading frames and corresponding proteins. However, a real exploitation of genome rearrangements resulting in a direct benefit for the individual without being previously subject to selection is achieved by antigenic variation. Random recombination of an expression locus with a linear array of repeated, structurally different peptide-encoding cassettes leads, for example, in the case of pilin variation to a vast repertoire of antigenically different proteins. Each of these examples of genome flexibility has been identified by the accurate analysis of DNA sequences. Availability of the complete genomes of both pathogenic *Neisseria* species will systematically speed up this process of identification and annotation.

However, availability of an increasing amount of prokaryotic and eukaryotic genome sequences will not replace actual biochemical and cell biological experiments. But it will provide life sciences with a new and very powerful access to address basic biological questions on how living matter is organized and functions at a molecular level. One striking finding of genomes analysed to date is the unprecedented high percentage of putative genes sharing no similarity to any other gene deposited in sequence databases. In the case of the *B. subtilis* genome (Kunst *et al.*, 1997) one-quarter of presumably functional open reading frames belongs to this group of unassigned sequences. The significance of such uncharacterized ORFs is clearly strengthend by their assignment to families of orthologues sharing members in different prokaryotes. Whole operons have been found conserved between sequenced genomes that code for as-yet elusive functions. Most excitingly, this could mean that present-day science is expecting to discover completely new areas of molecular biology dealing with phenotypes of unknown impact on human society.

References

Biswas, G.D., Blackman, E.Y. and Sparling, P.F. (1980). High-frequency conjugal transfer of a gonococcal penicillinase plasmid. *J. Bacteriol.* **143**, 1318–1324.

Biswas, G.D., Burnstein, K.L. and Sparling, P.F. (1986). Linearization of donor DNA during plasmid transformation in *Neisseria gonorrhoeae*. *J. Bacteriol.* **168**, 756–761.

Boyle-Vavra, S. and Seifert, H.S. (1993). Shuttle mutagenesis: two mini-transposons for gene mapping and for *lacZ* transcriptional fusions in *Neisseria gonorrhoeae*. *Gene* **129**, 51–57.

Boyle-Vavra, S. and Seifert, H.S. (1995). Shuttle mutagenesis: a mini-transposon for producing PhoA fusions with exported proteins in *Neisseria gonorrhoeae*. *Gene* **155**, 101–106.

Campbell, L.A. and Yasbin, R.E. (1984). Mutagenesis of *Neisseria gonorrhoeae*: absence of error-prone repair. *J. Bacteriol.* **160**, 288–293.

Chen, Wen-ping and Kuo, Tsong-teh (1993) A simple and rapid method for the preparation of Gram-negative bacterial genomic DNA. *Nucl. Acids Res.* **21**, 2260.

Drazek, E.S., Stein, D.C. and Deal, C.D. (1995) A mutation in the *Neisseria gonorrhoeae rfaD* homolog results in altered lipooligosaccharide expression. *J. Bacteriol.* **177**, 2321–2327.

Frosch, M. and Meyer, T.F. (1992) Transformation-mediated exchange of virulence determinants by co-cultivation of pathogenic *Neisseriae*. *FEMS Microbiol. Lett.* **79**, 345–349.

Fussenegger, M., Facius, D., Meier, J. and Meyer, T.F.(1996). A novel peptidoglycan-linked lipoprotein (ComL) that functions in natural transformation competence of *Neisseria gonorrhoeae*. *Mol. Microbiol.* **19**, 1095–1105.

Fussenegger, M., Rudel, T., Barten, R., Ryll, R. and Meyer, T.F.(1997). Transformation competence and type-4 pilus biogenesis in *Neisseria gonorrhoeae* – a review. *Gene* **192**, 125–134.

Gibbs, C.P., Reimann, B.Y., Schultz, E., Kaufmann, A., Haas, R. and Meyer, T.F. (1989). Reassortment of pilin genes in *Neisseria gonorrhoeae* occurs by two distinct mechanisms. *Nature* **338**, 651–652.

Goodman, S.D. and Scocca, J.J. (1988). Identification and arrangement of the DNA sequence recognized in specific transformation of *Neisseria gonorrhoeae*. *Proc. Natl. Acad. Sci. USA* **85**, 6982–6986.

Gotschlich, E.C. (1994). Genetic locus for the biosynthesis of the variable portion of *Neisseria gonorrhoeae* lipooligosaccharide. *J. Exp. Med.* **180**, 2181–2190.

Haas, R., Kahrs, A.F., Facius, D., Allmeier, H., Schmitt, R. and Meyer, T.F. (1993). Tn*Max* – a versatile mini-transposon for the analysis of cloned genes and shuttle mutagenesis. *Gene* **130**, 23–31.

Holmes, D.S. and Quigley, M. (1981). A rapid boiling method for the preparation of bacterial plasmids. *Anal. Biochem.* **114**, 193.

Johnston, D.M., Isbey, S.F., Snodgrass, S.F., Apicella, M., Zhou, D. and Cannon, J.G. (1996). A strategy for constructing mutant strains of *Neisseria gonorrhoeae* containing no antibiotic resistance markers using a two gene cassette with selectable and counterselectable markers, and its use in constructing a *pgm* mutant for use in human challenge trials. Abstracts of the 10th International Pathogenic *Neisseria* Conference 37 (Abstract).

Jonsson, A.B., Nyberg, G. and Normark, S. (1991). Phase variation of gonococcal pili by frameshift mutation in *pilC*, a novel gene for pilus assembly. *EMBO J.* **10**, 477–488.

Kahrs, A.F., Bihlmaier, A., Facius, D. and Meyer, T.F. (1994). Generalized transposon shuttle mutagenesis in *Neisseria* gonorrhoeae: a method for isolating epithelial cell invasion-defective mutants. *Mol. Microbiol.* **12**, 819–831.

Kahrs, A.F., Odenbreit, S., Schmitt, W., Heuermann, D., Meyer, T.F. and Haas, R. (1995). An improved Tn*Max* mini-transposon system suitable for sequencing, shuttle mutagenesis and gene fusions. *Gene* **167**, 53–57.

Kathariou, S., Stephens, D.S., Spellman, P. and Morse, S.A. (1990). Transposition of Tn*916* to different sites in the chromosome of *Neisseria meningitidis*: a genetic tool for meningococcal mutagenesis. *Mol. Microbiol.* **4**, 729–735.

Koomey, M.J., Gill, R.E. and Falkow, S. (1982). Genetic and biochemical analysis of gonococcal IgA1 protease: cloning in *Escherichia coli* and construction of mutants of gonococci that fail to produce the activity. *Proc. Natl. Acad. Sci. USA* **79**, 7881–7885.

Koomey, M., Gotschlich, E.C., Robbins, K., Bergstrom, S. and Swanson, J. (1987). Effects of *recA* mutations on pilus antigenic variation and phase transitions in *Neisseria gonorrhoeae*. *Genetics* **117**, 391–398.

Kunst, F. *et al.* (1997). The complete genome sequence of the Gram-positive bacterium *Bacillus subtilis*. *Nature* **390**, 249–256.

Kupsch, E.M., Knepper, B., Kuroki, T., Heuer, I. and Meyer, T.F. (1993). Variable opacity (Opa) outer membrane proteins account for the cell tropisms displayed by *Neisseria gonorrhoeae* for human leukocytes and epithelial cells. *EMBO J.* **12**, 641–650.

Kupsch, E.-M., Aubel, D., Gibbs, C.P., Kahrs, A.F., Rudel, T. and Meyer, T.F. (1996) Construction of Hermes shuttle vectors: a versatile system useful for genetic complementation of transformable and non-transformable *Neisseria* mutants. *Mol. Gen. Genet.* **250**, 558–569.

Laible, G., Hakenbeck, R., Sicard, M.A., Joris, B. and Ghuysen, J.M. (1989) Nucleotide sequences of the *pbpX* genes encoding the penicillin-binding proteins 2x from *Streptococcus pneumoniae* R6 and a cefotaxime-resistant mutant, C506. *Mol. Microbiol.* **3**, 1337–1348.

Mehr, I.J. and Seifert, H.S. (1997). Random shuttle mutagenesis: gonococcal mutants deficient in pilin antigenic variation. *Mol. Microbiol.* **23**, 1121–1131.

Meyer, T.F., Pohlner, J. and van Putten, J.P.M. (1994). Biology of the pathogenic *Neisseriae. Curr. Top. Microbiol. Immunol.* **192**, 283–317.

Nassif, X., Puaoi, D. and So, M. (1991). Transposition of Tn*1545*-Δ3 in the pathogenic *Neisseriae*: a genetic tool for mutagenesis. *J. Bacteriol.* **173**, 2147–2154.

Nassif, X., Beretti, J.L., Lowy, J., Stenberg, P., O'Gaora, P., Pfeifer, J., Normark, S. and So, M. (1994). Roles of pilin and PilC in adhesion of *Neisseria meningitidis* to human epithelial and endothelial cells. *Proc. Natl. Acad. Sci. USA* **91**, 3769–3773.

Projan, S.J., Monod, M., Narayanan, C.S. and Dubnau, D. (1987). Replication properties of pIM13, a naturally occurring plasmid found in *Bacillus subtilis*, and of its close relative pE5, a plasmid native to *Staphylococcus aureus*. *J. Bacteriol.* **169**, 5131–5139.

Robertson, B.D. and Meyer, T.F. (1992). Genetic variation in pathogenic bacteria. *Trends Genet.* **8**, 422–427.

Rudel, T., Boxberger, H.J. and Meyer, T.F. (1995a). Pilus biogenesis and epithelial cell adherence of *Neisseria gonorrhoeae pilC* double knock-out mutants. *Mol. Microbiol.* **17**, 1057–1071.

Rudel, T., Scheuerpflug, I. and Meyer, T.F. (1995b). *Neisseria* PilC protein identified as type-4 pilus tip-located adhesin. *Nature* **373**, 357–359.

Sandlin, R.C., Apicella, M.A. and Stein, D.C. (1993). Cloning of a gonococcal DNA sequence that complements the lipooligosaccharide defects of *Neisseria gonorrhoeae* 1291$_d$ and 1291$_e$. *Infect. Immun.* **61**, 3360–3368.

Sarkari, J., Pandit, N., Moxon, E.R. and Achtmann, M. (1994). Variable expression of the Opc outer membrane protein in *Neisseria meningitidis* is caused by size variation of a promoter containing poly-cytidine. *Mol. Microbiol.* **13**, 207–217.

Schneider, H., Hammack, C.A., Apicella, M.A. and Griffiss, J.M. (1988). Instability of expression of lipooligosaccharides and their epitopes in *Neisseria gonorrhoeae*. *Infect. Immun.* **56**, 942–946.

Seifert, H.S. (1996). Questions about gonococcal pilus phase- and antigenic variation. *Mol. Microbiol.* **21**, 433–440.

Seifert, H.S. (1997). Insertionally inactivated and inducible *recA* alleles for use in *Neisseria. Gene* **188**, 215–220.

Sharetzsky, C., Edlind, T.D., LiPuma, J.J. and Stull, T.L. (1991). A novel approach to insertional mutagenesis of *Haemophilus influenzae*. *J. Bacteriol.* **173**, 1561–1564.

Stein, D.C., Silver, L.E., Clark, V.L. and Young, F.E. (1983). Construction and characterization of a new shuttle vector, pLES2, capable of functioning in *Escherichia coli* and *Neisseria gonorrhoeae*. *Gene* **25**, 241–247.

Stein, D.C., Gregoire, S. and Piekarowicz, A. (1988). Restriction of plasmid DNA during transformation but not conjugation in *Neisseria gonorrhoeae*. *Infect. Immun.* **56**, 112–116.

Stern, A., Brown, M., Nickel, P. and Meyer, T.F. (1986) *Opacity* genes in *Neisseria gonorrhoeae*: control of phase and antigenic variation. *Cell* **47**, 61–71.

Stibitz, S., Black, W. and Falkow, S. (1986). The construction of a cloning vector designed for gene replacement in *Bordetella pertussis*. *Gene* **50**, 133–140.

Trieu-Cuot, P., Gerbaud, G., Lambert, T. and Courvalin, P. (1985). In vivo transfer of genetic information between Gram-positive and Gram-negative bacteria. *EMBO J.* **4**, 3583–3587.

van der Ende, A., Hopman, C.T., Zaat, S., Essink, B.B., Berkhout, B. and Dankert, J. (1995) Variable expression of class 1 outer membrane protein in *Neisseria meningitidis* is caused by variation in the spacing between the −10 and −35 regions of the promoter. *J. Bacteriol.* **177**, 2475–2480.

Yang, Q.L. and Gotschlich, E.C. (1996). Variation of gonococcal lipooligosaccharide structure is due to alterations in poly-G tracts in *lgt* genes encoding glycosyl transferases. *J. Exp. Med.* **183**, 323–327.

6 Genetic Methods in Clostridia

DI Young, VJ Evans, JR Jefferies, KCB Jennert, ZEV Phillips, A Ravagnani and M Young

Institute of Biological Sciences, University of Wales, Aberystwyth, UK

◆◆

CONTENTS

◆◆◆◆◆◆ I. INTRODUCTION

The genus *Clostridium* encompasses a diverse range of organisms (Stackebrandt and Rainey, 1997). Five new genera and 11 new species combinations have been proposed, based on 16S rRNA/DNA sequences of about 100 different organisms (Collins *et al.*, 1994). Gene transfer has only been documented in a handful of organisms and is most advanced in *Clostridium acetobutylicum*, *Clostridium beijerinckii* and *Clostridium perfringens* all of which belong to Group I of Collins *et al.* (1994). In these three "model" organisms, genetic methods are being employed to investigate some problems of fundamental biological interest, two of which are considered briefly at the end of this chapter (see also Rood and Cole, 1991; Young and Cole, 1993). Gene transfer has also been documented in several other clostridia, including *Clostridium difficile*, *Clostridium saccharo-butyl-acetonicum-liquefaciens* (strain P262), *Clostridium saccharoperbutylacetonicum* (strain NI-4 and its derivatives), *Clostridium botulinum* and *Clostridium tetani*, and is referred to, as appropriate, in the text.

Owing to the extreme diversity of the genus, the gene transfer methods described below will probably require optimization, refinement and even further development, for use with organisms and strains other than those for which they were specifically devised. It is worth stressing that in any experiment involving gene transfer to clostridia, the physiological state of the recipient is of cardinal importance. Since gene transfer frequencies are generally rather low (often below 10^{-6} per recipient), the methods and media employed must allow recovery of as many potential

METHODS IN MICROBIOLOGY, VOLUME 29
ISBN 0–12–521529–0

transformants or transconjugants as possible. Optimal survival of the recipient is of the utmost importance. Brief exposure to oxygen during manipulation at the laboratory bench may be perfectly acceptable for isolating and routinely subculturing some of these obligately anaerobic organisms, but this is usually a recipe for failure when attempting gene transfer. Unless the organism under investigation is unusually tolerant of exposure to oxygen, all manipulations should be carried out in an anaerobic workstation.

◆◆◆◆◆◆ II. TRANSPOSON MUTAGENESIS

Procedures for undertaking classical mutagenesis and selecting mutant strains have been extensively documented (Jones, 1993; Sebald, 1993) and they are not further considered here.

Provided complete loss of function can be tolerated, mutated genes are most readily isolated and characterized if they are generated by insertional mutagenesis using, for example, a transposon. Several clostridia, including *C. perfringens*, harbour transposable genetic elements (reviewed by Rood and Cole, 1991; Young, 1993; Lyras and Rood, 1997), but they have not been adapted for use as mutagenic agents. Extensive use has, however, been made of two closely related conjugative transposons, Tn*916* (*tetM*; Tc^R) and Tn*1545* (*ermAM, tetM, aphA-3*; Em^R, Tc^R, Km^R), which are the best characterized examples of this widely distributed class of genetic element (reviewed by Clewell and Flannagan, 1993; Scott and Churchward, 1995) (see Chapter 4, III). A variety of organisms may be used as donors, but transfer is most efficient with a donor such as *Enterococcus faecalis* BM4110, which harbours four or five copies of Tn*1545* (Woolley *et al.*, 1989).

One drawback to the use of Tn*916* and Tn*1545* is that both elements show a distinct preference for A+T-rich target sites with sequence similarity to the transposon ends (Trieu-Cuot *et al.*, 1993). In the mesophilic clostridia intergenic regions tend to be more A+T-rich than coding sequences and as a result most transposon insertions are phenotypically silent (Wilkinson *et al.*, 1995b). Moreover, if a near-identical sequence is present in the genome of the organism under study, it will be a hotspot for transposon insertion. This behaviour has limited the use of Tn*916* in several organisms, including *C. difficile* (Mullany *et al.*, 1991). Another drawback is that transconjugants harbouring multiple copies of the transposon are frequently encountered (e.g. Gawron-Burke and Clewell, 1984; Woolley *et al.*, 1989).

As with any conjugative gene transfer protocol (see also section III.D), media and culture conditions that can be used to select transconjugants (i.e. counter-select both recipient and donor) are essential. It may also be important to enumerate donor and recipient selectively, so that transfer frequencies can be determined. Protocol 1 is a typical protocol for transferring Tn*1545* to *C. beijerinckii*; it routinely gives transfer frequencies of the order of 10^{-4} per recipient.

Protocol I. Typical protocol for transferring Tn*1545* to *C. beijerinckii*

1. Grow the *E. faecalis* BM4110 donor aerobically at 37°C in BHI (Brain Heart Infusion, Difco) overnight to stationary phase (it is not necessary to use antibiotics to select for maintenance of the conjugative transposon in the donor). Dilute back the overnight culture tenfold in the same medium and grow as above until the recipient is ready. Alternatively, the overnight culture may be diluted with an equal volume of pre-warmed BHI, just before use.

2. Grow the *C. beijerinckii* NCIMB 8052 recipient anaerobically at 37°C in CBM (O'Brien and Morris, 1971) overnight. Dilute back the following morning, if necessary, and grow at 37°C to mid–late exponential phase ($OD_{600nm} = 0.6$).

3. Take the donor culture into the anaerobic workstation and mix together equal volumes of the donor and the recipient (*ca*. 10^9 cfu ml^{-1} of the donor and *ca*. 10^7 cfu ml^{-1} of the recipient. Harvest the bacterial mixture by filtration (positive pressure) onto a 0.45µm nitrocellulose membrane (Millipore).

4. Place the membrane, bacteria uppermost, on a non-selective RCM (reinforced clostridial medium – Oxoid) agar plate spread with 0.2 ml of a solution of catalase (10 mg ml^{-1}, Sigma) and incubate anaerobically at 37°C for at least 6 h (overnight incubation works well).

5. Place the membrane with the bacteria in a boiling tube. Add 0.5–1.0 ml holding buffer (HB = 25 mM potassium phosphate, 1 mM MgSO$_4$, pH 7.5) and vortex mix to resuspend the bacteria.

6. Select transconjugants containing Tn*1545* by plating serial (10^0–10^{-2}) dilutions on modified CBM in which either D$^+$-xylose or L$^+$-arabinose (0.5%) is substituted for D$^+$-glucose (0.5%) containing 10 µg erythromycin per millilitre. Transconjugants usually appear after overnight incubation, although they may continue to appear for up to 48 h after plating. *E. faecalis* is unable to grow on D$^+$-xylose or L$^+$-arabinose.

7. Count the recipient by plating serial (10^{-4}–10^{-6}) dilutions on modified CBM in which either D$^+$-xylose or L$^+$-arabinose (0.5%) is substituted for D$^+$-glucose (0.5%) and incubating anaerobically at 37°C for 24 h.

8. Count the donor by plating serial (10^{-6}–10^{-8}) dilutions on BHI and incubating aerobically at 37°C overnight.

When adapting this protocol for use with other clostridia, transfer frequencies may be enhanced by modifying one or more of the following:

- Growth medium for the donor and/or the recipient
- Growth stage of the donor and/or the recipient at the time of mixing
- Ratio of donor to recipient cells in the mating mixture
- Non-selective medium for growth of the mating mixture
- Time of incubation on the filter
- Medium for selection of transconjugants.

Tn1545 transfer also occurs if the donor/recipient mixture obtained at stage 3 is harvested by centrifugation, the bacterial pellet resuspended in 0.5–1.0 ml HB and plated directly on CBM containing 10 μg erythromycin per millilitre. As might be expected when the Em^s recipient comes into contact with erythromycin before mating, the transfer frequency is reduced, compared with that obtained by the procedure outlined above, but this method is particularly useful for generating transposon insertion libraries, since each colony is the result of an independent conjugative transposition event.

The streptococcal conjugative transposons have found widespread use in clostridia (see Chapter 4, VI). For example, use was made of the relatively large size of Tn1545 (25.3 kbp – Caillaud *et al.*, 1987). for constructing a physical map of the *C. beijerinckii* chromosome (Wilkinson and Young, 1995). Since as little as a 5% increase in fragment size could be detected after pulsed-field gel electrophoresis, the macro-restriction fragments into which the element had inserted in a small bank of Tn1545 mutants were readily identified. Both Tn1545 and Tn916, and a derivative of the related element, Tn925, have been used in *C. beijerinckii, C. acetobutylicum* and *C. saccharo-butyl-acetonicum-liquefaciens* to isolate mutants with defects in solvent formation (Bertram *et al.*, 1990; Babb *et al.*, 1993; Mattsson and Rogers, 1994; Wilkinson *et al.*, 1995b). This led to the identification of tRNA^{thrACG} as a possible regulator of solventogenesis in *C. acetobutylicum* (Sauer and Dürre, 1992). Tn916 has also been used in *C. perfringens* to isolate mutants defective in toxin production (Awad and Rood, 1997) and its utility as a shuttle vector for gene cloning in *C. difficile* has also been explored (Mullany *et al.*, 1994). Transfer to *C. tetani* and *C. botulinum* has also been obtained (Volk *et al.*, 1988; Lin and Johnson, 1991). Two oligonucleotides, TnLE (5'CCTTGATAAAGT-GTGATAAGT3') and TnRE (5'CGTGAAGTATCTTCCTACAGT3'), complementary to sequences near the transposon ends, may be employed in inverse PCR and ligation-mediated PCR reactions to isolate junction fragments containing host DNA. The resulting products will serve as a starting point for isolation and characterization of the disrupted gene (Evans *et al.*, 1998). The entire sequence of Tn916 has been determined (Flannagan *et al.*, 1994) and mini-derivatives of Tn1545 are available (Poyart-Salmeron *et al.*, 1989, 1990; Nassif *et al.*, 1991), although they have not yet been adapted for use in clostridia.

An experimental system based on a thermosensitive plasmid would be an extremely useful adjunct to the tools currently available, since it would permit efficient delivery of smaller transposons lacking pronounced target site preference, such as Tn917 and its derivatives (Youngman, 1987) or mini-Tn10 derivatives (Petit *et al.*, 1990) (see also Chapter 4, VI), into clostridia. Unfortunately, plasmid pG+Host, used extensively for this purpose in other Gram-positive bacteria (see Maguin *et al.*, 1992), does not show thermosensitive replication in *C. beijerinckii*, at least (D. I. Young and M. Young, unpublished observations).

◆◆◆◆◆◆ III. GENE TRANSFER PROCEDURES

A. Plasmid Vectors

There are no particular difficulties associated with the isolation of plasmid (or chromosomal) DNA from most clostridia and standard methods usually give very satisfactory results (Woolley *et al.*, 1989; Williams *et al.*, 1990) (see Chapter 2, VI). Some of the wide variety of plasmid vectors available for use in clostridia, are shown in Table 1. Generally speaking, the selectable markers used in *C. perfringens* vectors are derived from naturally occurring conjugative plasmids and transposons found within this rather clearly defined species. Endogenous plasmids or transposons carrying selectable antibiotic-resistance markers have not been found in either *C. acetobutylicum* or *C. beijerinckii*, and therefore most vectors have been constructed using genes of heterologous origin, especially from enterococci.

For all three model organisms both high and low copy number plasmids are available, as are suicide (non-replicative) vectors permitting selection of the products of either Campbell-type or double crossover (gene replacement) recombination into the host chromosome. (A more detailed discussion of gene replacement technology is in Chapter 10, II, III and IV.) For further information about these as well as the more specialized vectors that are currently available, the reader should consult one of the reviews dealing specifically with this topic (Minton *et al.*, 1993a, 1993b; Rood, 1997).

B. Protoplast Transformation

Before the development of electrotransformation methods (see section III.C), procedures had been painstakingly devised for making L-forms and for regenerating walled bacilli from wall-less protoplasts of a variety of organisms, including strains of *C. perfringens*, *C. acetobutylicum*, *C. saccharobutyl-acetonicum-liquefaciens*, *C. saccharoperbutylacetonicum*, *Clostridium pasteurianum*, *Clostridium tertium* and *Clostridium thermohydrosulfuricum*. Protoplast transformation was also reported in some instances (e.g. Reid *et al.*, 1983; Heefner *et al.*, 1984; Squires *et al.*, 1984; Lin and Blaschek, 1984; Soutschek-Bauer *et al.*, 1985; Mahony *et al.*, 1988; Reysset *et al.*, 1988; Truffaut *et al.*, 1989). These rather exacting methods are not easily transferred from one laboratory to another; they are seldom used today and will not be further considered here. This topic has been succinctly summarized by Reysset and Sebald (1993), whose laboratory enjoyed considerable success in practising this unforgiving art (see also Chapter 1, III.D.7 in this volume).

C. Electrotransformation

When bacteria are very briefly exposed to a high strength electric field, the cell membrane undergoes transient breakdown, after which it re-forms (see Chapter 1, III.D.5). This process, called electroporation, provides a window of opportunity for ingress and egress of material, to and from the cell. It has been used successfully to introduce plasmids into a very wide spectrum of organisms for which no alternative gene transfer methods are currently

Table I. Selected plasmids for use in clostridia

Plasmid	Size (kbp)	Clostridium		E. coli		Comments	Reference
		Replicon	Selection	Replicon	Selection		
C. acetobutylicum vectors							
pSYL2	8.7	pCBU2	Em[a]	pBR322	Tc	pCBU2 is a cryptic *C. butyricum* plasmid	Luczak *et al.* (1985); Collins *et al.* (1985)
pSYL7	9.2	pJU122	Em[a]	pBR322	Tc	pJU122 is a cryptic *C. perfringens* plasmid	Squires *et al.* (1984); Lee *et al.* (1992a)
pFNK1[b]	2.4	pIM13	Em[a]	none	none	pIM13 is an Em[R] plasmid from *B. subtilis*	Lee *et al.* (1992b); Mermelstein *et al.* (1992)
C. beijerinckii vectors							
pAT187	10.6	pAMβ1	Km[c]	pBR322	Ap	Contains *oriT* for mobilization from *E. coli*	Trieu-Cuot *et al.* (1987); Williams *et al.* (1990)
pCTC1	7.2	pAMβ1	Em[a]	pMTL20[d]	Ap	Contains *oriT* for mobilization from *E. coli*	Williams *et al.* (1990); Minton *et al.* (1993b)
pCTC511	7.8	pCB101	Em[a]	pMTL20[d]	Ap	Contains *oriT* for mobilization from *E. coli*	Williams *et al.* (1990)
pMTL30	4.4	none	Em	pMTL20[d]	Ap XG[e]	Integrative vector containing *oriT* permitting mobilization from *E. coli* – for single crossover gene disruption	Collins *et al.* (1985); Williams *et al.* (1990); Brehm *et al.* (1992); Minton *et al.* (1993b)
pSRW1	3.2	none	none	pMTL20[d]	Ap XG[e]	Integrative vector containing *oriT* permitting mobilization from *E. coli* – for double crossover gene replacement	S. R. Wilkinson and M Young, unpublished
C. perfringens vectors							
pJIR418	7.4	pIP404	Em Cm	pUC18	Em Cm XG[e]	Fully sequenced, contains *C. perfringens catP* and *ermBP* genes	Sloan *et al.* (1992)
pHR106	7.9	pJU122	Cm	pSL100	Ap Tc	pJU122 is a cryptic *C. perfringens* plasmid	Squires *et al.* (1984); Allen and Blaschek (1988); Roberts *et al.* (1988); Scott and Rood (1989)
pAK201	8.0	pHB101	Cm	pBR322	Cm	pHB101 is a small *C. perfringens* caseinase plasmid	Kim and Blaschek (1989); Allen and Blaschek (1990)

[a] Em[R] is conferred by the MLS[R] gene (*ermAM*) from pAMβ1.
[b] Km[R] is conferred by an *aphA-3* gene from *Campylobacter coli*.
[c] pFNK1 is a *B. subtilis* / *C. acetobutylicum* shuttle vector containing the pUC9 polylinker.
[d] Constructed using the pMTL20 plasmid backbone (Chambers *et al.*, 1988).
[e] Blue-white selection with X-gal in *E. coli*.

available (Chassy *et al.*, 1988; Luchansky *et al.*, 1988; see also Chapter 1, III.D.5). Electrotransformation is often therefore the method of choice, when attempting gene transfer to a genetically uncharacterized organism. However, it has three drawbacks. Firstly, specialized equipment is required for generating the high voltage electric field. Several instruments are commercially available, and the Gene Pulser™ (Bio-Rad) seems to enjoy particular popularity. Secondly, the organism to be electroporated must be able to tolerate suspension in a medium of extremely low ionic strength; arcing will otherwise occur when the pulse is generated. Bacteria that are particularly prone to lyse under these conditions are difficult to electrotransform. Thirdly, for organisms with active restriction endonucleases, the DNA must be protected by appropriate methylation (see below).

The protocols employed for different organisms are often rather similar. Two examples are given below, for use with *C. acetobutylicum* and *C. perfringens*. Other clostridia, including *C. beijerinckii* NCIMB 8052, *C. botulinum* and *C. saccharoperbutylacetonicum* NI-4 and its derivatives have also been electrotransformed using similar methods (Oultram *et al.*, 1988; Reysset, 1993; Zhou and Johnson, 1993). An excellent summary of the available methods for electrotransformation of clostridia has been provided by Phillips-Jones (1995).

The following protocol (Protocol 2) for electrotransformation of *C. acetobutylicum* ATCC 824 is based on the method used by Mermelstein *et al.* (1992). Although this organism contains an active restriction endonuclease (*Cac*824I) it is electrotransformable if the DNA is first protected by appropriate methylation; this is effected by propagation in an *E. coli* strain expressing the gene which encodes the Φ3T-I methyltransferase (Mermelstein *et al.*, 1992; Mermelstein and Papoutsakis, 1993).

Protocol 2. Electrotransformation of *C. acetobutylicum*

1. Grow the bacteria anaerobically at 37°C in CSM at pH 6.6 (Roos *et al.*, 1985) to late exponential phase (OD_{600nm} = 0.6) and use 6 ml to inoculate 60 ml RCM (Difco). Continue incubation as before until stationary phase (OD_{600nm} = 1.2)
2. Harvest the culture by centrifugation, resuspend in 1.8 ml ETB (270 mM sucrose, 5 mM Na phosphate, pH 7.4) and place on ice.
3. Place 0.6 ml samples in pre-cooled 0.4 cm electroporation cuvettes, add DNA (10–500 ng dissolved in H_2O), mix by pipetting and incubate on ice for 10 min.
4. Apply a single pulse (2000 V = 5 kV cm^{-1}, 25 μF capacitor, time constant *ca.* 13 ms) with the electroporation apparatus.
5. Transfer the cell suspension immediately to 10 ml pre-warmed RCM and incubate anaerobically at 37°C for 4 h to allow phenotypic expression of plasmid-encoded antibiotic resistance markers before plating on the selective medium (agar-solidified RCM containing appropriate antibiotics).

Several methods have been described for electrotransformation of different strains of *C. perfringens* (Allen and Blaschek, 1988, 1990; Kim and Blaschek 1989; Scott and Rood, 1989; Phillips-Jones, 1990). Protocol 3 is based on that described in a recent review (Phillips-Jones, 1995)

Protocol 3. Electrotransformation of *C. perfringens*

1. Grow the bacteria anaerobically at 37°C in TGY medium (30 g tryptone, 20 g glucose, 1 g L-cysteine, 10 g yeast extract per litre) to early stationary phase.
2. Harvest the culture by centrifugation, resuspend in 0.1 volume cold electroporation solution – use either 15% (v/v glycerol) or SMP buffer (7 mM Na phosphate, pH 7.4 containing, per litre, 92.4 g sucrose, 0.2 g $MgCl_2.6H_2O$). Place the bacterial suspension on ice.
3. Pipette 0.8 ml of the cell suspension in 0.4 cm electroporation cuvettes pre-cooled on ice. Add 1–5 µg DNA, dissolved in 10 µl TE buffer (10 mM Tris.HCl; 1 mM EDTA, pH 8.0) mix by pipetting and leave on ice for 10 min.
4. Apply a single pulse (2.5 kV = 6.25 kV cm^{-1}, 25 µF, time constant *ca.* 8.3 ms) with the electroporation apparatus.
5. Replace the cuvette on ice and incubate for a further 10 min.
6. Add the electrotransformed cells to 4.8 ml pre-warmed, anaerobic TGY medium and incubate anaerobically for 3 h to allow phenotypic expression of plasmid-encoded antibiotic resistance markers.
7. Plate on BHI (Brain Heart Infusion – Oxoid) medium solidified with agar and containing appropriate antibiotics.

D. Conjugation

Conjugative plasmids are well documented from *C. perfringens* (reviewed by Rood and Cole, 1991; Young, 1993), but they have not been adapted for use in gene transfer. Conjugative gene transfer methods have however found extensive use in *C. beijerinckii*. Initially, plasmid pAMβ1 was employed in intergeneric matings with *B. subtilis* donors to transfer small non-conjugative plasmids as co-integrate molecules (Oultram *et al.*, 1987). This method was inefficient and pAMβ1 showed structural instability in the donor (van der Lelie and Venema, 1987). It has now been superseded by more efficient methods which exploit the extraordinarily wide host range of the IncP plasmids and the lack of specificity of their conjugation apparatus (Thomas and Smith, 1987; Krishnapillai, 1988).

Trieu-Cuot *et al.* (1987) demonstrated conjugative mobilization of plasmids from *E. coli* to several different aerobic Gram-positive bacteria. The donor harboured an IncP helper plasmid (Chapter 2, IV.D) together with the transferable vector, pAT187 (see Table 1). This latter contained a cognate *oriT* region, required for conjugative mobilization (Chapter 2, V.B), together with the broad host range replication genes of pAMβ 1 and an *aphA-3* gene conferring Km^R, which is selectable in a wide range of different organisms. Transfer of pAT187, as well as a variety of other plasmids, to *C. beijerinckii* was demonstrated subsequently (Williams *et al.*, 1990). Since a single DNA strand appears to be transferred (Young *et al.*, 1993), restriction endonucleases, which may be present in the recipient, do not normally represent a significant barrier to gene transfer by conjugation. Protocol 4 gives details for undertaking conjugative mobilization of plasmid pCTC1 (Ap^R, Em^R – see Table 1), from an *E. coli* donor containing the IncPβ plasmid, R702 (Km^R), to *C. beijerinckii*; it routinely gives transfer frequencies of the order of 10^{-5} per recipient.

Protocol 4. Conjugative mobilization of plasmid pCTC1

1. Grow the *E. coli* donor aerobically at 37°C in BHI (Brain Heart Infusion, Difco) containing 50 µg ampicillin per millilitre to maintain selection for the mobilizable plasmid, pCTC1. It is not necessary to maintain selection for the IncP helper plasmid.
2. Grow the *C. beijerinckii* NCIMB 8052 recipient anaerobically at 37°C in CBM (O'Brien and Morris, 1971) overnight. Dilute back the following morning, if necessary, and grow at 37°C to mid–late exponential phase (OD$_{600nm}$ = 0.6).
3. Take the donor culture into the anaerobic workstation. To remove residual antibiotics, centrifuge and resuspend the bacterial pellet in an equal volume of BHI. Harvest the bacteria again by centrifugation.
4. Resuspend the washed donor bacteria in an equal volume of BHI and mix together 1.8 ml (*ca.* 2 x 10^9 cfu) of the donor and 0.2 ml (*ca.* 2 x 10^7 cfu) of the recipient. Harvest the bacterial mixture by filtration (positive pressure) on to a 0.45 µm nitrocellulose membrane (Millipore).
5. Place the membrane, bacteria uppermost, on a non-selective RCM (reinforced clostridial medium – Oxoid) agar plate spread with 0.2 ml of a solution of catalase (10 mg ml^{-1}, Sigma) and incubate anaerobically at 37°C for 6 h (overnight incubation may also be employed).
6. Place the membrane with the bacteria in a boiling tube. Add 0.5–1.0 ml holding buffer (HB = 25 mM potassium phosphate, 1 mM MgSO$_4$, pH 7.5) and vortex mix to resuspend the bacteria.
7. Select transconjugants containing pCTC1 by plating serial (10^0–10^{-2}) dilutions on CBM containing 10 µg erythromycin and 10 µg trimethoprim per millilitre. The trimethoprim counter-selects the donor. Transconjugants usually appear after overnight incubation, although they may continue to appear for up to 48 h after plating.
8. Count the recipient by plating serial (10^{-4}–10^{-6}) dilutions on CBM containing 10 µg trimethoprim per millilitre (to counter-select the donor) and incubating anaerobically at 37°C for 24–48 h.
9. Count the donor by plating serial (10^{-5}–10^{-7}) dilutions on BHIB containing 50 µg ampicillin per millilitre and incubating aerobically at 37°C overnight.

Genetic Methods in Clostridia

A variety of different *E. coli* donors are available. Strains SM10 and S17-1 (Simon *et al.*, 1983a, 1983b) contain the IncP helper plasmid functions integrated into the bacterial chromosome. Others containing different IncP helper plasmids (e.g. RK2) are useful for mobilizing plasmids conferring Km^R (Thomas and Smith, 1987; Trieu-Cuot *et al.*, 1987). Derivatives harbouring the lambda *cI857*(Ts) prophage are available from the authors; they are particularly useful for plasmid transfer to organisms able to grow at 42°C since this permits extremely efficient counter-selection against the donor (Young *et al.*, 1993). Although this method has proved extremely successful with *C. beijerinckii*, it has not found widespread use in other clostridia.

◆◆◆◆◆◆ IV. REPORTER GENES AND EXPRESSION VECTORS

Several reporter gene systems have been developed for use in clostridia. Both the *C. perfringens catP* gene and the *E. coli gusA* gene have been employed successfully in *C. perfringens* (Matsushita *et al.*, 1994; Melville *et al.*, 1994; Bullifent *et al.*, 1995). Several other candidates have also been explored, but they catalyse oxygen-dependent reactions, which may limit their general utility in these anaerobic bacteria. Examples include: a *Pseudomonas xylE* gene, which was used to isolate promoters from *C. beijerinckii* (Minton *et al.*, 1993a, 1993b) and the *lux* system of *Vibrio fischeri*, which was adapted for use in *C. perfringens* (Phillips-Jones, 1993). (For further details on reporter systems see Chapter 4, X and XI.) Aequorin (GFP, or green fluorescent protein) and its derivatives (Chalfie *et al.*, 1994; Hein and Tsien, 1996; Yang *et al.*, 1996) also suffer from the same potential limitation since cyclisation of the tyr-ser-gly chromophore is also an oxygen-dependent reaction (Heim *et al.*, 1994).

Finally, an expression vector, pMTL500F, has been developed for use in *C. beijerinckii* (Minton *et al.*, 1993a). Gene expression is driven by the *C. pasteurianum* ferredoxin promoter and ribosome binding site. A staphylococcal *cat* gene inserted in this vector was expressed to a level where the gene product (chloramphenicol acetyl transferase) represented about 7% of the total soluble protein.

◆◆◆◆◆◆ V. CASE STUDIES

In this final section we very briefly outline how the available genetic tools are being employed to analyse two problems of both fundamental and applied importance in clostridia.

A. Solvent Production by Clostridia

Several of the sugar-fermenting clostridia, including *C. acetobutylicum* and *C. beijerinckii*, produce solvents at the end of the exponential growth phase in batch culture and this has been exploited for the industrial scale production of acetone and butanol from renewable feedstocks such as molasses and corn starch, etc. (reviewed by Jones and Woods, 1986: Mitchell, 1998). The nature of the switch from volatile fatty acid production to solvent production, which coincides with the end of exponential growth, is still the subject of ongoing research. A close linkage exists between solventogenesis and endospore formation in these organisms (Jones and Woods, 1986; Woolley and Morris, 1990) suggesting that both processes are co-ordinately regulated. One aspect of this shared control has recently been unravelled. Gene disruption technology has been employed to inactivate the *spo0A* gene in *C. beijerinckii* (Brown *et al.*, 1994; Wilkinson *et al.*, 1995b). The resulting strain was asporogenic and asolventogenic, indicating that both processes are controlled by the Spo0A transcription factor (Hoch, 1993). (Note that there is confusion in the pre-1995 literature, in which the NCIMB 8052 strain of *C. beijerinckii* was incorrectly designated *C. acetobutylicum* – Keis *et al.*, 1995; Wilkinson *et al.*, 1995a.) Combined physical and genetic maps have been constructed for both of these organisms. The nucleotide sequence of the *C. acetobutylicum* ATCC 824 genome is currently being determined (http://pandora.cric.com/htdocs/sequences/clostridium/clospage.html) and this will provide an invaluable resource for further research over the coming years.

The genes concerned with solvent formation and their protein products have been extensively characterized (reviewed by Chen, 1993; Papoutsakis and Bennett, 1993; Dürre *et al.*, 1995). Functional analysis of these genes "*in clostridio*" is now underway (Wilkinson and Young, 1994; Green and Bennett, 1996; Green *et al.*, 1996). There have been several reports of metabolic engineering in *C. acetobutylicum* to investigate the effects on solvent formation of either increased gene dosage or gene disruption, using multicopy replicative plasmids and integrational plasmids, respectively. Acetone production by the wild-type strain was enhanced by introducing a synthetic operon encoding the relevant enzymes on a multicopy plasmid (Mermelstein *et al.*, 1993). In another example genes concerned with volatile fatty acid (butyrate or acetate) production during exponential growth have been disrupted (Green *et al.*, 1996). Inactivation of *pta* reduced phosphotransacetylase and acetate kinase activity and significantly reduced acetate production. Similarly, inactivation of *buk* reduced butyrate kinase activity and significantly reduced butyrate production. In both cases, however, there was some residual production of the volatile fatty acid whose biosynthetic machinery had been disrupted. This was not unexpected, since the enzymes concerned with butyrate and acetate production in this organism have broad substrate specificities (reviewed by Chen, 1993; Papoutsakis and Bennett, 1993). The important role played by the *add/adh* gene product (a broad spectrum alcohol/aldehyde dehydrogenase) in solvent production was also established by gene disruption (Green and Bennett, 1996).

Several of the genes encoding enzymes specifically concerned with solvent formation reside on a 210 kbp plasmid in *C. acetobutylicum* strains ATCC 824 and ATCC 4259 (Cornillot and Soucaille, 1996; Cornillot *et al.*,

1997). Loss of the plasmid engenders loss of solvent-forming ability (strain degeneration) and solvent formation can be restored, but not enhanced above that shown by the wild type, by reintroducing the missing genes on a multicopy plasmid (Nair and Papoutsakis, 1994). Degeneration of solvent production also occurs in *C. beijerinckii*, in which the genes concerned with solvent production are not plasmid encoded (Wilkinson and Young, 1995). In this organism Tn*1545* mutagenesis has been employed to identify, isolate and characterize genes whose disruption ameliorates degeneration (Kashket and Cao, 1993, 1995). One mutant class harbours a Tn*1545* insertion that truncates the *fms* gene, encoding peptide deformylase, and in these organisms, the enhanced stability of solvent formation is a consequence of a reduced bacterial growth rate (Evans *et al.*, 1998).

B. Pathogenic Determinants in Clostridia

There is a substantial body of work on the pathogenic determinants (essentially toxins and extracellular hydrolytic enzymes) of several proteolytic, amino-acid fermenting clostridia associated with animal and human diseases (Rood *et al.*, 1997). At present there is little or no evidence for the production by these organisms of determinants such as invasins, adhesins or variable surface antigens, specifically associated with a pathogenic lifestyle, which has promoted the suggestion that many of these clostridia are probably opportunistic pathogens normally associated with putrefaction (Rood and Cole, 1991). Genes encoding many of the toxins and extracellular hydrolytic enzymes they elaborate have been isolated, characterized and manipulated for analytical and therapeutic purposes. Discussion of the extensive literature on this topic is beyond the scope of this chapter, but many of the most interesting aspects are dealt with in a recent publication (Rood *et al.*, 1997).

Genetic analysis has contributed significantly to our understanding of the molecular basis of pathogenicity in several clostridia, most notably *C. perfringens*, *C. difficile*, *C. botulinum* and *C. tetani* (Rood and Cole, 1991; Aktories *et al.*, 1997; Henderson *et al.*, 1997; Moncrief *et al.*, 1997). However, at the present time, functional analysis *in vivo* is only possible in *C. perfringens* for which there is now a quite detailed physical and genetic map (Canard and Cole, 1989; Katayama *et al.*, 1995). Many of the toxin genes in this organism are encoded by plasmids (Katayama *et al.*, 1996) and the typing of new clinical isolates is based on differences in the profiles of toxins they produce. Current ideas about the probable roles of the different toxins in virulence and pathogenesis have their basis in the toxin typing scheme. The important role thus inferred for α toxin (a protein with both phospholipase C and sphingomyelinase activities) in *C. perfringens*-mediated gas gangrene has recently been confirmed by gene disruption experiments (Ninomiya *et al.*, 1994; Awad *et al.*, 1995; see also section III.A above). Disruption of the *pfoA* gene encoding the θ toxin (a thiol-activated, pore-forming cytolysin, also known as perfringolysin O, which uses cholesterol as receptor), rather unexpectedly showed that it does not appear to play an important role in tissue necrosis associated with gas gangrene (Awad *et al.*, 1995). Some *C. perfringens* strains produce an enterotoxin and are responsible for outbreaks of a relatively mild form of food poisoning. There is a wealth of compelling evidence connecting enterotoxin production with this condition (Kokai-Kun and McClane, 1997), but this has yet to be confirmed by disruption of the toxin-encoding *cpe* gene. The *cpe* gene is variously located on the bacterial chromosome or on a plasmid in different strains of *C. perfringens* and it may be associated with a mobile genetic element (Cornillot *et al.*, 1995; Lyras and Rood, 1997). Sporulation-specific regulation of *cpe* expression in both *C. perfringens* and *B. subtilis* has been confirmed using GusA fusions to the *cpe* promoter (Melville *et al.*, 1994)

    C. difficile is the causative agent of pseudomembranous colitis (PMC) (reviewed by Johnson and Gerding, 1997). This condition appears to be associated with the production by this organism of two related toxins, denoted A and B, whose mechanism of action is well understood. Both toxins glucosylate a conserved threonine residue in proteins belonging to the Rho family, involved in regulation of the actin cytoskeleton of eukaryotic cells (Just *et al.*, 1994, 1995; Aktories *et al.*, 1997). However, the direct involvement of toxins A and B in PMC has not yet been confirmed by functional analysis *in vivo* and this must await further development and refinement of genetic analysis in this organism, which is still relatively poorly developed (Mullany *et al.*, 1990, 1991, 1994).

    Finally, strains of *C. botulinum* and *C. tetani* encode a family of potent neurotoxins, whose molecular biology and mode of action have been well documented (Henderson *et al.*, 1997; Schiavo and Montecucco, 1997). In some instances the genes encoding these toxins appear to be associated with mobile genetic elements including plasmids, bacteriophages and possibly also transposons (reviewed by Henderson *et al.*, 1997). Genetic analysis is in its infancy in the neurotoxin-producing clostridia (Volk *et al.*, 1988; Lin and Johnson, 1991; Zhou and Johnson, 1993) and allelic replacement methods may be needed to determine whether these molecules do indeed have a biological role in the organisms that produce them.

References

Aktories, K., Selzer, J., Hofmann, F. and Just, I. (1997). In: *The Clostridia: Molecular Biology and Pathogenesis*, (J. I. Rood, B. A. McClane, G. J. Songer and R. W. Titball, eds), pp. 393–407. Academic Press, San Diego.

Allen S.P. and Blaschek H.P. (1988). *Appl. Environ. Microbiol* **54**, 2322–2324.

Allen S.P. and Blaschek H.P. (1990). *FEMS Microbiol. Lett.* **70**, 217–220.

Awad, M. M. and Rood, J. I. (1997). *Microb. Path.* **22**, 275–284.

Awad, M. M., Bryant, A. E., Stevens, D. L. and Rood, J. I. (1995). *Mol. Microbiol.* **15**, 191–202.

Babb, B. L., Collett, H. J., Reid, S. J. and Woods, D. R. (1993). *FEMS Microbiol. Lett.* **114**, 343–348.

Bertram, J., Kuhn, A. and Dürre, P. (1990). *Arch. Microbiol.* **153**, 373–377.

Brehm, J. K., Pennock, A., Bullman, H. M. S., Young, M., Oultram, J. D. and Minton, N. P. (1992). *Plasmid* **28**, 1–13.

Brown, D. P., Ganova-Raeva, L., Green, B. D., Wilkinson, S. R., Young, M. and Youngman, P. (1994). *Mol. Microbiol.* **14**, 411–426.

Bullifent, H. L., Moir, A. and Titball, R.W. (1995). *FEMS Microbiol. Lett.* **131**, 99–105.

Caillaud, F., Carlier, C. and Courvalin, P. (1987). *Plasmid* **17**, 58–60.

Canard, B. and Cole, S. T. (1989). *Proc. Natl. Acad. Sci. USA* **86**, 6676–6680.

Chalfie, M., Tu, Y., Euskirchen, G., Ward, W. W. and Prasher, D. C. (1994). *Science* **263**, 802–805.

Chambers, S. P., Prior, S. E., Barstow, D. A. and Minton, N. P. (1988). *Gene* **68**, 139–149.

Chassy B. M., Mercenier, A. and Flickinger, J. (1988). *Trends Biotechnol.* **6**, 303–309.

Chen, J.-S. (1993). In: *The Clostridia and Biotechnology*, (D.R. Woods, ed.), pp. 51–76. Butterworth, Boston.

Clewell, D. B. and Flannagan, S. E. (1993). In: *Bacterial Conjugation*, (D.B. Clewell, ed.), pp. 369–393. Plenum, New York.

Collins, M. D., Lawson, P. A., Willems, A., Cordoba, J. J., Fernandez-Garayzabal, J., Garcia, P., Cai, J., Hippe, H. and Farrow, J. A. E. (1994). *Int. J. Syst. Bacteriol.* **44**, 812–826.

Collins, M. E., Oultram, J. D. and Young, M. (1985). *J. Gen. Microbiol.* **131**, 2097–2105.

Cornillot, E. (1996). PhD Thesis, INSA, Toulouse, France.

Cornillot, E. and Soucaille, P. (1996). *Nature* **380**, 489.

Cornillot, E., Saint-Joanis, B., Daube, G., Katayama, S., Granum, P. E., Canard, B. and Cole, S. T. (1995). *Mol. Microbiol.* **15**, 639–647.

Cornillot, E., Nair, R.V., Papoutsakis, E.T., and Soucaille, P. (1997). *J. Bacteriol.* **179**, 5442–5447.

Dürre, P., Fischer, R. J., Kuhn, A., Lorenz, K., Schreiber, W., Sturzenhofecker, B., Ullmann, S., Winzer, K. and Sauer, U. (1995). *FEMS Microbiol. Rev.* **17**, 251–262.

Evans, V. J., Liyanage, H., Ravagnani, A., Young, M. and Kashket, E. R. (1998). Truncation of peptide deformylase reduces the growth rate and stabilizes solvent production in *Clostridium beijerinckii* NCIMB 8052. *Appl. Environ. Microbiol.* **64**, 1780–1785.

Flannagan, S. E, Zitzow, L. A. and Clewell, D. B. (1994). *Plasmid* **32**, 350–354.

Gawron-Burke, C. and Clewell, D. B. (1984). *J. Bacteriol.* **159**, 214–221.

Green, E. M. and Bennett, G. N. (1996). *Appl. Biochem. Biotechnol.* **57**, 213–221.

Green, E. M., Boynton, Z. L., Harris, L., M., Rudolph, F. B., Papoutsakis, E. T. and Bennett, G. N. (1996). *Microbiology UK* **142**, 2079–2086.

Heefner, D. L, Squires, C. H., Evans, R. J., Kopp, B. J. and Yarus, M. J. (1984). *J. Bacteriol.* **159**, 460–464.

Heim, R., Prasher, D. C. and Tsien, R. Y. (1994). *Proc. Natl. Acad. Sci. USA* **91**, 12501–12504.

Hein, R. and Tsien, R. Y. (1996). *Curr. Biol.* **6**, 178–182.

Henderson, I., Davis, T., Elmore, M. and Minton, N. P. (1997). In: *The Clostridia: Molecular Biology and Pathogenesis* (J. I. Rood, B. A. McClane, G. J. Songer and R. W. Titball, eds.), pp. 261–294. Academic Press, San Diego.

Hoch, J. A. (1993). *Ann. Rev. Microbiol.* **47**, 441–465.

Johnson, S. and Gerding, D. N. (1997). In: *The Clostridia, Molecular Biology and Pathogenesis* (J. I. Rood, B. A. McClane, G. J. Songer and R. W. Titball, eds), pp. 117–140. Academic Press, San Diego.

Jones, D. T. (1993). In: *The Clostridia and Biotechnology* (D. R. Woods, ed.), pp. 77–97. Butterworth-Heinemann, Boston.

Jones, D. T. and Woods, D. R. (1986). *Microbiol. Revs.* **50**, 484–524.

Just, I., Fritz, G., Aktories, K., Giry, M., Popoff, M. R., Boquet, P., Hegenbarth, S. and von Eichel-Streiber, C. (1994). *J. Biol. Chem.* **269**, 10706–10712.

Just, I., Selzer, J., Wilm, M., von Eichel-Streiber, C., Mann, M. and Aktories, K. (1995). *Nature* **375**, 500–503.

Kashket, E.R. and Cao, Z.Y. (1993). *Appl. Environ. Microbiol.* **59**, 4198–4202.

Kashket, E.R. and Cao, Z.Y. (1995). *FEMS Microbiol. Rev.* **17**, 307–315.

Katayama, S., Dupuy, B., Garnier, T. and Cole, S. T. (1995). *J. Bacteriol.* **177**, 5680–5685.

Katayama, S., Dupuy, B., Daube, G., China, B. and Cole, S.T. (1996). *Mol. Gen. Genet.* **251**, 720–726.

Keis, S., Bennett, C. F., Ward, V. K. and Jones, D. T. (1995). *Int. J. Syst. Bacteriol.* **45**, 693–705.

Kim, A. Y. and Blaschek, H. P. (1989). *Appl. Environ. Microbiol.* **55**, 360–365.

Kokai-Kun, J. F. and McClane, B. A. (1997). In: *The Clostridia, Molecular Biology and Pathogenesis* (J. I. Rood, B. A. McClane, G. J. Songer and R. W. Titball, eds.), pp. 325–357. Academic Press, San Diego.

Krishnapillai V. (1988). *FEMS Microbiol. Rev.* **54**, 223–238.

Lee, S. Y., Bennett, G. N. and Papoutsakis, E. T. (1992a). *Biotechnol. Lett.* **14**, 427–432.

Lee, S. Y., Mermelstein, L. D., Bennett, G. N. and Papoutsakis, E. T. (1992b). *Ann. NY Acad. Sci.* **665**, 39–51.

Lin, W. J. and Johnson, E. A. (1991). *Appl. Environ. Microbiol.* **57**, 2946–2950.

Lin, Y. L. and Blaschek, H. P. (1984). *Appl. Environ. Microbiol.* **48**, 737–742.

Luchansky, J. B, Muriana, P. M. and Klaenhammer, T. R. (1988). *Mol. Microbiol.* **2**, 637–646.

Luczak, H., Schwarzmoser, H. and Staudenbauer, W. L. (1985). *Appl. Microbiol. Biotechnol.* **23**, 114–122.

Lyras, D. and Rood, J. I. (1997). In: *The Clostridia, Molecular Biology and Pathogenesis* (J. I. Rood, B. A. McClane, G. J. Songer and R. W. Titball, eds), pp. 73–92. Academic Press, San Diego.

Maguin, E., Duwat, P., Hege, T., Ehrlich, S. D. and Gruss, A. (1992). *J. Bacteriol.* **174**, 5633–5638

Mahony, D. E., Mader, J. A. and Dubel, J. R. (1988). *Appl. Environ. Microbiol.* **54**, 264–267.

Matsushita, C., Matsushita, O., Koyama, M. and Okabe, A. (1994). *Plasmid* **31**, 317–319.

Mattsson, D. M. and Rogers, P. (1994). *J. Indust. Microbiol.* **13**, 258–268.

Melville, S. B, Labbe, R. and Sonenshein, A. L. (1994). *Infect. Immun.* **62**, 5550–5558.

Mermelstein, L. D. and Papoutsakis, E. T. (1993). *Appl. Environ. Microbiol.* **59**, 1077–1081.

Mermelstein, L. D., Welker, N. E., Bennett, G. N. and Papoutsakis, E. T. (1992). *BioTechnology* **10**, 190–195.

Mermelstein, L. D., Papoutsakis, E. T., Petersen, D. J. and Bennett, G.N. (1993). *Biotechnol. Bioeng.* **42**, 1053–1060.

Minton, N. P., Brehm, J. K., Swinfield, T.-J., Whelan, S. M., Mauchline, M. L., Bodsworth, N. and Oultram, J. D. (1993a). In: *The Clostridia and Biotechnology*, (D.R. Woods, ed.), pp. 119–150. Butterworth, Boston.

Minton, N. P., Swinfield, T.-J., Brehm, J. K., Whelan, S. M. and Oultram, J. D. (1993b). In: *Genetics and Molecular Biology of Anaerobic Bacteria* (M. Sebald, ed.), pp. 120–140. Springer-Verlag, New York.

Mitchell, W. J. (1998). Physiology of carbohydrate to solvent conversion by Clostridia. *Adv. Microb. Physiol.* **39**, 31–130.

Moncrief, J. S., Lyerly, D. M. and Wilkins, T. D. (1997). In: *The Clostridia: Molecular Biology and Pathogenesis* (J. I. Rood, B. A. McClane, G. J. Songer and R. W. Titball, eds), pp. 369–392. Academic Press, San Diego.

Mullany, P., Wilks, M., Lamb, I., Clayton, C., Wren, B. and Tabaqchali, S. (1990). *J. Gen Microbiol.* **136**, 1343–1349.

Mullany, P., Wilks, M. and Tabaqchali, S. (1991). *FEMS Microbiol. Lett.* **79**, 191–194.

Mullany, P., Wilks, M., Puckey, L. and Tabaqchali, S. (1994). *Plasmid* **31**, 320–323.

Nair, R. V. and Papoutsakis, E. T. (1994). *J. Bacteriol.* **176**, 5843–5846.

Nassif, X., Puaoi, D. and So, M. (1991). *J. Bacteriol.* **173**, 2147–2154.

Ninomiya, M., Matsushita, O., Minami, J., Sakamoto, H., Nakano, M. and Okabe, A. (1994). *Infect. Immun.* **62**, 5032–5039.

O'Brien, R. W. and Morris, J.G. (1971). *J. Gen. Microbiol.* **68**, 307–318.

Oultram, J. D., Davies, A. and Young, M. (1987). *FEMS Microbiol. Lett.* **42**, 113–119.

Oultram, J. D., Loughlin, M., Swinfield, T. J., Brehm, J. K., Thompson, D. E. and Minton, N. P. (1988). *FEMS Microbiol. Lett.* **56**, 83–88.

Papoutsakis, E. T. and Bennett, G. N. (1993). In: *The Clostridia and Biotechnology* (D.R. Woods, ed.), pp. 157–199. Butterworth, Boston.

Petit, M. A., Bruand, C., Jannière, L. and Ehrlich, S. D. (1990) *J. Bacteriol.* **172**, 6736–6740

Phillips-Jones, M. K. (1990). *FEMS Microbiol. Lett.* **66**, 221–226.

Phillips-Jones, M. K. (1993). *FEMS Microbiol. Lett.* **106**, 265–270.

Phillips-Jones, M. K. (1995). In: *Electroporation Protocols for Micro-organisms* (J.A. Nickoloff, ed.), pp. 227–235. Humana Press Inc, Totowa, New Jersey.

Poyart-Salmeron, C., Trieu-Cuot, P., Carlier C. and Courvalin, P. (1989). *EMBO J.* **8**, 2425–2433.

Poyart-Salmeron, C., Trieu-Cuot, P., Carlier, C. and Courvalin, P. (1990). *Mol. Microbiol.* **4**, 1513–1521.

Reid, S. J., Allcock, E. R., Jones, D. T. and Woods, D. R. (1983) *Appl. Environ. Microbiol.* **45**, 305–307.

Reysset, G. (1993). In: *Genetics and Molecular Biology of Anaerobic Bacteria* (M. Sebald, ed.), pp. 111–119. Springer-Verlag, New York.

Reysset, G. and Sebald, M. (1993). In: *The Clostridia and Biotechnology* (D. R. Woods, ed.), pp. 151–156. Butterworth-Heinemann, Boston.

Reysset, G., Hubert, J., Podvin, L. and Sebald, M. (1988). *Biotechnol. Tech.* **2**, 199–204.

Roberts, I., Holmes, W. M. and Hylemon, P. B. (1988). *Appl. Environ. Microbiol.* **54**, 268–270.

Rood, J. I. (1997). In: *The Clostridia: Molecular Biology and Pathogenesis* (J. I. Rood, B. A. McClane, G. J. Songer, and R. W. Titball, eds.), pp. 65–72. Academic Press, San Diego.

Rood, J. I. and Cole, S. T. (1991). *Microbiol. Rev.* **55**, 621–648.

Rood, J. I., McClane, B. A., Songer, G. J. and Titball, R. W. (1997). *The Clostridia: Molecular Biology and Pathogenesis*, 533pp. Academic Press, San Diego.

Roos, J. W., McLaughlin, J. K. and Papoutsakis, E. T. (1985). *Biotechnol. Bioeng.* **27**, 681–694.

Sauer, U. and Durre, P. (1992). *FEMS Microbiol. Lett.* **100**, 147–153.

Schiavo, G. and Montecucco, C. (1997). In: *The Clostridia: Molecular Biology and Pathogenesis* (J. I. Rood, B. A. McClane, G. J. Songer and R. W. Titball, eds.), pp. 295–322. Academic Press, San Diego.

Scott, J. R. and Churchward, G. G. (1995). *Ann. Rev. Microbiol.* **49**, 367–397.

Scott, P. T. and Rood, J. I. (1989). *Gene* **82**, 327–333.

Sebald, M. (1993). In: *Genetics and Molecular Biology of Anaerobic Bacteria* (M. Sebald, ed.), pp. 64–97. Springer-Verlag, New York.

Simon, R., Priefer, U. and Puhler, A. (1983a). *BioTechnology* **1**, 784–791.

Simon, R., Priefer, U. and Puhler, A. (1983b). In: *Molecular Genetics of Bacteria-Plant Interaction* (A. Puhler, ed.), pp. 98–106. Springer-Verlag, Berlin.

Sloan, J., Warner, T. A., Scott, P. T., Bannam, T. L., Berryman, D. I. and Rood, J. I. (1992). *Plasmid* **27**, 207–219.

Soutschek-Bauer, E., Hartl, L. and Staudenbauer, W.L. (1985). *Biotechnol. Lett.* **7**, 705–710.

Squires, C. H., Heefner, D. L., Evans, R. J., Kopp, B. J. and Yarus, M. J. (1984). *J. Bacteriol.* **159**, 465–471.

Stackebrandt, E. and Rainey, F. A. (1997). In: *The Clostridia, Molecular Biology and Pathogenesis* (J. I. Rood, B. A. McClane, G. J. Songer and R. W. Titball, eds.), pp. 3–19. Academic Press, San Diego.

Thomas C. M. and Smith C. A. (1987). *Ann. Rev. Microbiol.* **41**, 77–101.

Trieu-Cuot, P., Carlier, C., Martin, P. and Courvalin, P. (1987). *FEMS Microbiol. Lett.* **48**, 289–294.

Trieu-Cuot, P., Poyart-Salmeron, C., Carlier, C. and Courvalin, P. (1993). *Mol. Microbiol.* **8**, 179–185.

Truffaut, N., Hubert, J. and Reysset, G. (1989). *FEMS Microbiol. Lett.* **58**, 15–19.

van der Lelie, D. and Venema, G. (1987). *Appl. Environ. Microbiol.* **53**, 2458–2463.

Volk, W. A., Bizzini, B., Jones, K. R. and Macrina F. L. (1988). *Plasmid* **19**, 255–259.

Wilkinson, S. R. and Young, M. (1994). *Microbiology UK* **140**, 89–95.

Wilkinson, S. R. and Young, M. (1995). *J. Bacteriol.* **177**, 439–448.

Wilkinson, S. R., Young, M., Goodacre, R., Morris, J. G., Farrow, J. A. E. and Collins, M. D. (1995a). *FEMS Microbiol. Lett.* **125**, 199–204.

Wilkinson, S. R., Young, D. I., Morris, J. G. and Young, M. (1995b). *FEMS Microbiol. Rev.* **17**, 275–285.

Williams, D. R., Young, D. I. and Young, M. (1990). *J. Gen. Microbiol.* **136**, 819–826.

Woolley, R.C. and Morris, J.G. (1990). *J. Appl. Bacteriol.* **69**, 718–728.

Woolley, R. C., Pennock, A., Ashton, R. J., Davies, A. and Young, M. (1989). *Plasmid* **22**, 169–174.

Yang, T. T., Kain, S. R., Kitts, P., Kondepudi, A., Yang, M. M. and Youvan, D. C. (1996). *Gene* **173**, 19–23.

Young, M. (1993). In: *The Clostridia and Biotechnology* (D. R. Woods, ed.), pp. 99–117. Butterworth, Boston.

Young, M. and Cole, S. T. (1993). In: Bacillus subtilis *and other Gram-Positive Bacteria: Biochemistry, Physiology and Molecular Genetics* (A. L. Sonenshein, J. A. Hoch and R. Losick, eds), pp. 35–52. ASM, Washington, D.C.

Young, D. I., Williams, D. R. and Young, M. (1993). In: *DNA Transfer and Gene Expression in Microorganisms* (E. Balla, G. Berencsi and A. Szentirmai, eds). Intercept, Andover, U.K.

Youngman, P. J. (1987). In: *Plasmids, a Practical Approach.* (K. G. Hardy, ed.), pp. 79–103. IRL Press, Oxford.

Zhou, Y. and Johnson, E. A. (1993). *Biotechnol. Lett.* **15**, 121–126.

7 Genetic Methods in *Borrelia* and Other Spirochaetes

Patricia Rosa, Brian Stevenson and Kit Tilly
Laboratory of Microbial Structure and Function, Rocky Mountain Laboratories, Hamilton, Montana, USA

◆◆◆

CONTENTS

◆◆◆◆◆◆ I. INTRODUCTION

Spirochaetes are a phylogenetically ancient bacterial group and one of the few for which classification by morphology is consistent with molecular typing based upon rRNA sequence comparisons (Paster *et al.*, 1984, 1991; Woese, 1987). This group includes the genera *Treponema*, *Spirochaeta*, *Borrelia*, *Serpulina*, *Leptonema*, *Leptospira* and *Brevinema*; untyped spirochaetes from various sources may constitute additional unrecognized genera. Although a unique spiral shape and periplasmic flagella for motility are universal among the spirochaetes, they are a diverse group of bacteria with regard to many other characteristics (Johnson, 1977). They include both free-living and host-associated bacteria, pathogens and non-pathogens, and their genomes range widely in size, structure and base composition. Several spirochaetes represent important human and animal pathogens, with histories of disease that extend for centuries. None of the spirochaetes has been the subject of extensive genetic analysis and there is a corresponding paucity in current genetic tools and methods with which to manipulate them. In this chapter we describe some recent experiments that begin to address this shortcoming. We focus primarily on *Borrelia burgdorferi*, the causative agent of Lyme disease (Burgdorfer *et al.*, 1982), as an example of the development of genetic methods for a bacterium that was isolated only 17 years ago. We start with a description of the complex genome of *B. burgdorferi*, and the potential promise and problems that its unusual features present for genetic studies. Next we summarize a set of experiments that represent the first genetic manipulations of this spirochaete. Finally, we describe genetic studies and tools that have been used with other spirochaetes, including *Leptospira*, *Treponema* and *Serpulina*.

◆◆◆◆◆◆ II. *BORRELIA* GENOMICS

A. *Borrelia burgdorferi* and Lyme Disease

Borrelia burgdorferi is the causative agent of Lyme disease, a multisystemic ailment of humans that is spread by the bite of certain species of *Ixodes* ticks (Burgdorfer *et al.*, 1982; Steere *et al.*, 1983). Described as a single species when first isolated (Johnson *et al.*, 1984), these bacteria have recently been subdivided into the disease-associated B. *burgdorferi* (*sensu stricto*), B. *garinii* and B. *afzelii*, plus several related species that have not been shown to cause disease in humans (Fukunaga *et al.*, 1996). These bacteria are phylogenetically closely related and are often referred to as B. *burgdorferi sensu lato* or simply B. *burgdorferi*, a terminology we will use throughout this chapter. Relapsing fever is caused by several different species of *Borrelia* including B. *hermsii*, (Barbour and Hayes, 1986), which we will refer to as "relapsing fever *Borrelia*".

B. Plasmids

In addition to an approximately one megabase linear chromosome (Baril *et al.*, 1989; Ferdows and Barbour, 1989; Casjens and Huang, 1993), isolates of B. *burgdorferi* generally contain numerous different linear and circular plasmids (e.g. Barbour, 1988, 1993; Hyde and Johnson, 1988; Simpson *et al.*, 1990; Stålhammar-Carlemalm *et al.*, 1990; Dunn *et al.*, 1994; Xu and Johnson, 1995; Casjens *et al.*, 1997b; Stevenson *et al.*, 1997), as do other members of this genus (Hayes *et al.*, 1988; Perng and LeFebvre, 1990; Kitten and Barbour, 1992; Barbour, 1993). Linear DNA, although unusual among the eubacteria, has also been observed in other bacterial species (reviewed in Hinnebusch and Tilly, 1993). Specific procedures for the purification of B. *burgdorferi* plasmids have been described (Barbour, 1988), although we have found that Qiagen midi plasmid purification kits (Qiagen, Chatsworth, California, USA) can be used efficiently to purify both the linear and circular plasmids of B. *burgdorferi* with essentially no chromosomal contamination (Casjens *et al.*, 1997b; Stevenson *et al.*, 1997). Caesium chloride gradients may be used to separately purify linear and supercoiled circular plasmids (Barbour and Garon, 1987). A technique that distinguishes linear and circular DNA, such as two-dimensional gel electrophoresis, is essential for accurate mapping of plasmid-encoded genes (Barbour, 1993; Marconi *et al.*, 1993; Sadziene *et al.*, 1993; Samuels and Garon, 1993; Stevenson *et al.*, 1996, 1997).

The naturally occurring linear and circular plasmids may prove to be useful for constructing cloning vectors, although there have not yet been any reports of success. Some *B. burgdorferi* plasmid fragments are unstable when cloned into *E. coli* (Dunn *et al.*, 1994; Stevenson *et al.*, 1996), which may be an obstacle to shuttle vector construction.

Developing vectors based upon *B. burgdorferi* linear plasmids may be especially complicated, since the mechanisms of linear DNA replication in these bacteria are unknown. There is circumstantial evidence that a circular intermediate may be involved in linear plasmid replication, since a linear *B. burgdorferi* plasmid, 1p56, shares extensive homology with the circular cp32 plasmids of these bacteria (Zückert and Meyer, 1996; Casjens *et al.*, 1997b). Also, a linear *B. hermsii* plasmid converted to a stable circular form during laboratory cultivation (Ferdows *et al.*, 1996).

Borrelia burgdorferi may contain both high and low copy number plasmids. A small circular plasmid, cp8.3, apparently has a very high copy number, a feature that aided in the cloning and sequencing of this plasmid (Dunn *et al.*, 1994). Three other *B. burgdorferi* plasmids, linear plasmids 1p16 and 1p49 and circular plasmid cp26, have copy numbers of approximately one each per chromosome (Hinnebusch and Barbour, 1992). The number of chromosomes per bacterium is unknown for *B. burgdorferi*, but is assumed to be similar to that for *B. hermsii*, with between 4 and 16 copies per cell, depending upon growth conditions (Kitten and Barbour, 1992).

With the exception of the high copy number cp8.3 (Dunn *et al.*, 1994), all circular and linear *B. burgdorferi* plasmids that have been sequenced contain the open reading frame *orfC* (also called *cdsM*) (Barbour *et al.*, 1996; Zückert and Meyer, 1996; Casjens *et al.*, 1997b; Fraser *et al.*, 1997; Stevenson *et al.*, 1997; Stevenson *et al.*, 1998), which is similar to the *parA* and *sopA* genes found in the *E. coli* plasmids P1 and F, respectively (Ogura and Hiraga, 1983; Abeles *et al.*, 1984) (see Chapter 2, IV.A.2 in this volume). The ParA/SopA proteins are ATPases that are required for efficient partitioning of these low copy number plasmids (Motallebi-Veshareh *et al.*, 1990; Watanabe *et al.*, 1992). The sequences of the plasmid *orfC* homologues that have been identified within individual bacteria are heterogeneous (Barbour *et al.*, 1996; Zückert and Meyer, 1996; Casjens *et al.*, 1997b; Stevenson *et al.*, 1997; Stevenson *et al.*, 1998), which may account for the compatibility of the many *B. burgdorferi* plasmids. All characterized plasmids contain another gene of unknown function, *orf3* (also called *orfD* or *cdsN*), which varies among different plasmids within a bacterium (Dunn *et al.*, 1994; Barbour *et al.*, 1996; Zückert and Meyer, 1996; Casjens *et al.*, 1997b; Stevenson *et al.*, 1998), suggesting that its product may also play a role in plasmid segregation or replication. The *orfC* and *orf3* sequences of resident plasmids in a particular isolate may need to be determined before attempting to introduce a chimera based upon a *B. burgdorferi* plasmid. *B. burgdorferi* plasmids contain additional open reading frames encoding proteins with unknown functions (Dunn *et al.*, 1994; Barbour *et al.*, 1996; Porcella *et al.*, 1996; Stevenson *et al.*, 1996, 1997; Zückert and Meyer, 1996; Casjens *et al.*, 1997b; Fraser *et al.*, 1997), some of which may be essential for plasmid maintenance and might be required in *cis* on novel vectors.

Genetic Methods in Borrelia

Predicting sequence relationships and compatibility groups on the basis of plasmid size is problematic. The linear plasmid that includes the *ospAB* operon ranges between 49 and 58 kb (Samuels *et al.*, 1993), and a 100 kb dimer of this plasmid has been observed in several isolates (Munderloh *et al.*, 1993; Marconi *et al.*, 1996). We have also identified *B. burgdorferi* containing stable dimers of the 26 kb circular plasmid that carries *ospC* (Tilly *et al.*, 1998). Conversely, a single bacterium of *B. burgdorferi* may contain as many as six different circular plasmids that are all 32 kb in size (Porcella *et al.*, 1996; Stevenson *et al.*, 1996; Zückert and Meyer, 1996; Casjens *et al.*, 1997b). However, all of the 32 kb circular plasmids that have been found within individual bacteria carry different *orfC* and *orf3* genes (Zückert and Meyer, 1996; Stevenson *et al.*, 1998), which might account for their compatibility. (See Chapter 2, IV.D for general methods to test for incompatability.)

C. Evidence of Natural Genetic Exchange

Many bacteria are able to take up DNA from external sources and they often integrate the new DNA into their genomes, characteristics that may be exploitable for the introduction of recombinant genes. Restriction site mapping of the chromosomes and sequencing of chromosomally-encoded genes from a large number of *B. burgdorferi* idolates indicate that chromosomal rearrangements are rare events (Dykhuizen *et al.*, 1993; Ojaimi *et al.*, 1994; Casjens *et al.*, 1995). Sequence comparisons of the plasmid-encoded *ospAB* genes from a large number of isolates indicate that the *ospAB* locus is also very stable (Dykhuizen *et al.*, 1993). In contrast, the *ospC* locus, carried on an approximately 26 kb circular plasmid, does exhibit considerable sequence variation between isolates, and dendrograms of *ospC* sequences are often inconsistent with trees based upon chromosomal or *ospAB* genes (Stevenson and Barthold, 1994; Jauris-Heipke *et al.*, 1995; Livey *et al.*, 1995; Gibbs *et al.*, 1996). The variations within the *ospC* genes appear to be due to recombination with small fragments of DNA from external sources (Livey *et al.*, 1995; Gibbs *et al.*, 1996; Tilly *et al.*, 1997), indicating that natural genetic exchange may occur between *B. burgdorferi*. The mechanism(s) responsible for the uptake of DNA by these bacteria is not yet known.

D. Bacteriophages

Bacteriophages have been used in other bacteria as transduction systems or for constructing phagemid vectors (see Chapter 3, V). *B. burgdorferi* appears to harbour several types of prophages, since three types of morphologically distinct bacteriophage particles have been observed in cultures of the bacteria (Hayes *et al.*, 1983; Neubert *et al.*, 1993; Schaller and Neubert, 1994). Bacteriophages have also been seen in cultures of the relapsing fever spirochete *B. hermsii* (Barbour and Hayes, 1986). Ciprofloxacin was found to induce lytic bacteriophages in some *B. burgdorferi* cultures (Schaller and Neubert, 1994). All the bacteriophages described were associated with *B. burgdorferi* freshly isolated from infected ticks and mammals, and it is possible that lytic phage genomes were lost or became defective during laboratory cultivation.

Restriction site mapping comparisons of the chromosomes from numerous isolates indicate their lengths and physical maps are conserved among all isolates (Ojaimi *et al.*, 1994; Casjens *et al.*, 1995), suggesting that any prophage genomes within *B. burgdorferi* are integrated into the same location of all these chromosomes, are very small, or are episomal. As mentioned previously, *B. burgdorferi* may contain six or more different cp32 plasmids, members of a circular plasmid family that are all 32 kb in size and appear to contain homologous sequences and gene orders throughout their entireties (Porcella *et al.*, 1996; Stevenson *et al.*, 1996; Zückert and Meyer, 1996; Casjens *et al.*, 1997b). Such conservation of size and sequence is a common characteristic of bacteriophage families such as the lambdoid phages (Casjens and Earnshaw, 1980; Casjens *et al.*, 1992), in part because the size constraints of the capsid require maintenance of a constant genome size. It remains to be seen whether the cp32 plasmids or any other *B. burgdorferi* plasmids are actually prophage genomes.

E. Recombination Mechanisms

The natural ability of bacteria to recombine homologous DNA has been used in many bacterial systems to integrate DNA into specific target sites. *B. burgdorferi* has a *recA* homologue (Dew-Jager *et al.*, 1995) and the bacteria appear to be fully capable of homologous recombination, since deletion events between the homologous *ospA* and *ospB* genes of a single bacterium have yielded chimeric genes (Rosa *et al.*, 1992). This ability of *B. burgdorferi* has been exploited in the targeted allelic exchange experiments described in Section III below.

The relapsing fever borreliae apparently use homologous recombination in a mechanism of avoiding clearance by mammalian immune systems. These bacteria contain non-transcribed *vmp* genes that can recombine via a gene conversion-like mechanism into an expression locus, allowing the synthesis of a large repertoire of different Vmp surface proteins and survival of these bacteria in the face of antibodies directed against previously synthesized Vmp proteins (Meier *et al.*, 1985; Plasterk *et al.*, 1985; Barbour *et al.*, 1991; Restrepo and Barbour, 1994). A *vmp*-like system (*vls*) in *B. burgdorferi* may function similarly (Zhang *et al.*, 1997).

The telomeres of linear borrelial DNAs appear to be especially recombinogenic and have unique properties. Both the *vmp* and *vls* recombination sites are located near the telomeres of linear plasmids. Furthermore, recombination among the telomeres of the linear chromosome and some linear plasmids of *B. burgdorferi* has been reported (Casjens *et al.*, 1997a).

Other *B. burgdorferi* DNA sites appear to be resistant to homologous recombination. Despite extensive sequence homologies, the multiple cp32 plasmids appear to be extremely stable, as recombination among these plasmids during growth of bacteria in culture has never been observed (Casjens *et al.*, 1997b).

F. Analysis of the *B. burgdorferi* Genome

The Institute for Genomic Research (TIGR) is presently sequencing the entire genome of *B. burgdorferi* isolate B31, including all of its plasmids, and the data were released and annotated in 1998 (http://www.tigr.org/tdb/mdb/bbdb/bb_bg.html) (Fraser *et al.*, 1997). The complete sequence of this genome will undoubtedly help researchers develop more and better genetic methods for exploring the biology of *B. burgdorferi*. For example, there is evidence of a DNA restriction/modification system in *B. burgdorferi* (Norton Hughes and Johnson, 1990) and the complete sequence may identify the genes encoding these enzymes.

To date, most "genetic studies" in *B. burgdorferi* have been limited to characterizing the "genetic material" (chromosome and plasmids) or cloning, sequencing and expressing genes in *E. coli*. With the availability of the entire genomic sequence, genetic studies will assume a different character. Lyme disease researchers will be challenged to identify the functions of a large number of gene products and to understand the molecular basis of particular phenotypes. These studies will require the ability to routinely manipulate *B. burgdorferi* with a wide array of genetic methods. In the following section we describe the preliminary steps that have been taken towards this goal.

◆◆◆◆◆◆ III. TRANSFORMATION OF *B. BURGDORFERI*

A. Development of a Selectable Marker

Genetic manipulation of *B. burgdorferi* was first achieved following a careful set of experiments by D. Scott Samuels and co-workers. Initially, Samuels demonstrated that *B. burgdorferi* is quite susceptible to coumermycin A_1, an inhibitor of the DNA gyrase B subunit (growth inhibition at 0.2 µg ml^{-1}), whereas it is relatively resistant to DNA gyrase A subunit inhibitors, such as ciprofloxacin, nalidixic acid and oxolinic acid (Samuels and Garon, 1993). Next, spontaneous coumermycin-resistant mutants were isolated by selecting for growth in the presence of the antibiotic (Samuels *et al.*, 1994b). Although selected at sub-inhibitory concentrations of coumermycin (0.1 µg ml^{-1}), these mutants had 100–300-fold higher resistance to the antibiotic than wild-type *B. burgdorferi*. Resistance to coumermycin in *B. burgdorferi*, as in *E. coli* and other bacteria, arose through mutations in the *gyrB* gene, encoding the B subunit of DNA gyrase; resistant mutants had single point mutations in their *gyrB* genes and the level of resistance varied with the substituted amino acid (Samuels *et al.*, 1994b).

B. Electroporation and Transformation

Samuels *et al.* next utilized *gyrB*-conferred resistance to coumermycin (*gyrB^r*) as a marker to monitor the transformation of *B. burgdorferi*

(Samuels *et al.*, 1994a). There is a significant advantage to this approach in developing transformation in a bacterium for which no genetic tools were previously available. By utilizing a characterized endogenous gene from a spontaneous mutant as a selectable marker, the phenotype of the desired transformant and its ability to grow and form colonies at a particular level of drug are already established. When a method for transformation does not yet exist, expression of a foreign gene at an adequate level to confer a selectable phenotype is an additional unknown.

DNA containing the *gyrB*[r] gene from spontaneous coumermycin-resistant bacteria was introduced into wild-type *B. burgdorferi* by electroporation and transformants were selected for growth in the presence of antibiotic. Homologous recombination between donor DNA and the endogenous chromosomal *gyrB* gene resulted in a single nucleotide change, confirming that mutations at this residue in *gyrB* conferred coumermycin resistance. The number of antibiotic-resistant colonies following electroporation with donor DNA from coumermycin-resistant mutants was significantly higher than the background level of spontaneous resistance, demonstrating that transformation was successful. A silent mutation was also added to the transforming *gyrB*[r] gene to allow differentiation between recombinants and spontaneous mutants. Site-directed mutations in *gyrB* could be demonstrated following transformation with total genomic DNA from spontaneous mutants or with a PCR-amplified fragment spanning the mutation in the *gyrB*[r] gene.

Transformation of *B. burgdorferi* by electroporation was accomplished using fairly standard conditions (see Samuels, 1995, for a detailed protocol) (see also Chapter 1, III.D.5 and Figure 4 in this volume). Approximately ten transformants were obtained per microgram of total genomic DNA from coumermycin-resistant *B. burgdorferi*, whereas a PCR-amplified fragment of the *gyrB*[r] gene from resistant bacteria routinely yielded approximately 200 transformants per microgram of DNA (Samuels *et al.*, 1994a). A difference at least that great is expected because of the complexity of the two DNA substrates and the corresponding molarity of *gyrB*[r]. In more recent experiments, the transformation efficiency has reached 10^4 transformants per microgram of DNA, using either recombinant plasmids purified from *E. coli* or PCR fragments (Rosa *et al.*, 1996). Neither the number of recipient bacteria nor the amount of donor DNA appears to be in excess with the current protocol, so the transformation efficiency can be improved by increasing either parameter in the electroporation and both should be taken into account when comparing separate experiments. With the current method, up to 1 in 10^5 bacteria can be transformed. Short oligonucleotides, either single- or double-stranded and spanning the necessary mutation in *gyrB*, have also been used to transform *B. burgdorferi* (Samuels and Garon, 1997). While they are considerably less efficient than PCR fragments on a molar basis, oligonucleotides represent a convenient source of transforming DNA if a positive selection for the targeted mutation exists. Since borrelial DNA, recombinant plasmids purified from *E. coli*, synthetic oligonucleotides and PCR fragments have all been used successfully to transform *B. burgdorferi*, the modification state of the DNA does not appear to be critical.

C. Gene Inactivation by Allelic Exchange

After developing a selectable marker and demonstrating transformation (Samuels *et al.*, 1994a), the next step was the targeted insertion of *gyrB*^{r} at a new site. In other bacteria *gyrB*^{r} is a dominant mutation (Maxwell, 1993), and this would have to be true in borreliae as well. If a similar approach were taken to develop another selectable marker based upon a mutation in a different endogenous gene, it must also confer a dominant phenotype in order to be useful in allelic exchange.

The *gyrB*^{r} gene was successfully targeted to a non-essential site on a circular plasmid via homologous recombination with flanking sequences (Rosa *et al.*, 1996; see also Chapters 10, II, III and IV and 14, IV.A). Subsequently, the same approach was used to inactivate several structural genes on this plasmid (Tilly *et al.*, 1997; Tilly *et al.*, 1998). These experiments confirmed that *gyrB*^{r} was dominant. Although the borrelial chromosome and plasmids can be present at multiple copies per spirochaete (Kitten and Barbour, 1992), we found no examples of transformants in which the selectable marker was present in only some copies of the plasmids, confirming the feasibility of gene inactivation by allelic exchange in *B. burgdorferi*.

To date, targeted insertion of *gyrB*^{r} has been successful at all four plasmid sites with which allelic exchange has been attempted. In each case, however, insertion of *gyrB*^{r} into the plasmid via flanking sequences was inefficient relative to recombination with the chromosomal *gyrB* gene. Less than 0.5% of the coumermycin-resistant transformants had undergone allelic exchange at the targeted site; the majority of the transformants resulted from conversion of the wild-type chromosomal *gyrB* locus to *gyrB*^{r}. The relative inefficiency of recombination via plasmid flanking sequences required that large numbers of transformants be analysed in order to identify mutants that had undergone allelic exchange at the targeted gene. Colonies were directly screened by PCR, using primers that flanked the targeted insertion site. Transformants with targeted insertion on the plasmid yielded PCR fragments that were 2 kb (the size of the *gyrB*^{r} gene) larger than the wild-type fragment. Up to 96 colonies could be screened at a time using a Perkin Elmer model 9600 PCR machine, with size analysis of the amplification products on agarose gels. Only candidate colonies were grown up and analysed by Southern and Western blots to confirm targeted insertion of *gyrB*^{r} at the plasmid loci.

The screen is a labour-intensive, but not prohibitive, feature of using a selectable marker for which an endogenous copy of the gene is a second (and apparently preferred) site for allelic exchange. There was no evidence of illegitimate recombination resulting in insertion of *gyrB*^{r} at an unrelated site. The amount of plasmid-derived flanking sequence in the transforming DNA (approximately 1.5 kb on each side) may have been insufficient to efficiently mediate homologous recombination around a 2 kb segment of heterologous DNA (*gyrB*^{r}). Consistent with this explanation, targeted insertion of *gyrB*^{r} into the plasmid increased approximately 20-fold when the flanking DNA was extended to more than 10 kb on either side (Rosa *et al.*, 1996). Also, other transformation systems show preferential incorporation of small mismatches over large heterologous sequences (Pasta and Sicard, 1996).

D. Transformation into Infectious vs. Non-infectious Bacteria

Most of the transformations have been done with a single isolate of *B. burgdorferi sensu stricto*, the type strain B31 (ATCC 35210), and efficient transformation has only been accomplished with non-infectious, attenuated variants. In fact, experiments to date have used an uncloned B31 population, rendering transformants non-isogenic because of the heterogeneity of the recipient. Obviously, any comparisons between wild-type and mutant bacteria require that a clonal population be transformed.

Borrelia burgdorferi is maintained in nature in an infectious cycle between ticks and mammals. We and others are interested in the roles that particular gene products play in transmission between the tick vector and the mammalian host, and adaptation to these very different environments. We have inactivated several plasmid-encoded genes in the non-infectious B31 variant, with the intention of moving these mutations into an infectious background in order to analyse their phenotypes in the natural infectious cycle. So far we have not succeeded in transforming infectious *B. burgdorferi* by electroporation. A major technical hurdle seems to be related to the change in surface properties that accompanies loss of infectivity. Non-infectious *B. burgdorferi* variants are easily resuspended in the sucrose/glycerol solution in which electroporation is performed, whereas infectious *B. burgdorferi* form large insoluble masses that are not accessible to electroporation. Despite extensive efforts, we have not identified a solution compatible with electroporation in which infectious *B. burgdorferi* remain viable and do not clump. We are currently attempting to circumvent this problem with alternative methods of gene transfer in order to inactivate genes in infectious *B. burgdorferi*.

◆◆◆◆◆◆ IV. OTHER SPIROCHAETES

A number of non-borrelial genera of spirochaetes have been subjected to various degrees of genetic analysis. *Leptospira* species are carried by animals and some cause acute illness in both animals and humans (Vinetz, 1997), while others are non-pathogens. The genus *Treponema* includes a wide variety of species, ranging from *T. pallidum*, the non-cultivable causative agent of syphilis, to *T. denticola*, which is easily cultivable and associated with periodontal disease but has not been proven to be a pathogen. *Serpulina* species were originally thought to be treponemes, on the basis of their morphology, but were recently shown, on the basis of ribosomal RNA gene sequencing, to comprise a separate genus (Paster *et al.*, 1991; Stanton *et al.*, 1991). *Serpulina hyodysenteriae* is an anaerobic, but cultivable, bacterium that causes swine dysentery (Harris *et al.*, 1972). Other genera include *Spirochaeta*, which is composed of free-living bacteria, none of which is known to cause disease (Johnson, 1977) and *Brevinema*, a mouse and shrew isolate (Defosse *et al.*, 1995).

Genetic tools that can be used for laboratory manipulation of these diverse spirochaetes range from better than those found for *B. burgdorferi* to completely absent. Elegant experiments using *S. hyodysenteriae* have demonstrated both generalized transduction (Humphrey *et al.*, 1997) and allelic exchange following electroporation (ter Huurne *et al.*, 1992; Rosey *et al.*, 1995) whereas *T. pallidum* cannot survive or grow *in vitro*, creating a huge obstacle in all genetic manipulations. We will describe observations with potential for the development of tools in a number of spirochaetes, then concentrate on the two species (*T. denticola* and *S. hyodysenteriae*) in which transformation has been demonstrated. We are not aware of any published genetic studies of the other spirochaetes.

A. *Leptospira*

Leptospira interrogans appears to have a segmented genome, with a 4.5 Mb chromosome and at least one mini-chromosome, of 350 kb (Taylor *et al.*, 1991; Zuerner *et al.*, 1993). Both DNA species carry genes presumed to be essential (e.g. in *L. interrogans*, the *asd* gene, encoding aspartate β-semi-aldehyde dehydrogenase, is located on the small species and many essential genes, such as *dnaA* and the genes encoding ribosomal RNAs, are located on the large species). In contrast to *B. burgdorferi*, in which the restriction and genetic maps of isolates and genospecies are highly conserved, the restriction and genetic maps of *L. interrogans* serovars differ significantly (Herrmann *et al.*, 1991; Zuerner *et al.*, 1993). Recombination among repetitive sequences found in *L. interrogans* and other leptospires (Van Eys *et al.*, 1988; Zuerner and Bolin, 1988; Woodward and Sullivan, 1991; Pacciarini *et al.*, 1992) may account for some of the genome instability (Zuerner *et al.*, 1993).

Although no successful transformations have been reported in any *Leptospira* species, several experiments demonstrating potentially useful observations have been described. Phages of *L. biflexa* (a saprophytic leptospire) with double-stranded DNA genomes were isolated from the Paris sewers (Saint Girons *et al.*, 1990), although they appeared to have only lytic life cycles. *L. borgpetersenii* possesses a putative transposable element, with inverted repeats at its termini, an open reading frame coding for a protein related to previously characterized transposases, and duplications in the genomic DNA flanking its termini (Zuerner, 1994). This transposon is located at different sites in different serovars, although transposition within a strain has not been described. The phylogeny (among *Leptospira* species) of an intervening sequence in the large ribosomal RNA gene differs from that for *rrs*, encoding the small ribosomal RNA, a genetically unlinked marker (Ralph and McClelland, 1994), providing evidence for past horizontal transfer of DNA. Finally, the genomic DNA appears to have 4-methyl-C in the sequence GTAC, which may mean that DNA used for gene transfer experiments should be likewise modified in order to avoid restriction (Ralph *et al.*, 1993; see Chapter 1, IV.B).

B. *Treponema*

Treponema pallidum, the agent of syphilis, is uncultivable *in vitro*, severely limiting the possibility of genetic studies in this important pathogen. The complete genomic sequence was determined by a collaboration between The Institute for Genomic Research (http://www.tigr.org) and the University of Texas (http://utmmg.med.uth.tmc.edu/treponema/tpall.html), which may allow investigators to define essential nutrients for *in vitro* cultivation (Fraser *et al.*, 1998). A gene encoding a putative DNA adenine methylase has been identified, correlating well with the evidence from restriction enzyme sensitivities that the DNA has methyl adenine in the site GATC (Stamm *et al.*, 1997). A phage inducible by mitomycin C was described in *T. phagedenis* (Masuda and Kawata, 1979), but transduction has not been demonstrated.

Genetic studies in *T. denticola* have proceeded further, facilitated by its ability to grow in culture medium. Natural plasmids have been described (Ivic *et al.*, 1991; Caudry *et al.*, 1994) and the sequence of one resembled that of a class of plasmids found in Gram-positive bacteria (MacDougall *et al.*, 1992). Transformation via electroporation has been used in two ways. First, the *flgE* gene, encoding the flagellar hook protein, was inactivated by targeted insertion of a cassette conferring erythromycin resistance on recipient bacteria (Li *et al.*, 1996). Erythromycin-resistant transformants were obtained at a frequency of 0.9 transformants per microgram of transforming DNA, despite having only 250 bp of flanking sequences to provide homology for allelic exchange (Li *et al.*, 1996). A similar strategy was used to inactivate the *dmcA* gene, encoding a methyl-accepting chemotaxis protein (Kataoka *et al.*, 1997). In a second approach, a plasmid derived from the broad host range plasmid RSF1010 (pKT210) was introduced into *T. denticola* by electroporation (Li and Kuramitsu, 1996). The plasmid appeared to replicate as an episome, conferring chloramphenicol resistance on transformants (Li and Kuramitsu, 1996). These experiments provide the groundwork for further gene inactivations, with the additional prospect of complementing mutants with gene copies cloned into pKT210. Unfortunately, the lack of an animal model for *T. denticola* growth means that the effects of mutations on growth in a natural environment cannot be tested.

C. *Serpulina*

Several methods for gene transfer into *S. hyodysenteriae* have been developed. Furthermore, determining the importance of targeted genes in the disease process was possible because infectious strains are transformable and there is an animal model.

I. Electroporation

Two markers have been used in targeted gene inactivation by electroporation. First, although the bacteria are resistant to relatively high levels of kanamycin, a gene conferring kanamycin resistance (*kan*) was inserted into the gene for a haemolysin and used to transform *S. hyodysenteriae* by electroporation (ter Huurne *et al.*, 1992). The efficiency of transformation in this experiment was 0.2–0.8 transformants per microgram of transforming DNA. The mutants had reduced virulence in mice and pigs (ter Huurne *et al.*, 1992; Hyatt *et al.*, 1994), marking the first genetic test of the importance of a gene for infection by a spirochaete.

A second set of experiments used the *kan* gene and a gene conferring chloramphenicol resistance (*cat*) to inactivate the *flaA1* and *flaB1* genes, which encode flagellar components (Rosey *et al.*, 1995). These authors found transformation efficiencies of 20–40 transformants per microgram of transforming DNA. The improved transformation efficiency may have resulted from several differences in their protocols, including using fresh rather than frozen electrocompetent bacteria, and differences in electroporation cuvettes, voltages, and electroporation buffers. Circular plasmid DNA was used in the transformations, as was true for the earlier experiments, and all of the constructs had relatively short (<1 kb) flanking sequences (ter Huurne *et al.*, 1992; Rosey *et al.*, 1995). Surprisingly, the *flaA1* and *flaB1* mutants and a double mutant had normal-looking flagella but were attenuated in a mouse model of swine dysentery (Rosey *et al.*, 1995, 1996; Kennedy *et al.*, 1997).

2. Bacteriophage-mediated generalized transduction

Bacteriophages were first identified in the spirochaete *S. hyodysenteriae* in 1978 (Ritchie *et al.*, 1978). This study inspired Humphrey and co-workers (Humphrey *et al.*, 1995) to demonstrate that bacteriophages could be induced from cultures by mitomycin C treatment. Bacteriophage-associated DNA appeared to be similar to a previously characterized extrachromosomal DNA population (Combs *et al.*, 1992). Upon further characterization, both the bacteriophage DNA and the extrachromosomal DNA were found to be random bacterial DNA fragments of a uniform size (Humphrey *et al.*, 1997; Turner and Sellwood, 1997), suggesting that the bacteriophage might be packaging host DNA within its head. Bacteriophages derived from a strain carrying the *flaA1::cat* mutation were able to transfer that mutation into a strain carrying a *nox::kan* mutation (inactivating the gene for NADH oxidase), demonstrating the feasibility of using the bacteriophages for generalized transduction (Humphrey *et al.*, 1997). Cultures treated with mitomycin C and mixed with untreated bacteria were also able to transfer genes. Gene transfer via generalized transduction is very efficient, so methods like mapping random transposon insertions and determining linkage by co-transduction mapping might become feasible. The bacteriophage genome has yet to be identified, although, presumably, it is integrated into the *S. hyodysenteriae* chromosome, since the bacteria have no plasmids. Another unique feature of this system is the absence of surface exclusion by lysogenic conversion, in which a prophage alters the bacterial surface to prevent attachment of superinfecting bacteriophage (Robbins *et al.*, 1965), since all isolates of *S. hyodysenteriae* tested appear to be lysogens yet they are competent recipients for transduction.

◆◆◆◆◆◆ V. PERSPECTIVES

The contrast between the tools available for *S. hyodysenteriae*, with multiple selectable markers, two methods of transferring genes into the bacteria, and transformable infectious strains, and our primary example, *B. burgdorferi*, with a single problematic marker and a method of gene transfer that works poorly with infectious bacteria, demonstrates that status as a human pathogen is not a guarantee of rapid progress. To facilitate genetic studies of *B. burgdorferi*, additional genetic markers, shuttle vectors, methods of mutagenesis, and improved culture medium and plating protocols would be useful (see Chapters 1, 2, 3 and 4 of this volume for general discussions of these). Some or all of these are also missing in the other spirochaete systems.

A complementary approach that has and will provide additional insights is reverse genetics. These techniques include methods to assay regulation by reconstructing systems in *E. coli*, such as *phoA* fusions to isolate sequences carrying signals for protein secretion (Giladi *et al.*, 1993). A second approach is to design reporter systems for use in the spirochaete, as in assaying chloramphenicol acetyl transferase activity after transient transfection with *B. burgdorferi* promoters fused to a *cat* gene (Sohaskey *et al.*, 1997). A third approach is constructing cDNA libraries from bacteria grown in various conditions and performing subtractive hybridizations to identify genes induced by the variable stimuli. Despite the number of investigators interested in spirochaetes and the myriad techniques possible, it will be a long time before the potential for original genetic studies of these diverse bacteria is exhausted.

Genetic Methods in Borrelia

References

Abeles, A. L. Snyder, K. M. and Chattoraj, D. K. (1984). P1 plasmid replication: replicon structure. *J. Mol. Biol.* **173**, 307–324.

Barbour, A. G. (1988). Plasmid analysis of *Borrelia burgdorferi*, the Lyme disease agent. *J. Clin. Microbiol.* **26**, 475–478.

Barbour, A. G. (1993). Linear DNA of *Borrelia* species and antigenic variation. *Trends Microbiol.* **1**, 236–239.

Barbour, A. G. and Garon, C. F. (1987). Linear plasmids of the bacterium *Borrelia burgdorferi* have covalently closed ends. *Science* **237**, 409–411.

Barbour, A. G. and Hayes, S. F. (1986). Biology of *Borrelia* species. *Microbiol. Rev.* **50**, 381–400.

Barbour, A. G., Burman, N., Carter, C. J., Kitten, T. and Bergstrom, S. (1991). Variable antigen genes of the relapsing fever agent *Borrelia hermsii* are activated by promoter addition. *Mol. Microbiol.* **5**, 489–493.

Barbour, A. G., Carter, C. J., Bundoc, V. and Hinnebusch, J. (1996). The nucleotide sequence of a linear plasmid of *Borrelia burgdorferi* reveals similarities to those of circular plasmids of other prokaryotes. *J. Bacteriol.* **178**, 6635–6639.

Baril, C., Richaud, C., Baranton, G. and Saint Girons, I. (1989). Linear chromosome of *Borrelia burgdorferi*. *Res. Microbiol.* **140**, 507–516.

Burgdorfer, W., Barbour, A. G., Hayes, S. F., Benach, J. L., Grunwaldt, E. and Davis, J. P. (1982). Lyme disease – a tick-borne spirochetosis? *Science* **216**, 1317–1319.

Casjens, S. and Earnshaw, W. (1980). DNA packaging by the double-stranded DNA bacteriophages. *Cell* **21**, 319–331.

Casjens, S. and Huang, W. M. (1993). Linear chromosomal physical and genetic map of *Borrelia burgdorferi*, the Lyme disease agent. *Mol. Microbiol.* **8**, 967–980.

Casjens, S., Hatfull, G. and Hendrix, R. (1992). Evolution of dsDNA tailed-bacteriophage genomes. *Semin. Virol.* **3**, 383–397.

Casjens, S., DeLange, M., Ley, H. L., III, Rosa, P. and Huang, W. M. (1995). Linear chromosomes of Lyme disease agent spirochetes: genetic diversity and conservation of gene order. *J. Bacteriol.* **177**, 2769–2780.

Casjens, S., Murphy, M., DeLange, M., Sampson, L., van Vugt, R. and Huang, W. M. (1997a). Telomeres of the linear chromosomes of Lyme disease spirochaetes: nucleotide sequence and possible exchange with linear plasmid telomeres. *Mol. Microbiol.* **26**, 581–596.

Casjens, S., van Vugt, R., Tilly, K., Rosa, P. A. and Stevenson, B. (1997b). Homology throughout the multiple 32-kilobase circular plasmids present in Lyme disease spirochetes. *J. Bacteriol.* **179**, 217–227.

Caudry, S., Klitorinos, A., Gharbia, S. E., Pssara, N., Siboo, R., Keng, T. and Chan, E. C. S. (1995). Distribution and characterization of plasmids in oral anaerobic spirochetes. *Oral Microbiol. Immunol.* **10**, 8–12.

Combs, B. G., Hampson, D. J. and Harders, S. J. (1992). Typing of Australian isolates of *Treponema hyodysenteriae* by serology and by DNA restriction endonuclease analysis. *Vet. Microbiol.* **31**, 273–285.

Defosse, D. L., Johnson, R. C., Paster, B. J., Dewhirst, F. E. and Fraser, G. J. (1995). *Brevinema andersonii* gen. nov., sp. nov., an infectious spirochete isolated from the short-tailed shrew (*Blarina brevicauda*) and the white-footed mouse (*Peromyscus leucopus*). *Int. J. Syst. Bacteriol.* **45**, 78–84.

Dew-Jager, K., Yu, W.-Q. and Huang, W. M. (1995). The *recA* gene of *Borrelia burgdorferi*. *Gene* **167**, 137–140.

Dunn, J. J., Buchstein, S. R., Butler, L.-L., Fisenne, S., Polin, D. S., Lade, B. N. and Luft, B. J. (1994). Complete nucleotide sequence of a circular plasmid from the Lyme disease spirochete, *Borrelia burgdorferi*. *J. Bacteriol.* **176**, 2706–2717.

Dykhuizen, D. E., Polin, D. S., Dunn, J., Wilske, B., Preac-Mursic, V., Dattwyler, R. J. and Luft, B. J. (1993). *Borrelia burgdorferi* is clonal: implications for taxonomy and vaccine development. *Proc. Natl. Acad. Sci. USA* **90**, 10163–10167.

Ferdows, M. S. and Barbour, A. G. (1989). Megabase-sized linear DNA in the bacterium *Borrelia burgdorferi*, the Lyme disease agent. *Proc. Natl. Acad. Sci. USA* **86**, 5969–5973.

Ferdows, M. S., Serwer, P., Griess, G. A., Norris, S. J. and Barbour, A. G. (1996). Conversion of a linear to a circular plasmid in the relapsing fever agent *Borrelia hermsii*. *J. Bacteriol.* **178**, 793–800.

Fraser, C. M., Casjens, S., Wai Mun Huang, Sutton, G. G., Clayton, R., Lathigra, R., White, O., Ketchum, K. A., Dodson, R., Hickey, E. K., Gwinn, M., Dougherty, B., Tomb, J-F., Fleischmann, R. D., Richardson, D., Peterson, J., Kerlavage, A. R., Quackenbush, J., Salzberg, S., Hanson, M., van Vugt, R., Palmer, N., Adams, M. D., Gocayne, J., Weidman, J., Utterback, T., Watthey, L., McDonald, L., Artiach, P., Bowman, C., Garland, S., Fujii, C., Cotton, M. D., Horst, K., Roberts, K., Hatch, B., Smith, H.O., Venter, J. C. (1997) Genomic sequence of a Lyme disease spirochaete, *Borrelia burgdorferi*. *Nature* **390**, 580–586.

Fraser, C. M., Norris, S. J., Weinstock, G. M., White, O., Sutton, G. G., Dodson, R., Gwinn, M., Hickey, E. K., Clayton, R., Ketchum, K. A., Sodergren, E., Hardham, J. M., McLeod, M. P., Salzberg, S., Peterson, J., Khalak, H., Richardson, D., Howell, J. K., Chidambaram, M., Utterback, T., McDonald, L., Artiach, P., Bowman, C., Cotton, M. D., Fujii, C., Garland, S., Hatch, B., Horst, K.,

Roberts, K., Sandusky, M., Weidman, J., Smith, H. O., Venter, J. C. (1998) Complete genome sequence of *Treponema pallidum*, the syphilis spirochete. *Science* **281**, 375–378, 387–388.

Fukunaga, M., Okada, K., M., N., Konishi, T. and Sato, Y. (1996). Phylogenetic analysis of *Borrelia* species based on flagellin gene sequences and its application for molecular typing of Lyme disease borreliae. *Int. J. Syst. Bacteriol.* **46**, 898–905.

Gibbs, C. P., Livey, I. and Dorner, F. (1996). The role of recombination in OspC variation in Lyme disease *Borrelia*. *Acta Dermatovenerologica Alpina, Pannonica et Adriatica* **5**, 179–183.

Giladi, M., Champion, C. I., Haake, D. A., Blanco, D. R., Miller, J. F., Miller, J. N. and Lovett, M. A. (1993). Use of the "blue halo" assay in the identification of genes encoding exported proteins with cleavable signal peptides: cloning of a *Borrelia burgdorferi* plasmid gene with a signal peptide. *J. Bacteriol.* **175**, 4129–4136.

Harris, D. L., Glock, R. D., Christensen, C. R. and Kinyon, J. M. (1972). Swine dysentery. I. Inoculation of pigs with *Treponema hyodysenteriae* (new species) and reproduction of the disease. *Vet. Med. Small Anim. Clin.* **67**, 61–64.

Hayes, L. J., Wright, D. J. M. and Archard, L. C. (1988). Segmented arrangement of *Borrelia duttonii* DNA and location of variant surface antigen genes. *J. Gen. Microbiol.* **134**, 1785–1793.

Hayes, S. F., Burgdorfer, W. and Barbour, A. G. (1983). Bacteriophage in the *Ixodes dammini* spirochete, etiological agent of Lyme disease. *J. Bacteriol.* **154**, 1436–1439.

Herrmann, J. L., Baril, C., Bellenger, E., Perolat, P., Baranton, G. and Saint Girons, I. (1991). Genome conservation in isolates of *Leptospira interrogans*. *J. Bacteriol.* **173**, 7582–7588.

Hinnebusch, J. and Barbour, A. G. (1992). Linear- and circular-plasmid copy numbers in *Borrelia burgdorferi*. *J. Bacteriol.* **174**, 5251–5257.

Hinnebusch, J. and Tilly, K. (1993). Linear plasmids and chromosomes in bacteria. *Mol. Microbiol.* **10**, 917–922.

Humphrey, S. B., Stanton, T. B. and Jensen, N. S. (1995). Mitomycin C induction of bacteriophages from *Serpulina hyodysenteriae* and *Serpulina innocens*. *FEMS Microbiol. Lett.* **134**, 97–101.

Humphrey, S. B., Stanton, T. B., Jensen, N. S. and Zuerner, R. L. (1997). Purification and characterization of VSH-1, a generalized transducing bacteriophage of *Serpulina hyodysenteriae*. *J. Bacteriol.* **179**, 323–329.

Hyatt, D. R., ter Huurne, A. A. H. M., van der Zeijst, B. A. M. and Joens, L. A. (1994). Reduced virulence of *Serpulina hyodysenteriae* hemolysin-negative mutants in pigs and their potential to protect pigs against a challenge with a virulent strain. *Infect. Immun.* **62**, 2244–2248.

Hyde, F. W. and Johnson, R. C. (1988). Characterization of a circular plasmid from *Borrelia burgdorferi*, etiologic agent of Lyme disease. *J. Clin. Microbiol.* **26**, 2203–2205.

Ivic, A., MacDougall, J., Russell, R. R. B. and Penn, C. W. (1991). Isolation and characterization of a plasmid from *Treponema denticola*. *FEMS Microbiol. Lett.* **78**, 189–194.

Jauris-Heipke, S., Liegl, G., Preac-Mursic, V., Robler, D., Schwab, E., Soutschek, E., Will, G. and Wilske, B. (1995). Molecular analysis of genes encoding outer surface protein C (OspC) of *Borrelia burgdorferi* sensu lato: relationship to *ospA* genotype and evidence of lateral gene exchange of *ospC*. *J. Clin. Microbiol.* **33**, 1860–1866.

Johnson, R. C. (1977). The spirochetes. *Ann. Rev. Microbiol.* **31**, 89–106.

Johnson, R. C., Schmid, G. P., Hyde, F. W., Steigerwalt, A. G. and Brenner, D. J. (1984). *Borrelia burgdorferi* sp. nov.: etiologic agent of Lyme disease. *Int. J. Syst. Bacteriol.* **34**, 496–497.

Kataoka, M., Li, H., Arakawa, S. and Kuramitsu, H. (1997). Characterization of a methyl-accepting chemotaxis protein gene, *dmcA*, from the oral spirochete *Treponema denticola*. *Infect. Immun.* **65**, 4011–4016.

Kennedy, M. J., Rosey, E. L. and Yancey, R. L., Jr (1997). Characterization of *flaA⁻* and *flaB⁻* mutants of *Serpulina hyodysenteriae*: both flagellin subunits, FlaA and FlaB, are necessary for full motility and intestinal colonization. *FEMS Microbiol. Lett.* **153**, 119–128.

Kitten, T. and Barbour, A. G. (1992). The relapsing fever agent *Borrelia hermsii* has multiple copies of its chromosome and linear plasmids. *Genetics* **132**, 311–324.

Li, H. and Kuramitsu, H. K. (1996). Development of a gene transfer system in *Treponema denticola* by electroporation. *Oral Microbiol. Immunol.* **11**, 161–165.

Li, H., Ruby, J., Charon, N. and Kuramitsu, H. (1996). Gene inactivation in the oral spirochete *Treponema denticola*: construction of an *flgE* mutant. *J. Bacteriol.* **178**, 3664–3667.

Livey, I., Gibbs, C. P., Schuster, R. and Dorner, F. (1995). Evidence for lateral transfer and recombination in OspC variation in Lyme disease *Borrelia*. *Mol. Microbiol.* **18**, 257–269.

MacDougall, J., Margarita, D. and Saint Girons, I. (1992). Homology of a plasmid from the spirochete *Treponema denticola* with the single-stranded DNA plasmids. *J. Bacteriol.* **174**, 2724–2728.

Marconi, R. T., Samuels, D. S. and Garon, C. F. (1993). Transcriptional analyses and mapping of the *ospC* gene in Lyme disease spirochetes. *J. Bacteriol.* **175**, 926–932.

Marconi, R. T., Casjens, S., Munderloh, U. G. and Samuels, D. S. (1996). Analysis of linear plasmid dimers in *Borrelia burgdorferi* sensu lato isolates: implications concerning the potential mechanism of linear plasmid replication. *J. Bacteriol.* **178**, 3357–3361.

Masuda, K. and Kawata, T. (1979). Bacteriophage-like particles induced from the Reiter treponeme by mitomycin C. *FEMS Microbiol. Lett.* **6**, 29–31.

Maxwell, A. (1993). The interaction between coumarin drugs and DNA gyrase. *Mol. Microbiol.* **9**, 681–686.

Meier, J. T., Simon, M. I. and Barbour, A. G. (1985). Antigenic variation is associated with DNA rearrangements in a relapsing fever *Borrelia*. *Cell* **41**, 403–409.

Motallebi-Veshareh, M., Rouch, D. A. and Thomas, C. M. (1990). A family of ATPases involved in active partitioning of diverse bacterial plasmids. *Mol. Microbiol.* **4**, 1455–1463.

Munderloh, U. G., Park, Y.-J., Dioh, J. M., Fallon, A. M. and Kurtti, T. J. (1993). Plasmid modifications in a tick-borne pathogen, *Borrelia burgdorferi*, cocultured with tick cells. *Insect Mol. Biol.* **1**, 195–203.

Neubert, U., Schaller, M., Januschke, E., Stolz, W. and Schmieger, H. (1993). Bacteriophages induced by ciprofloxacin in a *Borrelia burgdorferi* skin isolate. *Zentralbl. Bakteriol.* **279**, 307–315.

Norton Hughes, C. A. and Johnson, R. C. (1990). Methylated DNA in *Borrelia* species. *J. Bacteriol.* **172**, 6602–6604.

Ogura, T. and Hiraga, S. (1983). Partition mechanism of F plasmid: two plasmid gene-encoded products and a cis-acting region are involved in partition. *Cell* **32**, 351–360.

Ojaimi, C., Davidson, B. E., Saint Girons, I. and Old, I. G. (1994). Conservation of gene arrangement and an unusual organization of rRNA genes in the linear chromosomes of the Lyme disease spirochaetes *Borrelia burgdorferi*, *B. garinii* and *B. afzelii*. *Microbiology* **140**, 2931–2940.

Pacciarini, M. L., Savio, M. L., Tagliabue, S. and Rossi, C. (1992). Repetitive sequences cloned from *Leptospira interrogans* servovar hardjo genotype hardjoprajitno and their application to serovar identification. *J. Clin. Microbiol.* **30**, 1243–1249.

Pasta, F. and Sicard, M. A. (1996). Exclusion of long heterologous insertions and deletions from the pairing synapsis in pneumococcal transformation. *Microbiology* **142**, 695–705.

Paster, B. J., Stackebrandt, E., Hespell, R. B., Hahn, C. M. and Woese, C. R. (1984). The phylogeny of the spirochetes. *Syst. Appl. Microbiol.* **5**, 337–351.

Paster, B. J., Dewhirst, F. E., Weisburg, W. G., Tordoff, L. A., Fraser, G. J., Hespell, R. B., Stanton, T. B., Zablen, L., Mandelco, L. and Woese, C. R. (1991). Phylogenetic analysis of the spirochetes. *J. Bacteriol.* **173**, 6101–6109.

Perng, G. C. and LeFebvre, R. B. (1990). Expression of antigens from chromosomal and linear plasmid DNA of *Borrelia coriaceae. Infect. Immun.* **58**, 1744–1748.

Plasterk, R. H. A., Simon, M. I. and Barbour, A. G. (1985). Transposition of structural genes to an expression sequence on a linear plasmid causes antigenic variation in the bacterium *Borrelia hermsii. Nature* **318**, 257–263.

Porcella, S. F., Popova, T. G., Akins, D. R., Li, M., Radolf, J. D. and Norgard, M. V. (1996). *Borrelia burgdorferi* supercoiled plasmids encode multi-copy tandem open reading frames and a lipoprotein gene family. *J. Bacteriol.* **178**, 3293–3307.

Ralph, D. and McClelland, M. (1994). Phylogenetic evidence for horizontal transfer of an intervening sequence between species in a spirochete genus. *J. Bacteriol.* **176**, 5982–5987.

Ralph, D., Que, Q., Van Etten, J. L. and McClelland, M. (1993). *Leptospira* genomes are modified as 5′-GTAC. *J. Bacteriol.* **175**, 3913–3915.

Restrepo, B. I. and Barbour, A. G. (1994). Antigen diversity in the bacterium *B. hermsii* through "somatic" mutations in rearranged *vmp* genes. *Cell* **78**, 867–876.

Ritchie, A. E., Robinson, I. M., Joens, L. A. and Kinyon, J. M. (1978). A bacteriophage for *Treponema hyodysenteriae. Veterin. Rec.* **102**, 34–35.

Robbins, P. W., Keller, J. M., Wright, A. and Bernstein, R. L. (1965). Enzymatic and kinetic studies on the mechanism of O-antigen conversion by bacteriophage ϵ^{15}. *J. Biol. Chem.* **240**, 384–390.

Rosa, P. A., Schwan, T. and Hogan, D. (1992). Recombination between genes encoding major outer surface proteins A and B of *Borrelia burgdorferi. Mol. Microbiol.* **6**, 3031–3040.

Rosa, P., Samuels, D. S., Hogan, D., Stevenson, B., Casjens, S. and Tilly, K. (1996). Directed insertion of a selectable marker into a circular plasmid of *Borrelia burgdorferi. J. Bacteriol.* **178**, 5946–4953.

Rosey, E. L., Kennedy, M. J., Petrella, D. K., Ulrich, R. G. and Yancey, R. J., Jr. (1995). Inactivation of *Serpulina hyodysenteriae flaA1* and *flaB1* periplasmic flagellar genes by electroporation-mediated allelic exchange. *J. Bacteriol.* **177**, 5959–5970.

Rosey, E. L., Kennedy, M. L. and Yancey, R. J., Jr (1996). Dual *flaA1 flaB1* mutant of *Serpulina hyodysenteriae* expressing periplasmic flagella is severely attenuated in a murine model of swine dysentery. *Infect. Immun.* **64**, 4154–4162.

Sadziene, A., Wilske, B., Ferdows, M. S. and Barbour, A. G. (1993). The cryptic *ospC* gene of *Borrelia burgdorferi* B31 is located on a circular plasmid. *Infect. Immun.* **61**, 2192–2195.

Saint Girons, I., Margarita, D., Amouriaux, P. and Baranton, G. (1990). First isolation of bacteriophages for a spirochaete: potential genetic tools for *Leptospira. Res. Microbiol.* **141**, 1131–1138.

Samuels, D. S. (1995). Electrotransformation of the spirochete *Borrelia burgdorferi*. In *Methods in Molecular Biology* (J. A. Nickoloff, ed.), pp. 253–259. Humana Press, Totowa, New Jersey.

Samuels, D. S. and Garon, C. F. (1993). Coumermycin A_1 inhibits growth and induces relaxation of supercoiled plasmids in *Borrelia burgdorferi*, the Lyme disease agent. *Antimicrob. Agents Chemother.* **37**, 46–50.

Samuels, D. S. and Garon, C. F. (1997). Oligonucleotide-mediated genetic transformation of *Borrelia burgdorferi*. *Microbiology* **143**, 519–522.

Samuels, D. S., Marconi, R. T. and Garon, C. F. (1993). Variation in the size of the *ospA*-containing linear plasmid, but not the linear chromosome, among three *Borrelia* species associated with Lyme disease. *J. Gen. Microbiol.* **139**, 2445–2449.

Samuels, D. S., Mach, K. E. and Garon, C. F. (1994a). Genetic transformation of the Lyme disease agent *Borrelia burgdorferi* with coumarin-resistant *gyrB*. *J. Bacteriol.* **176**, 6045–6049.

Samuels, D. S., Marconi, R. T., Huang, W. M. and Garon, C. F. (1994b). *gyrB* mutations in coumermycin A_1-resistant *Borrelia burgdorferi*. *J. Bacteriol.* **176**, 3072–3075.

Schaller, M. and Neubert, U. (1994). Bacteriophages and ultrastructural alterations of *Borrelia burgdorferi* induced by ciprofloxacin. *J. Spirochetal Tick-Borne Dis.* **1**, 37–40.

Simpson, W. J., Garon, C. F. and Schwan, T. G. (1990). Analysis of supercoiled circular plasmids in infectious and non-infectious *Borrelia burgdorferi*. *Microb. Pathogen.* **8**, 109–118.

Sohaskey, C. D., Arnold, C. and Barbour, A. G. (1997). Analysis of promoters in *Borrelia burgdorferi* by use of a transiently expressed reporter gene. *J. Bacteriol.* **179**, 6837–6842.

Stålhammar-Carlemalm, M., Jenny, E., Gern, L., Aeschlimann, A. and Meyer, J. (1990). Plasmid analysis and restriction fragment length polymorphisms of chromosomal DNA allow a distinction between *Borrelia burgdorferi* strains. *Zentralbl. Bakteriol.* **274**, 28–39.

Stamm, L. V., Greene, S. R., Barnes, N. Y., Bergen, H. L. and Hardham, J. M. (1997). Identification and characterization of a *Treponema pallidum* subsp. *pallidum* gene encoding a DNA adenine methyltransferase. *FEMS Microbiol. Lett.* **155**, 115–119.

Stanton, T. B., Jensen, N. S., Casey, T. A., Tordoff, L. A., Dewey, F. E. and Paster, B. J. (1991). Reclassification of *Treponema hyodysenteriae* and *Treponema innocens* in a new genus, *Serpula* gen. nov., as *Serpula hyodysenteriae* comb. nov. and *Serpula innocens* comb. nov. *Int. J. Syst. Bacteriol.* **41**, 50–58.

Steere, A. C., Grodzicki, R. L., Kornblatt, A. N., Craft, J. E., Barbour, A. G., Burgdorfer, W., Schmid, G. P., Johnson, E. and Malawista, S. E. (1983). The spirochetal etiology of Lyme disease. *N. Engl. J. Med.* **308**, 733–740.

Stevenson, B. and Barthold, S. W. (1994). Expression and sequence of outer surface protein C among North American isolates of *Borrelia burgdorferi*. *FEMS Microbiol. Lett.* **124**, 367–372.

Stevenson, B., Tilly, K. and Rosa, P. A. (1996). A family of genes located on four separate 32-kilobase circular plasmids in *Borrelia burgdorferi* B31. *J. Bacteriol.* **178**, 3508–3516.

Stevenson, B., Casjens, S., van Vugt, R., Porcella, S. F., Tilly, K., Bono, J. L. and Rosa, P. (1997). Characterization of cp18, a naturally truncated member of the cp32 family of *Borrelia burgdorferi* plasmids. *J. Bacteriol.* **179**, 4285–4291.

Stevenson, B., Casjens, S. and Rosa, P. (1998). Evidence of past recombination events among the genes encoding the Erp antigens of *Borrelia burgdorferi*. *Microbiology* **144**, 1869–1879.

Taylor, K. A., Barbour, A. G. and Thomas, D. D. (1991). Pulsed-field gel electrophoretic analysis of leptospiral DNA. *Infect. Immun.* **59**, 323–329.

ter Huurne, A. A. H. M., van Houten, M., Muir, S., Kusters, J. G., van der Zeijst, B. A. M. and Gaastra, W. (1992). Inactivation of a *Serpula (Treponema) hyodysenteriae* hemolysin gene by homologous recombination: importance of this hemolysin in pathogenesis of *S. hyodysenteriae* in mice. *FEMS Microbiol. Lett.* **92**, 109–114.

Tilly, K., Casjens, S., Stevenson, B., Bono, J. L., Samuels, D. S., Hogan, D. and Rosa, P. (1997). The *Borrelia burgdorferi* circular plasmid cp26: conservation of plasmid structure and targeted inactivation of the *ospC* gene. *Mol. Microbiol.* **25**, 361–373.

Tilly, K., Lubke, L. and Rosa, P. (1998). Characterization of circular plasmid dimers in *Borrelia burgdorferi*. *J. Bacteriol* **180**, 5676–5681.

Turner, A. K. and Sellwood, R. (1997). Extracellular DNA from *Serpulina hyodysenteriae* consists of 6.5 kbp random fragments of chromosomal DNA. *FEMS Microbiol. Lett.* **150**, 75–80.

Van Eys, G. J. J. M., Zaal, J., Schoone, G. J. and Terpstra, W. J. (1988). DNA hybridization with hardjobovis-specific recombinant probes as a method for type discrimination of *Leptospira interrogans* serovar *hardjo*. *J. Gen. Microbiol.* **134**, 567–574.

Vinetz, J. M. (1997). Leptospirosis. *Curr. Opin. Infect. Dis.* **10**, 357–361.

Watanabe, E., Wachi, M., Yamasaki, M. and Nagai, K. (1992). ATPase activity of SopA, a protein essential for active partitioning of F plasmid. *Mol. Gen. Genet.* **234**, 346–352.

Woese, C. R. (1987). Bacterial evolution. *Microbiol. Rev.* **51**, 221–271.

Woodward, M. J. and Sullivan, G. J. (1991). Nucleotide sequence of a repetitive element isolated from *Leptospira interrogans* serovar *hardjo* type *hardjo-bovis*. *J. Gen. Microbiol.* **137**, 1101–1109.

Xu, Y. and Johnson, R. C. (1995). Analysis and comparison of plasmid profiles of *Borrelia burgdorferi* sensu lato strains. *J. Clin. Microbiol.* **33**, 2679–2685.

Zhang, J.-R., Hardham, J. M., Barbour, A. G. and Norris, S. J. (1997). Antigenic variation in Lyme disease borreliae by promiscuous recombination of VMP-like sequence cassettes. *Cell* **89**, 1–20.

Zückert, W. R. and Meyer, J. (1996). Circular and linear plasmids of Lyme disease spirochetes have extensive homology: characterization of a repeated DNA element. *J. Bacteriol.* **178**, 2287–2298.

Zuerner, R. L. (1994). Nucleotide sequence analysis of IS*1533* from *Leptospira borgpetersenii* : identification and expression of two IS-encoded proteins. *Plasmid* **31**, 1–11.

Zuerner, R. L. and Bolin, C. A. (1988). Repetitive sequence element cloned from *Leptospira interrogans* serovar hardjo type hardjo-bovis provides a sensitive diagnostic probe for bovine leptospirosis. *J. Clin. Microbiol.* **26**, 2495–2500.

Zuerner, R. L., Herrmann, J. L. and Saint Girons, I. (1993). Comparison of genetic maps for two *Leptospira interrogans* serovars provides evidence for two chromosomes and intraspecies heterogeneity. *J. Bacteriol.* **175**, 5445–5451.

8 Genetic Methods for *Bacteroides* Species

Abigail A. Salyers, Nadja Shoemaker, Andrew Cooper, John D'Elia and Joseph A. Shipman
Department of Microbiology, University of Illinois, Urbana, Illinois, USA

CONTENTS

◆◆◆◆◆◆ ## I. IMPORTANCE OF *BACTEROIDES* AND RELATED BACTERIA

Bacteroides spp. are Gram-negative obligate anaerobes. Although *Bacteroides* have a Gram-negative cell envelope, they are not members of the *Escherichia coli* phylogenetic group but belong to a phylogenetic group that is more distantly related to *E. coli* than the Gram-positive bacteria (Weisburg *et al.*, 1985). Given this, it is not surprising that *E. coli* plasmids, even the supposedly broad host range ones, do not replicate in *Bacteroides* and *E. coli* promoters do not work in *Bacteroides*. The *Bacteroides* phylogenetic group contains a number of important members.

Bacteroides is one of the numerically predominant genera of bacteria in the human colon. *Bacteroides* spp. such as *Bacteroides fragilis* and *Bacteroides thetaiotaomicron* are opportunistic human pathogens. *Bacteroides* infections have become a problem due to increasing antibiotic resistance in this group. Most of the resistance genes are on mobile elements. These mobile elements include not only plasmids but also integrated gene transfer elements such as conjugative transposons and NBUs (Salyers *et al.*, 1995). Another clinically important genus in the *Bacteroides* phylogenetic group is *Porphyromonas*. *Porphyromonas gingivalis* and other *Porphyromonas* species have been implicated in periodontal disease. Some of the vectors developed for use in *Bacteroides* have been used successfully for the genetic manipulation of *P. gingivalis* (Dyer *et al.*, 1992; Park and McBride, 1993; Genco *et al.*, 1995).

In the rumen and intestine of cattle, *Prevotella ruminicola*, another member of the *Bacteroides* phylogenetic group, is an important member of the bacterial microflora. *Bacteroides* genetic tools have been used as a basis for genetic engineering of *P. ruminicola* (Shoemaker *et al.*, 1991; Gardner *et al.*, 1996). Finally, a number of environmentally important bacteria such as cytophagas and flexibacters are members of the *Bacteroides* group. Recently, *Bacteroides* genetic tools have been used successfully for the genetic manipulation of *Cytophaga johnsoni* (McBride and Baker, 1996; McBride and Kempf, 1996).

◆◆◆◆◆◆ II. MOVING DNA INTO *BACTEROIDES*

A. Conjugation vs. Transformation

The first demonstrations of gene transfer from *E. coli* to *Bacteroides* were done using conjugation. Although *E. coli* plasmids generally do not replicate in *Bacteroides* and vice versa, the broad host range IncP plasmids such as R751 or RK2 are capable of transferring themselves or mobilizing other plasmids into *Bacteroides* (see Chapter 2, III.C–IV.B). These plasmids were used to mobilize the first *E. coli–Bacteroides* shuttle vectors and have subsequently formed the backbone of *Bacteroides* genetic manipulations. Using electroporation, we and others have tried to introduce plasmids into *Bacteroides* with very limited success. The usual experience has been that plasmids isolated from a particular *Bacteroides* strain can be electroporated back into the same strain with relatively high frequency, but that when the same plasmid was isolated from *E. coli* no electroporants were obtained. The most likely explanation for the poor success of transformation is that most *Bacteroides* strains contain restriction enzymes that interfere with the survival of double-stranded foreign DNA (see Chapter 1, IV.B for further discussion).

Smith (1985) has developed a transformation protocol for one strain of *B. fragilis*, *B. fragilis* 638, which introduces DNA into the strain at frequencies high enough to make transposon mutagenesis and gene disruption possible. *Bacteroides uniformis* strain 0061 can be electroporated, but the frequency is very low. Thus, for most strains of *Bacteroides* conjugation is the only means of introducing foreign DNA. Unfortunately, although conjugation works for many *Bacteroides* strains, there are strains that do not serve as recipients for transfer of cloned DNA for unknown reasons. Thus, if you want to introduce DNA into a *Bacteroides* strain to study some biological phenomenon, you may have to try several strains before finding one amenable to genetic manipulation.

The protocol for conjugation is a simple one (Shoemaker *et al.*, 1986b). Both the *E. coli* donor and the *Bacteroides* recipient are grown in broth medium to an OD_{650} of less than 0.2 (about 10^8 per ml). Then, donor and recipient are mixed in a microfuge tube to a final volume of 1 to 1.5 ml, and the bacteria pelleted by centrifugation in a microfuge. The ratio of donor to recipient is 1 : 5 for *E. coli* to *Bacteroides* matings and 1 : 1 for *Bacteroides* to *Bacteroides* or *Bacteroides* to *E. coli* matings. The pellet is resuspended in 0.2 ml of medium and the suspended cells are placed on a Millipore filter on an agar plate. (*Warning*: If the drop spreads rapidly to the edge of the filter rather than staying in the centre, you have filters with detergent in them and they are probably killing your cells). The mixture is incubated aerobically overnight if *E. coli* is the donor. For matings in which *Bacteroides* is the donor, the filters are incubated overnight under anaerobic conditions. The medium can be either pre-reduced trypticase–yeast extract–glucose medium or supplemented brain heart infusion medium.

The filters are then placed in a glass tube containing about 2 to 3 ml of medium, depending on the filter diameter, and vortexed to remove the bacteria from the filter. Bacteria are then plated on selective media. In the case of *E. coli* to *Bacteroides* matings, gentamicin (200 µg ml^{-1}) is used to select against the *E. coli* donors and the antibiotic corresponding to the resistance on the element being transferred is used to select for *Bacteroides* transconjugates. Some selectable markers that work in *Bacteroides* spp. are described in Table 1. Antibiotic concentrations normally used for selecting *Bacteroides* transconjugants are: tetracycline (Tc), 3 µg ml^{-1}; erythromycin (Erm), 10 µg ml^{-1}; chloramphenicol (Cm), 15 µg ml^{-1}. In *Bacteroides* to *E. coli* matings, aerobic incubation on LB agar is an effective selection against *Bacteroides* donors. For *Bacteroides* to *Bacteroides* matings, rifampicin resistance (Rif, 10 µg ml^{-1}) can be used to select for recipients and against donor (Salyers and Shoemaker, 1997). Tetracycline must be included in medium used to grow the donor as an inducer of the mating (Stevens et al, 1993).

Spontaneous trimethoprim-resistant mutants (Tp = 100–200 µg ml^{-1}) are usually thymidine auxotrophs. If such a mutant is used as a donor, simply leaving thymidine out of the medium can be used to eliminate donor cells after mating. If a Tp^r mutant is used as a recipient, Tp selection performs the same function of eliminating donors. Since spontaneous resistance to rifampicin or trimethoprim arises at a relatively high frequency (10^{-7}), combining these selections is advisable to obtain the cleanest background. Many of these same selections can be used for *Porphyromonas* or *Prevotella* strains with one exception. The gentamicin selection is not applicable because most members of these genera are sensitive to gentamicin.

B. Conjugation Caveats

Anyone who plans to work on a strain of *Bacteroides* that has not previously been manipulated genetically should understand that *Bacteroides* strains can differ considerably in how amenable they are to genetic manipulation. As already mentioned, we have encountered *Bacteroides* strains into which we were not able to introduce even a shuttle plasmid, usually the highest frequency genetic event. For this reason, it is desirable to start with several strains to give yourself a better chance of finding one that is a good recipient of DNA transfer. You should also take the time to

Table 1. Origin and characteristics of selectable marker genes, reporter genes and promoters that are useful in *Bacteroides* genetics

Gene or promoter	Origin and characteristics
'IS4351-ermF	3 kbp *Eco*RI-*Cla*I fragment from Tn*4351* or Tn*4400*; Missing one end of IS*4351* but still contains an intact transposase gene; IS*4351* promoter runs *ermF* expression
cat	775 bp Tn*9 cat* gene; Useful as reporter gene or selectable marker (Smith *et al.*, 1992)
'IS4351-*cat*	2.5 kbp fragment from pFD308 with *Bam*HI site filled in. Missing one end of IS*4351* but still contains an intact transposase gene; IS*4351* promoter runs *cat* expression in both *Bacteroides* and *E. coli* (Smith *et al.*, 1992)
P_{xyl}-*cat*	1.9 kbp *Xba*I-*Nar*I fragment containing the xylanase gene promoter from *P. ruminicola* cloned upstream of *cat* on pFD325 (Whitehead *et al.*, 1991; Smith, unpublished)
xyaA	1.1 kb partially filled in *Kpn*I-*Hind*III fragment containing the *B. ovatus* β-xylosidase gene (Whitehead, 1997; Xing *et al.*, submitted)
uidA	β-glucuronidase (GUS) gene on a 1.9 kbp *Bam*HI-*Sst* I fragment from pBT101 (Clonetech Labs, Palo Alto, CA); Useful as reporter gene (Feldhaus *et al.*, 1991)
ermG	1.3 kbp PCR product from amplification of *ermG* of CTn 7853; No IS*4351* homology (Cooper *et al.*, 1996)
tetQ	2.7 kbp *Sst*I fragment from pNFD13-2 (Nikolich *et al.*, 1992)
cfxA	2.3 kbp *Hind*III fragment from pFD351 (Parker and Smith, 1993)
$oriT_{RK2}$	750 bp *Hae*II fragment from pDG5 containing the *oriT* of RK2 (Guiney and Yakobson, 1983). Mobilized by RK2 or RP4 but not by R751
P_{chuR}	443 bp *Sph*I fragment from pSQW 7 containing the promoter of *chuR*; Constitutively expressed at a moderate level (Cheng *et al.*, 1992)
P_{susA}	376 bp PCR fragment from the *B. thetaiotaomicron* chromosome containing the promoter of *susA*; Maltose-induced expression at a moderate to low level (D'Elia and Salyers, 1996)
P_{susB}	1.8 kbp fragment from the *B. thetaiotaomicron* chromosome containing the promoter of *susB*; Maltose-induced expression at a high level (D'Elia and Salyers, 1996)
P_{csuBC}	2 kbp *Eco*RV-*Pst*I fragment from *B. thetaiotaomicron* containing the promoter of the *csuBC* operon; Chondroitin sulphate-induced expression at a high level (Feldhaus *et al.*, 1991)
P_{xyl}	730 bp *Sau*3A fragment from *P. ruminicola* containing the promoter of a xylanase gene; constitutively expressed at a high level in *Bacteroides* but xylan-regulated expression in *P. ruminicola* (Whitehead *et al.*, 1991)
IS4351	Promoters at both ends firing outward; constitutive expression at a moderate level (Smith *et al.*, 1992)

optimize the ratio of donor to recipient and check the effect of growth phase of donor and recipient on transfer efficiency because these parameters may vary from strain to strain.

When working with a new strain, it is important to check the antibiotic resistance profile of the strain (see examples in Chapter 2, II.A, Table 1). It is also critical to determine the frequency of transfer into that strain of a shuttle plasmid. If you want to do transposon mutagenesis or gene disruption, a shuttle plasmid transfer frequency of at least 10^{-3}–10^{-4} per recipient is necessary because transposition and homologous recombination events will be 3 to 4 orders of magnitude below this. An exception to this rule, of course, applies in cases where the shuttle plasmid does not replicate in the recipient. *E. coli–Bacteroides* shuttle plasmids are unstable in *P. gingivalis*, so in this case frequencies of transposon insertion or of insertion of a homologous DNA segment are actually as high or higher than transfer of the shuttle plasmid (Dyer *et al.*, 1992).

We have found that the growth phase of both donor and recipient – but especially of the donor – is critical for successful conjugation experiments. Unless very young donor cells are used (OD 650 < 0.2), transfer rates plummet. The recipients can be grown to a somewhat higher OD, but it is advisable to use early to mid-exponential phase cells. We do not know why this is so, but it is a fact of life that needs to be observed.

In matings where *E. coli* is the donor, cells must be grown aerobically and will need to have some oxygen exposure during mating, if only for a short time. The IncP plasmids used to mobilize DNA require oxygen for high frequency transfer. Nitrate does not replace oxygen, so energy limitation does not appear to be the factor that affects mating frequencies under anaerobic conditions. No one has determined the reason for this oxygen dependence but it means that completely anaerobic matings are not going to work well. This is a problem in the case of *Porphyromonas* or *Prevotella* spp., because these genera are very sensitive to oxygen. Nonetheless, the short period of oxygen exposure needed to maximize the mating frequency can still be achieved.

If the recipient strain can survive a short period of oxygen exposure, especially in a mixture with actively respiring *E. coli* cells, the period during which a Gaspak jar interior is gradually depleted of oxygen is sufficient to stimulate transfer. If the recipient strain will not tolerate this much exposure, a small amount of oxygen can be injected into the jar or tube containing the mating mixture. *E. coli* donors can protect oxygen-sensitive recipients from low amounts of oxygen. To determine the minimum oxygen exposure necessary, simply do some control matings with an *E. coli* donor and an *E. coli* recipient to determine the minimum amount of oxygen exposure needed to boost mating frequencies under the conditions being employed.

C. *Bacteroides* Promoters

So far, no one has defined a consensus *Bacteroides* promoter. One thing we can be pretty certain about, however, is that it does not look much like the consensus promoter of *E. coli*, because *E. coli* genes with perfectly good consensus promoters do not express at all in *Bacteroides* spp. (Smith *et al.*, 1992b; Salyers and Shoemaker, 1997). *Bacteroides* genes will sometimes be expressed at low level in *E. coli* but expression is usually not from the true *Bacteroides* promoter but from some AT-rich region upstream of the gene. People who publish *Bacteroides* sequences should not underline *E. coli* consensus promoters and treat them like true *Bacteroides* promoters!

There are now a number of sequences of upstream regions of *Bacteroides* genes in the databases. Unfortunately, most of them are expressed in a regulated manner and thus are unlikely to have a good consensus promoter. Also, the ones that are expressed constitutively are fairly weak. For these reasons, sequence comparison is not an effective way to deduce promoter structure. An alternative approach would be to mutate known promoter regions to determine where the promoter lies. So far, no one seems to have cared enough about what a *Bacteroides* promoter looks like to undertake such a study. There are, however, several DNA segments known to contain promoters of different strengths and these segments can be used to drive the expression of heterologous genes in *Bacteroides*. Some *Bacteroides* promoters that have been used for this purpose are listed and described briefly in Table 1. The xylanase promoter, the *chuR* promoter and the promoters coming out of the ends of *IS4351* are expressed constitutively, whereas the others listed in Table 1 are regulated, e.g. *csuBC* promoter by chondroitin sulphate and *susA/susB* promoters by maltose.

Some of these promoters are species specific. We and others have found examples of promoter regions that drive gene expression in one *Bacteroides* species but not in another. Also, the promoter may be regulated differently in different species. The promoter region of a xylanase gene from *P. ruminicola*, which was regulated in *P. ruminicola*, was expressed constitutively in *B. thetaiotaomicron* (Whitehead *et al.*, 1991). The *IS4351* promoters seem to work in a variety of *Bacteroides* species. Some regulated promoters such as the promoters of *csuBC* can be used only in strains that grow on chondroitin sulphate, the inducer to which the promoter responds.

More serious than the lack of sufficient information about *Bacteroides* promoters is the absence of information about *Bacteroides* ribosome binding sites. More than once, when we have tried to produce *Bacteroides* proteins in *Bacteroides*, from genes under the control of heterologous promoters, we have obtained disappointingly low protein yields even in cases where we know that transcription levels are high. Commonly, such examples involve genes that are part of operons, where translational coupling presumably makes a good ribosome binding site unnecessary. The *Bacteroides* ribosome binding site could be the same as that used by *E. coli* because the 3' ends of 16SrRNA are virtually identical (Weisburg *et al.*, 1985), but this has not been shown conclusively by mutation of candidate ribosome binding sites.

D. Reporter genes

β-Glucuronidase (*uidA*, GUS) and *ermF* were the first reporter genes to be used to test gene expression in *Bacteroides* (Smith *et al.*, 1992b; Feldhaus *et al.*, 1991). These and other reporter genes are listed in Table 1. GUS has proved useful but is not a very sensitive indicator of promoter activity and cannot detect weak promoter activity. At least GUS activity can be quantitated. There is no way to quantitate *ermF* activity unless you use Northern blotting to detect message levels. Smith *et al.* (1992b) has shown that *cat* works well as a reporter group in *Bacteroides* spp. In his hands, chloramphenicol transacetylase levels are higher than β-glucuronidase levels when *cat* and *uidA* are under control of the same promoter. Another reporter group that should prove to be very useful is the *xyaA* gene from *Bacteroides ovatus* (Whitehead *et al.*, 1997). This gene encodes a β-xylosidase. Many *Bacteroides* spp. do not have this activity and β-xylosidase may be a more sensitive indicator of gene expression than any reporter group previously used in *Bacteroides*.

A limitation associated with the highly reducing medium commonly used to grow *Bacteroides* is that detection methods based on X-gal or similar compounds do not work. The blue colour requires oxidation of the X group after hydrolysis, and cysteine in the medium prevents this reaction from occurring. A visible colour reaction that would work with *Bacteroides* medium would be very useful because it would allow plate screens of mutants for different levels of gene expression.

◆◆◆◆◆◆ III. TRANSPOSON MUTAGENESIS

A. Uses of Transposon Mutagenesis

Transposon mutagenesis can be an excellent method for finding new genes, especially genes for which there is no assay available (Chapter 4). For example, we located genes for five starch-associated outer membrane proteins we had not known about previously by screening transposon-generated mutants of *B. thetaiotaomicron* for mutants that could no longer grow on starch. By cloning DNA adjacent to the transposon insertion (Chapter 4, IX) and sequencing it, we found a region containing the genes that encoded the outer membrane proteins (Tancula et al., 1992; Reeves et al., 1996). To find mutants with a particular phenotype, it is usually necessary to screen thousands of colonies. Thus, high frequency transfer – as well as persistence and patience on the part of the scientist doing the screen – is imperative to maximize the number of transposon-generated mutants obtained from each mating.

B. Tn*4351*, Tn*4400* and other *Bacteroides* Transposons

Two *Bacteroides* transposons have been used for transposon mutagenesis, Tn*4351* and Tn*4400*. These two transposons are virtually identical to each other (Figure 1). They carry an *ermF* gene that confers erythromycin resistance on *Bacteroides* and a *tetX* gene that works only in aerobically grown *E. coli* and cannot be used as a selectable marker in *Bacteroides*. Tn*4551* (Figure 1) is a third *Bacteroides* transposon that carries *ermF* (Smith *et al.*,

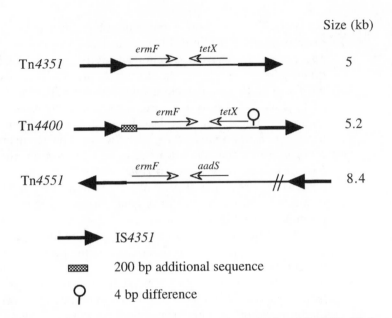

Figure 1. Diagramatic representation of Tn*4351*, Tn*4400* and Tn*4551*. All three of these transposons have the same IS*4351* direct repeats at their ends, although IS*4351* is reversed relative to *ermF* in Tn*4551*. All three transposons contain *ermF*. In Tn*4400* there is a 200 bp insertion that separates *ermF* from the end of IS*4351*. Tn*4351* and Tn*4400*, but not Tn*4551* carry *tetX*, a gene that confers resistance only on aerobically grown *E. coli*. Tn*4351* and Tn*4400* are virtually identical except for the 200 bp insertion in Tn*4400* and a 4 bp difference upstream of *tetX*. This 4 bp difference occurs at a site for insertion of the conjugative transposon XBU4400. Tn*4551* carries a streptomycin-resistance gene, *addS*, which is not expressed in *Bacteroides*.

1992a). Tn*4551* is probably not a good candidate for use in transposon mutagenesis because it is 8.4 kbp in size and does not carry the *tetX* gene, which provides a convenient selection in *E. coli*. Also, it has the same IS elements as Tn*4351* and thus will have all the same problems as Tn*4351* with none of the benefits. A fourth transposon, Tn*4399*, has been characterized (Hecht *et al.*, 1989). This interesting transposon carries no resistance markers but has a region that allows it to be mobilized. Like Tn*4551*, Tn*4399* is probably not a good candidate for use in transposon mutagenesis because it has a proclivity for AT-rich regions and is thus not as random in its insertion as is desirable for a transposon. The transfer origin of Tn*4399* has been identified (Murphy and Malamy, 1995)

The first Tn*4351* delivery vehicle was R751::Tn*4351* which was introduced into *Bacteroides* by conjugation. R751 is a large self-transmissible plasmid that transfers at high frequency from *E. coli* to *Bacteroides*. This high frequency helps to maximize the number of transposon insertion mutants obtained in each mating. This delivery vehicle has an important drawback, however. Tn*4351* co-integrates R751 about half the time when it integrates itself (Shoemaker *et al.*, 1986a). The integrated R751 is flanked by at least one copy of IS*4351* on one side and a copy of Tn*4351* on the other. Since the copy of IS*4351* can occur on either side, it is necessary to determine the nature and orientation of the flanking IS elements before proceeding to clone adjacent DNA. This can be time consuming, not to mention confusing.

Recently, we have constructed a delivery vehicle for Tn*4351* that should make cloning of adjacent chromosomal regions easier, pEP*4351* (Figure 2). This delivery vehicle also gives higher frequencies of transposon insertions than R751::Tn*4351*. pEP*4351* is based on the small plasmid, pEP*185.2* (Miller and Mekalanos, 1988). This plasmid requires the R6K *pir* gene product for replication, so it must be grown in an *E. coli* strain that carries this gene. Also, it is only mobilized by IncPα plasmids. R751 is not a member of this IncP subgroup, so it is necessary to use RK2 for mobilization. A strain created by W. Metcalf, strain *E. coli* BW19851, contains the *pir* gene and the transfer functions of RP4 in its chromosome. This strain was derived from *E. coli* S17–1 (Metcalf *et al.*, 1994).

As with R751, pEP*185.2* co-integrates with Tn*4351* in about half of the transconjugants that received a transposon insertion. If this occurs, the flanking DNA can be cloned easily by cutting chromosomal DNA with a restriction enzyme that cuts a site in the multiple cloning region adjacent to Tn*4351*, ligating the DNA, then transforming *E. coli* to select for the circle that contains the plasmid plus adjacent chromosomal DNA (Figure 3). pEP*4351* is a relatively low copy number plasmid (15 copies per cell). The copy number can be increased to 250 per cell if the plasmid is grown in *E. coli* BW18815 *uidA::pir-116*, a strain that contains a mutated allele of the *pir* gene. It may still be necessary to reclone the chromosomal junction region into a higher copy number plasmid such as pUC19 for sequencing. In our

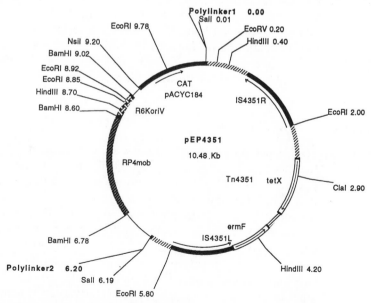

Polylinker1: 0/T7.Kpnl.Apal.Drall.Xhol.Sall.
Polylinker2: 6.2/Accl.Hincll.Clal.Hindlll.EcoRV.EcoRl.PsthSmal.Sstll.Sstl.T3.

Figure 2. Map of the new transposon delivery vector pEP*4351*. Tn*4351* has been cloned into pEP185.2, a *pir*-dependent plasmid (Cooper *et al.*, 1997).

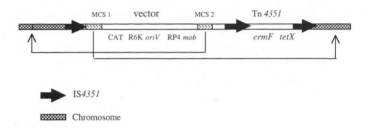

Figure 3. Diagram of an insertion of pEP4351 into the *Bacteroides* genome, in cases where the transposon co-integrates the carrier plasmid. Usually, there is one copy of IS4351 at one end of the inserted plasmid and a copy of Tn4351 at the other. The order can be the reverse of the one shown in the figure. Strategies for cloning adjacent chromosomal DNA are indicated by the vertical arrows (see text). MCS1 and MCS2 are multiple cloning sites in the vector.

experience, the low copy number of pEP4351 can actually be an advantage because it makes it more likely that DNA segments toxic to *E. coli* when cloned in high copy number can be cloned intact in this vector. If such a segment is encountered, it may be necessary to amplify the chromosomal segment by PCR in order to sequence it.

Things are not always as simple, unfortunately, as Figure 3 makes it seem. We have already encountered a case in which a tandem duplication of the plasmid occurred, so that the strategy outlined in the previous paragraph yielded mostly the original plasmid, not the desired flanking DNA. For this reason, it is desirable to check transposon-generated mutants of interest by Southern hybridization before trying the cloning strategy shown in Figure 3. If there is a tandem duplication, it will be necessary to extract the desired band from a gel before carrying out the ligation and transformation steps. In fact, this additional step is desirable even in cases where there is not a tandem duplication because it increases the likelihood of obtaining the desired clone.

C. Transposition Caveats

Tn4351 has proved to be very useful, but it has an annoying propensity that needs to be taken into account. In our experience, Tn4351 has a pronounced affinity for regions between open reading frames. So far, nearly half of the transposon insertions we have cloned and sequenced have proved to be in intergenic regions. Sometimes the polar effects of such an insertion give a similar phenotype to that of an insertion within the adjacent gene. Since, however, Tn4351 has promoters coming out of both its IS4351 ends, a transposon insertion in a promoter region can actually activate expression of a gene that is not normally expressed under the growth conditions being used. We have already seen a couple of mutants in which we suspect the phenotype is due to gain of function rather than loss of function. Since Tn4400 has the same IS elements as Tn4351, Tn4400 presumably has this same property as well. There is a real need for another *Bacteroides* transposon that can be used for mutagenesis.

If you clone and sequence DNA adjacent to a Tn*4351* insertion and find that the transposon appears to have inserted an intergenic region, it is imperative to disrupt the downstream open reading frame immediately to check whether a knock-out of the gene reproduces the phenotype of the transposon-generated mutant. If not, you may have a gain of function mutation – a fact that will change your interpretation of the phenotype.

D. Conjugative Transposons

With the possible exception of Tn*4399*, the transposons described in the previous section are conventional transposons. *Bacteroides* spp. also harbour numerous conjugative transposons as well as integrated elements they mobilize (Salyers *et al.*, 1995; see Chapter 4, III). Tn*4399*, Tn*4555*, NBU1 and NBU2 are examples of mobilizable integrated elements. The conjugative transposons and the mobilizable integrated elements are quite different from conventional transposons. In particular they have a circular transposition intermediate and are transferred by conjugation. We are encouraging everyone to use CTn or CT designations for conjugative transposons to alleviate the confusion caused by giving one of these transposons a transposon name, since it is too late to change the transposon names. We are now calling the conjugative transposons we previously designated as Tc^r ERL, Tc^{r}Em^r DOT, Tc^{r}Em^r 12256 and Tc^{r}Em^{r}7853, CTnERL, CTnDOT, CTn12256 and CTn7853, respectively. The conjugative transposons and the NBUs are not useful for transposon mutagenesis because they are fairly site specific. Also, the conjugative transposons are too large.

◆◆◆◆◆◆ IV. GENE DISRUPTIONS

A. Uses of Gene Disruption

The ability to disrupt cloned genes is essential for determining their function in the intact organism (a more detailed description of gene disruption is given in Chapter 10, II, III and IV). For example, we thought from biochemical studies that a particular starch-degrading enzyme, a neopullulanase, was the most important – if not the only – starch degrading enzyme in *B. thetaiotaomicron*. Yet when we disrupted the gene that encoded this enzyme, we found that it was not essential for growth on starch (D'Elia and Salyers, 1996). Moreover, eliminating this enzyme activity from our extracts allowed us to identify and characterize a much lower activity enzyme that was essential for growth.

B. Suicide Vectors

There are now many suicide (or insertional) vectors available for generating gene disruptions or introducing DNA into the chromosome for other purposes. Some of these are listed in Table 2. They carry a variety of resistance markers: *tetQ*, *ermF*, *ermG*, *cat* (with a *Bacteroides* promoter) or *cfxA*.

Table 2. Suicide (insertional) vectors we have found useful for making gene disruptions in *Bacteroides* spp. In each case, the base vector is described, followed by suicide vectors made from this base

Base vector	*E.coli* replication origin	Mobilization origin type	Selection in *E. coli*[a]
pUC19::*oriT*$_{RK2}$ High copy number (> 100 copies per cell)	pUC19 (ColE1)	*oriT* of RK2 Mobilized by RK2 and RP4, but not by R751	Ap (50 µg ml^{-1})

Vector	*Bacteroides* Selectable marker (reporter gene)	References
pLYL001	*tetQ*	Reeves *et al.*, 1996
pGERM	*ermG*	N. Shoemaker, unpublished
pNLY3	IS4351-cat	N. Shoemaker, unpublished
pCQW1 (GUS fusions)	*ermF* (*uidA*)	Feldhaus *et al.*, 1991
pGJL1 (GUS fusions)	*tetQ* (*uidA*)	L.-Y. Li, unpublished
pANS910 (β-xylosidase fusions)	*tetQ* (*xyaA*)	Xing *et al.*, submitted

Base vector	*E.coli* replication origin	Mobilization origin type	Selection in *E. coli*[a]
pACYC184::*oriT* Low copy number (20 copies per cell)	pACYC184 Compatible with pUC19-based vectors	*oriT*$_{RK2}$ Mobilized by RK2 and RP4 but not by R751	Tc (10 µg ml^{-1}); Cm (20 µg ml^{-1})

Vector	*Bacteroides* Selectable marker (reporter gene)	References
pNLY2	IS4351-cat	Reeves *et al.*, 1997

Base vector	*E.coli* replication origin	Mobilization origin type	Selection in *E. coli*[a]
pJRD215 Low copy number (< 10 copies per cell)	RSF1010 Compatible with pUCY19 and pACYC184	RSF1010 Mobilized by RK2, RP4, and R751	Kn(50 µg ml^{-1})

Vector	*Bacteroides* Selectable marker (Reporter gene)	References
pNJR6	'IS4351-ermF$_{Tn4400}$	Stevens *et al.*, 1990
pBT-1	*tetQ*	Tancula *et al.*, 1992
pNJR14	*cfxA*	N. B. Shoemaker, unpublished
pBT-GUS (GUS fusions)	*tetQ* (*uidA*)	L. Hooper, unpublished

Table 2. contd

Base vector	E.coli replication origin	Mobilization origin type	Selection in E. coli[a]
pEP185.2 Low copy number (8–10 copies per cell; Miller and Mekalanos, 1988)	R6K Requires *pir* to replicate	*oriT* of RP4 Mobilized by RK2 and RP4 but not by R751	Cm (15 µg ml⁻¹)

Vector	Bacteroides Selectable marker (reporter gene)	References
pEP4351[b] (Contains Tn4351)	*ermF*	Cooper *et al.*, 1997
pEPE[c]	*ermG*	J. Wang, unpublished
pEPCm[c]	'IS4351-cat	J. Wang, unpublished

[a]Antibiotic abbreviations and concentrations: Ap, ampicillin (50 µg ml⁻¹); Cm, chloramphenicol (15 µg ml⁻¹ in *Bacteroides*, 20 µg ml⁻¹ in *E. coli*; Em, erythromycin (10 µg ml⁻¹); Kn, kanamycin (50 µg ml⁻¹); Tc, tetracycline (3 µg ml⁻¹ in *Bacteroides*, 10 µg ml⁻¹ in *E. coli*).

[b]Vector used for transponson mutagenesis.

[c]Vector used for insertional gene disruption. Target DNA segment cloned into these vectors.

In our experience, the P_{xyl}-*cat* (Table 1) is not very good for generating chromosomal insertions because *cat* is too weak a resistance in *Bacteroides* spp. Since P_{xyl}-*cat* is expressed in *E. coli*, it is necessary to use concentrations high enough to eliminate the *E. coli* donors. The other form of *cat*, 'IS4351-cat, confers a stronger resistance phenotype in *Bacteroides* spp. and can be used for selection in single copy. We have also experienced background problems with *cfxA*, presumably due to the low level of β-lactamase production found naturally in many *Bacteroides* strains while *tetQ*, *ermF* and *ermG* give strong, clean selections even in single copy.

There is a safety problem that should be addressed at this point. There are not many drugs that are useful for treating *Bacteroides* infections. Clindamicin, cefoxitin, imipenen and metronidazole are currently drugs of choice. Chloramphenicol can also be used, since most *Bacteroides* strains are still sensitive to it, but side-effects make this drug less than optimum. *ermF* and *ermG* both confer resistance to clindamicin as well as erythromycin, and *cfxA* confers resistance to cefoxitin. The *cat* gene confers resistance to chloramphenicol. Thus, the genetic tools and mutants used in genetic studies of *Bacteroides* should be treated with respect and precautions taken to make sure that people working with these strains are not colonized with them. *tetQ* is not so much of a concern because tetracycline resistance is now so widespread in *Bacteroides*, even in community isolates, and this is one reason for preferring *tetQ* as a genetic tool. Although most modern *Bacteroides* strains are resistant to tetracycline, there are plenty of susceptible isolates of *Bacteroides* that were isolated before *tetQ* swept through the *Bacteroides* spp. By contrast, people who work with *Porphyromonas* or *Prevotella* need to worry about *tetQ* as well as the other resistance genes. Because tetracycline resistance is still relatively uncommon in these genera tetracycline is one of the drugs being considered as a therapeutic agent for treating people with periodontal disease.

The availability of several different selectable markers has made it possible to create double or even triple mutants. A word of caution is in order if you are planning such a construction. Many of the available vectors have homologous portions, so that the second vector may insert in the first vector rather than in the desired gene. Careful inspection of the plasmids to be used eliminates this problem. If you are forced to use two plasmids with DNA sequence homology, be sure to check the insertions by Southern blot to find mutants that have the desired secondary insertion event.

◆◆◆◆◆◆ V. SHUTTLE PLASMIDS

A. Characteristics of Shuttle Plasmids

Some of the most useful shuttle plasmids are listed in Table 3 (see Chapter 2, II.B for a general discussion of shuttle vectors). All of these are based on one of two *Bacteroides* cryptic plasmids, pBI143 or pB8-51. These plasmids both have a copy number of 10–20 per cell in *Bacteroides*, are compatible

Table 3. Examples of shuttle vectors used in *Bacteroides* genetic manipulations

Base vector	E. coli replication origin	Bacteroides replication origin	Mobilization origin type	Selection in E. coli[d]
pFD160 (Smith, 1985) High copy number (> 100 copies per cell)	pUC19	pBI143 Mobilized by RK2, RP4, R751, *Bacteroides* CTs[a]	pBI143	Ap (50 μg ml[-1])

Vector	Bacteroides Selectable marker (reporter gene)	References
pAFD1	'IS4351-*ermF*	Stevens, unpublished
pLYL01	*tetQ*	Li *et al.*, 1995
pNLY1	'IS4351-*cat*	N. B. Shoemaker, unpublished
pLYL05	*cfxA*	L-Y. Li, unpublished
pMJF2[b]	'IS4351-*ermF* (*uidA*)	Feldhaus *et al.*, 1991
pMJF3	*tetQ* (*uidA*)	Feldhaus *et al.*, 1991

Table 3. continued

Base vector	E. coli replication origin	Bacteroides replication origin	Mobilization origin type	Selection in E. coli[d]
pVAL Moderate copy number (50 copies per cell)	pBR328	pB8-51 Mobilized by RK2, RK4, R751 and *Bacteroides* CTns	pB8-51	Ap(50 μg ml)$^{-1}$ Tc (50 μg ml^{-1})

Vector	Bacteroides Selectable marker (reporter gene)	References
pVAL1	'IS4351-ermF	Valentine *et al.*, 1988
pT-COW[c]	tetQ	Gardner *et al.*, 1996
pTC-COW[c]	tetQ, IS4351-cat	Gardner *et al.*, 1996
pC-COW	'IS4351-cat	N. B. Shoemaker, unpublished
pRDB6	tetQ, cfxA	N. B. Shoemaker, unpublished

Base vector	E. coli replication origin	Bacteroides replication origin	Mobilization origin type	Selection in E. coli[d]
pNJR1 (Shoemaker *et al.*, 1988) Low copy number (8–10 copies per cell)	pJRD215 (RSF1010)	pB8–51 Non-mobilizable due to disruption of mobilization gene by cloning	RSF1010 Mobilized by RK2, RP4 and R751	Kn (50 μg ml^{-1})

Vector	Bacteroides Selectable marker (reported gene)	References
pNJR5	'IS4351-ermF	Shoemaker *et al.*, 1989
pNJR12	tetQ	Stevens *et al.*, 1992
pNJR17	'IS4351-cat	N. B. Shoemaker, unpublished
pNJR24	Pxyl-cat	A. M. Stevens, unpublished

[a]CT, conjugative transposon.
[b]Plasmid pLYL02 is the same as pMJF2 except that *uidA* is in the opposite orientation.
[c]Designed to introduce DNA into *P. ruminicola*. The *tetQ* gene works both in *Bacteroides* spp. and in *P. ruminicola*. *P. ruminicola* B$_1$4 strain used in these studies was already Cm^r.
[d]Antibiotic abbreviations as in Table 2.

with each other and do not share regions of high sequence similarity. They are also mobilized both by IncP plasmids (R751 and RK2) (Chapter 2, IV) and by *Bacteroides* conjugative transposons. Thus, they provide a transfer origin as well as a replication origin. Neither one replicates in *E. coli*, so an *E. coli* replication origin needs to be provided. pB1143 has been completely sequenced and appears to replicate by a rolling circle mechanism (Chapter 2, III.A.3) (Smith *et al.*, 1995). Nothing is known about the replication mechanism of pB8-51.

Of the two plasmids, pB8-51 appears to have the broader host range and replicates in *Prevotella ruminicola* and *Porphyromonas* spp. as well as in *Bacteroides*. pBI143 seems to be restricted to the *Bacteroides* spp. The narrower host range of pBI143 could actually be an advantage in some settings. For example, although we were able to introduce DNA into *P. ruminicola* from *Bacteroides* donors, we were never able to transfer plasmids from *E. coli* directly to *P. ruminicola*. Thus, to get DNA into *P. ruminicola*, it was necessary to mate the plasmid into a *Bacteroides* strain, then move it from *Bacteroides* to *P. ruminicola*. If it is necessary to use *Bacteroides* as a donor in matings to genetically engineer *P. ruminicola*, pBI143 could act as a suicide vector for non-*Bacteroides* recipients.

B. Shuttle Plasmid Caveats

The fact that the copy number of all known *Bacteroides* plasmids is relatively low makes it less likely that cloned DNA will be toxic to the cell. Yet, copy number effects can be encountered. We have found that promoter regions cloned on a plasmid often titrate enough regulatory protein to affect the expression of the chromosomal copy of the gene (D'Elia and Salyers, 1996). In such cases, complementation might not be observed because of decreased expression of other genes in the same operon. Another example of titration of a regulatory protein involved chondroitin sulphate utilization genes. Multiple copies of the promoter of the *csuBC* operon caused one of the chondroitin sulphate utilization genes to be derepressed during growth on glucose, a condition under which it was normally not detectable (Hwa and Salyers, 1992).

Another deleterious effect of a multicopy plasmid can occur if the plasmid contains an intact copy of the transposon or an *ermF* allele whose expression is driven by IS*4351*, as most extant *ermF* alleles on shuttle plasmids are. If there is a copy of Tn*4351*, IS*4351* or Tn*4400* in the chromosome, the presence of such a plasmid in *trans* can make the transposon or IS element unstable and loss of the insertion can result. Presumably this is due to the fact that transposase from the plasmid copy of IS*4351* or IS*4400* acts *in trans* on the chromosomal IS elements to increase their transposition frequency.

◆◆◆◆◆◆ VI. RECOMBINATION-DEFICIENT *BACTEROIDES* STRAIN

The *recA* gene of *B. fragilis* was first cloned and sequenced by Goodman and Wood (1990). Recently, we used this sequence to create a *recA* disruption mutant of *B. thetaiotaomicron* 4000. To do this, it was necessary to PCR amplify and reclone the *recA* gene from *B. thetaiotaomicron* because the sequence of the *B. thetaiotaomicron recA* gene was different enough from that of the *B. fragilis recA* gene to lower the recombination frequency below our level of detection (Cooper *et al.*, 1997). In our hands, the *recA* knock-out mutant was more sensitive to oxygen than the wild type. This sensitivity can be alleviated to some extent by using brain heart infusion medium without added vitamin K, resazurin or cysteine, all of which are generators of superoxide. Malamy's group has created a *recA* knock-out strain of *B. fragilis* and did not encounter this problem (personal communication), possibly because they were using brain heart infusion medium.

◆◆◆◆◆◆ VII. GENOMICS

So far, no member of the *Bacteroides* group has had its genome sequenced, but the genome of a strain of *Porphyromonas gingivalis* has now been sequenced and sequence information is being made available in the databases. *Bacteroides* spp. were passed over in the sequencing sweepstakes because of their large genome sizes (5–6 Mb). We have been doing a small-scale random genome sequencing project. We created a library of fragments in the 4–6 kbp length range and sequenced into the fragment from both directions. Any open reading frames are translated and used to search the databases. So far, we have sequenced 112 clones. Of the 224 sequences (one from each side of the cloned region), over 30 had significant amino acid homology with sequences in the database ($P < 10^{-8}$). A list of these homologues is given in Table 4.

It is dangerous to assume that database matches identify genes. The main utility of such matches is to generate hypotheses about what the gene might encode. Some of the matches make sense based on what we know about *Bacteroides* physiology, but others such as the match to a nitrogen fixation protein from *H. influenzae* are unexpected and could well be misleading. One striking finding so far is the number of homologues to SusC, an outer membrane protein of *B. thetaiotaomicron* that is essential for utilization of starch. SusC has the classic outer membrane protein characteristics, including a number of possible transmembrane segments and a terminal phenylalanine (Reeves *et al.*, 1996). SusC may be a specialized porin for intermediate chain length oligomers of glucose (G4 to G7). Thus, the homologues might be outer membrane proteins. So far, none of the *susC* homologue genes has been disrupted in *B. thetaiotaomicron*. Recently, we found still another SusC homologue next to a transposon insertion that affected the ability to utilize starch. The SusC family appears to be quite large.

The main conclusion from this very low throughput sequencing project is that partial genome sequencing on a more ambitious level might serve as a good way to search for interesting genes. We will be unable to follow up any but a few of these homologues. Anyone interested in obtaining clones and more information should contact the authors. The completed genome sequence for *Porphyromonas gingivalis* should be a very valuable resource for those interested in the physiology of bacteria in this group.

Table 4. List of homologues to random DNA sequences from the *B. thetaio-taomicron* chromosome. The sequences were translated then used to search the databases. The homologues are listed in order of decreasing similarity

Homologue	Origin of homologous gene	Log [*P* value]
Clp protease	*Synechocystis* sp.	−89
Homoserine *O-trans* succinylase	*Escherichia coli*	−75
Methylmalonyl-CoA mutase β-subunit	*Porphyromonas gingivalis*	−74
UvrA protein	*Micrococcus luteus*	−66
Exopolysaccharide biosynthesis enzyme	*Erwinia amylovora*	−48
Uridylate kinase	*Escherichia coli*	−46
Renin-binding protein	*Sus scrofa*	−45
K⁺-transporting ATPase	*Escherichia coli*	−41
α-Galactosidase	*Glycine max*	−39
DNA gyrase A subunit	*Staphylococcus aureus*	−39
Arabinosidase	*Bacteroides ovatus*	−38
Nitrogen fixation protein	*Haemophilus influenzae*	−35
GMP synthase	*Bacillus subtilis*	−34
Topoisomerase	*Haemophilus influenzae*	−31
Glucose kinase	*Staphylococcus xylosus*	−28
RecQ	*Escherichia coli*	−26
RprX (sensor protein)	*Bacteroides fragilis*	−25
L-Lactate permease	*Escherichia coli*	−25
Ribokinase	*Escherichia coli*	−24
StySKI methylase	*Salmonella enterica*	−24
ATP-dependent RNA helicase	*Escherichia coli*	−21
N-acetylmuramoyl-L-alanine amidase	T7 bacteriophage	−21
Glucosamine-6-phosphate isomerase	*Escherichia coli*	−18
Exp8 protein	*Streptococcus pneumoniae*	−17
Arylsulphatase	*Escherichia coli*	−16
Zinc/cadmium resistance	*Saccharomyces cervisiae*	−16
Alanyl-tRNA synthetase	*Escherichia coli*	−13
Vir protein	*Bordetella pertussis*	−12
VsrD protein	*Burkholderia solanase*	−10
LytS protein	*Staphylococcus aureus*	−8
SusC homologues (8 different ones)	*Bacteroides thetaiotaomicron*	−3 to −34

Acknowledgments

We acknowledge the people who provided us with information about unpublished plasmids. These include: Jun Wang and Lhing-Yew Li from University of Illinois, Urbana, IL; Ann M. Stevens from Virginia Polytechnic

Institute, Blacksburg, VA; C. Jeff Smith from East Carolina University, Greenville, NC; and Laura Hooper from Washington University, St Louis, MO. Maps of these plasmids are available from the authors on request. Some of the work described in this chapter was supported by grant AI17876 and grant AI22383 from the US National Institutes of Health.

References

Cheng, Q., Hwa, V. and Salyers, A. A. (1992). A locus that contributes to colonization of the intestinal tract by *Bacteroides thetaiotaomicron* contains a single regulatory gene (*chuR*) that links two polysaccharide utilization pathways. *J. Bacteriol.* **174**, 7185–7193.

Cooper, A. J., Kalinowski, A. P., Shoemaker, N. B. and Salyers, A. A. (1997) Construction and characterization of a *Bacteroides thetaiotaomicron recA* mutant: Transfer of *Bacteroides* integrated conjugative elements is RecA independent. *J. Bacteriol.* **179**, 6221–6227.

Cooper, A. J., Shoemaker, N. B. and A. Salyers, A. A. (1996). The erythromycin resistance gene from the *Bacteroides* conjugative transposon Tc^{r}Em^r 7853 is nearly identical to *ermG* from *Bacillus sphaericus*. *Antimicrob. Agents Chemother.* **40**, 506–508.

D'Elia, J. and Salyers, A. A. (1996). Contribution of a neopullulanase, a pullulanase and an α-glucosidase to the growth of *Bacteroides thetaiotaomicron* on starch. *J. Bacteriol.* **178**, 7173–7179.

Dyer, D. W., Bilalis, G., Michel, J. H. and Malek, R. (1992). Conjugal transfer of plasmid and transposon DNA from *Escherichia coli* into *Porphyromonas gingivalis*. *Biochem. Biophys. Res. Comm.* **186**, 1012–1019.

Feldhaus, M. J., Cheng, Q., Hwa, V. and Salyers, A. A. (1991) Use of *Escherichia coli* β-glucuronidase (GUS) gene as a reporter gene for investigation of *Bacteroides* promoters. *J. Bacteriol.* **173**, 4540–4543.

Gardner, R. G., Russel, J. B., Wilson, D. B., Wang, G.-R. and Shoemaker, N.B. (1996). Use of a modified *Bacteroides-Prevotella* shuttle vector to transfer a reconstructed β-1,4-D-endogluconase gene into *Bacteroides uniformis* and *Prevotella ruminicola* B$_1$4. *Appl. Environ. Microbiol.* **62**, 196–202.

Genco, C. A., Schifferlem, R. E., Njoroge, T. Fong, R. Y. and Cutler, C. W. (1995). Resistance of a Tn*4351*-generated polysaccharide mutant of *Porphyromonas gingivalis* to polymorphonuclear leukocyte killing. *Infect. Immun.* **63**, 393–401.

Goodman, H. J. K. and Woods, D. R. (1990). Molecular analysis of the *Bacteroides fragilis recA* gene. *Gene* **94**, 77–82.

Guiney, D. G. and Yakobson, E. (1983). Location and nucleotide sequence of the transfer origin of the broad host range plasmid RK2, *Proc, Natl. Acad. Sci.*, USA **80**, 3593–3598.

Hecht, D. W., Thompson, J. S. and Malamy, M. H. (1989). Characterization of the termini and transposition products of Tn*4399*, a conjugal mobilizing transposon of *Bacteroides fragilis*. *Proc Natl. Acad. Sci.*, USA **86**: 5340–5344.

Li, L.-Y., Shoemaker, N. B. and Salyers, A. A. (1995). Localization and characterization of the transfer origin of a *Bacteroides* conjugative transposon. *J. Bacteriol.* **177**, 4992–4999.

McBride, M. J and Baker, S. A. (1996). Development of techniques to genetically manipulate members of the genera *Cytophaga, Flavobacterium, Flexibacter* and *Sporocytophaga*. *Appl. Environ. Microbiol.* **62**, 3017–3022.

McBride, M. J. and Kempf, M. J. (1996) Development of techniques for genetic manipulation of the gliding bacterium, *Cytophaga johnsonae*. *J. Bacteriol.* **178**, 583–590.

Metcalf, W. W., Jiang, W. and Wanner, B. L. (1994) Use of the rep technique for allele replacement to construct new *Escherichia coli* hosts for maintenance of R6K-gamma origin plasmids at different copy numbers. *Gene* **138**, 1–7.

Miller, V. and Mekalanos, J. J. (1988). A novel suicide vector and its use in construction of insertion mutations: Osmoregulation of outer membrane proteins and virulence determinants in *Vibrio cholerae* requires *taxR*. *J. Bacteriol.* **170**, 2575–2583.

Murphy, C. G. and Malamy, M. H. (1995). Requirements for strand- and site-specific cleavage within the *oriT* of Tn*4399*, a mobilizing transposon from *Bacteroides fragitis*. *J. Bacteriol.* **177**, 3158–3165.

Nikolich, M. P., Shoemaker, N. B. and Salyers, A. A. (1992). A *Bacteroides* tetracycline resistance gene represents a new class of ribosome protection tetracycline resistance. *Antimicrob. Agents Chemother.* **36**, 1005–1012.

Park, Y. and McBride, C. (1993). Characterization of the *tpr* gene product and isolation of a specific protease-deficient mutant of *Porphyromonas gingivalis* W83. *Infect. Immun.* **61**, 4139–4146.

Parker, A. C. and Smith, C. J. (1993). Genetic and biochemical analysis of a novel Ambler class A β-lactamase responsible for cefoxitin resistance in *Bacteroides* species. *Antimicrob. Agents Chemother.* **37**, 1028–1036.

Reeves, A., D'Elia, J. D., Frias, J., and Salyers, A. A. (1996). A *Bacteroides thetaiotaomicron* outer membrane protein that is essential for utilization of maltooligosaccharides and starch. *J. Bacteriol.* **178**, 823–830.

Reeves, A., Wang, G.-R., and Salyers, A. A. 1997. Characterization of four outer membrane proteins involved in the utilization of starch by *Bacteroides thetaiotaomicron*. *J. Bacteriol.* **179**, 643–649.

Salyers, A. A. and Shoemaker, N. B. (1997). Genetic methods for *Bacteroides* species. In: *Genetic Engineering*, Vol. 19 (J. K. Serlow, ed.), pp. 89–99. Plenum Press, New York.

Salyers, A. A., Shoemaker, N. B., Li, L. Y. and Stevens, A. M. (1995) Conjugative transposons: an unusual and diverse set of integrated gene transfer elements. *Microbiol. Rev.* **59**, 579–590.

Shoemaker, N., Getty, C., Gardner, J. F. and Salyers, A. A. (1986a). Tn*4351* transposes in *Bacteroides* and mediates the integration of R751 into the *Bacteroides* chromosome. *J. Bacteriol.* **165**, 929–936.

Shoemaker, N. B., Getty, C., Guthrie, E. P. and Salyers, A. A. (1986b). Two *Bacteroides* plasmids, pBFTM10 and pB8-51, contain transfer regions that are recognized by broad host range IncP plasmids and by a conjugative *Bacteroides* tetracycline resistance element. *J. Bacteriol.* **166**, 959–965.

Shoemaker, N. B., Barber, R. and Salyers, A. A. (1989). Cloning and characterization of a *Bacteroides* conjugal tetracycline resistance element using shuttle cosmid vector. *J. Bacteriol.* **171**, 1294–1302.

Shoemaker, N. B., Anderson, K. L., Smithson, S. L., Wang, G. R. and Salyers, A. A. (1991). Conjugal transfer of a shuttle vector from the human colonic anaerobe, *Bacteroides uniforms* to the ruminal anaerobe, *Prevotella* (formerly *Bacteroides*) *ruminicola* $B_1$4: *Appl. Environ Microbiol.* **57**, 2114–2121.

Smith, C. J. (1985). Development and use of cloning systems for *Bacteroides fragilis* cloning of a plasmid-encoded resistance determinant. *J. Bacteriol.* **164**, 294–301.

Smith, C. J., Owen, C. and Kirby, L. (1992a). Activation of a cryptic streptomycin-resistance gene in the *Bacteroides erm* transposon, Tn*4551*. *Mol. Microbiol.* **6**, 2287–2297.

Smith, C. J., Rogers, M. B. and McKee, M. L. (1992b). Heterologous gene expression in *Bacteroides fragilis*. *Plasmid* **27**, 141–154.

Smith, C. J., Rollins, L. A. and Parker, A. C. (1995). Nucleotide sequence determination and genetic analysis of the *Bacteroides* plasmid, pBI143. *Plasmid* **34**, 211–222.

Stevens, A. M., Shoemaker, N. B. and Salyers, A. A. (1990). Genes on a *Bacteroides* conjugal tetracycline resistance element which mediate production of plasmid-like forms from unlinked chromosomal DNA may be involved in transfer of the resistance element. *J. Bacteriol.* **172**, 4271–4279.

Stevens, A. M., Sanders, J. M., Shoemaker, N. B. Salyers, A. A. (1992). Genes involved in production of plasmidlike forms by a *Bacteroides* conjugal chromosomal element share significant amino acid homology with two component regulatory systems. *J. Bacteriol.* **174**, 2935–2942.

Stevens, A. M., Shoemaker, N. B., Li, L.-Y. and Salyers, A. A. (1993). Tetracycline regulation of genes on *Bacteroides* conjugative transposons. *J. Bacteriol.* **175**, 6134–6141.

Tancula, E., Feldhaus, M. J., Bedzyk, L. A. and Salyers, A. A. (1992). Location and characterization of genes involved in binding of starch to the surface of *Bacteroides thetaiotaomicron*. *J. Bacteriol.* **174**, 5609–5616.

Valentine, P. J., Shoemaker, N. B. and Salyers, A. A. (1988). Mobilization of *Bacteroides* plasmids by *Bacteroides* conjugal elements, *J. Bacteriol.* **170**, 1319–1324.

Weisburg, W. G., Oyaizu, Y., Oyaizu, H. and Woese, C. R. (1985). Natural relationship between *Bacteroides* and Flavobacteria. *J. Bacteriol.* **164**, 230–236.

Whitehead, T. R. (1997). Development of a bifunctional xylosidase-arabinosidase gene as a reporter gene for Gram-negative anaerobes. *Curr. Microbiol.* **33**, 1–6.

Whitehead, T. R., Cotta, M. A. and Hespell, R. B. (1991). Introduction of the *Bacteroides ruminicola* xylanase gene into the *Bacteroides thetaiotaomicron* chromosome for production of xylanase activity. *Appl. Environ. Microbiol.* **57**, 277–282.

9 Genetic Methods in Mycobacteria

Graham F. Hatfull

Department of Biological Sciences, University of Pittsburgh, Pittsburgh, USA

◆◆◆

CONTENTS

Genetic Methods in
Mycobacteria

◆◆◆◆◆◆ I. INTRODUCTION

The mycobacteria represent a diverse collection of organisms involved in numerous pathologies of animals and Man. Perhaps the most important of these is *Mycobacterium tuberculosis*, the causative agent of human tuberculosis, a disease that kills nearly 3 million people each year (Bloom and Murray, 1992). *M. tuberculosis* infects about one-third of the world's population, and although most of these latent infections are not lethal, a number of factors – including co-infection with HIV – can lead to induction of active tuberculosis (Bloom and Murray, 1992).

Since the introduction of antibiotics, there has been a general decline in the incidence of tuberculosis in the developed world (Snider *et al.*, 1994). However, the treatment of tuberculosis with antibiotics is not simple; several drugs must be taken in combination and for an extensive period of time, usually at least six months. This makes the treatment not only expensive, but difficult to monitor, and lack of compliance (fuelled by the side effects of some of these antibiotics) provides a breeding ground for antibiotic-resistant organisms. This has now culminated in the appearance of multidrug resistant tuberculosis (MDRTB) which is resistant to all or most of the effective antituberculosis antibiotics and which has a mortality rate (~50%) similar to that for TB in the pre-antibiotic era.

The mycobacteria are readily identified by their acid-fast staining pattern which results from the unique characteristics of their cell walls.

However, they can be further classified into two general groups, the slow-growing mycobacteria and fast-growing mycobacteria, with most of the major pathogens belonging to the first group (Wayne, 1984). Indeed, these do grow extremely slowly, with doubling times usually in the range of 16–24 h. The so-called fast-growing mycobacteria still grow relatively slowly, with doubling times typically in the range of 3–4 h.

The mycobacteria have been intensively studied for many years, particularly with regard to their general physiology, pathogenicity, interactions with the host and host responses. However, genetic approaches have been employed only more recently, and the development of genetic techniques in the mycobacteria has proven particularly challenging. While many factors have contributed to this, the most notable are the extremely slow growth rate, the propensity for mycobacterial cells to grow as clumps, and the virulence of these organisms (Hatfull, 1996). However, over the past ten years enormous strides have been made and today there is a reasonably facile and generally available genetic system, bringing new hope to the prospects of more rapid diagnostic methods, effective vaccines and novel treatments for the control of tuberculosis.

In this chapter, I will discuss the current state of mycobacterial genetics and the application of these methods to both fast- and slow-growing mycobacteria. For the most part, detailed protocols will not be presented but are available elsewhere (Jacobs, 1991; Parish and Stoker, 1997).

◆◆◆◆◆◆ II. GENERAL PROPERTIES

A. Mycobacterial Strains

Although there are more than forty species of *Mycobacterium*, only a few have become the focus of detailed genetic studies. Among the slow-growers, several strains of *M. tuberculosis* have been widely used, particularly *M. tuberculosis* H37Rv which has also been used for non-genetic methods of characterization and for the establishment of animal model systems. However, the pathogenicity of *M. tuberculosis* requires it to be grown under BL3 containment which substantially restrains its use. This is also the case for *Mycobacterium bovis*, which is a very close relative of *M. tuberculosis*, and while primarily associated with bovine tuberculosis is also a pathogen of Man (O'Reilly and Daborn, 1995). The vaccine strain, bacille Calmette-Guérin (BCG) – an avirulent derivative of *M. bovis* – has become a popular laboratory strain, in that it has been inoculated as live bacteria into over 2 billion people and thus BL3 containment is usually not required. *Mycobacterium leprae*, the causative agent of leprosy, has not been successfully grown in defined culture and there is no genetic system for studying this organism.

Among the fast-growing mycobacteria, *Mycobacterium smegmatis* is the most commonly used for genetic studies. While it is apparent that the genomes of *M. smegmatis* and *M. tuberculosis* have substantial differences, many genes are shared, and understanding the genetics of *M. smegmatis* should contribute substantially to understanding tuberculosis. The faster growth-rate of *M. smegmatis* does offer substantial advantages to the microbial geneticist, with bacterial colonies appearing in 3–4 days rather than the 3–4 weeks needed for the slow-growing mycobacteria. Numerous strains of *M. smegmatis* have now been described, but the high efficiency transformation strain *M. smegmatis* mc^{2}155 – a derivative of ATCC607 – is one of the most important (Snapper *et al.*, 1990).

As the genetics of mycobacteria continues to develop, it is likely that additional species will be added to this short list. Notable additions are *Mycobacterium aurum* (Hermans *et al.*, 1991) and *Mycobacterium marinum*, a pathogen of fish and frogs, in which gene replacement has been demonstrated (Ramakrishnan *et al.*, 1997).

B. Growth of Mycobacteria

With the exception of *M. leprae*, most strains of mycobacteria can be grown in defined media. Optimal growth of *M. tuberculosis* and BCG occurs at 37°C but only a narrow temperature range is tolerated, with a sharp reduction in growth rate below 32°C or above 39°C (Wayne, 1994). *M. smegmatis* has more relaxed temperature requirements and will grow at room temperature or up to 50°C. While growing mycobacteria is usually fairly simple, a few special requirements are noteworthy. For example, propagation of a dispersed culture of *M. tuberculosis* requires the addition of Tween-80 (0.05%) and OADC or ADC supplements containing albumin (Wayne, 1994); *M. smegmatis* does not require this but may benefit from it; Tween-80 can also be added to solid media to facilitate replica plating (A. Belanger, and G. F. Hatfull, unpublished). Antibiotics can be added to either liquid or solid media, although drugs like tetracycline that are unstable for extended periods at 37°C are not useful for the slow-growing mycobacteria. Thus the main challenge in growing mycobacterial cultures (other than pathogenicity) is their growth rate, which demands considerable patience with the slow-growers, excellent sterile technique, and attention to the problems of dehydration, especially when using solid media.

C. Mycobacterial Genomes

Mycobacteria contain a single circular chromosome in the range of 2–5 Mbp and are characteristically G+C rich (62–70%; Clark-Curtiss, 1990). The analysis of mycobacterial genomes has been advanced by the construction of cosmid libraries, and ordered cosmid maps have been described for both *M. leprae* (Eiglmeier *et al.*, 1993) and *M. tuberculosis* (Philipp *et al.*, 1996). These have proven particularly useful for genome sequencing projects and at the time of writing, the DNA sequence has been determined for cosmids representing over 50% of the *M. leprae* genome and over 80% of the *M. tuberculosis* H37Rv genome. Genome sequence determination is also in progress for a recent clinical isolate (CSU#93) of *M. tuberculosis*, using a shotgun strategy. Currently, this project is at the stage of closure of the remaining gaps between contiguous sequences. Completion of both *M. tuberculosis* sequencing projects is expected in the very near future and the analysis and comparison should be extremely enlightening. They will certainly have a profound influence on mycobacterial genetics. With the realization that whole bacterial genome sequencing is technically feasible, it is reasonable to expect other mycobacterial genomes to be sequenced within the next few years. Mycobacterial strains at the top of the priority list are *M. avium*, *M. smegmatis* and BCG.

The rather high G+C content of mycobacterial genomes demands particular attention in DNA sequencing projects due to the predominance of compressions and replication problems. In situations where random shotgun and assembly strategies are used, the high redundancy (seven- to eightfold determination of each nucleotide position) helps to abrogate these difficulties, although it is still important that both DNA strands are determined throughout; in automated sequencing reactions, errors can be minimized by using dRhodamine dye-terminators. The G+C content does, however, prove useful when identifying protein-coding genes in mycobacterial DNA, since the very heavy bias towards G-C bases in the third positions of codons provides a powerful predictive tool (Hatfull, 1994).

◆◆◆◆◆◆ III. GENETIC EXCHANGE

A. Recombinational Properties

A fully effective genetic system requires the ability to exchange markers by homologous recombination. However, the propensity to undergo homologous recombination is a key feature that distinguishes *M. tuberculosis* (and its close relatives) from *M. smegmatis*. The recombinational properties of *M. smegmatis* appear to be generally similar to those of most other bacteria, and exchange of different alleles of otherwise homologous DNA segments occurs at a reasonable frequency (Husson *et al.*, 1990). A substantial number of gene replacement events have been described at assorted loci in *M. smegmatis*. Homologous recombination has also been demonstrated in *Mycobacterium intracellulare* (Marklund *et al.*, 1995).

The recombinational properties of *M. tuberculosis* and its relatives – including BCG – are quite peculiar and are characterized by a low ratio of recombination events between homologous sequences relative to illegitimate recombination events (Kalpana *et al.*, 1991; Aldovani *et al.*, 1993; Reyrat *et al.*, 1995; Balasubramanian *et al.*, 1996). In one example, electroporation of a 40 kb linear DNA fragment of *M. tuberculosis* containing a kanamycin-resistance marker within the *leuD* gene, into *M. tuberculosis* yielded many kanamycin-resistant transformants, but only 6% were leucine auxotrophs (Balasubramanian *et al.*, 1996). This unusual recombinational behaviour has been reported by several investigators, with the proportion of homologous events ranging from virtually undetectable up to 10%. At such frequencies, it could be possible to screen for the desirable, but infrequent, homologous events although specialized strategies have been described for selecting or identifying gene replacement products (Norman et al., 1995; Baulard et al., 1996; Pelicic et al., 1996a). Nevertheless, the defect in homologous recombination in M. tuberculosis presents a serious barrier to the development of methods for efficient exchange of markers and genetic mapping by transduction and conjugation.

It is possible that the recombinational properties of M. tuberculosis result from the peculiarities of its recA gene which contains an intein (Davis et al., 1992; Colston and Davis, 1994). However, it also seems plausible that the defect could result from a fundamental requirement for DNA replication in homologous recombination (although not in illegitimate recombination), and that homologous exchange may be limited to only those cells that are active in DNA replication during the course of the experiment. With a generation time of 24 h, this proportion may be quite small, although little information is available regarding the timing of DNA replication and cell division. If this latter explanation is correct and the frequency of homologous recombination fundamentally results from the growth rate, then one solution may be to provide longer incubation times prior to imposing selection, such that there is a greater chance that the part of the chromosome in question is replicated in all of the cells. The observation that a high frequency of homologous recombination was observed following counter-selection of a resident plasmid carrying a segment of chromosomal DNA (Pelicic et al., 1997) is consistent with this, since all of the cells must have undergone several rounds of replication prior to the selection.

B. Transformation

In spite of conflicting early reports about the natural competence of mycobacteria for DNA transformation (Tsukamura et al., 1960; Norgard and Imaeda, 1978), it appears as though at least those strains in common use are not naturally transformable at any detectable frequency. The first clear demonstration of DNA uptake was in spheroplasts of M. smegmatis and BCG that can take up mycobacteriophage DNA and produce plaques in an infectious centre assay (Jacobs et al., 1987). Subsequently, it was shown that electroporation is an effective and efficient method for uptake of both phage and plasmid DNA to give up to 10^5 plaques per microgram of DNA and 10^7 colonies per microgram of DNA respectively (Snapper et al., 1990).

Most mycobacteria including various strains of M. tuberculosis and BCG are readily transformable by electroporation. The most striking departure from the conditions usually employed for bacterial electroporation (Chapter 1, II.D.5) is that transformation of these strains (and also M. intracellulare) is several orders of magnitude higher when electroporated at 37°C than at 0°C (Wards and Collins, 1996). In contrast, the optimal temperature for electroporation of M. smegmatis is 0°C. However, M. smegmatis ATCC607 yields only very few transformants (1–10 per microgram of

DNA) after electroporation, although those that do arise appear to contain mutations that now render them highly transformable (Snapper *et al.*, 1990). Once cured of the newly introduced plasmid, these strains are useful for transformation experiments, and one such strain (mc^{2}155) is in widespread use. However, the nature of the mutation that gives rise to this high efficiency transformation (*ept*) phenotype is not known. Electroporation has also been used successfully to introduce DNA into a variety of other mycobacterial strains including *M. intracellulare* (Marklund *et al.*, 1995) and *M. marinum* (Ramakrishnan *et al.*, 1977).

Essentially all of the plasmid vectors that have been described for the mycobacteria (see below) are shuttle vectors that can replicate in both *E. coli* and mycobacteria. A particularly useful method for transferring extrachromosomal plasmids between these strains is by electroduction (Baulard *et al.*, 1992). For example, a colony of *M. smegmatis* can be picked from solid media, mixed with electrocompetent *E. coli*, electroporated and plated on selective media that also contains a counter-selection against the *M. smegmatis*. This has the advantage of being simple and provides a considerable time-saving, since it avoids the need to grow mycobacterial cultures and subsequent isolation of DNA.

C. Transduction

Mycobacteriophage I3, first described by Raj and Ramakrishnan (1970) is the only mycobacteriophage reported to be capable of generalized transduction (Hatfull and Jacobs, 1994). Transduction of several auxotrophic markers in *M. smegmatis* was described at a frequency of 10^{-6} of plated cells (Raj and Ramakrishnan, 1970, 1971) and subsequently for streptomycin and isoniazid resistance (Saroja and Gopinathan, 1973). However, there is still no clear demonstration of co-transduction or use of I3 for genetic mapping. In addition, no high frequency transduction mutants have been described nor has transduction of markers in any mycobacterial strain other than *M. smegmatis*. Nevertheless, I3-mediated transduction should prove a useful genetic tool and hopefully will be further developed in the near future.

D. Conjugation

Conjugation between mycobacterial strains and between mycobacteria and other bacteria has been reported. Mizuguchi and Tokunaga (1971) demonstrated that mixing two strains of *M. smegmatis* (Jucho and Lacticola) containing different auxotrophic markers produced prototrophic recombinants at a frequency of about 10^{-5} presumably due to chromosomal transfer. Subsequent studies showed that *M. smegmatis* strains can be sorted into five groups with characteristic mating behaviours (Mizuguchi *et al.*, 1976). Recent studies confirm these findings and show that *M. smegmatis* mc^{2}155 acts as a donor with some other *M. smegmatis* strains (Keith Derbyshire, personal communication). Presumably, mc^{2}155 contains a mobilizable element within the genome that initiates chromosomal transfer within mating pairs. However, the nature and location of this element is not known. Chromosomal transfer has not been demonstrated in mycobacterial strains other than *M. smegmatis* (Greenberg and Woodley, 1984).

Conjugal transfer of plasmids (Chapter 2, V) in mycobacteria has also been described. In particular, there is evidence that transfer of plasmids can occur between strains of *M. fortuitum* and *M. smegmatis* and this was used to isolate plasmid pJAZ38, although this does not itself appear to be a conjugative plasmid (Gavigan *et al.*, 1997). Other studies have shown that plasmids RSF1010 (Chapter 2, V.B) and pMY10 can be transferred from *E. coli* to *M. smegmatis* by conjugation (Gormley and Davies, 1991; Gavigan *et al.*, 1995).

◆◆◆◆◆◆ IV. CONSTRUCTING MYCOBACTERIAL RECOMBINANTS

A. Selectable Markers

Construction of recombinant bacteria requires genetically selectable markers for selection of events that occur at relatively low frequencies. A variety of selectable markers are available for the mycobacteria, with antibiotic resistance being the most common. The *aph* genes of transposons Tn*5* and Tn*903*, which confer resistance to kanamycin, have been widely used in slow- and fast-growing mycobacteria (Hatfull, 1993). The hyg^R gene conferring resistance to hygromycin is also useful in fast- and slow-growing strains but is problematic in *E. coli* due to a high level of resistance (Radford and Hodgson, 1991). Tetracycline resistance is a useful marker in *M. smegmatis* but not for the slow-growing strains due to its poor stability at 37°C. Chloramphenicol resistance conferred by the *cat* gene can be used, but spontaneous resistant mutants occur at high frequency (10^{-4}; Jacobs *et al.*, 1987). Other available antibiotic-resistant markers include spectinomycin resistance (Guilhot *et al.*, 1994), sulphonamide resistance (Gormley and Davies, 1991) and gentamycin resistance (Gormley and Davies, 1991).

A variety of other selectable markers are available that do not involve antibiotic resistance genes. For example, a number of auxotrophic mutants of both fast- and slow-growing mycobacteria have been isolated and the corresponding genes identified by complementation (Hinshelwood and Stoker, 1992; Guilhot *et al.*, 1994; McAdam *et al.*, 1995; Balasubramanian *et al.*, 1996; Pavelka and Jacobs, 1996; Pelicic *et al.*, 1996b). In principle, these are excellent potential selectable markers, but their utility is limited to the specific strains containing the auxotrophic mutation, and – as described above – there are not yet easy, simple and effective means of moving these markers from one strain to another.

A second type of alternative selectable marker is heavy metal resistance, and mercury resistance has been demonstrated in *M. smegmatis* and BCG, using the *mer* genes derived from *Pseudomonas aeruginosa* and *Serratia marcescens* (Baulard *et al.*, 1995). Resistance or immunity to bacteriophages can also be used and has been demonstrated for the immunity gene (71) of mycobacteriophage L5 in *M. smegmatis* and BCG (Donnelly-Wu *et al.*, 1993) and the multicopy resistance gene (*mpr*) in *M. smegmatis* (Barsom and Hatfull, 1996). Further development of non-antibiotic-based selectable markers is of importance since they are helpful for constructing recombinants of mycobacterial pathogens and strains that are naturally antibiotic resistant, as well as live recombinant vaccine strains (Stover *et al.*, 1991).

Two counter-selectable genetic markers are also available. One utilizes the observation that the wild-type allele of *rpsL* is dominant over a streptomycin-resistant allele of *rpsL* (Sander *et al.*, 1995). Thus, if both are present in a partial diploid, the strain is phenotypically streptomycin sensitive, and streptomycin resistance can be used to select for loss of the wild-type allele (Pavelka and Jacobs, 1996). The second example utilizes the *Bacillus subtilis sacB* gene which confers sensitivity to sucrose in mycobacteria, such that its loss can be selected for by plating on solid media containing sucrose (Pelicic *et al.*, 1996a, 1996b, 1997). These counter-selective genes are probably most effective – and provide the strongest selection – when used in combination (Pelicic *et al.*, 1997).

B. Plasmids

I. Extrachromosomal plasmids

Most strains of *M. smegmatis* and *M. tuberculosis* appear to be normally plasmid-free. Other strains such as *M. fortuitum* and the *M. avium* complex contain a wide variety of plasmids, both large and small, in a variety of combinations (Falkingham and Crawford, 1994). While few of these plasmids are well understood, some have been very useful for constructing extrachromosomal plasmid vectors. Since most of these replicate at fairly low copy numbers in mycobacteria and it is time consuming and difficult to recover DNA in substantial quantities, most vectors are constructed as shuttle plasmids that replicate in *E. coli* as well as in mycobacteria.

Extrachromosomal shuttle plasmids have been constructed using the replication origins from pAL5000 (Ranes *et al.*, 1990; Snapper *et al.*, 1990), pMSC262 (Goto *et al.*, 1991), RSF1010 (Hermans *et al.*, 1991), pNG2 (Radford and Hodgson, 1991) and pLR7 (Beggs *et al.*, 1995). The most extensively developed plasmids are those containing the pAL5000 origin and these replicate in both fast- and slow-growing strains. The copy numbers of these are relatively low (3–10 copies per cell) and are not well maintained in the absence of selection at least in *M. smegmatis* (Lee *et al.*, 1991; Stolt and Stoker, 1996). Plasmid derivatives have been described that contain a variety of features such as polylinker sequences (e.g. pMV261, Stover *et al.*, 1991), blue–white screens for insertions (e.g. pMD31, Donnelly-Wu *et al.*, 1993), expression signals (e.g. hsp60 promoter, Stover *et al.*, 1991) and reporter genes (e.g. *lacZ*, Barletta *et al.*, 1991). Little is known about the compatibility of the various extrachromosomal plasmid vectors that have been described.

Extrachromosomal plasmid vectors have the advantage that they can be easily transformed into mycobacteria, phenotypic effects easily analysed or selected, and the recombinant plasmids readily recovered, typically by electroduction into *E. coli* and plasmid purification. The potential disadvantages are that there are commonly pronounced phenotypic effects when even wild-type mycobacterial genes are present on extrachromosomal plasmids (Banerjee *et al.*, 1994; Barsom and Hatfull, 1996), and that the plasmids may not be well maintained in the absence of selection, particularly if the presence of recombinant plasmids confers even a modest growth disadvantage (Burlein *et al.*, 1994).

2. Integrating plasmids

Integrating vectors transform mycobacteria by stable integration into the chromosome. Several mycobacterial integration-proficient vectors have been described, most of which utilize phage integration systems (see Chapter 3, III and IX). For example, several vectors have been constructed that carry the *attP* site and integrase gene (*int*) of mycobacteriophage L5 which promote integration into the chromosomal attachment site (*attB*) for the phage (Lee *et al.*, 1991; Stover *et al.*, 1991). The *attB* site overlaps a conserved tRNA^{gly} gene which is present in *M. smegmatis* and *M. tuberculosis* and probably many other mycobacterial species (Lee *et al.*, 1991). These vectors may therefore be useful in a broad selection of strains. Similar vectors have been constructed from the integration apparatus of mycobacteriophages FRAT1 (Haeseleer *et al.*, 1993) and Ms6 (Anes *et al.*, 1992) and from the integrative *Streptomyces* plasmid pSAM1 (Martin *et al.*, 1991).

These integration-proficient vectors transform both fast- and slow-growing mycobacteria with high efficiencies and yield transformants in which a single copy of the plasmid is integrated at the *attB* site. Provided that the phage-encoded excisionase gene is not included on the plasmid, the integrated sequences are well maintained even in the absence of continued selection (Lee *et al.*, 1991). These vectors also overcome the phenotypic effects resulting from extrachromosomal plasmids but have the disadvantage of being difficult to recover from recombinant strains (Pascopella *et al.*, 1994).

It is worth noting that the replication origins of extrachromosomal plasmids such as those derived from pAL5000 are not tolerated when inserted into the *M. smegmatis* genome. This is illustrated by the construction of plasmids containing both the *attP-int* segment of L5 and the pAL5000 *oriM*; these do not efficiently transform *M. smegmatis*, and the few transformants recovered arise from mutations either within *oriM* – such that they now integrate – or within the *attP-int* segment – such that they now replicate extrachromosomally (Hatfull, 1996).

C. Other Vectors

Several other important vector systems have been described. For example, shuttle cosmids have been widely used, which combine the packaging signals of bacteriophage lambda with shuttle plasmids (both extrachromosomal and integrating; Pascopella *et al.*, 1994). Ligation of large segments of DNA (35–45 kb) can be selected by packaging into lambda particles *in vitro* and recovering the plasmids in *E. coli* (see Chapter 3, VII). DNA prepared from pools of such transformants can then be introduced into mycobacteria by electroporation.

Shuttle phasmids have also been developed, which replicate as plasmids in *E. coli* but as phages in mycobacteria (Jacobs *et al.*, 1987; Snapper *et al.*, 1988; Bardarov *et al.*, 1997). Typically, these are constructed by treatment of mycobacteriophage DNA with ligase, followed by partial digestion with Sau3A and isolation of fragments in the 35–45 kb size range. This fraction is then ligated with an *E. coli* cosmid (linearized with BamHI), packaged into lambda particles *in vitro*, and used to transduce *E. coli*. DNA is prepared from a pool of colonies, and introduced into mycobacteria by electroporation, followed by plating for plaque production. Shuttle phasmids have been generated from lytic mycobacteriophages such as TM4, D29, and Bxb1 as well as the temperate phages L1 and L5 (Snapper *et al.*, 1988). Inclusion of restriction sites (such as Pac I that cuts very infrequently in mycobacterial DNA) at the junctions of the cosmid moiety of these shuttle vectors enables simple substitution of the cosmid and introduction of other genes of choice (Bardarov *et al.*, 1997).

◆◆◆◆◆◆ V. TRANSPOSONS AND TRANSPOSITION

A. Transposons

More than a dozen different transposons have been identified in a variety of mycobacterial species. However, particular transposons are associated with one or a small number of related mycobacterial species. For example, IS6110 and IS1081 are restricted to the *M. tuberculosis* complex (Thiery *et al.*, 1990; Collins and Stevens, 1991), IS1096, IS6120 and IS1137 are found in *M. smegmatis* (Cirillo *et al.*, 1991; Guilhot *et al.*, 1992; Garcia *et al.*, 1994), IS900, IS901, IS1245 and IS1110 in the *M. avium* complex (Green *et al.*, 1989; Kunze *et al.*, 1991; Hernandez Perez *et al.*, 1994; Guerrero *et al.*, 1995), and IS1395 in *Mycobacterium xenopi* (Picardeau *et al.*, 1996). However, the elements that have been shown experimentally to transpose can move in a broad range of mycobacterial species (McAdam *et al.*, 1994).

In general, all of the mycobacterial transposons identified to date are of the class I type, being either IS elements or composite transposons that carry antibiotic resistance genes (see Chapter 4, II). Typically, they transpose at low frequencies, although one exception is IS1110 which transposed at high frequencies in at least one strain background and exhibited different patterns among individual unselected colonies of *M. avium* (Hernandez Perez *et al.*, 1994). The mycobacterial transposons generally exhibit little target specificity and generate short target duplications upon transposition, common features of bacterial transposons. Whereas many of these elements probably generate simple insertions through conservative transposition events, the composite element, Tn*610* (and its IS*6100* constituents) uses a strictly replicative mechanism, such that the products are replicon fusions (Martin *et al.*, 1990; see Chapter 4). IS*900* also generates replicon fusions by replicative transposition (England *et al.*, 1991; Dellagostin *et al.*, 1993). The mechanism by which these elements move and the products that are generated determine their usefulness for the development as genetic tools.

B. Transposon Mutagenesis

Efficient transposon mutagenesis requires simple strategies for transposon delivery; several approaches have been described, including shuttle mutagenesis (see Chapter 4, VI and VIII). Initial efforts used electroporation to introduce transposons on non-replicating, non-integrating, plasmid vectors followed by selection of a transposon-associated marker (Martin *et al.*, 1990). While transposition can be demonstrated, the combined frequencies of transposition and electroporation result in only a relatively small number (<100) of progeny in a typical experiment (McAdam *et al.*, 1995). A second approach is to use plasmids that are temperature sensitive for replication such that transposition of an element from the plasmid on to the chromosome can be selected at the temperature that is non-permissive for plasmid replication (Guilhot *et al.*, 1994). This was successfully used to transpose Tn*611* (a kanamycin-resistant derivative of the composite element, Tn*610*) on to the chromosome of *M. smegmatis*, and a large number of mutants were generated, including 15 different auxotrophic phenotypes (Guilhot *et al.*, 1994). Unfortunately, since Tn*610* transposes by a replicative process, the progeny also contain a complete copy of the plasmid replicon on the *M. smegmatis* chromosome, and since the origin of pAL5000 replication is not well tolerated on the chromosome, they are likely to be unstable at lower temperatures. The temperature-sensitive replicons are only weakly thermosensitive in slow-growing strains such as *M. tuberculosis* and thus less useful. However, inclusion of the counter-selective marker, *sacB*, provides a much stronger positive selection for transposition events (Pelicic *et al.*, 1997; see also Chapter 4, VI). This was used to demonstrate transposition of Tn*5367* (a kanamycin-resistant derivative of IS*1096* normally absent in *M. tuberculosis*) on to the chromosome of *M. tuberculosis*, and the possibility of generating large representative libraries of insertion mutants (Pelicic *et al.*, 1997).

An alternative transposon delivery strategy is to use conditionally replicating bacteriophages (Chapter 4, VI). Bardarov *et al.* (1997) constructed a series of shuttle phasmids (see above) in which mycobacteriophages TM4 or D29 carry either Tn*5367* (kanamycin-resistant derivative of IS*1096*) or mini-Tn*10*(kan). Thermosensitive mutants of these phages were generated which grow at 30°C but fail to grow or lyse their host at 38.5°C. These phages can be propagated at the permissive temperature in *M. smegmatis* and then used to infect either the fast- or slow-growing strains at the higher temperature, followed by selection of the transposon. For example, phage phAE94, which carries Tn*5367* in a TM4 backbone, yields large numbers of insertion mutants following infection of

M. tuberculosis (Bardarov *et al.*, 1997). The potential advantages of this system are that Tn*5367* shows little target specificity and large numbers of independent transposition events can be generated.

These plasmid and phage transposon delivery systems are now reasonably well developed and should be useful for generating many types of mutants of interest, particularly attenuated avirulent derivatives of *M. tuberculosis* that represent potential new vaccine candidates. In addition to the auxotrophic mutants of *M. smegmatis* (Guilhot *et al.*, 1994), insertions were isolated in a variety of *M. tuberculosis* genes including those in polyketide synthase, transketolase and ferritin H (Bardarov *et al.*, 1997). Moreover, with the impending availability of the entire *M. tuberculosis* genome sequence, these efficient transposition systems should help in defining the complete set of non-essential mycobacterial genes.

◆◆◆◆◆◆ VI. MYCOBACTERIOPHAGES

A large number of mycobacteriophages have been isolated and described, largely through efforts to use them for typing mycobacterial strains (Redmond, 1963; Hatfull and Jacobs, 1994; Mizuguchi 1984; Snider *et al.*, 1984). These phages vary considerably in their host ranges, with some forming plaques on a broad range of mycobacterial species, while others are restricted to small subgroups of mycobacteria. Phages of varying plaque morphologies have been identified including both lytic and temperate phages. However, only a handful of phages have been described in molecular detail, although these are evidently rich sources of genetic information and tools.

A. Mycobacteriophage L5

L5 is perhaps the best characterized of the mycobacteriophages (Hatfull, 1994). It was initially isolated in Japan by Doke (1960) from a lysogenic strain of *M. smegmatis*. It is a temperate phage and forms turbid plaques on *M. smegmatis*, from which stable lysogens can be isolated. It appears to have a broad host range and infects both *M. smegmatis* and BCG although it does so under rather different conditions, requiring high concentrations of calcium for infection of the slow-growing strains (Fullner and Hatfull, 1997). L5 appears to be virtually identical to phage L1 which was isolated at the same time by Doke (1960), with the main difference being that L1 does not form plaques at 42°C. Phage D29, which was isolated independently from L1 and L5 (Froman *et al.*, 1954) is a close relative and has extensive similarity at the sequence level. While less well characterized, phage FRAT 1 (Haeseleer *et al.*, 1992) is likely to also fall within this group of phages.

While L5 shares no sequence relationship with phage λ or any of its close relatives there are many features in common. For example, they are morphologically similar and have similar sized genomes with cohesive termini. Both have an attachment site (*attP*) close to the centre of the genome which is used as a site for integration into the host chromosome (Lee *et al.*, 1991). The overall gene organization is also similar, with the structure and assembly genes to the left of *attP* and DNA replication and control functions to its right (Hatfull and Sarkis, 1993). However, L5 differs from λ in a number of respects. For example, the L5 genome contains three tRNA genes, DNA polymerase and ribonucleotide reductase genes and appears to negatively regulate host gene expression during lytic growth (Hatfull and Sarkis, 1993). These are features more commonly associated with lytic bacteriophages.

The formation of L5 lysogens is accompanied by integration of the genome into the mycobacterial chromosome to form a prophage. Integration involves site-specific recombination between the phage *attP* site and the bacterial attachment site, *attB*, which overlaps a tRNAgly gene present in both *M. smegmatis* and *M. tuberculosis* (Lee *et al.*, 1991; see Chapter 3, III and IX). The phage-encoded integrase protein catalyses this reaction (Lee and Hatfull, 1993), although a host integration factor (mIHF) is also required (Pedulla *et al.*, 1996). A putative excisionase gene is probably closely linked to *attP* but has not yet been identified (Lee *et al.*, 1991). This information has been used to construct integration-proficient vectors that contain the L5 *attP* site and *int* gene and transform both fast- and slow-growing strains at high efficiency to produce stable single-copy transformants (Lee *et al.*, 1991).

The lysogenic state of L5 is maintained by a repressor, gp71, which acts as a transcriptional regulator (Donnelly-Wu *et al.*, 1993). However, it does so in a rather unusual way, and acts at more than two dozen binding sites present throughout the genome (Brown *et al.*, 1997). One of these overlaps an early lytic promoter, P_{left}, which transcribes the right arm genes (Nesbit *et al.*, 1995), such that the repressor interferes with transcription initiation. The other sites are located between genes in short intergenic intervals, and the binding of gp71 to these sites prevents transcription elongation (Brown *et al.*, 1997). Although it is clear that the left arm genes are transcribed at a later time than the right arm genes, the signals for late gene expression have not yet been identified.

Phage L5 has yet to be fully exploited as a cloning vehicle. L5-derived shuttle phasmids have been isolated (Snapper *et al.*, 1988) but they appear to have limited cloning capacity as a result of a requirement for nearly all of the genome for phage production (the only non-essential region is at the right end) and because the size of the L5 genome (52.3 kb) is close to the packaging limit of the phage. L5 recombinants containing the firefly luciferase gene (*FFlux*) have been isolated using homologous recombination in *M. smegmatis* and although these were generated at rather low frequencies (10^{-6}), this results in part from the requirement for deletions to occur that compensate for the size of the additional DNA (Sarkis *et al.*, 1995); further exchanges occur at a much higher frequency (10^{-3}; G.F. Hatfull, unpublished; see also Chapter 2, VIII).

B. Mycobacteriophage D29

Mycobacteriophage D29 is clearly a close relative of L5, although this has become much clearer now that the genome sequences of both phages are available (Ford *et al.*, 1997). D29 infects both fast- and slow-growing mycobacteria and does not have the same stringent requirement for high calcium concentrations for infection of BCG that L5 has (Fullner and Hatfull, 1997); it has also been shown to adsorb to *M. leprae* (David *et al.*, 1984). However, D29 is not a temperate phage and forms clear plaques, apparently as a result of a genomic deletion that removes part of the putative repressor gene (Ford *et al.*, 1997).

Comparison of the D29 and L5 genomes strongly suggests that D29 is a very recent derivative of a temperate parent (Ford *et al.*, 1997). The genomes are sufficiently different to show that L5 itself was not the parent, but D29 has the other characteristics of a temperate phage. For example, it not only has an *attP* site and integrase gene, but these are fully active and can mediate site-specific integration both *in vivo* and *in vitro* (Ford *et al.*, 1997); it also has a large number of gp71-binding sites throughout the genome. Moreover, D29 is able to lysogenize *M. smegmatis* if the L5 repressor gene, *71*, is present (Ford *et al.*, 1997).

Like L5, D29 has proven to be a useful genetic agent, and has been used for transposon delivery (Bardarov *et al.*, 1997) and for the construction of luciferase reporter phages (Pearson *et al.*, 1996). Because D29 is a very efficient and effective killer of its hosts it is also the phage of choice when selecting for transformants containing L5 *71* as a selectable marker (Donnelly-Wu *et al.*, 1993). A segment of D29 DNA was reported to act as an origin of replication (Lazraq *et al.*, 1994) but since this contains the D29 *attP* and *int* gene, it is likely that plasmids containing it are able to transform *M. smegmatis* via integration, rather than autonomous replication.

C. Mycobacteriophage TM4

Phage TM4 was initially isolated by induction from a strain of *M. avium* (Timme and Brennan, 1984) but has a broad host range. It is morphologically similar to L5 and D29 and contains a genome of approximately 50 kb. It is not obviously temperate in either *M. smegmatis* or *M. tuberculosis*, although the kinetics of light output following infection of *M. smegmatis* by a TM4-derived luciferase reporter phage does suggest that it lyses its host much slower than does either L5 or D29 (Sarkis *et al.*, 1995). The genome has been recently sequenced (M. Ford, R. Hendrix and G. F. Hatfull, unpublished) and shows that it is not related to L5 or D29 at the DNA sequence level, although some of the TM4 structural proteins have similarity to L5 and phage r1t. TM4 does not appear to encode an integrase protein and an *attP* site has not been identified.

It has proven rather easy to isolate shuttle phasmids with phage TM4 presumably because there are substantial regions of the genome that are non-essential, although it is not yet known which these are. TM4-derived shuttle phasmids have been useful for transposon delivery (Bardarov *et al.*, 1997) and for constructing luciferase reporter phages (Jacobs *et al.*, 1993).

D. Other Mycobacteriophages

Mycobacteriophage I3 is of some interest due its ability to act as a generalized transducing phage (see Chapter 3, V). The genome is approximately 145 kb in length and does not appear to contain cohesive termini (Gopinathan, 1993; Hatfull and Jacobs, 1994). Unlike L5, D29 and TM4, I3 has a contractile tail that contracts up to 60% following DNA injection into the cell, and a large hexagonal head (Kozloff *et al.*, 1972). I3 phage DNA is unusual in that it contains random, short, single-stranded regions, 6–10 bases long (Reddy and Gopinathan, 1986). Another phage of some interest is DS6A mainly because of its host range, which is restricted to species of the *M. tuberculosis* complex.

◆◆◆◆◆◆ VII. GENE EXPRESSION

A. RNA Polymerase

Mycobacterial RNA polymerases have the same overall structure as other bacterial RNA polymerases, and contain large β and β′ subunits, two protomers of the α subunit and associated sigma factors (Wiggs *et al.*, 1979). RNA polymerase isolated from *M. smegmatis* is active in promoting transcription, but has a strong requirement for supercoiling of the DNA substrate, particularly for transcription from promoters of mycobacterial origin (Levin and Hatfull, 1993). Promoter-specific initiation is absolutely dependent upon the presence of sigma factors, and two, MysA and MysB, that are present in exponentially-growing *M. smegmatis* cells have been identified (Predich *et al.*, 1995). Close homologues of these sigma factor genes are also present in *M. tuberculosis* (Doukhan *et al.*, 1995). *In vitro* transcription reactions using purified mycobacterial RNA polymerase are useful to the mycobacterial molecular biologist since they provide a means of showing that the 5′ ends of mRNAs isolated *in vivo* reflect transcription initiation as opposed to RNA processing or degradation (Levin and Hatfull, 1993).

Both fast- and slow-growing mycobacteria contain additional sigma factors that have specialized functions. For example, both *M. smegmatis* and *M. tuberculosis* contain SigE, a sigma factor thought to mediate responses to various environmental conditions (Wu *et al.*, 1997). The *sigE* gene is not essential for viability in *M. smegmatis*, but if absent, *M. smegmatis* cells have reduced survival following heat shock or exposure to acidic pH, detergent or oxidative stress (Wu *et al.*, 1997). Another sigma factor, SigF, with sequence similarity to the SigF sporulation sigma factors in *Streptomyces coelicolor* and *Bacillus subtilis*, is present in *M. tuberculosis* but absent from fast-growers such as *M. smegmatis*; SigF is induced during stationary phase, nitrogen starvation and cold shock and is proposed to play a role in adaptation of the bacterium to host responses in infection (DeMaio *et al.*, 1996). Finally, a mutation in the *rpoV* gene encoding a principal sigma factor in *M. bovis* is responsible for loss of virulence in that particular strain (Collins *et al.*, 1995).

B. Promoters

The complexity of the RNA polymerase in mycobacteria is likely to be reflected in the promoter signals required for transcription initiation. This has contributed to the difficulty in defining what specific signals are required for recognition by each of the holoenzyme forms. However, in general, the hexamer sequence at −10 from the start site is reasonably well conserved and is similar to that of *E. coli* promoters (Bashyam *et al.*, 1996). In contrast, the sequences corresponding to the −35 hexamer are hard to define, and a great variety of sequences appear to be tolerated, perhaps reflecting the action of multiple sigma factors (Bashyam *et al.*, 1996). This represents a feature shared with promoters of *Streptomyces*. It is noteworthy, however, that the P_{left} promoter of phage L5 has a 5′-TTGACA hexamer at −35, corresponding to the consensus *E. coli* sequence (Nesbit *et al.*, 1995) and this is an extremely active promoter in both *M. smegmatis* and BCG (Brown *et al.*, 1997). Similarly, one of the promoters upstream of the *M. tuberculosis recA* gene has a 5′-TTGTCA at the −35 region although its spacing from the putative −10 was only 9 bp, rather than the optimum of 17 bp in *E. coli* promoters; a second promoter upstream of recA has sequences similar to those of σ-32 responsive *E. coli* promoters (Movahedzadeh *et al.*, 1997). Consensus sequences have been proposed for the −10 and −35 sequences of *M. paratuberculosis* promoters which are very different from the canonical *E. coli* sequences (Bannantine *et al.*, 1997), although only a small number of promoters were identified and there is as yet no direct evidence that these sequences are important for transcription.

The majority of promoters isolated from *M. smegmatis* are also active in *M. tuberculosis* and vice versa, reflecting a fundamental similarity in the transcription apparatus of the fast- and slow-growing mycobacteria (Das Gupta *et al.*, 1993). However, some promoters do behave differently and only one of the two promoters upstream of the BCG hsp60 gene that are active in BCG (Stover *et al.*, 1991) is also active in *M. smegmatis* (Levin and Hatfull, 1993). Thus, differences between both specificity and strength of individual promoters in slow- and fast-growing mycobacteria are likely (Timm *et al.*, 1994a; Bashyam *et al.*, 1996). Most mycobacterial promoters have rather weak activity in *E. coli* (Bashyam *et al.*, 1996).

The complexity of mycobacterial promoter structure unfortunately makes it difficult to predict the locations of promoters in mycobacterial DNA sequences. This will undoubtedly change as we learn more about mycobacterial promoters, and the availability of the *M. tuberculosis* genome sequence should be helpful. Meanwhile, it is important that transcription start sites are identified empirically before assigning promoter identity to specific DNA sequences upstream of genes. Since a number of relatively simple methods for isolation of RNA from mycobacteria are available (Levin and Hatfull, 1993; Bashyam and Tyagi, 1994) this can be accomplished reasonably easily by a combination of primer-extension and S1 nuclease mapping.

C. Terminators

Very little is known about transcription termination in mycobacteria and there are few reports of direct mapping of the 3′ ends of transcription units. However, Rho-independent stem–loop type terminator-like structures are present in mycobacterial DNA, and these are sufficiently similar to well-characterized *E. coli* terminators that they must perform this function. Some of the more impressive of these types of RNA structures are found in the mycobacteriophages, and L5 has one putative terminator with a 17 bp perfectly matched stem followed by seven U bases (Hatfull and Sarkis, 1993). Little is known about termination at other signals or factor-dependent transcription termination, apart from the phage-specific example of repressor-mediated polarity of downstream gene expression (Brown *et al.*, 1997).

D. Reporter Genes

Characterization of mycobacterial transcription is simplified by the development of a plethora of reporter gene systems. More than half-a-dozen reporter genes have been reported, each of which have their advantages and disadvantages. The popular *E. coli* reporter gene, *lacZ*, has been utilized in a variety of circumstances to identify promoters in plasmid libraries and to characterize promoter strengths (Barletta *et al.*, 1992; Donnelly-Wu *et al.*, 1993; Timm *et al.*, 1994a, 1994b; Dellagostin *et al.*, 1995; Jain *et al.*, 1997). It has the advantage that there is no endogenous β-galactosidase in mycobacteria and *lacZ*-expressing strains can be identified as blue colonies on solid media containing X-Gal. A potential disadvantage is that cells must be efficiently lysed for quantitative assays, usually by sonication, which complicates and introduces some variability into the assay. The *E. coli phoA* gene has also been used as a reporter gene in mycobacteria and has the advantage that PhoA activity is dependent on protein export (Timm *et al.*, 1994b). For example, fusions of the mycobacterial exported BlaF protein to PhoA have significant alkaline phosphatase activity (Timm *et al.*, 1994b). A third reporter gene is *xylE* from *Pseudomonas*, which can be detected as a bright yellow colour when mycobacterial colonies expressing XylE are sprayed with a catechol solution (Curcic *et al.*, 1994). The *xylE* product, catechol 2, 3-dioxygenase can also be measured quantitatively in sonicated extracts (Curcic *et al.*, 1994). Tyagi and colleagues (Das Gupta *et al.*, 1993; Bashyam *et al.*, 1996) have also utilized the chloramphenicol acetyl-transferase (CAT) gene in both slow- and fast-growing mycobacteria for the isolation of promoters, using chloramphenicol resistance as a selection for promoter activity.

The luciferase genes of bacteria and of fireflies have been used effectively as reporters in mycobacteria. While these are not as useful in assays on solid media, they are simple to measure quantitatively. To measure FFlux activity, no lysis of the cells is required and an aliquot of culture can be placed directly into a luminometer followed by automated injection of the luciferase substrate and measurement of light output (Jacobs *et al.*, 1993). Luciferase activity is dependent upon intracellular ATP such that metabolically inactive bacteria do not produce light, and live and dead bacteria can be easily distinguished (Jacobs *et al.*, 1993; Sarkis *et al.*, 1995). FFlux activity has also been used to determine antimycobacterial activities of antibiotics in whole animal systems (Hickey *et al.*, 1996; Shawar *et al.*, 1997; see also Chapter 3, VIII). Recombinants of *M. smegmatis* expressing bacterial luciferase have been described for use as a reporter of environmental changes (Gordon *et al.*, 1994).

The green fluorescent protein (GFP) of *Aequorea victoria* has unique properties as a reporter gene in mycobacteria (Dhandayuthapani *et al.*, 1995; Kremer *et al.*, 1995). Expression of *gfp* in both *M. smegmatis* and BCG can be detected by their green fluorescence following irradiation and used as part of a promoter-probe vector (Kremer *et al.*, 1995). Moreover, gene expression following infection of macrophages can be monitored by fluorescence and infected cells studied and sorted by flow cytometry. Macrophages infected with GFP-expressing mycobacteria can also be viewed by scanning confocal microscopy. Thus, GFP provides a number of nice advantages and opportunities for studying mycobacterial gene expression in infected cells.

◆◆◆◆◆◆ VIII. RESOURCES

The previous discussion illustrates the growing level of sophistication in mycobacterial genetics and the impressive advances that have been made since the uptake of phage DNA was described ten years ago (Jacobs *et al.*, 1987). Many manipulations are now fairly routine, and anyone with some general background in microbial genetics, can obtain the various strains, plasmids and phages from either the American Type Culture Collection or individual investigators and conduct experiments in mycobacterial genetics. There is also a large collection of research materials available at Colorado State University through a National Institutes of Health funded contract, including bacterial cells, proteins, lipids, nucleic acids and carbohydrates, monoclonal antibodies and polyclonal sera (see www.cvmbs.colostate.edu/microbiology/tb/top.htm). Of course, if using *M. tuberculosis* rather than *M. smegmatis* or BCG, the appropriate containment facilities must be available. Nonetheless, notwithstanding the notable lack of facile methods for moving specific genetic markers between strains of slow-growing mycobacteria, and the relatively slow pace of individual experiments, there are few serious limitations to mycobacterial genetics.

The manner in which the experimental approaches are applied is likely to change substantially within the coming year, as the *M. tuberculosis* genome sequences are completed, and as new mycobacterial genome projects are initiated. Fortunately, this information is readily available on the Internet. The sequencing project in progress at the Sanger Center can be found at www.sanger.ac.uk/pathogens/ and the project at The Institute for Genomic research (TIGR) is at www.tigr.org; the TIGR sequences can also be searched by using the

BLAST function at NCBI (www.ncbi.nlm.nih.gov/BLAST/). *M. leprae* cosmid sequences determined by Genome Therapeutics Inc. are at www.cric.com/htdocs/sequences/leprae/index.html.

A key resource for the mycobacterial geneticist is the MycDB database established by Stewart Cole and Staffan Bergh (Bergh and Cole, 1994). This consolidates a vast array of information on the mycobacteria into a single database which can be accessed at kiev.physchem.kth.se/MycDB.html. The database contains DNA sequences of mycobacteria, macrorestriction maps, information of clones, antigens and antibodies as well as publications and their authors. There is also a comprehensive directory of colleagues in mycobacterial research.

References

Aldovani, A., Husson, R. H. and Young, R. A. (1993). The *uraA* locus and homologous recombination in *Mycobacterium bovis* BCG. *J. Bacteriol.* **175**, 7282–7289.

Anes, E. Portugal I. and Moniz-Pereira, J. (1992). Insertion into the *Mycobacterium smegmatis* genome of the *aph* gene through lysogenization with the temperate mycobacteriophage Ms6. *FEMS Microbiol. Lett.* **74**, 21–25.

Balasubramanian, V., Pavelka, M.S., Jr, Bardarov, S. S., Martin, J., Weisbrod, T. R., McAdam, R. A., Bloom, B. R. and Jacobs, W. R., Jr. (1996). Allelic exchange in *Mycobacterium tuberculosis* with long linear recombination substrates. *J. Bacteriol.* **178**, 273–279.

Banerjee, A., Dubnau, E., Quemard, A., Balasubramanian, V., Um, K. S., Wilson, T., Collins, D., de Lisle, G. and Jacobs, W. R., Jr (1994). *inhA*, a gene encoding a target for isoniazid and ethionamide in *Mycobacterium tuberculosis*. *Science* **263**, 27–230.

Bannantine, J. P., Barletta, R. G., Thoen, C. O. and Andrews, R. E., Jr (1997). Identification of *Mycobacterium paratuberculosis* gene expression signals. *Microbiology* **143**, 921–928.

Bardarov, S., Kriakov, J., Carriere, C., Yu, S., Vaamonde, C., McAdam, R. A., Bloom, B. R., Hatfull, G. F. and Jacobs, W. R., Jr (1997). Conditionally replicating mycobacteriophages: A system for transposon delivery to *Mycobacterium tuberculosis*. *Proc. Natl. Acad. Sci. USA* **94**, 10961–10966.

Barletta, R. G., Kim, D. D., Snapper, S. B., Bloom, B. R. and Jacobs, W. R., Jr (1992). Identification of expression signals of the mycobacteriophages Bxb1, L1 and TM4 using the *Escherichia–Mycobacterium* shuttle plasmids pYUB75 and pYUB76 designed to create translational fusions to the *lacZ* gene. *J. Gen. Microbiol.* **138**, 23–30.

Barsom, E. K. and Hatfull, G. F. (1996). Characterization of a *Mycobacterium smegmatis* gene that confers resistance to phages L5 and D29 when overexpressed. *Mol. Microbiol.* **21**, 159–170.

Bashyam, M. D. and Tyagi, A. (1994). An efficient and high-yielding method for isolation of RNA from mycobacteria. *Biotechniques* **17**, 834–836.

Bashyam, M. D., Kaushal, D., Dasgupta, S. K. and Tyagi, A. K. (1996). A study of mycobacterial transcripitonal apparatus: identification of novel features in promoter elements. *J. Bacteriol.* **178**, 4847–4853.

Baulard, A., Jourdan, C., Mercenier, A. and Locht, C. (1992). Rapid mycobacterial plasmid analysis by electroduction between *Mycobacterium* spp. and *Escherichia coli*. *Nucl. Acids Res.* **20**, 4105.

Baulard, A., Escuyer, V., Haddad, N., Kremer, L., Locht, C. and Berche, P. (1995).

Mercury resistance as a selective marker for recombinant mycobacteria. *Microbiology* **141**, 1045–1050.

Baulard, A., Kremer, L. and Locht, C. (1996). Efficient homologous recombination in fast-growing and slow-growing mycobacteria. *J. Bacteriol.* **178**, 3091–3098.

Beggs, M. L., Crawford, J. T. and Eisenach, K. D. (1995). Isolation and sequencing of the replication region of *Mycobacterium avium* plasmid pLR7. *J. Bacteriol.* **177**, 4836–4840.

Bergh, S. and Cole, S. T. (1994). MycDB: an integrated mycobacterial database. *Mol. Microbiol.* **12**, 517–534.

Bloom, B. R. and Murray, C. J. L. (1992). Tuberculosis: commentary on a reemergant killer. *Science* **257**, 1055–1064.

Brown, K. L., Sarkis, G. J., Wadsorth, C. and Hatfull, G. F. (1997). Transcriptional silencing by the mycobacteriophage L5 repressor. *EMBO J.* **16**, 5914–5921.

Burlein, J. E., Stover, C. K., Offutt, S. and Hanson, M. S. (1994). Expression of foreign genes in mycobacteria. In *Tuberculosis: Pathogenesis, Protection and Control* (B. R. Bloom, ed.), pp. 239–252. American Society for Microbiology, Washington, DC.

Cirillo, J. D., Barletta, R. G., Bloom, B. R. and Jacobs, W. R., Jr (1991). A novel transposon trap for mycobacteria: isolation and characterization of IS*1096*. *J. Bacteriol.* **173**, 7772–7780.

Clark-Curtiss, J. (1990). Genome structure of mycobacteria. In *Molecular Biology of Mycobacteria* (J. McFadden, ed.), pp. 77–96. Academic Press, London.

Collins, D. M. and Stephens, D. M. (1991). Identification of an insertion sequence, IS1081, in *Mycobacterium bovis. FEMS Microbiol. Lett.* **83**, 11–16.

Collins, D. M., Kawakami, R. P., De Lisle, G. W., Pascopella, L., Bloom, B. R. and Jacobs, W. R., Jr (1995). Mutation of the principal s factor causes loss of virulence in a strain of the *Mycobacterium tuberculosis* complex. *Proc. Natl. Acad. Sci. USA* **92**, 8036–8040.

Colston, M. J. and Davis, E. O. (1994). Homologous recombination, DNA repair, and mycobacterial *recA* genes. In *Tuberculosis: Pathogenesis, Protection and Control* (B. R. Bloom, ed.), pp. 217–226. American Society for Microbiology, Washington, DC.

Curcic, R., Dhandayuthapani, S. and Deretic, V. (1994). Gene expression in mycobacteria: transcriptional fusions based on *xylE* and analysis of the promoter region of the response regulator *mtrA* from *Mycobacterium tuberculosis. Mol. Microbiol.* **13**, 1057–1064.

Das Gupta, S. K., Bashyam, M. D. and Tyagi, A. K. (1993). Cloning and assessment of mycobacterial promoters by using a plasmid shuttle vector. *J. Bacteriol.* **175**, 5186–5192.

David, H.L., Clement, F., Clavel-Seres, S. and Rastogi, N. (1984). Abortive infection of *Mycobacterium leprae* by the mycobacteriophage D29. *Intl. J. Lepr. Myco. Dis.* **52**, 515–523.

Davis, E. O., Jenner, P. J., Brooks, P. C., Colston, M. J. and Sedgwick, S. G. (1992). Protein splicing in the maturation of *M. tuberculosis* RecA protein: a mechanism for tolerating a novel class of intervening sequences. *Cell* **71**, 201–210.

Dellagostin, O. A., Wall, S., Norman, E., O'Shaughnessy, T., Dale, J. W. and McFadden, J. J. (1993). Construction and use of integrative vectors to express foreign genes in mycobacteria. *Mol. Microbiol.* **10**, 983–993.

Dellagostin, O. A., Esposito, G., Eales, L. J., Dale, J. W. and McFadden, J. (1995). Activity of mycobacterial promoters during intracellular and extracellular growth. *Microbiology* **141**, 1785–1792.

DeMaio, J., Zhang, Y., Ko, C., Young, D. B. and Bishai, W. R. (1996). A stationary-phase stress-response sigma factor from *Mycobacterium tuberculosis. Proc. Natl. Acad. Sci. USA* **93**, 2790–2794.

Dhandayuthapani, S., Via, L. E., Thomas, C. A., Horowitz, P. M., Deretic, D. and Deretic, V. (1995). Green fluorescent protein as a marker for gene expression and cell biology of mycobacterial interactions with macrophages. *Mol. Microbiol.* **17**, 901–912.

Doke, S. (1960). Studies on mycobacteriophages and lysogenc mycobacteria. *J. Kunamoto Med. Soc.* **34**, 1360–1371.

Donnelly-Wu, M., Jacobs,W. R. and Hatfull, G. F. (1993). Superinfection immunity of mycobacteriophage L5 : Applications for genetic transformation of mycobacteria. *Mol Microbiol.* **7**, 407–417.

Doukhan, L., Predich, M., Nair, G., Dussurget, O., Mandie-Mulec, I., Cole, S. T., Smith, D. R. and Smith, I. (1995). Genomic organization of the mycobacterial sigma gene cluster. *Gene* **165**, 67–70.

Eiglmeier, K., Honore, N., Woods, S. A., Caudron, B. and Cole, S. T. (1993). Use of an ordered cosmid library to deduce the genomic organization of *Mycobacterium leprae. Mol. Microbiol.* **7**, 197–206.

England, P. M., Wall, Q. and McFadden, J. J. (1991). IS*900*-promoted stable integration of a foreign gene into mycobacteria. *Mol. Microbiol.* **4**, 1771–1777.

Falkingham, J. O. and Crawford, J. T. (1994). Plasmids. In *Tuberculosis: Pathogenesis, Protection and Control* (B. R. Bloom, ed.), pp. 185–198. American Society for Microbiology, Washington, DC.

Ford, M. E., Sarkis, G. J., Belanger, A. E., Hendrix, R. W. and Hatfull, G. F. (1997). Genome structure of mycobacteriophage D29: Implications for phage evolution *J. Mol. Biol.* **279**, 143–164.

Froman, S., Will, D.W. and Bogen, E. (1954). Bacteriophage active against virulent *Mycobacterium tuberculosis* I. Isolation and activity. *Am. J. Pub. Hlth.* **44**, 1326–1333.

Fullner, K. J. and Hatfull, G. F. (1997). Mycobacteriophage L5 infection of *Mycobacterium bovis* BCG: Implications for phage genetics in the slow-growing mycobacteria. *Mol. Microbiol.* **26**, 755–766.

Garcia, M. J., Guilhot, C., Lathigra, R., Menendez, M. C., Domenech, P., Moreno, C., Gicquel, B. and Martin, C. (1994). Insertion sequence IS1137, a new IS3 family element from *Mycobacterium smegmatis. Microbiology* **140**, 2821–2828.

Gavigan, J. A., Guilhot, C., Gicquel, B. and Martin, C. (1995). Use of conjugative and thermosensitive cloning vectors for transposon delivery to *Mycobacterium smegmatis. FEMS Microbiol. Lett.* **127**, 35–59.

Gavigan, J. A., Ainsa, J. A., Perez, E., Otal, I. and Martin, C. (1997). Isolation by genetic labeling of a new mycobacterial plamisd, pJAZ38, from *Mycobacterium fortuitum. J. Bacteriol.* **179**, 4115–4122.

Gopinathan, K. P. (1993). Molecular biology of mycobacteria and mycobacteriophages. *J. Indian Inst. Sci.* **73**, 31–45.

Gordon, S., Parish, T., Roberts, I. S. and Andrew, P. W. (1994). The application of luciferase as a reporter of environmental regulation of gene expression in mycobacteria. *Lett. Appl. Microbiol.* **19**, 336–340.

Gormley, E. P. and Davies, J. (1991). Transfer of plasmid RSF1010 by conjugation from *Escherichia coli* to *Streptomyces lividans* and *Mycobacterium smegmatis. J. Bacteriol.* **173**, 6705–6708.

Goto, Y., Taniguchi, H., Udou, T., Mizuguchi, Y. and Tokunaga, T. (1991). Development of a new host vector system in mycobacteria. *FEMS Microbiol. Lett.* **83**, 277–282.

Green, E. P., Tizard, M. L. V., Moss, M. T., Thompson, D. J., Winterbourne, D. J., McFadden, J. and Hermon-Taylor, J. (1989). Sequence and characteristics of IS*900*, an insertion element idenitifed in a human Crohn's disease isolate of *Mycobacterium paratuberculosis. Nucl. Acids Res.* **17**, 9063–9073.

Genetic Methods in Mycobacteria

Greenberg, J. and Woodley, C. L. (1984). Genetics of mycobacteria. In *The Mycobacteria: A Sourcebook* (G. P. Kubica and L. G. Wayne, eds), pp. 629–639. Marcel Dekker, New York.

Guerrero, C., Bernasconi, C., Burki, D., Bodmer, T. and Telenti, A. (1995). A novel insertion element from *Mycobacterium avium*, IS1245, is a specific target for analysis of strain relatedness. *J. Clin. Microbiol.* **33**, 304–307.

Guilhot, C., Gicquel, B., Davis, J. and Martin, C. (1992). Isolation and analysis of IS6120, a new insertion sequence from *Mycobacterium smegmatis*. *Mol. Microbiol.* **6**, 107–113.

Guilhot, C., Otal, I., Van Rompaey, I., Martin, C. and Gicquel, B. (1994). Efficient transposition in mycobacteria: construction of *Mycobacterium smegmatis* insertional mutant libraries. *J. Bacteriol.* **176**, 535–539.

Haeseleer, F., Pollet, J. F., Bollen, A. and Jacobs, P. (1992). Molecular cloning and sequencing of the attachment site and integrase gene of the temperate mycobacteriophage FRAT1. *Nucl. Acids Res.* **20**, 1420.

Haeseleer, F., Pollet, J-F., Haumont, M., Bollen, A. and Jacobs, P. (1993). Stable integration and expression of the *Plasmodium falciparum* circumsporozoite protein coding sequence in mycobacteria. *Mol. Biochem. Parasitol.* **57**, 117–126.

Hatfull, G. F. (1993). Genetic transformation of the mycobacteria. *Trends Microbiol.* **1**, 310–314.

Hatfull, G. F. (1994). Mycobacteriophage L5: A toolbox for tuberculosis. *ASM News* **60**, 255–260.

Hatfull, G. F. (1996). The molecular genetics of *Mycobacterium tuberculosis*. *Curr. Top. Microbiol. Immunol.* **215**, 29–47.

Hatfull, G. F. and Jacobs, W. R., Jr (1994). Mycobacteriophages: cornerstones of mycobacterial research. In *Tuberculosis: Pathogenesis, Protection and Control.* (B. R. Bloom, ed.), pp. 165–183. American Society for Microbiology, Washington, DC.

Hatfull, G. F. and Sarkis, G. J. (1993). DNA sequence, structure and gene expression of mycobacteriophage L5: A phage system for mycobacterial genetics. *Mol. Microbiol.* **7**, 395–405.

Hermans, J., Martin, C., Huijberts, G. N. M., Goosen, T. and de Bont, J. A. M. (1991). Transformation of *Mycobacterium aurum* and *Mycobacterium smegmatis* with the broad host range Gram-negative cosmid vector pJRD215. *Mol. Microbiol.* **5**, 1561–1566.

Hernandez Perez, M., Fomukong, N. G., Hellyer, T., Brown, I. N. and Dale, J. W. (1994). Characterization of IS1110, a highly mobile genetic element from *Mycobacterium avium*. *Mol. Microbiol.* **12**, 717–724.

Hickey, M. J., Arain, T. M., Shawar, R. M., Humble, D. J., Langhorne, M. H., Morgenroth, J. N. and Stover, C. K. (1996). Luciferase *in vivo* expression technology: use of recombinant mycobacterial reporter strains to evaluate antimycobacterial activity in mice. *Antimicrob. Agents Chemother.* **40**, 400–407.

Hinshelwood, S. and Stoker, N. G. (1992). Cloning of mycobacterial histidine synthesis genes by complementation of a *Mycobacterium smegmatis* auxotroph. *Mol. Microbiol.* **6**, 2887–2895.

Husson, R. A., James, B. E. and Young, R. A. (1990). Gene replacement and expression of foreign DNA in mycobacteria. *J. Bacteriol.* **172**, 519–524.

Jacobs, W. R., Jr (1991). Genetics systems for Mycobacteria. *Meth. Enzymol.* **155**, 153–160.

Jacobs, W. R. Jr., Tuckman, M. and Bloom, B. R. (1987). Introduction of foreign DNA into mycobacteria using a shuttle phasmid. *Nature* **327**, 532–535.

Jacobs, W. R., Jr, Barletta, R. G., Udani, R., Chan, J., Kalkut, G., Sosne, G., Kieser, T., Sarkis, G. J., Hatfull, G. F. and Bloom, B. R. (1993). Rapid assessment of drug susceptibilities of *Mycobacterium tuberculosis* by means of luciferase reporter phages. *Science* **260**, 819–822.

Jain, S., Kaushal, D., DasGupta, S. K. and Tyagi, A. K. (1997). Construction of shuttle vectors for genetic manipulation and molecular analysis of mycobacteria. *Gene* **190**, 37–44.

Kalpana, G. V., Bloom, B. R. and Jacobs, W. R., Jr (1991). Insertional mutagenesis and illegitimate recombination in mycobacteria. *Proc. Natl. Acad. Sci. USA* **88**, 5433–5437.

Kozloff, L. M., Raj, C. V. S., Rao, R. N., Chapman, V. A. and DeLong, S. (1972). Structure of a transducing mycobacteriophage. *J. Virol.* **9**, 390–393.

Kremer, L., Baulard, A., Estaquier, J., Poulain-Godefroy, O. and Locht, C. (1995). Green fluorescent protein as a new expression marker in mycobacteria. *Mol. Microbiol.* **17**, 913–922.

Kunze, Z. M., Wall, S., Appelberg, R., Silva, M. T., Portaels, F. and McFadden, J. J. (1991). IS*901*, a new member of a widespread class of atypical insertion sequences, is associated with pathogenicity in *Mycobacterium avium*. *Mol. Microbiol.* **5**, 2265–2272.

Lazraq, R., Houssaini-Iraqui, M., Clavel-Séres, S, and David, H. L. (1991). Cloning and expression of the origin of replication of mycobacteriophage D29 in *Mycobacterium smegmatis*. *FEMS Microbiol. Lett.* **80**, 117–120.

Lee, M. H. and Hatfull, G. F. (1993). Mycobacteriophage L5 integrase-mediated site-specific recombination *in vitro*. *J. Bacteriol.* **175**, 6836–6841.

Lee, M. H., Pascopella, L., Jacobs, W. R. and Hatfull, G. F. (1991). Site-specific integration of mycobacteriophage L5: integration-proficient vectors for *Mycobacterium smegmatis*, *Mycobacterium tuberculosis*, and bacille Calmette-Guérin. *Proc. Natl. Acad. Sci. USA* **88**, 3111–3115.

Levin, M. E. and Hatfull, G. F. (1993). *Mycobacterium smegmatis* RNA polymerase: DNA supercoiling, action of rifampicin and mechanism of rifampicin resistance. *Mol. Microbiol.* **8**, 277–285.

Marklund, B. I., Speert, D. P. and Stokes, R. W. (1995). Gene replacement through homologous recombination in *Mycobacterium intracellulare*. *J. Bacteriol.* **177**, 6100–6105.

Martin, C., Timm, J., Rauzier, R., Gomez-Lus, R., Davis, J. and Gicquel, B. (1990). Transposition of an antibiotic resistance element in mycobacteria. *Nature* **345**, 739–743.

Martin, C. P., Mazodier, E., Mediola, M. V., Gicquel, B., Smokvina, T., Thompson, C. J. and Davis, J. (1991). Site-specific integration of the *Streptomyces* plasmid pSAM2 in *Mycobacterium smegmatis*. *Mol. Microbiol.* **5**, 2499–2502.

McAdam, R., Guilhot, C. and Gicquel, B. (1994). Transposition in mycobacteria. In *Tuberculosis: Pathogenesis, Protection and Control* (B. R. Bloom, ed.), pp. 199-216. American Society for Microbiology, Washington, DC.

McAdam, R. A., Weisbrod, T. R., Martin, J., Scuderi, J. D., Brown, A. M., Cirillo, J. D., Bloom, B. R. and Jacobs, W. R., Jr (1995). *In vivo* growth characteristics of leucine and methionine auxotrophic mutants of *Mycobacterium bovis* BCG generated by transposon mutagenesis. *Infect. Immun.* **63**, 1004–1012.

Mizuguchi, Y. (1984). Mycobacteriophages. In *The Mycobacteria: A Sourcebook* (G. P. Kubica and L. G. Wayne, eds), pp. 641–662. Marcel Dekker, New York, NY.

Mizuguchi, Y. and Tokunaga, T. (1971). Recombination between *Mycobacterium smegmatis* strains Jucho and Lacticola. *Jpn. J. Microbiol.* **15**, 359–366.

Mizuguchi, Y., Suga, K. and Tokunaga, T. (1976). Multiple mating types of *Mycobacterium smegmatis*. *Jpn. J. Microbiol.* **20**, 435–443.

Movahedzadeh, F., Colston, M. J. and Davis, E. O. (1997). Determination of DNA sequences required for regulated *Mycobacterium tuberculosis* RecA expression in response to DNA-damaging agents suggests that two modes of regulation exist. *J. Bacteriol.* **179**, 3509–3518.

Nesbit, C.E., Levin, M.E., Donnelly-Wu, M.K. and Hatfull, G.F. (1995). Transcriptional regulation of repressor synthesis in mycobacteriophage L5. *Mol. Microbiol.* **17**, 1045–1056.

Norgard, M. V. and Imaeda, T. (1978). Physiological factors affecting transformation of *Mycobacterium smegmatis. J. Bacteriol.* **133**, 1254–1262.

Norman, E., Dellagostin, O. A., McFadden, J. and Dale, J. W. (1995). Gene replacement by homologous recombination in *Mycobacterium bovis* BCG. *Mol. Microbiol.* **16**, 755–760.

O'Reilly, L. M. and Daborn, C. J. (1995). The epidemiology of *Mycobacterium bovis* infections in animals and man: a review. *Tuber. Lung Dis.* **76**, 1–46.

Parish, T. and Stoker, N. G. (1997). *Methods in Molecular Biology*, Vol. 101. *Mycobacteria Protocols.* (T. Parish and N. G. Stoker, eds). Humana Press, Totowa, New Jersey.

Pascopella, L., Collins, F. M., Martin, J. M., Lee, M. H., Hatfull, G. F., Bloom, B. R. and Jacobs, W. R., Jr (1993). Use of *in vivo* complementation in *Mycobacterium tuberculosis* to identify a genomic fragment associated with virulence. *Infect. Immun.* **62**, 1313–1319.

Pavelka, M. S. and Jacobs, W. R., Jr (1996). Biosynthesis of diaminopimelate, the precursor of lysine and a component of peptidoglycan, is an essential function of *Mycobacterium smegmatis. J. Bacteriol.* **178**, 6496–6507.

Pearson, R.E., Jurgensen, S., Sarkis, G.J., Hatfull, G.F. and Jacobs, W.R., Jr (1996). Construction of D29 shuttle phasmids and luciferase reporter phages for detection of mycobacteria. *Gene* **183**, 129–136.

Pedulla, M.L., Lee, M.H., Lever, D.C. and Hatfull, G.F. (1996). A novel host factor for integration of mycobacteriophage L5. *Proc. Natl. Acad. Sci., USA* **93**, 15411–15416.

Pelicic, V., Reyrat, J. M. and Gicquel, B. (1996a). Positive selection of allelic exchange mutants in *Mycobacterium bovis* BCG. *FEMS Microbiol. Lett.* **144**, 161–166.

Pelicic, V., Reyrat, J. M. and Gicquel, B. (1996b). Generation of unmarked directed mutations in mycobacteria, using sucrose counter-selectable suicide vectors. *Mol. Microbiol.* **20**, 919–925.

Pelicic, V., Jackson, M., Reyrat, J. M., Jacobs, W. R. Jr., Gicquel, B. and Guilhot, C. (1997). Efficient allelic exchange and transposon mutagenesis in *Mycobacterium tuberculosis. Proc. Natl. Acad. Sci., USA* **94**, 10955–10960.

Philipp, W. J., Poulet, S., Eiglmeier, K., Pascopella, L., Balasubramanian, V., Heym, B., Bergh, S., Bloom, B. R., Jacobs, W. R., Jr and Cole, S. T. (1996). An integrated map of the genome of the tubercle bacillus, *Mycobacterium tuberculosis* H37Rv, and comparison with *Mycobacterium leprae. Proc. Natl. Acad. Sci., USA* **93**, 3132–3137.

Picardeau, M., Varnerot, A., Rauzier, J., Gicquel, B. and Vincent, V. (1996). *Mycobacterium xenopi* IS*1395*, a novel insertion sequence expanding the IS*256* family. *Microbiology* **142**, 2453–2461.

Predich, M., Doukhan, L., Nair, G. and Smith, I. (1995). Characterization of RNA polymerase and two sigma-factor genes from *Mycobacterium smegmatis. Mol. Microbiol.* **15**, 355–366.

Radford, A. J. and Hodgson, A. L. M. (1991). Construction and characterization of a *Mycobacterium–Escherichia coli* shuttle vector. *Plasmid* **25**, 149–153.

Raj, C. V. S. and Ramakrishnan, T. (1970). Transduction of *Mycobacterium smegmatis. Nature* **228**, 280–281.

Raj, C. V. S. and Ramakrishnan, T. (1971). Genetic studies in mycobacteria

Isolation of auxotrophs and mycobacteriophages for *Mycobacterium smegmatis* and their use in transduction. *J. Indian Inst. Sci.* **53**, 126–140.

Ramakrishnan, L., Tran, H. T., Federspiel, N. A. and Falkow, S. (1997). A *crtB* homologue essential for photochromogenicity in *Mycobacterium marinum*: isolation, characterization, and gene disruption via homologous recombination. *J. Bacteriol.* **179**, 5862–5868.

Ranes, M. G., Rauzier, J., LaGranderie, M., Gheorghiu, M. and Gicquel, B. (1990). Functional analysis of pAL5000, a plasmid from *Mycobacterium fortuitum*: construction of a "mini" mycobacterium–*Escherichia coli* shuttle vector. *J. Bacteriol.* **172**, 2793–2797.

Reddy, A. B. and Gopinathan, K. P. (1986). Presence of random single-stranded gaps in mycobacteriophage I3. *Gene* **44**, 227–234.

Redmond, W. B. (1993). Bacteriophages of mycobacteria: a review. *Adv. Tuberc. Rev.* **12**, 191–229.

Reyrat, J. M., Berthet, F. X. and Gicquel, B. (1995). The urease locus of *Mycobacterium tuberculosis* and its utilization for the demonstration of allelic exchange in *Mycobacterium bovis* bacillus Calmette-Guérin. *Proc. Natl. Acad. Sci. USA* **92**, 8768–8772.

Sander, P., Meier, A. and Bottger, E. C. (1995). *rpsL*⁺: a dominant selectable marker for gene replacement in mycobacteria. *Mol. Microbiol.* **16**, 991–1000.

Sarkis, G. J., Jacobs, W. R., Jr. and Hatfull, G. F. (1995). L5 luciferase reporter mycobacteriophages: a sensitive tool for the detection and assay of live mycobacteria. *Mol. Microbiol.* **15**, 1055–1067.

Saroja, D. and Gopinathan, K. P. (1973). Transduction of isoniazid susceptibility-resistance and streptomycin resistance in mycobacteria. *Antimicrob. Agents Chemother.* **4**, 643–645.

Shawar, R. M., Humble, D. J., Van Dalfsen, J. M., Stover, C. K., Hickey, M. J., Steele, S., Mitscher, L. A. and Baker, W. (1997). Rapid screening of natural products for antimycobacterial activity by using luciferase-expressing strains of *Mycobacterium bovis* BCG and *Mycobacterium intracellulare*. *Antimicrob. Agents Chemother.* **41**, 570–574.

Snapper, S. B., Lugosi, L., Jekkel, A., Melton, R. E., Kieser, T., Bloom, B. R. and Jacobs, W. R., Jr (1988). Lysogeny and transformation in mycobacteria: stable expression of foreign genes. *Proc. Natl. Acad. Sci. USA* **85**, 6987–6991.

Snapper, S., Melton, R., Keiser, T. and Jacobs, W.R., Jr, (1990). Isolation and characterization of efficient plasmid transformation mutants of *Mycobacterium smegmatis*. *Mol. Microbiol.* **4**, 1911–1919.

Snider, D. E., Jones, W. D., Jr and Good, R. C. (1984). The usefulness of phage typing *Mycobacterium tuberculosis* isolates. *Am. Rev. Respir. Dis.* **130**, 1095–1099.

Snider, D. E., Jr, Raviglione, M. and Kochi, A. (1994). Global burden of tuberculosis. In *Tuberculosis: Pathogenesis, Protection and Control* (B. R. Bloom, ed.), pp. 3–11. American Society for Microbiology, Washington, DC.

Stolt, P. and Stoker, N. G. (1996). Functional definition of regions necessary for replication and incompatibility in the *Mycobacterium fortuitum* plasmid pAL5000. *Microbiology* **142**, 2795–2802.

Stover, C. K., de la Cruz, V. F., Fuerst, T. R., Burlein, J. E., Benson, L. A., Bennett, L. T., Bansal, G. P., Young, J. F., Lee, M. H., Hatfull, G. F., Snapper, S., Barletta, R. G., Jacobs, W. R., Jr and Bloom, B. R. (1991). New use of BCG for recombinant vaccines. *Nature* **351**, 456–460.

Thierry, D., Cave, M. D., Eisenach, K. D., Crawford, J. T., Bates, J. H., Gicquel, B. and van Embden, J. D. A. (1990). IS6110, an IS-like element of *Mycobacterium tuberculosis*. *Nucl. Acids Res.* **18**, 188.

Timm, J., Lim, E. M. and Gicquel, B. (1994a). *Escherichia coli*–mycobacterial shuttle vectors for operon and gene fusions to *lacZ*: the pJEM series. *J. Bacteriol.* **176**, 6749–6753.

Timm, J., Perilli, M. G., Duez, C., Trias, J., Orefici, G., Fattorini, L., Amicosante, G., Oratore, A., Joris, B., Frere, J. M., Pugsley, A. P. and Gicquel, B. (1994b). Transcription and expression analysis, using *lacZ* and *phoA* gene fusions, of *Mycobacterium fortuitum* β-lactamase gene cloned from a natural isolate and high-level β-lactamase producer. *Mol. Microbiol.* **12**, 491–504.

Timme, T.L. and Brennan, P.J. (1984). Induction of bacteriophage from members of the *Mycobacterium avium*, *Mycobacterium intracellulare*, *Mycobacterium scrofulaceum* serocomplex. *J. Gen. Microbiol.* **130**, 2059–2066.

Tsukamura, M., Hasimoto, M. and Noda, Y. (1960). Transformation of isoniazid and streptomycin resistance in *Mycobacterium avium* by the deoxyribonucleate derived from isoniazid- and streptomycin-double-resistant cultures. *Am. Rev. Respir. Dis.* **81**, 403–406.

Wards, B. J. and Collins, D. M. (1996). Electroporation at elevated temperatures substantially improves transformation efficiency of slow-growing mycobacteria. *FEMS Microbiol. Lett.* **145**, 101–105.

Wayne, L. G. (1984). Mycobacterial speciation. In *The Mycobacteria: A Sourcebook* (G. P. Kubica and L. G. Wayne, eds), pp. 25–65. Marcel Dekker, New York.

Wayne, J. G. (1994). Cultivation of mycobacterium tuberculosis for research purposes. In *Tuberculosis: Pathogenesis, Protection and Control* (B. R. Bloom, ed.), pp. 73–83. American Society for Microbiology, Washington, DC.

Wiggs, J.L., Bush, J.W. and Chamberlin, M.J. (1979). Utilization of promoter and terminator sites on bacteriophage T7 DNA by RNA polymerases from a variety of bacterial orders. *Cell* **16**, 97–109.

Wu, Q. L., Kong, D., Lam, K. and Husson, R. N. (1997). A mycobacterial extracytoplasmic function sigma factor involved in survival following stress. *J. Bacteriol.* **179**, 2922–2929.

10 Genetic Analysis in the Domain Archaea

William W. Metcalf
Department of Microbiology, University of Illinois, Urbana, Illinois, USA

◆◆

CONTENTS

◆◆◆◆◆◆ I. INTRODUCTION AND BACKGROUND

A. Discovery of Archaea

The recognition of the Archaea as a phylogenetic domain at the same hierarchical level as the Bacteria or Eukarya is a relatively recent development in the biological sciences (Woese *et al.*, 1990). This revision of the classical prokaryotic vs. eukaryotic phylogeny stems largely from the work of Carl Woese and his collaborators, and originated in the late 1970s from the study of what were then known as methane-producing bacteria. Utilizing 16S RNA as a molecular marker for determining phylogenetic relationships between organisms, Woese and his collaborators showed that these organisms were vastly different from the known bacterial species and proposed they should be considered as a separate group, designated the Archaebacteria (Balch *et al.*, 1977; Fox *et al.*, 1977; Woese *et al.*, 1978). Although this assertion was hotly contested for many years, its validity has been gradually substantiated by a variety of methods and is now widely accepted. The term Archaebacteria has been replaced by Archaea to emphasize the important point that these are not Bacteria (Woese *et al.*, 1990).

As might be assumed for a phylogenetic group of the highest level, the variety of organisms included in the Archaea is vast. During the twenty years since the Archaea were proposed as a separate lineage, a wide variety of fascinating organisms have been shown to be members of this domain. These include the previously mentioned methanogenic Archaea, the halophilic Archaea, and numerous thermophilic Archaea. In addition to these groups which have been amenable to laboratory growth in pure culture, as yet uncultivated members of the Archaea are present in most environments (Hershberger *et al.*, 1996; Jurgens *et al.*, 1997; MacGregor *et al.*, 1997; Olsen, 1994). In some environments, such as the Arctic, and Antarctic oceans, Archaea have been shown to be the numerically dominant organisms present (DeLong *et al.*, 1994).

Phylogenetically, the archaeal domain is split between two (or three) kingdom-level divisions based primarily on 16S RNA analysis (Woese, 1987). One group, designated the Euryarchaeota, includes the halophilic and methanogenic Archaea, and several others such as the sulphate-reducing thermophile *Archaeoglobus*, the cell-wall-less *Thermoplasma*, and the extreme thermophile *Pyrococcus*. The other group, designated the Crenarchaeota, is dominated by sulphur-metabolizing thermophilic organisms such as *Sulfolobus* and *Pyrodictium*, but also contains many as-yet uncultivated organisms including the numerically dominant Archaea detected in the polar oceans. A third kingdom level division, the Korarchaeota, has recently been proposed (Barns *et al.*, 1996).

B. Unique Nature of Archaea

In many ways the Archaea are similar to the Bacteria. All known Archaea are prokaryotic in morphology, which is to say they do not possess nuclei, and they are similar in size to the Bacteria. Further, the sizes of the known genomes in the Archaea are within the bacterial range, and circular in form, as are most bacterial genomes (Keeling *et al.*, 1994). Archaeal genes are commonly arranged in multi-gene transcriptional units, or operons, as are seen in the Bacteria (Brown *et al.*, 1989). Recent data, including analyses of the genome sequences of several Archaea, suggests that cell division in Archaea follows a bacterial model (Baumann *et al.*, 1996; Wang and Lutkenhaus, 1996). Finally, many aspects of archaeal metabolism are quite similar to their bacterial counterparts. It was for these reasons, especially the similarity in size and morphology, that the Archaea were so long considered Bacteria, and why the debate over the creation of the Domain Archaea raged so hotly.

Despite their similarities to the Bacteria, the Archaea have been shown to be specifically related to the Eukarya, and share many more common features with this group than with the Bacteria (Keeling and Doolittle, 1995; Olsen and Woese, 1996). A wide variety of cellular functions appear to be specifically related among the two domains. This similarity is particularly striking for the information-processing systems of the archaeal cell including transcription, translation, DNA replication and repair, amino-acylation of tRNAs, and tRNA splicing. While it is not the purpose of this review to discuss the relationship between the Archaea and Eukarya, many of these common features have important implications with regard to the methods used for genetic analysis. In particular the archaeal transcription and recombination machinery deserves discussion.

Archaeal RNA polymerases are multi-subunit enzymes similar to eukaryal RNA polymerases (Baumann *et al.*, 1995; Langer *et al.*, 1995;

Thomm, 1996). Thus, in addition to the main subunits that share homology with both the bacterial and eukaryal enzymes, archaeal RNA polymerases have extra subunits with homology to eukaryal subunits, and require initiation factors homologous to eukaryal TATA binding protein and TFIIB protein. The standard archaeal promoter consists of a distal promoter element similar to the eukaryal TATA box element and a downstream element where transcription initiates. This basic archaeal promoter appears to be similar in widely divergent members of this domain, and RNA polymerase from one species can correctly initiate transcription from promoters originating in other species *in vitro* (Hudepohl *et al.*, 1991; Thomm, 1996). Further, promoters from one archaeal species have been shown to be functional in other distantly related Archaea *in vivo* (de Macario *et al.*, 1996; Metcalf *et al.*, 1997). This feature has been important in the design of cassettes for the expression of foreign genes in members of this group, as will be discussed below. Unlike the eukaryal transcription system, archaeal mRNAs do not possess a 5' cap structure, nor do they possess long poly-A tails (short poly-A tails similar to those seen in bacteria have been demonstrated) (Brown and Reeve, 1985, 1986; Hennigan and Reeve, 1994). Also, translation of archaeal mRNAs seems to involve a ribosome binding site similar to the Shine–Dalgarno sequence observed in Bacteria. Most archaeal mRNAs have sequences complementary to the 3' end of their respective 16S RNAs. This, however, is not an absolute requirement, because some archaeal mRNAs start immediately adjacent to the first base of the protein coding sequence, and thus have no leader to base pair with 16S RNA (Betlach *et al.*, 1984; Pfeifer *et al.*, 1993).

DNA recombination is central to most genetic experiments and, as such, an understanding of the recombination machinery used by a particular organism can greatly enhance the probability for success in experiments involving that organism. Unfortunately, relatively little is known about DNA recombination in members of the Archaea. Based on DNA sequence analysis the archaeal recombination system appears to be similar to the eukaryal system. Accordingly, a homologue of the main eukaryal recombinase encoded by the *rad51* gene has been found in all Archaea examined to date (Sandler *et al.*, 1996). The archaeal gene has been designated *radA*. Homologues of the genes involved in the bacterial *recBCD* and *recF* pathways have not been identified despite examination of the complete genome sequences from four different Archaea. These data suggest that archaeal recombination will be more similar to that seen in Eukarya; however, to date, little biochemical or genetic evidence exists to support this supposition.

Despite their molecular similarity to Eukarya, and their superficial and metabolic similarity to Bacteria, it is inappropriate to categorize Archaea by features they do, or do not, share with the other two domains. Archaea possess a variety of unique features not found in the other domains. For example, all Archaea have ether-linked isoprenoid lipids, rather than the fatty acid ester lipids seen in both Bacteria and Eukarya (Gambacorta *et al.*, 1994; Langworthy, 1985). The metabolic diversity among Archaea is particularly striking. Thus, some members of the halophilic Archaea have evolved a novel photosynthetic metabolism utilizing the light-driven proton pump bacteriorhodopsin (Khorana, 1988). Certain thermophilic Archaea, like *Acidanius* and *Desulfurolobus*, are able to grow autotrophically by either sulphur oxidation under aerobic conditions, or by sulphur reduction under anaerobic conditions (Fischer *et al.*, 1983; Segerer *et al.*, 1985; Zillig *et al.*, 1985). The biochemistry of methanogenesis also represents a novel aspect of archaeal metabolism. Reduction of CO_2 to methane by the methanogenic Archaea involves no less than six novel enzyme co-factors, most of which are not found outside this domain (DiMarco *et al.*, 1990).

Another unique aspect of the Archaea is the adaptation to environmental extremes displayed by many members of this group. The extreme thermophiles exemplify this trait. Each of the dozen or so known organisms capable of growth above 95°C is an archaeon, as are the vast majority of those capable of growth above 70°C (Stetter, 1996b). Others are capable of growth under extremely acidic conditions. Members of the genus *Picrophilus* have a growth optimum at pH 0.7 at 60°C (Schleper *et al.*, 1995b). The *Halobacteriaceae* grow under conditions of high or even saturating salt concentrations, and will not grow at all in media with less than *ca*. 1M NaCl. Some species, such as *Natronococcus*, also require very high pH for optimal growth and, thus, are alkaliphilic in addition to halophilic (Tindall, 1991). The methanogenic Archaea are adapted to growth utilizing extremely poor energy sources. Members of this group use a very limited number of substrates, including H_2 plus CO_2, a variety of one-carbon compounds, and acetate as sole carbon and energy sources. Under physiological conditions the ΔG of many of these reactions is barely sufficient to support the production of ATP for growth (Muller *et al.*, 1993). Many methanogenic Archaea are capable of autotrophic growth under these extremely limiting conditions, making all their cellular material from CO_2. Amazingly, *Methanococcus jannaschii* has a doubling time of less than 20 min under these conditions (Jones *et al.*, 1983a).

C. Why study Archaea?

As can be seen from these brief descriptions, the Archaea include a variety of important and interesting organisms whose study is essential to numerous problems in modern biology. The unique metabolism found in various Archaea is one of the best reasons to study this group. Numerous unique enzymes have been found in the Archaea, particularly the methanogenic Archaea. Study of these has yielded a wealth of information on enzyme mechanisms, such as the role of metals in enzyme mediated catalysis (Thaur *et al.*, 1993). Study of bateriorhodopsin has been a mainstay of biophysicists, and our understanding of integral membrane proteins and the generation of ion gradients has been greatly enhanced by research on this protein (Knebs and Khorana, 1993). Methane production is relevant to the problems of global warming and production of fuel from biomaterials (all significant biological methane producers are Archaea) (Zinder, 1993). The extreme oxygen sensitivity of the methanogenic Archaea also suggests they are good model systems for the study of oxidative damage, a known factor in many diseases.

There is a great deal of interest today in the study of extreme environments. It is commonly believed that early life on the planet faced extreme conditions such as high temperature and anaerobiosis (Woese, 1977). Thus, study of the adaptations of Archaea to these environments may offer insight into the earliest life forms on Earth. Further, such extreme conditions are the norm on other planets and it has been suggested that knowledge of how life exists under these circumstances will be useful in the search for extra-terrestrial life (Stetter, 1996a). Adaptation to extreme environments necessarily requires proteins and enzymes that are stable under these conditions. There is a great deal of academic and commercial interest in study of heat stable enzymes (Adams, 1993). Enzymes that are stable under conditions of high or low pH, as well as enzymes stable under high ionic strengths are also actively being sought (Eisenberg, 1995). The Archaea represent one of the best sources for such enzymes. Last, but certainly not least, the specific relationship of the Archaea to the Eukarya suggests that much can be learned about eukaryotes from the study of the simpler Archaea.

D. Difficulties Inherent to the Genetic Analysis of Archaea

Genetic analysis among the Archaea has lagged far behind that in other more tractable organisms, such as *Escherichia coli* or *Saccharomyces cerevisiae*. There are a variety of reasons for this, beginning with the fact that most Archaea were not isolated in pure culture until quite recently. A great deal of basic information is required for design of genetic experiments, and in many cases this information is lacking. These include simple things like the media components required for growth in defined conditions, pH and temperature optima for growth, methods for storing strains, for obtaining pure cultures, and which treatments are mutagenic for a given organism. The ability to grow an organism as a clonal population derived from a single cell, usually by single colony isolation on solid medium, is an absolute requirement for genetic experimentation. This is often taken for granted by researchers using more familiar laboratory organisms, but it is by no means trivial to accomplish this for many Archaea. For example, agar is an inadequate solidifying agent at the temperatures and pH used for cultivation of many thermophiles (Rakhely and Kovacs, 1996). Absolute exclusion of oxygen and reduction of the medium below −300 mV is required for plating of the methanogenic Archaea (Balch *et al.*, 1979). Research into such basic areas of microbiology is not glamorous, is not easy to fund, and is often overlooked. For those of us working in this field the pioneering contributions of a number of researchers should not be forgotten. While these basic investigations are not described, the experiments presented here would not have been possible without them.

A second difficulty encountered in genetic analysis of Archaea stems from their unique nature. A critical requirement for genetic experimentation is the presence of selectable, or at least testable phenotypes, for use as markers to monitor exchange of DNA. Among the Bacteria this is most commonly accomplished by antibiotic selection. There are, however, few antibiotics that are effective in the Archaea (Bock and Kandler, 1985).

Antibiotics that target protein synthesis, such as streptomycin or tetracycline, are usually ineffective because of differences between archaeal and bacterial ribosomes and elongation factors. Antibiotics that target cell wall biosynthesis are ineffective because the Archaea do not possess peptidoglycan cell walls. Further, for those antibiotics that are effective there must be a resistance gene available that will function in the archaeon in question. Because of the differences in transcription between the domains, expression of such antibiotic resistance genes usually requires cloning the resistance gene behind an archaeal promoter (Gernhardt *et al.*, 1990). Additionally, the protein encoded by this gene must be stable and active under the growth condition employed, i.e. thermostable for use in thermophiles or halotolerant for use in halophiles. Such markers simply do not exist for most species. An alternative method is to use auxotrophic markers. This however, requires knowledge of the biochemical pathways used by the organism, a mutant defective in one of these pathways, and a cloned gene that complements the defect. Once again these are rarely available.

E. Progress in the Genetic Analysis of Archaea

Because of the lack of genetic methods available for use in Archaea the study of these organisms has largely been confined to physiological, biochemical or molecular biological approaches. Thus, although many individual enzymes have been studied, and the physiology of growth under certain extreme conditions is beginning to be understood, a great deal remains to be elucidated about the metabolism, physiology, biochemistry, and gene regulation of individual Archaea.

In the absence of traditional genetic methods for use in Archaea, a great deal of effort has been focused on the application of molecular biological techniques to the study of this group. An ever-growing body of knowledge relating to the genes putatively required for various functions has accumulated, based on studies of genes cloned into heterologous hosts such as *Escherichia coli*. This approach has culminated in the complete sequencing of four archaeal genomes, those of *Methanococcus jannaschii* (Bult *et al.*, 1996), Methanobacterium thermoautotrophicum (Smith *et al.*, 1997), *Archaeoglobus fulgidus* and *Pyrobaculum aerophilum*. At the time of writing several more archaeal genome sequencing projects are in progress. Examination of these data highlights the pressing need to develop genetic methods for analysis of Archaea. In each organism examined to date 50–60% of the putative genes present encode proteins that are not similar to any other proteins of known function. Also, many of the genes identified within these genomes were identifiable, but were not expected. For example, although *M. jannaschii* is known to fix carbon by the Ljungdahl–Wood pathway (and the genes required for this pathway identified), the key gene for fixation of carbon by the Calvin cycle, that encoding ribulose bisphosphate carboxylase (RUBISCO), is also present. Many other genes that must be present, such as those involved in the biosynthesis of certain amino acids, were not identifiable based on homology to known sequences (Bult *et al.*, 1996). Without methods for determining the functions of unknown genes, and for determining which genes are involved in known pathways these observations will remain unexplained. Further, although about 50% of the putative genes identified in each of these organisms encode proteins that are homologous to known proteins, this is not evidence that they actually perform the predicted function. Numerous examples are known of homologous proteins performing divergent functions. Experimental evidence is absolutely required to confirm the predictions made, based upon sequence homology. Genetic analysis is arguably the most direct method for obtaining this experimental evidence.

Over the last twenty years a gradual increase in our general understanding of Archaea has allowed development of genetic methods for use in various members of this domain. A variety of traditional and more modern molecular genetic methods have been adapted for use in selected members of the Archaea. The purpose of this review is to describe these methodologies. The material presented will be restricted to a discussion of *in vivo* genetic methods for use in Archaea, as opposed to molecular biological approaches utilizing heterologous hosts, as the latter have been reviewed extensively elsewhere (Brown *et al.*, 1989; Reeve, 1992). Particular attention will be paid to: (i) the ability to grow the organism in defined media and to isolate clonal populations; (ii) the ability to generate mutations in either a random or directed fashion; (iii) systems for gene transfer, i.e. transformation, transduction and conjugation; (iv) selectable markers that can be used in gene transfer experiments; and (v) vectors, especially plasmids, for gene cloning and expression in the archaeal host. Illustrative examples for many techniques will also be discussed.

Because of the large number of interesting species represented in this phylogenetic group it will be impossible to address the details of methods developed for each member of the group. Instead, methods developed for the genetic analysis of specific examples from each of three main groups – the methanogenic Archaea, the halophilic Archaea, and thermophilic sulphur metabolizing Archaea – will be presented. Specific examples from other organisms will also be discussed as relates to these case studies.

◆◆◆◆◆◆ II. RECOMBINATION EVENTS IN GENETIC EXPERIMENTATION

The hallmark of genetic experimentation is mutant analysis. By examining the phenotypic consequences of a mutation in a particular gene the normal function of that gene can be inferred. Thus, methods for production of mutations are essential to genetic analysis. Methods that allow targeting of *in vitro* constructed mutations to a specific gene are particularly desirable. This is especially true given the rapid increase in the availability of DNA sequence data. The ability to test hypotheses based on these sequences is of paramount importance to making sense of the mountains of data emerging from genome-sequencing projects (not to mention innumerable smaller-scale sequencing endeavours). Further, traditional genetic experimentation involving crosses between strains is as useful today as it ever was. In this way mutations can be combined in the same host to examine the effect on their respective phenotypes, to test linkage to other known markers, and to construct the desired strain backgrounds for further experimentation. Each of these techniques involves recombination between homologous sequences in the resident genome and incoming DNA. Because the types of recombination events employed in these experiments are similar regardless of the organism, whether it is archaeal, bacterial or eukaryal, a short description of selected recombination events that are commonly used is pertinent.

A. Recombination Involving Linear DNA

As shown in Figure 1, DNA can be transferred to a recipient molecule (such as the host chromosome) from linear donor molecules by homologous recombination on both sides of the exchanged marker. Because these recombination events are relatively infrequent, isolation of the desired recombinants requires some type of selection or screening procedure. This can be accomplished in a number of ways. If the donor DNA is capable of repairing a defect in the recipient, such as an auxotrophic mutation or inability to utilize a growth substrate or required nutrient, then recombinants can be selected on that basis. This is sometimes referred to as marker rescue. Alternatively, the donor DNA may confer resistance to some toxic compound. Often, antibiotic resistance genes are introduced into donor DNA to allow selection for the desired recombinant. When this modification is done *in vitro* using cloned genes it allows targeting mutations to specific genes. This method is commonly referred to as reverse genetics or gene replacement. Recombinants that acquire mutations preventing the uptake of toxic compounds or preventing the metabolism of benign compounds into toxic products can also be directly selected.

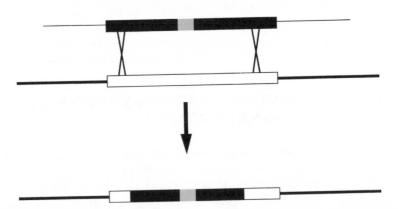

Figure 1. Recombination between linear donor and recipient molecules. Two homologous recombination events (crossed black lines), one each on either side of difference (grey shaded box) between linear donor (black rectangle) and recipient (white rectangle) DNA molecules results in the replacement of the recipient allele with the donor allele. The differing regions may represent a mutation on either the donor or recipient molecule. If the mutation lies on the donor it must confer a selectable phenotype (or at least a readily screened phenotype) to allow efficient recovery of the desired recombinant. If the mutation lies on the chromosome, then the donor allele must be selectable by correction of the mutant recipient phenotype (see text for details).

B. Recombination Involving Circular DNA

Several modifications of marker rescue and gene replacement experiments are possible using circular DNA molecules (Figures 2–4). Because commonly used cloning vectors are circular plasmids these techniques are especially useful. The key feature required for this type of genetic exchange is a plasmid that does not replicate in the target host. For most organisms standard *E. coli* plasmids suffice for this purpose, because most of these do not replicate outside of *E. coli*. For organisms more closely related to *E. coli* a variety of conditional replicons exist. To perform this type of experiment the target gene is cloned into an *E. coli* plasmid. After making the appropriate modifications the plasmid is then transformed into the target host. Because the plasmid cannot replicate in this heterologous host, the DNA and its associated selectable phenotype cannot be maintained in the absence of recombination into the host genome.

In its simplest form recombination with circular DNA is no different from recombination with linear DNA. Gene replacements or marker rescue events can be obtained with circular donor molecules by homologous recombination on both sides of a suitable selectable marker (Figure 2). Recombinants of this type must be selectable by nature of the homologous insert DNA as described above for recombination with linear fragments.

Because single recombination events are more frequent than double recombination events, a more likely result of transformation with a circular molecule is integration of the donor circle into the recipient molecule by a single homologous recombination event (Figure 3). Recombinants arising from this type of event carry a duplication of the homologous

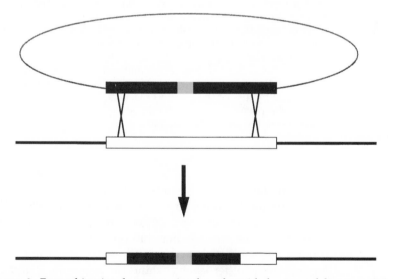

Figure 2. Recombination between circular plasmid donor and linear recipient molecules. The outcome of two homologous recombination events (crossed black lines), one each on either side of difference (grey shaded box) between circular donor (black rectangle) and linear recipient (white rectangle) DNA molecules is indistinguishable from that obtained when both donor and recipient are linear (see Figure 1 and text).

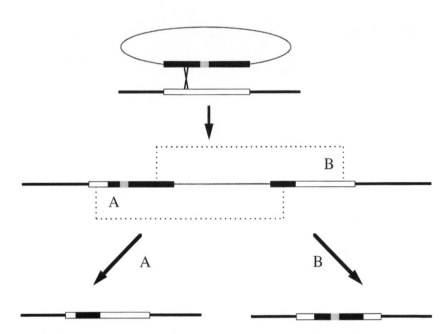

Figure 3. Recombination between circular plasmid donor and linear recipient molecules. The outcome of a single homologous recombination event (crossed black lines) between a circular plasmid donor (black rectangle) and a linear recipient (white rectangle) DNA molecules is integration of the donor molecule into the recipient. Partial diploid recombinants arising from this type of event carry a duplication of the donor and recipient alleles that differ in the grey shaded region. Partial diploid recombinants can be selected by a selectable phenotype conferred by the donor allele, or by repair of a recipient mutation, or by nature of a selectable marker carried on the plasmid backbone (grey line). The partial diploidy can be resolved by a second recombination event between the duplicated regions. Two types of resolution are possible (dotted lines labelled A or B). The A-type resolution simply reverses the original recombination event and restores the recipient phenotype. The B-type resolution is the result of a recombination event on the opposite side of the differing region and results in the replacement of the recipient allele with the donor allele. Recombinants that have undergone resolution can be recognized by loss of the plasmid markers. If a counter-selectable marker is present on the plasmid then recombinants that have undergone resolution can be selected directly (see text for details).

region. In these partial diploids, one of the duplicated regions is similar to the donor and the other to the recipient. A second homologous recombination event between the duplicated regions will resolve the partial diploidy. There are two possible resolutions. One simply reverses the original recombination event and restores the recipient phenotype (A in Figure 3). The other leaves behind the donor allele and occurs when the second recombination event is on the opposite side of the marker as the original recombination event that led to integration of the circle (B in Figure 3). There are several useful variations of this method depending on how the original integration event is selected.

If the selectable marker used in the experiment resides within the homologous region, then there is no selection for maintenance of the partial diploid state. These may resolve very quickly and be indistinguishable from double recombination events as depicted in Figures 1 and 2. Alternatively, the partial diploid may be quite stable and persist for some time. This is especially true if the mutation on the donor molecule causes a deleterious phenotype. In this case obtaining the desired recombinant may require the plasmid vector to carry a counter-selectable marker, i.e. a marker that allows selection for its loss. If this is not available then recombination using a linear donor would be preferable.

A second alternative is to place the selectable marker, and if possible also a counter-selectable marker, in the plasmid vector rather than in the homologous region. In this way a variety of mutations, including ones without a selectable phenotype, can be recombined onto the host chromosome. In this method, partial diploids are first selected using the plasmid encoded resistance marker after transformation with a circular donor molecule. Selective pressure is then removed allowing resolution of the partial diploid. Recombinants that have resolved the partial diploid duplication can be recognized by their loss of the original selectable marker, or can be selected if a counter-selectable marker was present on the plasmid vector. These recombinants can then be screened for the desired mutation by their phenotype or by using molecular biological methods. In some cases it may be desirable to maintain the partial diploid duplication indefinitely. Placing the selectable marker in the plasmid vector and maintaining selection can do this. This technique has been used to introduce reporter genes into the host genome while maintaining a wild-type copy of the gene under study.

Finally, integration of circular molecules by homologous recombination can be used as a method of gene disruption mutagenesis. This can be done by cloning internal gene fragments (i.e. a fragment lacking both amino- and carboxy-termini of the coding sequence) into non-replicating plasmids that carry a selectable marker for the target host. After transformation these plasmids can integrate into the target gene by homologous recombination. The plasmid marker is used to select recombinants. As shown in Figure 4 integration of the plasmid causes a gene disruption. Mutations caused by this type of integration event are unstable in the absence of selection for the plasmid marker because they carry duplicated regions of the target gene on both sides of the integrated plasmid. These duplicated regions can undergo homologous recombination to resolve the duplication just as in the preceding example. Therefore maintenance of selection is advisable when using this method. In addition to being useful for generation of mutations in specific genes, this method can be used for random insertional mutagenesis. If small random fragments are cloned, instead of small fragments from specific genes, then the resulting plasmids will integrate via recombination at random loci. If the cloned fragments are small enough so that the majority are internal to the gene from which they originate, then the majority of these random integration events will cause gene disruption mutations.

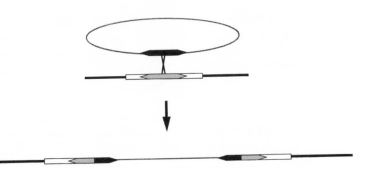

Figure 4. Recombination between circular plasmid donor and linear recipient molecules. The outcome of a single homologous recombination event (crossed black lines) between a circular plasmid donor that carries an internal gene fragment (black hexagon) and a linear recipient gene (white rectangle with shading to indicate homologous region) is integration of the donor molecule into the recipient. In this case the recipient gene is disrupted by the plasmid insertion. A marker on the plasmid backbone is required to select recombinants with the plasmid insertion. These insertions are unstable in the absence of selection because a second recombination event between the duplicated sequences flanking the plasmid insertion (grey and black hexagons) can reverse the integration event restoring the recipient phenotype (see text for details).

◆◆◆◆◆◆ III. GENETIC ANALYSIS OF METHANOGENIC ARCHAEA

The methanogenic Archaea are a diverse group of anaerobic organisms that produce methane as an end product of anaerobic respiration using a variety of one-carbon compounds as terminal electron acceptors. Although the existence of methane-producing organisms has been known since the late nineteenth century, pure cultures were not obtained until 1947. This is largely due to the extreme anaerobic conditions required for their growth. Thus, in addition to a complete absence of oxygen the redox potential of the medium must be –300 mV or below to allow growth of methanogenic Archaea (Balch *et al.*, 1979). This requirement has imposed severe limitations on the ability to grow and work with members of this group. It was not until development of the Hungate technique for anaerobic culture (Hungate, 1969) and its modification by Miller and Wolin (1974) and by Balch and Wolfe (Balch *et al.*, 1979) that these organisms became tractable for routine studies.

The result of these difficulties is that the basic requirements for genetic analysis of an organism, i.e. the ability to grow the organism and to isolate clonal population by plating on solid medium, were not met for any methanogenic archaeon until relatively recently (Jones *et al.*, 1983b; Kiener and Leisinger, 1983). For many methanogenic Archaea this is still not possible. Further, even for those organisms for which these basic requirements have been met, the methodologies are cumbersome and slow. To date it has not been possible to perform plating under aerobic conditions, followed by anaerobic incubation, as is possible for many anaerobes. Therefore, all manipulations must be carried out in an anaerobic glove box, or in sealed tubes, with rigorous exclusion of oxygen. A second complication for the routine culture of methanogenic Archaea is that many species exhibit optimum growth utilizing H_2 and CO_2 as substrates for carbon and energy. Due to the limited solubility of H_2 rapid growth is not achieved without pressurization of the H_2 and CO_2 atmosphere. This means that plates must be incubated in sealed, pressurized containers. However, each of these difficulties can be surmounted. A full description of the methodology for these anaerobic techniques is lengthy and has been presented elsewhere (Balch et al., 1979; Sowers and Schreier, 1995).

Most of the methanogenic Archaea can be placed into one of three large phylogenetic groups, the *Methanococcales*, the *Methanomicrobiales*, and the *Methanobacteriales* (Boone et al., 1993). The phylogenetic distance between these groups is roughly equivalent to that between *E. coli* and *Bacillus subtilus*. As might be expected from this, the metabolism and growth requirements of each group are substantially different, as are the methodologies developed for genetic analysis in each. The *Methanobacteriales* and *Methanococcales* both exhibit very limited substrate utilization. With few exceptions members of these groups are able to grow utilizing only H_2 and CO_2, or formate, as carbon and energy sources (Boone et al., 1993). The *Methanomicrobiales* have slightly broader metabolic capabilities and individual species within this group can utilize some or all of the following substrates as carbon and energy sources: H_2 and CO_2, methanol, methyl amines, methyl sulphides, acetate, and pyruvate (Boone et al., 1993; Bock et al., 1994; Rajagopal and LeGall, 1994).

Genetic analysis in the methanogenic Archaea is primitive at best, not withstanding the great effort that has gone into developing such methods. However, in the last few years these efforts have begun to produce truly workable genetic techniques for some members of this group. Most of this effort has gone into development of techniques for three genera; from the *Methanococcales*, the genus *Methanococcus*, from the *Methanomicrobiales*, the genus *Methanosarcina*, and from the *Methanobacteriales*, the genus *Methanobacterium*.

A. Genetic Analysis of Methanococcus

Genetic analysis in the genus *Methanococcus* has progressed further than in the other two methanogenic groups. The techniques available for use in this group include: reliable plating methods (Jones et al., 1983b); the ability to isolate auxotrophic mutants and to enrich for such mutations in a mutagenized population (Ladapo and Whitman, 1990); transformation with both chromosomal and plasmid DNA (Bertani and Baresi, 1987; Gernhardt et al., 1990), an *E. coli–Methanococcus* plasmid shuttle vector (Tumbula et al., 1997); use of reporter genes to monitor gene expression *in vivo* (Beneke et al., 1995; Cohen-Kupiec et al., 1997); and insertional inactivation of genes in both random and directed manner (Gernhardt et al., 1990; Blank et al., 1995).

Methanococci are relatively small, often motile, irregularly shaped cocci that inhabit a variety of environments from coastal sediments to extremely hot submarine volcanic vents (Boone *et al.*, 1993). Species of *Methanococcus* are known to use only H_2 and CO_2, or formate, as carbon and energy sources. Although the complete genome sequence of *Methanococcus jannaschii* has been determined (Bult *et al.*, 1996), the difficulties inherent to manipulation of this extreme thermophile make it a poor choice for genetic experimentation. Instead, two mesophilic species, *Methanococcus voltae* and *Methanococcus maripaludis*, have been singled out as the most useful for genetic studies. Both strains are among the fastest growing methanogenic Archaea with doubling times of 90 and 140 min respectively. Importantly, methods for growth of these species as clonal populations derived from colonies on solid medium have been described (Jones *et al.*, 1983b). Single cells of both organisms can be plated on solid media with efficiencies of *ca.* 90%.

A variety of methods have proven effective for generation of mutants in both *M. voltae* and *M. maripaludis*. These include mutagenesis by chemical treatment, by UV and gamma irradiation, as well as direct selection for spontaneous resistance to a variety of toxic compounds (Santoro and Konisky, 1987; Ladapo and Whitman 1990; Micheletti *et al.*, 1991; Yang *et al.*, 1995). A particularly useful method allows enrichment of non-growing mutants in a mutagenized population by inclusion of the toxic base analogues 6-azauracil and 8-azahypoxanthine in the medium (Ladapo and Whitman, 1990; Bowen *et al.*, 1995). This method is analogous to penicillin suicide enrichment in *E. coli* or inositol starvation suicide enrichment in *Saccharomyces* (Henry *et al.*, 1975). A number of useful mutants, including mutants auxotrophic for acetate, histidine, purines, vitamin $B_{12,}$ and coenzyme M (one of the unique co-factors involved in methanogenesis) have been generated by these methods (Bertani and Baresi, 1987; Ladapo and Whitman, 1990; Micheletti *et al.*, 1991).

Using these auxotrophically marked strains, transformation of *M. voltae* with genomic DNA has been demonstrated (Bertani and Baresi, 1987; Micheletti *et al.*, 1991). Two protocols have been used in these experiments, a natural transformation method and electroporation-mediated transformation. Neither method is very efficient with natural transformation yielding *ca.* 10 to 100 transformants per microgram of genomic DNA, and electroporation *ca.* 300 to 700 transformants per microgram of genomic DNA. While these efficiencies are not especially high they represent the only method demonstrated to date for transfer of chromosomal markers between strains of *Methanococcus*.

A major breakthrough in the genetic analysis of methanococci came with the finding that puromycin was an effective antibiotic, and that the puromycin acetyl-transferase gene, *pac*, from *Streptomyces alboniger* could confer resistance to this antibiotic (Possot *et al.*, 1988; Gernhardt *et al.*, 1990). This was accomplished by construction of an integration vector, pMip1, that carries the *pac* gene under the transcriptional control of a strong *M. voltae* promoter, that of the methyl-reductase operon (*pmcr*). The *pac* cassette in pMip1 is inserted into the *M. voltae hisA* gene cloned in the *E. coli* plasmid pUC18. Although pMip1 is unable to replicate in *M. voltae*, it can integrate into the chromosome by homologous recombination, as shown in Figures 2 and 3. Some of these integrants retain the plasmid backbone and arise by a single recombination event, resulting in a partial diploid for the *hisA* region. Others, however, have only the *pac* cassette inserted into the genomic *hisA* gene. The latter type may arise from double recombination events, one on either side of the *pac* marker, or by sequential integration and subsequent resolution of the partial diploid. These result in replacement of the wild-type *hisA* allele with the *pac*-disrupted allele. Importantly, these strains are histidine auxotrophs, demonstrating a role for the *hisA* gene in histidine biosynthesis in this organism. The construction of *hisA* mutants using pMip1 was the first demonstration of directed mutagenesis in any methanogenic archaeon (Gernhardt *et al.*, 1990).

The ability to select for the presence of the *pac* cassette has been vital for the development of further genetic methods, not only among the methanococci, but also in *Methanosarcina* (see below). Four types of genetic experiments have been made possible by this development. These are: (i) optimization of transformation protocols; (ii) directed mutagenesis of specific genes and testing of mutant phenotype following recombination onto the *Methanococcus* chromosome; (iii) construction of a plasmid shuttle vector for use in gene cloning in *M. maripaludis*; and (iv) construction of integration vectors with reporter genes under the control of *Methanococcus* promoters.

High frequency transformation greatly improves the chances for success of genetic experiments involving low frequency events such as homologous recombination. The initial experiments using the *pac* cassette involved natural transformation. The process was relatively inefficient. With a selectable marker to score for transformants it became possible to quantify easily the transformation efficiency and, thus, to optimize systematically the transformation protocol. The greatest improvements in the ability to transform methanococci were achieved through the use of protoplasts (Patel *et al.*, 1994; Tumbula *et al.*, 1994).

High efficiency transformation of methanococci relies on the ability to form protoplasts, and to regenerate these protoplasts into growing cells. This method has proven reliable for other Archaea as well (see below), and is a consequence of the unique cell wall structure of many members of this domain. *Methanococcus*, and many other Archaea, have a cell wall composed of a single repeating protein, or S-layer (Kandler and Konig, 1985). The S-layer of *Methanococcus* can be removed by careful resuspension in iso-osmotic media without divalent cations. Importantly, the protoplasts formed by this treatment can be regenerated (Patel *et al.*, 1993). In both *M. voltae* and *M. maripaludis* removal of the S-layer has been shown to enhance the ability of the cell to take up DNA (Patel *et al.*, 1994; Tumbula *et al.*, 1994). In *M. maripaludis* the efficiency of transformation is *ca.* 2×10^5 per microgram of DNA using an optimized PEG-mediated transformation method and an integration vector, pKAS102 (Sandbeck and Leigh, 1991), which is similar to pMip1 used in *M. voltae*. Transformation of *M. voltae* protoplasts by electroporation has increased the frequency of transformation in this host to a level of *ca.* 3×10^3 per microgram as scored by integration of pMip1 DNA into the chromosome. Although these efficiencies are not especially high, it is important to note that the numbers given here are based upon integration of non-replicating plasmids into the host genome. Thus the numbers reported represent the product of the efficiency of DNA transfer and the efficiency of DNA recombination.

In the course of these transformation studies several additional observations were made that are generally relevant to genetic experimentation in methanococci, and perhaps in other Archaea as well. In determining the conditions for protoplast transformation in *M. voltae* it was found that linear DNA was *ca.* 20 times more efficient at generating transformants than was covalently closed circular DNA (cccDNA) (Patel *et al.*, 1994). This was true despite the fact that linear DNA requires two recombination events to generate a recombinant, whereas circular DNA requires only a single event (see Figures 1 and 3). This result suggests that linear DNA is more efficient at recombination than cccDNA in *Methanococcus*. Linear DNA also gives much higher levels of transformation in eukaryotes such as *Saccharomyces* by stimulation of homologous recombination (Orr-Weaver *et al.*, 1981). The apparent similarity of the recombination proteins in Archaea and Eukarya suggests that linear DNA may stimulate recombination in Archaea as it does in Eukarya, and might be a more appropriate substrate for use in recombination experiments.

A second observation of general relevance was the finding of a *Pst* I-like restriction system in *M. maripaludis* (Tumbula *et al.*, 1994). Protection of DNA by *in vitro* methylation with *Pst* I methylase increases the transformation frequency in this host by a factor of at least fourfold. A systematic search for restriction endonucleases among the Archaea has not been reported; however, numerous examples are known (Schmid *et al.*, 1984; Prangishvili *et al.*, 1985; Thomm *et al.*, 1988; Lunnen *et al.*, 1989; Nolling and Vos, 1992). Analysis of the *M. jannaschii* genome suggests the presence of at least five restriction systems (Bult *et al.*, 1996). Although the efficiency of transformation in *M. maripaludis* is sufficient to surmount the restriction barrier, appropriate protection of DNA prior to transformation may be beneficial in other methanogenic Archaea.

The *pac* cassette has been used in a few cases to construct mutations in *Methanococcus* genes by recombination of *in vitro* constructed gene disruptions on to the *Methanococcus* chromosome, as depicted in Figures 1–3 (Gernhardt *et al.*, 1990). Given the utility of the method for defining the *in vivo* roles of known genes this method seems under-exploited. However, two modifications of this method have also been employed in the study of gene function in *Methanococcus*.

First, a derivative of the *E. coli* bacteriophage Mu transposon that carries the *pac* cassette has been constructed (Blank *et al.*, 1995). This can be used to mutagenize cloned *Methanococcus* genes in *E. coli*. These can subsequently be recombined on to the *Methanococcus* chromosome in order to test their phenotypic consequences in the native host. This shuttle mutagenesis with an *E. coli* transposon, similar to a system used in yeast (Seifert *et al.*, 1986), harnesses a powerful technique of bacterial genetics and provides a means for random insertional mutagenesis of cloned methanogenic archaeal genes (see also Chapter 4, section VIII). This method has been used to identify genes involved in nitrogen fixation by *M. maripaludis* (Blank *et al.*, 1995). Although not yet adopted, this method is also applicable to random mutagenesis of the entire *Methanococcus* genome. If a gene library is mutagenized instead of a specific cloned gene, then the resulting transposon insertions will represent random insertions in the *Methanococcus* chromosome. The insertion mutations can then be moved *en masse* into the native host to screen for those with specific phenotypes. This general method should be adaptable to any archaeon for which a transformation system and selectable marker exists.

A second modification using the *pac* cassette as a selectable marker for insertional mutagenesis of *Methanococcus* genes is also based on a technique used in Bacteria. A method of growing popularity in Bacteria is the disruption of genes by insertion of non-replicating plasmids by homologous recombination with cloned internal gene fragments (see Figure 4). This technique can be used for directed mutagenesis of a particular gene or for random mutagenesis by cloning in random small fragments of the target genome. Targeted gene disruption mutations have been made in either of two flagellin genes in *M. voltae* using this method (Jarnell *et al.*, 1996). A slight modification of this method in which polar mutations are created by integration of plasmids has been used to construct hydrogenase mutants of *M. voltae* (Berghofer and Klein, 1995). Random mutagenesis with small fragment libraries has recently been demonstrated in *M. maripaludis* (Kim and Whitman, 1997). The vector used for construction of the small fragment library was pMEB.2. This plasmid carries the *pac* cassette in the *E. coli* plasmid pUC18 (which does not replicate in *Methanococcus*) and has a variety of usable cloning sites (Gernhardt *et al.*, 1990). A number of mutants showing poor growth in minimal medium were isolated from puromycin-resistant colonies of *M. maripaludis* selected after transformation with this small fragment library. Integration vectors for use in *M. voltae* are described in Klein and Horner (1995). While not yet tested in *Methanococcus* species, a set of integration vectors constructed for use in *Methanosarcina* may also be useful for this type of experiment in *Methanococcus* (see Figure 5 and below).

Integration vectors have also been used in conjunction with reporter genes to measure transcription from various *Methanococcus* promoters *in vivo*. In *M. voltae*, pMip1 has been modified to carry a promoterless *uidA* gene, which can be placed under the control of the *M. voltae* prompter of choice (Beneke *et al.*, 1995). The *uidA* gene is an *E. coli* gene encoding the easily assayed β-glucuronidase protein (GUS), and is commonly used as a reporter gene in a variety of bacterial and eukaryal systems (Jefferson *et al.*, 1986). When inserted behind a *Methanococcus* promoter and recombined on to the *M. voltae* chromosome (at the *hisA* locus) this reporter gene allows *in vivo* quantification of promoter strength. The utility of this construct has been demonstrated by showing selenium-dependent expression of the selenocysteine-containing hydrogenase of this organism Beneke *et al.*, 1995). A similar plasmid for use in *M. maripaludis* uses a promoterless *E. coli lacZYA* operon, encoding the easily assayed β-galactosidase protein (β-GAL), in a modified pKAS102 integration vector (Cohen-Kupiec *et al.*, 1997). It has been used to measure repression of the nitrogenase operon by ammonia in *M. maripaludis*. One caveat must be presented for use of these constructs. The chromogenic 5-bromo-4-chloro-3-indolyl dyes, X-gluc and X-gal, commonly used for detection of GUS and β-GAL activity, do not form coloured products under the reducing conditions used for culture of methanogenic Archaea. Therefore, these cannot be used for screening colonies without resorting to some form of replica plating (the replica can be exposed to oxygen to allow colour development without killing the cells on the original plate). This unfortunate redox problem extends to virtually all reporter gene methods in common use. Because most commonly used chromogenic substrates, including the green fluorescent protein (GFP), are colourless under the strongly reducing conditions used to cultivate methanogenic Archaea, the direct use of *in vivo* detection systems for most reporter genes is prevented.

Finally, a M. *maripaludis*–E. *coli* plasmid shuttle vector has recently been described (Tumbula *et al.*, 1997). This plasmid is based on the cryptic pURB500 plasmid of M. *maripaludis* C5 (Woods *et al.*, 1985). The shuttle vector, designated pDLT44, combines the entire pURB500 replicon and the *pac* cassette for replication and selection in M. *maripaludis*, with the pMB1 replicon and β-lactamase gene for replication and selection in E. *coli*. The plasmid has *Xba* I and *Sac* I as usable cloning sites. Its use appears to be restricted to M. *maripaludis*. It does not replicate in M. *voltae*. By use of pDLT44 it should be possible to clone genes by complementation of mutations, and to create over-expression vectors or promoter probe vectors. Other methods involving replicating plasmids should now be adaptable to M. *maripaludis* as well.

B. Genetic Analysis of *Methanosarcina*

The *Methanosarcineae* are metabolically the most diverse of the methanogenic Archaea. They are capable of growth on a variety of substrates including H_2 and CO_2, methanol, methyl amines, methyl sulphides, acetate, and pyruvate (Boone *et al.*, 1993; Bock *et al.*, 1994; Rajagopal and LeGall, 1994). Further, they are the only methanogenic Archaea that have been shown capable of non-methanogenic growth (although this growth is of a limited nature) (Bock and Schonheit, 1995). As such, genetic analysis in *Methanosarcina* will allow experiments not possible in other methanogenic Archaea, namely direct mutagenic analysis of the methane-producing pathways. Because other methanogenic Archaea are obligate methanogens, mutations in the methanogenic pathway are expected to be lethal. Thus, obtaining such mutants in these other groups will require the use of conditionally lethal mutations, such as temperature-sensitive mutations. These are often difficult to obtain and hard to work with. However, in *Methanosarcina* it should be possible to obtain mutations that block use of certain methanogenic substrates, but not others, or that block the ability to produce methane altogether.

Three major obstacles to genetic analysis of *Methanosarcina* have recently been overcome. The first of these is the ability to grow the organism as single cells (Sowers *et al.*, 1993). Most natural isolates of this genus grow in large clumps of a few to a few thousand, or even tens of thousands, of cells. A thick layer of exopolysaccharide, known as methanochondroitin, encloses these clumps. Methanochondroitin is similar to chondroitin sulphate that is found in eukaryotic organisms. Growth of *Methanosarcina* in this clumped morphology prevents the simple isolation of clonal populations that is absolutely required for genetic analysis. However, it has recently been shown that production of methanochondroitin is regulated. In media of high osmotic strength the exopolysaccharide is not produced and the organism grows as a population of single cells (Sowers *et al.*, 1993). These cells possess a cell wall composed of an S-layer protein similar to that of methanococci or halobacteria. The single-celled *Methanosarcina* can be plated with high efficiency on solidified medium, forming colonies derived from individual cells, i.e. the clonal populations required for genetic manipulation.

A second major breakthrough in the genetic analysis of *Methanosarcina* was the development of transformation methods for this genus. Transformation of *Methanosarcina mazei* was first reported by de Macario *et al.* (1996), using CaCl$_2$-mediated transformation and an integration vector containing cloned *M. mazei* genes with the *Methanococcus pac* cassette as the selectable marker. In addition to providing the first method for introduction of foreign DNA to *Methanosarcina*, this experiment demonstrated that the *pac* gene under control of a *Methanococcus* promoter was expressed in *Methanosarcina* and that it provided resistance to the antibiotic puromycin. This method, and a variety of others, including PEG-mediated, lithium-acetate-mediated, electroporation, and natural transformation, give relatively low transformation frequencies in *Methanosarcina* (Metcalf *et al.*, 1997). However, a highly efficient liposome-mediated transformation protocol for *Methanosarcina* protoplasts has been developed (see also Chapter 1, II.D.3). Using an autonomously replicating shuttle vector (see below) the liposome-mediated transformation frequency can be as high as 2×10^8 transformants per microgram of DNA per 10^9 recipients, or about 20% of the recipient population. Further, the method is simple to perform, requiring a minimum number of steps (Metcalf *et al.*, 1997).

The third major development in the genetic analysis of *Methanosarcina* is the construction of a series of broad host range shuttle plasmids capable of replication in most *Methanosarcina* species tested, and in *E. coli* (Metcalf *et al.*, 1997) (see Chapter 2, II.B for discussion of general shuttle vectors). These plasmids are based on the naturally occurring plasmid, pC2A, from *Methanosarcina acetivorans* (Sowers and Gunsalas, 1988). The plasmids combine the pC2A replicon and a modified *pac* cassette (see below) for replication and selection in *Methanosarcina*, with the R6K plasmid replicon and β-lactamase gene for replication and selection in *E. coli*. Several plasmids were constructed providing a variety of cloning sites for introduction of DNA fragments of interest. Some of these also allow screening for recombinant plasmids in *E. coli* by loss of *lacZ* α-complementation (Ausebel *et al.*, 1992).

There is little published material on the mutagenesis of *Methanosarcina* species. A few reports describe the selection for mutants resistant to specific toxic compounds (Smith and Mah, 1981; Smith and Lequerica, 1985; Knox and Harris, 1988), but general methods, such as chemical or UV-induced mutagenesis, have not been developed. Given the availability of a shuttle vector that could be used to clone genes by complementation of mutants, there is an urgent need to adapt the methods used in other organisms for use in *Methanosarcina*. Because the *pac* cassette confers puromycin resistance upon *Methanosarcina* both directed and random insertional mutagenesis, as described above, should be possible in this genus.

The modified *pac* cassette used in the construction of the *Methanosarcina–E. coli* shuttle vectors should be quite useful for the *in vitro* construction of gene disruption mutations for use in reverse genetic experiments as described above (Metcalf *et al.*, 1997). This cassette was designed to utilize both the transcriptional and translational signals of the *M. voltae mcr* operon to direct expression of the *pac* gene. Accordingly, the DNA sequences between the *mcrB* and *pac* start codons of pMip1 were deleted creating a fusion of the *mcrB* ribosome binding site and *pac* gene. Numerous restriction sites to facilitate subsequent constructions flank the modified *pac* cassette. Because of the apparent similarities among all archaeal promoters identified to date, it is not surprising that the *M. voltae* promoter is active in *Methanosarcina*. Based on these findings it would not be surprising if the *pac* cassette functioned in other Archaea as well. Thus, this cassette may be generally useful among Archaea that are sensitive to puromycin.

A set of vectors designed to facilitate mutagenesis by gene disruption with non-replicating plasmids with cloned internal gene fragments has

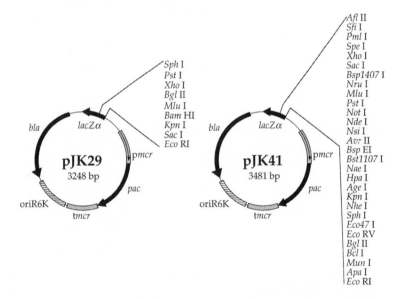

Figure 5. Integration vectors for use in methanogenic Archaea. Two representative plasmids from a set of integration vectors constructed for use in *Methanosarcina* are shown (Metcalf, Zhang, and Wolfe, unpublished). The plasmids can be used to disrupt genes by insertional mutagenesis as shown in Figure 4, or to insert other constructs such as reporter genes into the host chromosome via homologous recombination. These plasmids replicate in *E. coli* hosts that carry the *pir* gene required for replication using *ori*R6K (Stalker *et al.*, 1979). The plasmids will not replicate in non-*pir* hosts (which includes virtually all known organisms). The plasmids carry the *pac* gene for resistance to puromycin under the control of a strong constitutive promoter (*pmcr*) that is functional in *Methanosarcina* and *Methanococcus*. DNA from the target host can be cloned into these plasmids using the unique cloning sites shown to provide a region for homologous recombination. The *lacZα* gene allows blue–white screening for recombinant plasmids except for pJK41. Plasmids pJK31, pJK33, pJK35, pJK37, pJK39 are similar to pJK29 but carry different polylinkers derived from the pMTL series of cloning vectors (Chambers *et al.*, 1988). Plasmids are available from the author upon request.

been constructed (Figure 5; Metcalf, Zhang, and Wolfe, unpublished). These plasmids are based on the *E. coli* R6K replicon, which is not functional in *Methanosarcina*, and have convenient restriction sites for cloning. Most also allow *lacZα* complementation to screen for recombinant plasmids containing the desired internal gene fragment in *E. coli* prior to transformation in *Methanosarcina*. Each has the modified *pac* cassette for selection of plasmid integrants in *Methanosarcina*. One of these vectors has been used to construct *serC* mutations in *M. barkeri*, demonstrating the utility of these plasmids (Metcalf, Zhang, and Wolfe, unpublished).

C. Genetic Analysis of *Methanobacterium*

Few methods of genetic analysis are available for use in *Methanobacterium* species. This is particularly ironic, because in the areas of biochemistry and physiology *Methanobacterium* species are probably the best understood of the methanogenic Archaea. Species of *Methanobacterium* are similar to the methanococci in that they are restricted to growth utilizing H_2 and CO_2, or formate, as carbon and energy sources (Boone *et al.*, 1993). A few species are able to obtain energy by reduction of CO_2 coupled to oxidation of propanol or butanol; however, these compounds are not assimilated and CO_2 is still used as the source of carbon. Species of *Methanobacterium* are long rod-shaped organisms that inhabit mesophilic to moderately thermophilic environments. A principal characteristic of this group is the presence of a cell wall composed of a peptidoglycan designated pseudomurein due to its similarity to bacterial murein cell walls. However, antibiotics such as ampicillin and vancomycin do not inhibit synthesis of pseudomurein, nor is pseudomurein a substrate for degradative enzymes, such as lysozyme, that attack the murein cell walls of Bacteria (Kandler and Konig, 1985).

Many species of *Methanobacterium* species are easily mass-cultured (Leigh and Wolfe, 1983; Mukhopadhyay, 1993). Because of this much of the biochemistry of methanogenesis was first determined in members of this genus. A great deal of effort has gone into development of genetics for *Methanobacterium* as a natural complement to the well-characterized biochemistry. These efforts have yielded very good methods for plating of several *Methanobacterium* species on defined solid media (Kiener and Leisinger, 1983; Jain and Zeikus, 1987). Mutants of *Methanobacterium* are readily obtained by chemical mutagenesis using a variety of mutagens, and a method for selective enrichment of non-growing mutants using bacitracin has been developed (Kiener *et al.*, 1984; Rechsteiner *et al.*, 1986; Jain and Zeikus, 1987; Tanner *et al.*, 1989; Worrell and Nagle, 1990). With these methods a variety of useful auxotrophic strains have been obtained. Analysis of some of these mutants has provided insight into the metabolism of *Methanobacterium*. For example, biochemical analysis of glutamine auxotrophs of *M. ivanovii* conclusively demonstrated that ammonium ion is incorporated in this organism via the glutamine synthetase route (Bhatnagar *et al.*, 1986). *Methanobacterium* mutants resistant to a number of toxic compounds have also been selected (Kiener *et al.*, 1984; Hummel and

Bock, 1985; Nagle *et al.*, 1987). Auxotrophic mutants and mutants resistant to various toxic compounds, such as fluorouracil in *Methanobacterium thermoautotrophicum*, have proven useful in the demonstration of gene transfer between strains of this species (Worrell *et al.*, 1988; Meile *et al.*, 1990). In addition, a selectable marker for resistance to the antibiotic pseudomonic acid based on a mutant isoleucine-tRNA synthase gene (*ileS*) has been obtained for use in *M. thermoautotrophicum* (Jenal *et al.*, 1991).

Gene transfer between strains of *Methanobacterium* can occur by at least two methods. Transformation by high molecular weight chromosomal DNA has been demonstrated in *M. thermoautotrophicum* (Worrell *et al.*, 1988). Low levels of DNA (10 ng) saturate the process. Although the frequency of transformation cannot be quantitatively assessed in this type of experiment, it is apparently very low. Inexplicably, this DNA transfer only occurs on the surface of medium solidified by gellan gum. It does not occur in broth media, nor does it occur when agar is used as a solidifying agent. A second method for gene transfer among *Methanobacterium* strains is through the use of the generalized transducing phage ΨM1 (Meile *et al.*, 1990). ΨM1 is a lytic phage of *M. thermoautotrophicum* Marburg (Meile *et al.*, 1989). It is capable of transducing a variety of chromosomal markers at a frequency of 6×10^{-4} to 5×10^{-6} per plaque-forming unit. All markers tested to date could be transferred by ΨM1 indicating that this is a generalized transducing phage. This method could be of great use for strain constructions in this host, and for examining linkage groups of mutants. Potential difficulties with its use are the low burst size (~6) leading to low titre lysates, and relative instability of the lysates upon storage (lysates are stable for only two weeks at 4°C) (Meile *et al.*, 1995). To date ΨM1is the only known transducing phage among the Archaea.

Recently, the complete genome sequence of *Methanobacterium thermoautotrophicum* ΔH has been released (Smith *et al.*, 1997). If further methods can be developed for this organism, specifically transformation methods and plasmid vectors for gene cloning, our comprehension of this organism could be rapidly enhanced (see Chapter 1, IV for discussion of approaches). A number of plasmids have been identified among various *Methanobacterium* species (Meile *et al.*, 1983; Nolling *et al.*, 1991, 1992). Three of these have been completely sequenced (Bokranz *et al.*, 1990; Nolling *et al.*, 1992). To date, no replicating plasmid vectors, integration vectors, or methods for plasmid transformation have been reported for members of this group.

◆◆◆◆◆◆ IV. GENETIC ANALYSIS OF HALOPHILIC ARCHAEA

The halophilic Archaea have been known for centuries by the reddish appearance they impart to environments of high salinity, such as ponds used to produce salt by evaporation of sea water, and in salt-preserved foods. They were the first Archaea brought into pure culture for laboratory study, with some probable isolates dating to the late nineteenth century (these cultures have been lost and so cannot be verified) (Kushner,

1985; Tindall, 1991). The dominant feature of this group is their ability to grow under high, or even saturating salt conditions. Halophilic Archaea not only tolerate high salt, but require it. Most will not grow if the salt concentration falls below 1 M. Optimum NaCl concentration for most species is between 2.5 and 5.2 M. Phylogenetically distinct groups among the halophilic Archaea are defined by 16S RNA sequence comparison and by the environments they inhabit (Kamekura *et al.*, 1997). Members of the first group grow at pH values near neutrality and include the well-known genera *Halobacterium* (*Hb.*) and *Haloarcula* (*Ha.*). A second group with similar physiological traits includes the *Haloferax* (*Hf.*) species. Another group includes *Natronobacterium* and *Natronococcus* species, which are alkaliphilic in addition to halophilic. These organisms have been predominantly isolated from soda lakes, and require high pH for optimum growth.

In general, the halophilic Archaea grow aerobically by oxidation of a variety of compounds including amino acids, carbohydrates and fatty acids. A few species are also capable of anaerobic growth, either by fermentation or by anaerobic respiration with nitrate or fumarate as electron acceptors (Kushner, 1985; Rodriguez-Valera, 1995). One of the more interesting aspects of halobacterial metabolism is the production of bacteriorhodopsin by certain members of the group including *Hb. salinarium* (formerly *Hb. halobium*). This protein, which is synthesized in response to reduced oxygen tension and high light levels, functions as a light-driven proton pump (Shand and Betlach, 1991; Krebs and Khorana, 1993). The electrochemical gradient produced by bacteriorhodopsin allows synthesis of ATP, and indirectly the uptake of a variety of compounds by symport with Na^+, for which an $H+/Na+$ antiporter maintains a gradient. Under certain conditions, all energy required for growth can derived from this chlorophyll-independent photosynthesis (Hartmann *et al.*, 1980).

A. Growth, Mutagenesis and Selectable Markers

Unlike many Archaea, growth of halophilic archaeal species is relatively simple (Tindall, 1991; Rodriguez-Valera, 1995). Standard microbiological techniques for growth and plating on solidified media can be applied to members of this group. All halophilic Archaea are aerobic chemoheterotrophs, and although some have complex nutritional requirements many will grow in simple defined media. Certain species are also capable of anaerobic growth by fermentation, or by respiration with nitrate or fumarate as electron acceptors. Most strains grow well at 37°C. The only constraints placed upon these methods are those imposed by the requirement for high concentrations of Na^+ and Mg^{2+}. These sometimes limit the solubility of common media components such as phosphate buffers. Mass culture in stainless steel vessels is also to be avoided due to the corrosive nature of high-salt media.

A variety of mutagenesis methods have been developed for use in various halophilic Archaea. Chemical mutagenesis with ethylmethanesulphonate has proven very effective for random generation of auxotrophic mutants (Mevarech and Werczberger, 1985). Using this method hundreds of auxotrophic mutants of *Hf. volcanii* have been generated (Cohen *et al.*, 1992). In *Hb. salinarium* mutations that result in an obvious phenotypic characteristic are readily obtained without mutagenic treatment due to the high rate of spontaneous mutation in this species (Pfeifer *et al.*, 1981). Examples include the colony morphology displayed by gas vesicle mutants, or the colony colour displayed by bacteriorhodopsin or bacterioruberin mutants. This high spontaneous mutation rate (up to 1% for certain genetic loci) is due to an abundance of transposable insertion sequences, which frequently create insertions, deletions and other genetic rearrangements in this species (Sapienza *et al.*, 1982; DasSarma, 1989; Pfeifer and Blaseio, 1989). In addition to these traditional methods of mutant generation, the genetic malleability of the halophilic Archaea makes a variety of modern molecular methods possible for generation of both random and directed mutations (see below).

Several selectable markers are available for use in halophilic archaeal species. These include selection for complementation of the numerous auxotrophic mutants that are available, and antibiotic resistance markers, which are of more general use. The most commonly used antibiotics for this group are the DNA gyrase inhibitor novobiocin, and the HMG-CoA (hydroxymethylglutaryl-coenzyme A) reductase inhibitor mevinolin. The latter compound is effective against many Archaea because it prevents the synthesis of mevalonate, a key intermediate in the biosynthesis of the unique isoprenoid lipids found in all members of this domain. Mutant derivatives of the DNA gyrase gene (*gyrB*) from *Haloferax* strain Aa 2.2, and HMG-CoA reductase gene (alternatively designated *hmg* or *mev*) from *Hf. volcanii* provide resistance to these antibiotics when introduced into sensitive strains (Lam and Doolittle, 1989; Holmes and Dyall-Smith, 1990). Other antibiotics have also been used in certain circumstances including chloramphenicol, anisomycin and thiostrepton (a mutant 23S rRNA provides resistance in each case) (Mankin and Garrett, 1991; Mankin *et al.*, 1992), and trimethoprim (over-expression of dihydrofolate reductase provides resistance) (Zusman *et al.*, 1989).

B. Conjugation

A unique form of conjugation is displayed by *Hf. volcanii* that allows transfer of chromosomal markers and of shuttle vectors, but ironically not of indigenous plasmids (Mevarech and Werczberger, 1985; Rosenshine *et al.*, 1989). This form of mating is distinct from the plasmid-encoded conjugation systems known in Bacteria (see Chapter 2, V). In particular, DNA transfer appears to be bidirectional and no clear donor–recipient relationship exists. Conjugation requires cell–cell contact between live partners, and DNA transfer apparently occurs via cytoplasmic bridges that form between two or more cells. Further, transfer between different species within the *Haloferax* genus has been demonstrated, although transfer to other genera including *Halobacterium* and *Haloarcula* could not be demonstrated (Tchelet and Mevarech, 1994). A similar type of conjugation is seen in *Sulfolobus acidocaldarius* (Grogan, 1996). The conjugation system of *Haloferax* has been exploited to distinguish linkage groups of mutants with the same phenotype. Thus, when independent mutants with the same auxotrophic requirement are mated with members of the same linkage group no prototrophic exconjugates are obtained, whereas conjugation with members of other linkage groups does yield prototrophic exconjugates. Using this method it was shown that 14 adenine-requiring auxotrophs formed three separable linkage groups (Mevarech and Werczberger, 1985). An advantage to this method is that it is simple to perform. The method requires: (i) mixing the two cell populations on a filter; (ii) incubation of the filter on growth medium to allow conjugation; and (iii) resuspension of the cells and selection for exconjugates on appropriate media.

C. Transformation

High efficiency transformation of several halophilic archaeal species has been demonstrated using high molecular weight chromosomal DNA, nonreplicating plasmid integration vectors, autonomously replicating plasmid shuttle vectors, small linear DNA fragments and phage DNA (Cline and Doolittle, 1987; Cline et al., 1989a, 1989b; Conover and Doolittle, 1990; Mankin et al., 1992). Transformation of Hb. salinarium was first demonstrated by transfection with purified DNA from a lytic halophage ΦH1 (Cline and Doolittle, 1987). The observation of plaque formation after transfection allowed verification of DNA transfer without the need to first develop a selectable marker or plasmid vector. This generally useful method was also used to develop transformation in Sulfolobus (Schleper et al., 1992).

Like the Methanococcus and Methanosarcina species described above, the halophilic Archaea have an S-layer cell wall (Kandler and Konig, 1985). Suspension of these cells in low salt, iso-osmotic medium, followed by removal of Mg^{2+} by chelation with EDTA causes the loss of the S-layer and formation of protoplasts. These protoplasts can be induced to take up added DNA by addition of PEG-600 (Cline and Doolittle, 1989a, 1995; Chapter 1, II.D.7). Depending on the nature of the transformation experiment (i.e. whether the DNA is an autonomously replicating plasmid or whether recombination is required to generate transformants, and whether host restriction is a factor), the frequency of transformation ranges from ca. 10^2–10^7 transformants per microgram of DNA. Slight modifications of this technique have allowed transformation of Hb. salinarium, Hf. volcanii, Hf. mediterranei, Haloferax sp. Aa2.2, Ha. vallismortis, and Ha. hispanica (Cline et al., 1989a; Cline and Doolittle, 1992, 1995).

Because most of the halophilic Archaea possess restriction systems, DNA isolated from isogenic hosts transforms recipient cells at several orders of magnitude higher frequencies than DNA isolated from heterologous hosts (Schinzel and Burger, 1986; Blaseio and Pfeifer, 1990; Holmes et al., 1991; Cline and Doolittle, 1992). However, the high frequency of transformation obtained using protoplasts and PEG (see also Chapter 1, II.D.7) is usually adequate to surmount the restriction barrier. Hf. volcanii restricts DNA purified from E. coli hosts containing Dam methylase (Holmes et al., 1991). This particular restriction system can be bypassed using E. coli dam hosts to prepare transforming DNA.

An additional complication in transformation experiments involving selection for antibiotic resistances stems from the fact that the markers used for these selections are often derived from the same organism that is to be transformed (Lam and Doolittle, 1989; Holmes and Dyall-Smith 1990). As such, homologous recombination can occur between the transforming DNA and the genomic copy of the antibiotic resistance gene. In cases where plasmid maintenance is desired these marker rescue events can lead to false positives without the plasmid. False positives can also be obtained when integration is desired at a specific locus. In this case the resulting antibiotic-resistant integrants may reside at the native antibiotic resistance locus rather than at the desired target gene. These undesired recombinants have been shown to increase dramatically when restriction of the incoming DNA occurs (Blaseio and Pfeifer, 1990). To minimize this problem it is best to choose an antibiotic-resistance gene from a different species wherever possible.

High efficiency transformation has revolutionized the genetic analysis of halophilic Archaea. Gene replacement of the wild-type alleles with *in vitro* constructed mutant alleles is now a common experimental approach in both *Halobacterium* and *Haloferax* species. Another development that would not have been possible without transformation protocols has been the construction of a variety of autonomously replicating plasmid shuttle vectors.

D. Gene Replacement and Marker Rescue

Gene replacements in the halophilic Archaea have been made by several methods. Mutations made *in vitro* by insertion of antibiotic-resistance markers into cloned genes have been recombined on to the *Haloferax* chromosome after transformation with linear DNA or circular DNA (as in Figures 1 and 2). This method has been used to construct dihydrolipoamide dehydrogenase mutants in *Hf. volcanii* after disruption of the coding sequence with the mevinolin resistance marker (Jolley *et al.*, 1996). Transformation with circular DNA has been used to make *sopI* mutants of *Hb. salinarium* disrupted with the *gryB* resistance marker (Ferrando-May *et al.*, 1993). A more commonly used method in halophilic Archaea involves the formation and subsequent resolution of partial diploids by integration of non-replicating plasmids (as shown in Figure 3) (Mankin *et al.*, 1992; Krebs *et al.*, 1993a; Krebs, 1995; Woods and Dyall-Smith, 1997). In this case the non-replicating plasmids are often *E. coli* vectors that carry an appropriate selectable marker, usually resistance to mevinolin or novobiocin. Following transformation partial diploid recombinants are selected by virtue of this antibiotic resistance gene. In halophilic Archaea segregation of the partial diploids is relatively efficient, and can be accomplished simply by plating in the absence of antibiotic. Segregants are then identified by loss of the antibiotic-resistance marker, and subsequently screened for those with the desired mutation. In *Hb. salinarium* segregation of partial diploids is enhanced by insertion of the high-copy replicon pGRB1, or the closely related pHSB1 replicon, into the cloning vector. Plasmids carrying these replicons are notably unstable in *Hb. salinarium* and in the absence of antibiotic selection are rapidly lost in transformants (Mankin *et al.*, 1992; Krebs *et al.*, 1993a).

A wide variety of mutants have been constructed in both *Haloferax* and *Halobacterium* using this multi-step method. It has been particularly useful in the study of bacteriorhodopsin function and regulation (Krebs *et al.*, 1993a). Before gene replacement methods were available for use in halophilic Archaea mutant bacteriorhodopsin proteins had to be produced in *E. coli*, and studied after reconstitution in artificial liposomes. The advent of transformation and gene replacement methods now allows these studies to be conducted in the native host under physiological conditions appropriate to this protein. In addition, the bacteriorhodopsin gene (*bop*) promoter has been used to over-express other proteins after recombination into the chromosomal *bop* locus (Krebs *et al.*, 1993b). A second example of this technique is the study of 23S rRNA mutations (Mankin *et al.*, 1992). *In vitro* changes in the rDNA genes have been introduced into the genome of *Hb. salinarium* by selection for linked mutations conferring resistance to thiostrepton and anisomycin. Unlike most organisms this species has only a single copy of its 23S rRNA gene. This feature allows the unique opportunity to perform structure–function studies on mutant ribosomes purified from strains in which the wild-type rDNA gene has been replaced with rDNA containing specific site-directed mutations. This is not possible in organisms with multiple 23S RNA genes because these organisms produce mixtures of mutant and wild-type ribosomes when individual copies of their 23S RNA genes are mutated.

Marker rescue of chromosomal mutations by directly selecting for correction of the mutant phenotype has also been of great use in halophilic Archaea. This method was first demonstrated by transformation of auxotrophs to prototrophy using high molecular weight chromosomal DNA from otherwise isogenic strains (Cline *et al.*, 1989b). In these experiments the incoming DNA replaces the mutant allele by homologous recombination restoring the wild-type phenotype. An extremely useful application of this method is the mapping of genes on the chromosome by marker rescue of auxotrophic mutations. This can be done by transformation of the auxotrophs to prototrophy using an ordered cosmid library as donor DNA (Cohen *et al.*, 1992). Ordered cosmid libraries are available for *Halobacterium* sp. GRB and for *Hf. volcanii* (Charlebois *et al.*, 1991; St Jean *et al.*, 1994). In this method the ordered cosmid clones are grouped together into a number of small pools. These cosmid pools are then used to transform auxotrophic mutants. Only those pools that contain a clone with the wild-type version of the mutation being mapped will yield prototrophic transformants. The cosmids present in positive pools can then be screened individually to identify the particular clone carrying the gene of interest. Once a positive clone is identified, fine structure mapping can be done using small restriction endonuclease-generated DNA fragments purified by agarose gel electrophoresis. Importantly, these fragments do not require extensive purification. After electrophoresis in low melting point agarose individual bands can be excised, melted and used directly for transformation. Fragments as small as 3 kbp have been shown to give prototrophic recombinants. The simplicity of this experiment has allowed very rapid mapping of genes on the *Hf. volcanii* chromosome. In one study 139 mutations resulting in auxotrophies for various amino acids, purines, or pyrimidines were mapped on the *Hf. volcanii* genome by this approach (Cohen *et al.*, 1992).

E. Cloning Vectors

Plasmids are relatively common among the halophilic Archaea. These range from the megaplasmids that constitute a sizeable fraction of the *Haloferax* and *Halobacterium* genomes to a family of very small high copy plasmids from various *Halobacterium* species (Weidinger *et al.*, 1979; Charlebois *et al.*, 1987, 1991; Holmes and Dyall-Smith, 1990; Pfeifer and Ghahraman, 1993; St Jean *et al.*, 1994). Several of these plasmids have been modified to serve as halophilic Archaea–*E. coli* shuttle vectors (Lam and Doolittle, 1989; Blaseio and Pfeifer, 1990; Ng and DesSarma, 1993; Holmes *et al.*, 1991, 1994). These vary in their host range, copy number, selectable marker, compatibility with other shuttle vectors and useful cloning sites. Shuttle vectors have greatly expanded the range of experiments possible in halophilic Archaea. A vast array of techniques involving plasmids is commonplace in modern molecular genetics. Many of these have now been adapted for use in the halophilic Archaea. These include expression vectors, over-expression vectors, and vectors with reporter genes for detection of expression levels from cloned promoters.

Numerous examples using modified shuttle vectors to allow expression and over-expression of proteins in halophilic Archaea have been presented. Among the promoters used for expression and over-expression are the *bop* promoter (Ni *et al.*, 1990), the ferridoxin promoter (Danner and Soppa, 1996), the rRNA promoter from *Hb. cutirubrum* (Jolley *et al.*, 1996), and the *H. volcanii* tRNA^{Lys} promoter (Nieunlandt and Daniels, 1990). These plasmids have been particularly useful in defining the components of gas vesicles, and the gene that regulates their synthesis (Halladay *et al.*, 1992, 1993; Offner and Pfeifer, 1995; Offner *et al.*, 1996). They have also been used to study bacteriorhodopsin and genes involved in various metabolic pathways of *Haloferax* and *Halobacterium* (Ni *et al.*, 1990; Krebs *et al.*, 1991; Ruepp and Soppa, 1996).

The use of a reporter gene to quantify *in vivo* expression is of particular interest given the unique nature of archaeal RNA polymerase. What constitutes an archaeal promoter is an area of active research. A number of studies have used *in vitro* transcription systems to address this problem; however, these may not accurately reflect all the components required *in vivo*. The genetic manipulations made possible by the halophilic archaeal plasmids provide one of the few systems available for *in vivo* analysis of archaeal promoter sequences. These systems have allowed deletion analysis of promoters to define the regions required *in vivo* for gene expression, and introduction of specific mutations within the regions defined by deletion and *in vitro* studies to assess their effect on transcription (Gropp *et al.*, 1995; Yang *et al.*, 1996). A particularly elegant experiment used random oligonucleotides to saturate the distal promoter element with mutations, followed by quantification of *in vivo* expression from randomly picked clones and clones picked specifically for their high or low levels of expression (Danner and Sopper, 1996). Two reporter gene systems may be of general use. The dihydrofolate reductase gene provides resistance to trimethoprim in *Hf. volcanii* when it is over-expressed, with the degree of antibiotic resistance correlating to the level of over-expression (Danner and Sopper, 1996). A second system uses a modified tRNA^{Trp} that can be specifically detected by hybridization (Nieuwlandt and Daniels, 1990).

F. Transposons

One of the most powerful genetic techniques for identification of genes is the use of transposons simultaneously to disrupt and tag genes with selectable markers (see Chapter 4). This method has revolutionized bacterial genetics and finds widespread use in a number of eukaryotic systems as well. Transposable insertion sequences are extremely common in the halophilic Archaea. At least 15 unique elements have been identified (Charlebois and Doolittle, 1989). These elements are responsible for a variety of spontaneous mutations as has been described above. They are also responsible for the extensive rearrangements of the genome of *Hb. salinarium* (Sapienza *et al.*, 1982; Pfeifer and Blaseio, 1989). The *Hb. salinarium* transposable elements ISH2, ISH26 and ISH28 have been adapted to allow antibiotic-resistant selection for composite transposons made up of two copies of each element flanking the *Hf. volcanii hmg* gene for resistance to mevinolin (Dyall-Smith and Doolittle, 1994). These elements cannot be used in either *Hb. salinarium* or *Hf. volcanii* because of the presence of homologous sequences (the IS elements in *Hb. salinarium* and the *hmg* gene in *Hf. volcanii*). In these hosts homologous recombination occurs at a higher frequency than transposition resulting in the vast majority of antibiotic resistant transformants being due to insertion at homologous chromosomal loci rather than the desired transposition events. However, transformation of *Ha. hispanica* with plasmids carrying the composite IS26 and IS28 transposons results in drug-resistant transformants that arise from transposition of the element to random loci in the host genome. The IS2 composite transposon does not appear to be functional. Although not yet extensively used, these transposons should be of great use in genetic analysis of the halophilic Archaea.

◆◆◆◆◆◆ V. GENETIC ANALYSIS OF THE THERMOACIDOPHILE *SULFOLOBUS*

Sulphur metabolizing, thermophilic Archaea were first isolated in the late 1960s from hot springs in Yellowstone National Park, USA (Segener and Stetter, 1991). Since that time, the technology and methods for growth of organisms at high temperatures has improved dramatically. As a result numerous thermophilic Archaea have been isolated from both marine and terrestrial environments (Stetter, 1996b). Most of these organisms metabolize sulphur, including organisms that oxidize reduced sulphur compounds as an energy source and those that reduce oxidized sulphur compounds as electron acceptors for anaerobic respiration. Sulphur metabolism is not, however, universal among the thermophilic Archaea, nor is it obligatory in all organisms that possess this trait. The thermophilic Archaea are remarkable for their extremely high optimum growth temperatures, with several species capable of growth up to at least 110°C. A number of species are also acidophiles due to their production of sulphuric acid by oxidation of reduced sulphur compounds (referred to as thermoacidophiles). The thermophilic and hyperthermophilic Archaea have recently begun to receive a great deal of attention due to their unique metabolism and adaptation to both thermophilic and acidic conditions. Among this group the genus that has been studied in the greatest detail is *Sulfolobus*.

Sulfolobus acidocaldarius was described by Brock *et al*. (1972), and was among the first sulphur-metabolizing Archaea isolated. Although *Sulfolobus* species do not grow at the same extreme high temperatures as some other thermophilic Archaea, the optimum growth temperature for most strains is still quite high, *ca*. 75–85°C (Grogan, 1989; Segener and Stetter, 1991). *Sulfolobus* species have been isolated from a variety of terrestrial solfataric environments (sulphur-rich, geothermally heated environments) as thermophilic aerobes capable of autotrophic growth by oxidation of elemental sulphur. Marine isolates of this genus have not been reported. Two related organisms isolated from similar environments, *Acidanius* and *Desulfurolobus*, are remarkable for their ability to grow autotrophically either aerobically by sulphur oxidation or anaerobically by sulphur reduction using H_2 as the electron donor (Fischer *et al*., 1983; Segerer *et al*., 1985; Zillig *et al*., 1985). *Sulfolobus* and its relatives produce sulphuric acid as a by-product of sulphur oxidation. Consequently, *Sulfolobus*, and other organisms with similar metabolism are acidophilic. The pH required for optimum growth of *Sulfolobus* species is in the range of 2.0 to 4.0. The production of acid by organisms like *Sulfolobus* contributes greatly to the acidity of the soil and water surrounding the volcanic sites they inhabit. In addition to their ability to grow by oxidation of sulphur, some *Sulfolobus* species can obtain energy for growth by oxidation of ferrous iron. All *Sulfolobus* species are capable of heterotrophic growth with or without sulphur using a variety of substrates such as simple sugars and yeast extract. *Sulfolobus* species are small irregular cocci and can be motile depending on the growth conditions (Grogan, 1989; Segener and Stetter, 1991). As with many Archaea the cell wall of *Sulfolobus* species and their relatives is composed of a proteinaceous S-layer (Kandler and Konig, 1985).

Several species, including *S. acidocaldarius*, *S. solfataricus*, *S. islandicus* and *S. shibatae*, have been reported within the *Sulfolobus* genus (Segener and Stetter, 1991). The main criteria for the separation of these species are molecular, including antigenic cross-reactivity of their respective RNA polymerases, comparison of 16S RNA sequences, DNA–DNA hybridization and restriction analysis of genomic DNAs. Analysis of the 16S RNA sequences of *Sulfolobus* species places this genus in the Crenarchaeota (Olsen *et al*., 1985). As such, they are only distantly related to the methanogenic and halophilic Archaea, which are in the Euryarchaeota. *Sulfolobus* may be considered the model organism for the Crenarchaeota.

Much of what is known about this group (and about Archaea in general) has come from the study of this organism. For example, the RNA polymerase of *Sulfolobus* has been well studied biochemically and is a model for the function of archaeal polymerases (Langer *et al.*, 1995; Thomm, 1996). Reverse gyrase was first discovered in this genus (Kikuchi and Asai 1984). This intriguing enzyme introduces positive supercoils into DNA, as opposed to the negative supercoils introduced by bacterial and eukaryal topoisomerases. The enzyme is now known to be present in a variety of thermophiles and is believed to stabilize their DNA at high temperatures (Bouthier de la Tour *et al.*, 1990). A number of enzymes have been studied from *Sulfolobus* and several of these crystallized in an effort to understand the nature of thermostability of proteins (Hartl *et al.*, 1987; Pearl *et al.*, 1993; Aguilar *et al.*, 1997). There is also interest in *Sulfolobus* and related organisms due to their ability to solubilize metals from sulphidic ores (Huber and Stetter, 1991). This bioleaching of minerals at relatively high temperatures has great industrial potential.

In contrast to the numerous studies on the biochemistry and physiology of *Sulfolobus*, genetic techniques for this organism are just beginning to be developed. However, several promising results suggest that these studies will be fruitful in the near future. Accordingly, the physiology and growth requirements of several species have been characterized in detail (Grogan, 1989). Gene transfer has also been described in *Sulfolobus*, including both transformation and conjugation (Schleper *et al.*, 1995a; Grogan, 1996). Several plasmid vectors have recently been reported (Elferink *et al.*, 1996; Zillig *et al.*, 1996) (Cannio *et al.*, 1996), and progress has been made in developing a novel gene transfer agent based on the homing endonuclease-encoding 23S rRNA intron of *Desulfurococcus mobilis* (Aagaard *et al.*, 1996). In addition, the complete DNA sequence of *Sulfolobus solfataricus* P2 genome is currently being determined (Charlebois *et al.*, 1996). At the time of writing about one-third of the genome sequence has been completed. In conjunction with new genetic methods, and currently available techniques, these developments make *Sulfolobus* an excellent choice as a model organism for the study of thermophilic and acidic adaptation, and of Archaea in general.

A. Growth, Mutagenesis and Selectable Markers

The major difficulty in the study of thermophiles is the requirement of high temperatures for optimum growth. Desiccation of the medium, as a result of evaporation caused by the high incubation temperatures used for growth of these organisms, can be a serious problem. This is especially true for solid media. An additional problem with high temperature incubation on solid media is the instability of commonly used solidifying agents such as agar at high temperatures. These problems are compounded if the organism is a strict anaerobe (as are many thermophiles), as was described above for the methanogenic Archaea. These problems are less serious for *Sulfolobus* than for other thermophilic Archaea (Grogan, 1989). Because *Sulfolobus* is an aerobe no special anaerobic manipulations are required. Also growth does not require the extremely high temperatures needed for some thermophiles. Typically temperatures in the range of 75–80°C are used. Preparation of media with amorphous elemental sulphur can be quite laborious. Because *Sulfolobus* is capable of heterotrophic growth in the absence of sulphur this difficulty can also be avoided (unless, of course, one is studying sulphur metabolism). One slight complication in growth of *Sulfolobus* is the need to maintain acidic pH. Accordingly, some media components require separate sterilization to prevent their degradation at high temperature and low pH. An additional complication stems from the fact that non-growing cells are sensitive to acid. Thus the pH needs to be elevated prior to long-term storage of cultures (Grogan, 1989).

Methods for growth of *Sulfolobus* as colonies derived from single cells on medium solidified with gellan gum (Gelrite) have been developed (Grogan, 1989; Lindstrom and Sehlin, 1989; Rakhely and Kovacs, 1996). Unlike earlier methods using media solidified with agar or starch, these methods are reproducible and give high plating efficiencies. A very thorough characterization of the phenotypic characteristics of five wild-type *Sulfolobus* strains has been presented and provides a useful and required baseline for genetic studies on members of this genus. The characteristics examined in this study include optimal growth conditions, the substrates used for carbon, nitrogen and energy, growth inhibitors, motility, pigment production, and S-layer composition (Grogan, 1989).

The result of these basic studies is that most of the standard microbiological methods can be applied to *Sulfolobus*. This includes the generation of mutants by standard methods of chemical or UV mutagenesis (Grogan and Gunsalus, 1993; Grogan 1995). Numerous auxotrophic mutants have been obtained by screening mutagenized *S. acidocaldarius* populations by replica plating (Grogan and Gunsalus, 1993; Grogan 1996). These auxotrophic mutants provide a variety of selectable or testable phenotypes for use in genetic crosses between strains. Several toxic compounds have been used to select resistant mutants directly (Grogan, 1991; Kondo *et al.*, 1991; Grogan and Gunsalus, 1993). Thus mutants resistant to ethionine, fluorouracil, 5-fluoroorotic acid, and a variety of antibiotics have been isolated in *Sulfolobus*. These are of special interest due to their potential for use as selectable markers in gene transfer experiments.

The ability to generate auxotrophic mutants has been used in combination with biochemical analysis to elucidate the pathway for uracil biosynthesis in *Sulfolobus* (Grogan and Gunsalus, 1993). By obtaining pyrimidine auxotrophic mutants and determining their metabolic defects by enzymatic assays it was shown that a standard *de novo* pathway for uracil biosynthesis is used by *Sulfolobus*. Further, by selecting mutants resistant to various toxic nucleotide analogues, the presence of at least one salvage pathway was demonstrated. This approach, similar to classic genetic and biochemical studies of *E. coli* metabolism, has rarely been employed in the Archaea. This is especially unfortunate given the power of this technique. Much of our basic knowledge about metabolic pathways in Eukarya and Bacteria is derived from the use of this generally applicable method.

Like other Archaea, *Sulfolobus* species are resistant to most antibiotics. However, certain antibiotics including hygromycin, novobiocin, chloramphenicol, carbomycin, celesticetin, sparsomycin, puromycin, tetracycline and thiostrepton do inhibit growth of *Sulfolobus* (Grogan, 1991; Aagaard *et al.*, 1994; Cannio *et al.*, 1996). Selectable markers for many of these antibiotics are in common use. Unfortunately, all of these resistance genes are derived from mesophilic organisms, and are unlikely to be sufficiently stable at 75°C to be of use, even if they could be modified to allow expression in *Sulfolobus*. A potential selectable marker for use in *Sulfolobus* has been generated by selection for resistance to either celesticetin or chloramphenicol (Aagaard *et al.*, 1994). *Sulfolobus* mutants obtained by both selections carry the same mutational change in their 23S rRNA genes. This mutation confers resistance to carbomycin, celesticetin and chloramphenicol. It is expected to be a dominant selectable marker based on the presence of only a single 23S rDNA gene in *Sulfolobus*. Novobiocin-insensitive DNA gyrase genes have been used as selectable markers in a variety of organisms including some Archaea (the halophilic Archaea as described above). It seems possible that a similar marker might be developed for *Sulfolobus*.

Selection for 5-fluoroorotic acid resistance in the presence of uracil allows for direct selection of mutants that require uracil for growth. Mutants of this type are especially useful because they can be selected either for or against. Thus, in the forward direction, 5-fluoroorotic acid resistance can be selected, and in the reverse direction, growth in the absence of added uracil can be selected. The ability to select mutants in both forward and reverse directions has been heavily exploited for genetic experimentation in *Saccharomyces cerevisiae* using the *ura3* marker (Rothstein, 1991). In *Sulfolobus* 5-fluoroorotic-acid-resistant uracil auxotrophs have been obtained, and result from mutations in either the *pyrE* or *pyrF* genes. However, for use as selectable marker *pyrF* mutants appear to be preferable because *pyrE* mutants have a leaky phenotype (Kondo *et al.*, 1991; Grogan and Gunsalus, 1993).

B. Conjugation

Two types of conjugative gene transfer have been observed in various *Sulfolobus* species, one involving self-transmission of a conjugal plasmid (see Chapter 2, V for discussion of this in bacteria), and the other involving exchange of chromosomal mutations by a mechanism that resembles *Hf. volcanii* conjugation.

A large plasmid of *ca.* 45 kbp, designated pNOB8, was isolated from a *Sulfolobus islandicus* strain originating from solfataric fields in Japan. In the course of testing this isolate for phage production it was found that pNOB8 was rapidly transferred to recipient cells that did not have the plasmid (Schleper *et al.*, 1995a). This transfer requires the presence of donor cells, is not associated with any detectable phage particles and cannot be demonstrated with purified plasmid DNA under the conditions used for mating. Upon mixing a small number of pNOB8 donors with a vast excess of appropriate recipients, the plasmid is rapidly disseminated through the entire population. Microscopic examination of mating mixtures with fluorescently marked donor and recipient cells showed a rapid association between donor and recipient cells forming aggregates of 2 to 30 cells. These aggregates appear to be connected by cytoplasmic bridges. Further, electron microscopy revealed the presence of cellular appendages unlike either pili or flagella in the pNOB8 hosts. Recipients that have received pNOB8 via conjugation are fully competent to act as donors for subsequent transfer of the plasmid, suggesting that the transfer functions are localized to the plasmid. The finding of deletion derivatives of pNOB8 that are incapable of self-transmission supports this suggestion. Thus, transfer of pNOB8 occurs by a conjugation mechanism that is at least superficially similar to plasmid conjugation seen in the Bacteria.

Conjugal transfer of pNOB8 can occur between *S. solfataricus* and *S. islandicus* and between different strains within these species (Schleper *et al.*, 1995a). Plasmid transfer to *S. acidocaldarius*, *S. shibatae* and *Sulfolobus* sp. 59/2 could not be demonstrated. The pNOB8 plasmid has a relatively high copy number of 20 to 40 copies per chromosome. Transfer of pNOB8 is accompanied by a severe reduction in growth rate by recipient cells that has been attributed to the high copy number of the plasmid. The plating efficiency of pNOB8 exconjugates on solid medium is drastically reduced relative to parental strains. However, the few exconjugates obtained by plating show normal growth rates and a much reduced plasmid copy number (*ca.* 5 per genome equivalent). The reason for this lowering of copy number after selection by plating on solid medium is unclear; however, it must be a function of the host because exconjugants derived from low copy donors carry pNOB8 in the same high copy as the original isolate. This plasmid holds promise for the development of cloning vectors and as a gene transfer element as is described below.

Exchange of chromosomal markers via conjugation between various auxotrophic strains of *S. acidocaldarius* has also been demonstrated (Grogan, 1996). This process is apparently distinct from the conjugal transfer of pNOB8 in that no distinct donor–recipient relationship could be established for this method of gene transfer. Gene transfer requires cell–cell contact of living mating partners, and cannot be demonstrated using purified DNA or culture supernatants. Although many of the mechanistic details of this process remain to be elucidated, it appears in many ways to be similar to the conjugal exchange of chromosomal markers in *Hf. volcanii*. To date this form of conjugation represents the only demonstration of transfer of chromosomal markers in a thermophilic Archaeon. Probably because the development of this method is quite recent, it has yet to be widely applied. However, it should be quite useful for strain constructions and testing of genetic linkage groups in this species.

C. Transformation

The plaque-forming phage SSV1 from *S. shibatae* has been used to develop methods for transformation of *S. solfataricus* (Schleper *et al.*, 1992; Schleper and Zillig, 1995). Phage SSV1 is a lysogenic virus that can integrate into the *S. shibatae* genome in single copy, but also exists as a medium copy number plasmid of 15.6 kbp in lysogenic strains. Replication of the prophage is induced by treatment with UV radiation or mitomycin C, allowing the cccDNA form of the virus to be isolated in large amounts by a modified plasmid preparation protocol. UV induction also induces production of infectious particles. These spindle-shaped virions are capable of infecting *S. solfataricus* forming visible plaques in indicator lawns of sensitive cells (Schleper *et al.*, 1992). Plaque formation is the basis of a transfection assay for *Sulfolobus*. In this assay sensitive *Sulfolobus* strains are transformed with SSV1 cccDNA and plated on an indicator lawn of the same strain. Plaque formation indicates successful transformation, with the number of plaques indicating the number of transfection events. A similar method was used to develop transformation of *Hb. salinarium* (Cline and Doolittle, 1987). The most successful method developed to date uses electroporation in high-resistance medium (10% glycerol) to transform the recipient cells (Schleper *et al.*, 1992; Schleper and Zillig, 1995). With electroporation the transformation efficiency of *S. solfataricus* is *ca.* 10^6 transformants per microgram of SSV1 DNA. Transfection with SSV1 has been demonstrated with two strains of *S. solfataricus* and with *S. shibatae*, but could not be demonstrated with *S. acidocaldarius*. Interestingly, transfection of *S. solfataricus* is equally efficient with DNA isolated from either *S. solfataricus* or *S. shibatae*. This suggests that restriction does not occur in this host. In contrast, *S. acidocaldarius* has a restriction endonuclease, *Sau* I, that recognizes GGCC sequences (Prangishuili *et al.*, 1985). This activity may explain the lack of transfection of this host. Transformation of *S. acidocaldarius* has been demonstrated with the naturally occurring *Pyrococcus abysii* plasmid, pGT5, and its derivative pCVS1 (see below). Two methods, electroporation and CaCl$_2$-mediated transformation (for principles in bacteria see Chapter 1, II.D.1 and D.5), have been demonstrated in *S. acidocaldarius* (Aagaard *et al.*, 1996). The authors prefer the latter method. However, the PCR-based method used to screen for transformants (Chapter 2, VI.B.2) is not quantitative, and thus only relative efficiency of transformation can be determined with this method. The CaCl$_2$-mediated transformation method of *S. acidocaldarius* has been optimized in a relative fashion using the pGT5-based shuttle vector pCVS1. In these experiments the DNA was treated with *Hae* III methylase (GGCmC) prior to transformation to alleviate restriction by the *Sau* I endonuclease (Aagaard *et al.*, 1995).

D. Cloning Vectors

Screening of natural *Sulfolobus* isolates has led to the identification of a number of plasmids. Most of these exist in relatively high copy and have no known functions. The ability to transform *Sulfolobus* species has recently been used to allow the development of cloning vectors from several of these plasmids. The plasmids used for cloning vector constructions are the conjugal plasmid pNOB8 from *S. shibatae* (Schleper *et al.*, 1995a), the cryptic plasmid pRN1 from *S. islandicus* (Zillig *et al.*, 1994) and the cryptic plasmid pGT5 from the distantly related Euryarchaote *Pyrococcus abysii* (Erauso *et al.*, 1996). The origin of replication of phage SSV1 has also been used for construction of a cloning vector (Palm *et al.*, 1991).

The ability to promote its own transfer to non-plasmid-bearing strains makes pNOB8 an attractive candidate for cloning vector construction. Because pNOB8 is rapidly spread through the entire population following transformation a selectable marker is not essential with this vector. Also in characterizing the conjugation mechanism of this plasmid it had been noted that deletion derivatives of pNOB8 were commonly observed (Schleper *et al.*, 1995a). This indicates that many sites within the plasmid can be disrupted without interfering with plasmid replication or transfer. These sites represent possible regions for insertion of foreign genes. The thermostable β-galactosidase gene, *lacS*, from *S. solfataricus* was cloned into pNOB8 under the control of a strong constitutive promoter element to provide a marker to monitor transfer of pNOB8 derivatives (Elferink *et al.*, 1996). A stable *S. solfataricus lacS* mutant resulting from insertion of a transposable element was used as the recipient in these experiments (Schleper *et al.*, 1994). When transformed with the pNOB8::*lacS* plasmid the *lacS* mutant is complemented for production of β-galactosidase. In most of these complemented strains both the chromosomal *lacS* mutation (caused by insertion of a *Sulfolobus* transposable element) and the pNOB8::*lacS* could clearly be detected by hybridization. However, in a few instances the β-galactosidase-positive phenotype was the result of marker rescue rather than by complementation. In these cases the pNOB8::*lacS* plasmid was not retained, and the chromosomal mutation was presumably repaired by recombination with the cloned gene. Spreading of the pNOB8::*lacS* plasmid by conjugation was very efficient, with greater than 50% of the population carrying the plasmid after a few days' culture in liquid medium.

These data indicate the potential of pNOB8 as a cloning vector for *Sulfolobus*; however, despite the efficiency of plasmid transfer there are serious drawbacks to use of pNOB8 in this role (Elferink *et al.*, 1996). The most serious problem stems from the severe growth defect imposed upon strains carrying pNOB8 and its derivatives. Strains carrying pNOB8 and pNOB8::*lacS* show a reduction in growth rate of *ca.* 50% and a drastic reduction in plating efficiency. This problem appears to be aggravated by introduction of cloned genes into pNOB8. In the experiments described above isolated colonies of strains carrying the pNOB8::*lacS* plasmid could not be obtained. It is possible that the reduction of copy number observed upon serial plating of pNOB8 can be extended to pNOB8 derivatives (see above). If so this problem may be alleviated. A second potential problem with the use of pNOB8 is it large size, *ca.* 45 kbp. Because large segments of the plasmid have been deleted in naturally occurring derivatives of pNOB8 this drawback too can possibly be overcome.

A similar strategy is currently being explored for development of a cloning vector based on the small (5.5 kbp) cryptic plasmid pRN1 of *S. islandicus*. The complete DNA sequence of pRN1 has been determined to aid in design of potentially useful vectors (Keeling *et al.*, 1996), and the *lacS* gene has been cloned into this plasmid to serve as a marker in transformation experiments. In preliminary experiments reported by Zillig *et al.* (1996) β-galactosidase-positive transformants of the *S. solfataricus lacS* mutant have been obtained after electroporation with this plasmid.

A very promising cloning vector based on the phage SSV1 replicon has been reported by Cannio *et al.* (1996). This plasmid combines a 1.6 kbp fragment with the SSV1 replication region with an *E. coli* pMB1 replicon allowing replication in both *S. solfataricus* and *E. coli*. More importantly, the plasmid carries a thermostable mutant derivative of the *E. coli hph* gene which encodes resistance to hygromycin B. This antibiotic is active against *Sulfolobus* and sufficiently stable at high temperature to allow its use as selective agent. The mutant *hph* gene encoded on the shuttle vector is able to confer resistance upon *Sulfolobus* transformants. Thus, this plasmid represents the first selectable plasmid shuttle vector for a thermophilic Archaeon.

A second shuttle vector currently under development is based on the pGT5 plasmid from the Euryarchaeote *Pyrococcus abyssii* (Erauso *et al.*, 1996). This small plasmid has been combined with the *E. coli* plasmid pUC19 to create a shuttle vector designated pCSV1 (Aagaard *et al.*, 1996). Transformation of this shuttle vector into *P. furiosus* and *S. acidocaldarius* can be demonstrated using a PCR assay to detect the presence of the plasmid in transformed populations. Further, the plasmid was stably maintained for many generations. However, no selectable marker has been introduced to the plasmid, thus it is not yet possible to isolate clones carrying the shuttle vector. The broad host range of this plasmid (replication in both the Crenarchaeota and Euryarchaeota kingdoms) is somewhat surprising and suggests that future constructs based on pGT5 may have broad applications among the thermophilic Archaea.

E. Mobile Intron Constructs

A completely novel gene transfer agent based on the homing endonuclease encoding 23S rRNA intron of *Desulphurococcus mobilis* has been reported for use in *Sulfolobus* (Aagaard *et al.*, 1996). Large subunit rDNA introns that encode a special type of endonuclease have been identified in both Eucarya and Archaea (Lambowitz and Belfort, 1993). These introns are capable of promoting their own transfer to intron-less rRNA genes by the action of the endonucleases in a process that has been described as "homing". In eukaryotes homing generally occurs between the many copies of the rRNA gene within the same cell, although transformation of the intron from *Physarum polycephalum* to *Saccharomyces cerevisiae* can lead to its insertion into yeast rDNA by homing (Muscarella and Vogt, 1993). A similar experiment has demonstrated that the 23S rRNA intron of *Desulphurococcus mobilis* (encoding the homing endonuclease I-*Dmo* I) can promote its transfer to *S. acidocaldarius* rDNA by homing after electroporation of a non-replicating plasmid carrying the intron into this host (Aagaard *et al.*, 1995). The intron appears to provide a selective advantage over intron-less *S. acidocaldarius* strains. Further, the intron undergoes inter-cell transfer to intronless strains by an unknown mechanism (although, the conjugal exchange of host markers described for *S. acidocaldarius* could explain this observation). In combination these traits result in rapid transfer of the intron to most of the recipient population. These features suggest that the mobile intron might be modified to act as a gene transfer agent for *Sulfolobus*.

To test this the *D. mobilis* intron encoding the homing endonuclease I-*Dmo* I and a portion of the flanking exon regions has been cloned into an *E. coli* plasmid and modified by introduction of a cloning site immediately downstream of the I-*Dmo* I coding sequence (Aagaard *et al.*, 1996). The modified intron retains its ability to spread in *S. acidocaldarius* cultures after transformation of this construct by electroporation. DNA fragments as large as 276 bp have been cloned into this construct and shown to spread in the recipient population: however, larger fragments, such as the *lacS* coding sequence failed to be transferred.

◆◆◆◆◆◆ **VI. CONCLUSIONS**

Interest in Archaea has steadily increased since the observation in 1977 that this group represents a "third form of life" distinct from either the Eukarya or the Bacteria. This interest is reflected in the number of researchers investigating members of the domain, in the number of organisms isolated that belong to this domain, and in the number of publications relating to one or another aspect of their unique biochemistry, physiology, metabolism and phylogeny. In the twenty-two years since this discovery the methodologies for study of this unique group have been vastly improved. Methods for genetic analysis of Archaea have not been excluded from these improvements. It seems remarkable that the first report of gene transfer in any member of the domain Archaea, that of conjugation in *Hf. volcanii*, was only about thirteen years ago. The availability of model systems for genetic analysis of Eukarya and of Bacteria strongly influence the types of analyses that are being undertaken in Archaea today, and are in large part responsible for the rapid progress of the field. However, as has been the case with other aspects of archaeal research, it is important to remember that these are not Bacteria, nor are they Eukarya. The unique nature of the Archaea has forced modification of the classical genetic approaches used in the other two domains. Several of the methods described here may have general utility among other Archaea, in addition to the specific uses described here.

The unique S-layer cell walls of many Archaea suggest that generation of protoplasts may have broad applications in the development of transformation methods within this domain. The formation of protoplasts also suggests that liposomes will be generally useful as a method for DNA delivery (see Chapter 1, II.D.3 for discussion of this method in bacteria). Liposomes are expected to be particularly efficient at interaction with exposed lipid membranes of protoplasts. Further, in other systems liposomes have been used to deliver proteins and oligonucleotides in addition to double-stranded DNA. Thus, further applications involving liposomes and protoplasts can be envisioned. The use of plaque-forming phage as an assay method to develop transformation methods may also have broad application. Having such assays is particularly important in the Archaea due to a general lack of useful selectable markers to monitor uptake of DNA. The use of homing endonucleases to promote insertion of foreign DNA into target hosts could find broad application among the Archaea. At least two different endonucleases have been shown to promote transfer to heterologous hosts among the Archaea. The homing sites recognized by these endonucleases are highly conserved among Archaea (Maidak *et al.*, 1997), and the vector need not encode the endonucleases.

Electroporation of the endonuclease protein along with the intron sequence has been shown sufficient to promote homing of intron constructs (Aagaard *et al.*, 1995). Therefore, this method may possibly be applied to all Archaea. Lastly, the finding of plasmids that function in hosts as divergent as *Pyrococcus* and *Sulfolobus* is encouraging because of a lack of plasmid vectors developed for use in this domain. If similar broad host range vectors carrying selectable markers can be developed, they may find use in many Archaea. The finding that the *pac* gene expressed from the *Methanococcus* promoter is functional in widely divergent methanogenic Archaea provides hope that this marker may be used in this regard. Also the development of thermostable mutants of resistance markers for several antibiotics provides the possibility that these may be applied to thermophilic Archaea as well (Cannio *et al.*, 1996; Liao *et al.*, 1986).

It is widely believed that whole genome sequencing will revolutionize the biological sciences. The number of genome sequencing projects and the large commitment of funding to these projects attests to this belief. The Archaea have been well represented among these genome sequencing projects, as is appropriate to their phylogenetic status and their pivotal placement between the eukaryal and the bacterial domains. All indications are that they will continue to be well represented among future genome sequencing endeavours. It is important to remember that sequence data are only predictive. They do not in themselves constitute evidence of the presence of metabolic pathways, modes of gene regulation, physiology, biochemistry, etc., in the organisms in which these sequences are found. Confirmation of these predictions will require direct experimental evidence including growth and physiological studies, biochemical and genetic studies. Further, for those sequences that do not have homologues in the current databases, these methods are the only way to establish roles for many of the sequences obtained by the genome projects. Genetics directly relates sequence data to phenotypes and therefore is the most direct experimental method for verification of sequence-based predictions and assignment of function to new sequences. Use of the genetic method and the development of further genetic techniques are therefore of paramount importance to the study of Archaea. The methods described here provide an excellent start and suggest that study of the Archaea will continue to provide interesting science in the future.

Acknowledgements

I would like to thank Dr Stanley Maloy and Dr Ralph S. Wolfe for their critical reading of the manuscript, and J. K. Zhang for technical assistance.

References

Aagaard, C., Phan, H., Trevisanato, S. and Garrett, R. A. (1994). A spontaneous point mutation in the single 23S rRNA gene of the thermophilic arachaeon *Sulfolobus acidocaldarius* confers multiple drug resistance. *J. Bacteriol.* **176**, 7744–7747.

Aagaard, C., Dalgaard, J. Z. and Garrett, R. A. (1995). Intercellular mobility and

Genetic Analysis in the Domain Archaea

homing of an archaeal rDNA intron confers a selective advantage over intron-cells of *Sulfolobus acidocaldarius*. *Proc. Natl. Acad. Sci. USA* **92**, 12285–12289.

Aagaard, C., Leviev, I., Aravalli, R. N., Forterre, P., Prieur, D. and Garrett, R. A. (1996). General vectors for archaeal hyperthermophiles: strategies based on a mobile intron and a plasmid. *FEMS Microbiol. Rev.* **18**, 93–104.

Adams, M. W. (1993). Enzymes and proteins from organisms that grow near and above 100°C. *Ann. Rev. Microbiol.* **47**, 627–658.

Aguilar, C. F., Sanderson, I., Moracci, M., Ciaramella, M., Nucci, R., Rossi, M. and Pearl, L. H. (1997). Crystal structure of the beta-glycosidase from the hyperthermophilic archeon *Sulfolobus solfataricus*: resilience as a key factor in thermostability. *J. Mol. Biol.* **271**, 789–802.

Ausebel, F. M., Brent, R., Kingston, R. E., Moore, D. D., Seidman, J. G., Smith, J. A. and Struhl, K. (1992). *Current Protocols in Molecular Biology*. John Wiley, New York.

Balch, W. E., Magrum, L. J., Fox, G. E., Wolfe, R. S. and Woese, C. R. (1977). An ancient divergence among the bacteria. *J. Mol. Evol.* **9**, 305–311.

Balch, W. E., Fox, G. E., Magrum, L. J., Woese, C. R. and Wolfe, R. S. (1979). Methanogens: reevaluation of a unique biological group. *Microbiol. Rev.* **43**, 260–296.

Barns, S. M., Delwiche, C. F., Palmer, J. D. and Pace, N. R. (1996). Perspectives on archaeal diversity, thermophily and monophyly from environmental rRNA sequences. *Proc. Natl. Acad. Sci. USA* **93**, 9188–9193.

Baumann, P. and Jackson, S. P. (1996). An archaebacterial homologue of the essential eubacterial cell division protein FtsZ. *Proc. Natl. Acad. Sci. USA* **93**, 6726–6730.

Baumann, P., Qureshi, S. A. and Jackson, S. P. (1995). Transcription: new insights from studies on Archaea. *Trends Genet* **11**, 279–283.

Beneke, S., Bestgen, H. and Klein, A. (1995). Use of the *Escherichia coli uidA* gene as a reporter in *Methanococcus voltae* for the analysis of the regulatory function of the intergenic region between the operons encoding selenium-free hydrogenases. *Mol. Gen. Genet.* **248**, 225–228.

Berghofer, Y. and Klein, A. (1995). Insertional mutations in the hydrogenase *vhc* and *frc* operons encoding selenium-free hydrogenases in *Methanococcus voltae*. *Appl. Environ. Microbiol.* **61**, 1770–1775.

Bertani, G. and Baresi, L. (1987). Genetic transformation in the methanogen *Methanococcus voltae* PS. *J. Bacteriol.* **169**, 2730–2738.

Betlach, M., Friedman, J., Boyer, H. W. and Pfeifer, F. (1984). Characterization of a halobacterial gene affecting bacterio-opsin gene expression. *Nucl. Acids Res.* **12**, 7949–7959.

Bhatnagar, L., Jain, M. K., Zeikus, J. G. and Aubert, J. P. (1986). Isolation of auxotrophic mutants in support of ammonia assimilation via glutamine synthetase in *Methanobacterium ivanovii*. *Arch. Microbiol.* **144**, 350–354.

Blank, C. E., Kessler, P. S. and Leigh, J. A. (1995). Genetics in methanogens: transposon insertion mutagenesis of a *Methanococcus maripaludis nifH* gene. *J. Bacteriol.* **177**, 5773–5777.

Blaseio, U. and Pfeifer, F. (1990). Transformation of *Halobacterium halobium*: development of vectors and investigation of gas vesicle synthesis. *Proc. Natl. Acad. Sci. USA* **87**, 6772–6776.

Bock, A. K. and Kandler, O. (1985). Antibiotic sensitivity of archaebacteria. In *Archaebacteria* (C. R. Woese and R. S. Wolfe, eds), pp. 525–544. Academic Press, New York.

Bock, A. K. and Schonheit, P. (1995). Growth of *Methanosarcina barkeri* (Fusaro) under nonmethanogenic conditions by the fermentation of pyruvate to acetate: ATP synthesis via the mechanism of substrate level phosphorylation. *J. Bacteriol.* **177**, 2002–2007.

Bock, A. K., Prieger-Kraft, A. and Schonheit, P. (1994). Pyruvate – a novel substrate for growth and methane formation in *Methanosarcina barkeri*. *Arch. Microbiol.* **161**, 33–46.

Bokranz, M., Klein, A. and Meile, L. (1990). Complete nucleotide sequence of plasmid pME2001 of *Methanobacterium thermoautotrophicum* (Marburg). *Nucl. Acids Res.* **18**, 363.

Boone, D. R., Whitman, W. B. and Rouviere, P. (1993). Diversity and taxonomy of methanogens. In *Methanogenesis* (J. G. Ferry, ed.), pp. 35–80. Chapman & Hall, New York.

Bouthier de la Tour, C., Portemer, C., Nadal, M., Stetter, K. O., Forterre, P. and Duguet, M. (1990). Reverse gyrase, a hallmark of the hyperthermophilic archaebacteria. *J. Bacteriol.* **172**, 6803–6808.

Bowen, T. L. and Whitman, W. B. (1987). Incorporation of exogenous purines and pyrimidines by *Methanococcus voltae* and isolation of analog-resistant mutants. *Appl. Environ. Microbiol.* **53**, 1822–1826.

Bowen, T. L., Ladapo, J. A. and Whitman, W. B. (1995). Selection for auxotrophic mutants of *Methanococcus* In *Methanogens* (K. R. Sowers and H. J. Schreier, eds), pp. 405–407. Cold Spring Harbor Laboratory Press, Plainview, New York.

Brock, T. D., Brock, K. M., Belly, R. T. and Weiss, R. L. (1972). *Sulfolobus*: a new genus of sulphur-oxidizing bacteria living at low pH and high temperature. *Arch. Mikrobiol.* **84**, 54–68.

Brown, J. W. and Reeve, J. N. (1985). Polyadenylated, noncapped RNA from the archaebacterium *Methanococcus vannielii*. *J. Bacteriol.* **162**, 909–917.

Brown, J. W. and Reeve, J. N. (1986). Polyadenylated RNA isolated from the archaebacterium *Halobacterium halobium*. *J. Bacteriol.* **166**, 686–688.

Brown, J. W., Daniels, C. J. and Reeve, J. N. (1989). Gene structure, organization, and expression in archaebacteria. *Crit. Rev. Microbiol.* **16**, 287–338.

Bult, C. J., White, O., Olsen, G. J., Zhou, L., Fleischmann, R. D., Sutton, G. G., Blake, J. A., FitzGerald, L. M., Clayton, R. A., Gocayne, J. D., Kerlavage, A. R., Dougherty, B. A., Tomb, J. F., Adams, M. D., Reich, C. I., Overbeek, R., Kirkness, E. F., Weinstock, K. G., Merrick, J. M., Glodek, A., Scott, J. L., Geoghagen, N. S. M. and Venter, J. C. (1996). Complete genome sequence of the methanogenic archaeon, *Methanococcus jannaschii*. Science **273**, 1058–1073.

Cannio, R., Contursi, P., Rossi, M. and Bartolucci, S. (1996) An *E. coli/Sulfolobus* shuttle vector carrying the SSV1 viral ARS and a mutant of the hygromycin phosphotransferase gene as a genetic marker. *Thermophiles '96*, Athens, GA, 244.

Chambers, S. P., Prior, S. E., Barstow, D. A. and Minton, N. P. (1988). The pMTL nic-cloning vectors. I. Improved pUC polylinker regions to facilitate the use of sonicated DNA for nucleotide sequencing. *Gene* **68**, 139–149.

Charlebois, R. L. and Doolittle, W. F. (1989). Transposable elements and genome structure in halobacteria. In *Mobile DNA* (D. E. Berg and M. M. Howe, eds), pp. 297–305. American Society for Microbiology, Washington, DC.

Charlebois, R. L., Lam, W. L., Cline, S. W. and Doolittle, W. F. (1987). Characterization of pHV2 from *Halobacterium volcanii* and its use in demonstrating transformation of an archaebacterium. *Proc. Natl. Acad. Sci. USA* **84**, 8530–8504.

Charlebois, R. L., Schalkwyk, L. C., Hofman, J. D. and Doolittle, W. F. (1991). Detailed physical map and set of overlapping clones covering the genome of the archaebacterium *Haloferax volcanii* DS2. *J. Mol. Biol.* **222**, 509–524.

Charlebois, R. L., Gaasterland, T., Ragan, M. A., Doolittle, W. F. and Sensen, C. W. (1996). The Sulfolobus solfataricus P2 genome project. *FEBS Lett.* **389**, 88–91.

Cline, S. W. and Doolittle, W. F. (1987). Efficient transfection of the archaebacterium *Halobacterium halobium*. *J. Bacteriol.* **169**, 1341–1344.

Cline, S. W. and Doolittle, W. F. (1992). Transformation of members of the genus *Haloarcula* with shuttle vectors based on *Halobacterium halobium* and *Haloferax volcanii* plasmid replicons. *J. Bacteriol.* **174**, 1076–1080.

Cline, S. W., F., P. and Doolittle, W. F. (1995). Transformation of halophilic Archaea.

In *Halophiles* (S. DasSarma and E. M. Fleischmann, eds), pp. 197–204. Cold Spring Harbor Laboratory Press, Plainview, New York.

Cline, S. W., Lam, W. L., Charlebois, R. L., Schalkwyk, L. C. and Doolittle, W. F. (1989a). Transformation methods for halophilic archaebacteria. *Can. J. Microbiol.* **35**, 148–152.

Cline, S. W., Schalkwyk, L. C. and Doolittle, W. F. (1989b). Transformation of the archaebacterium *Halobacterium volcanii* with genomic DNA. *J. Bacteriol.* **171**, 4987–4991.

Cohen, A., Lam, W. L., Charlebois, R. L., Doolittle, W. F. and Schalkwyk, L. C. (1992). Localizing genes on the map of the genome of *Haloferax volcanii*, one of the Archaea. *Proc. Natl. Acad. Sci. USA* **89**, 1602–1606.

Cohen-Kupiec, R., Blank, C. and Leigh, J. A. (1997). Transcriptional regulation in Archaea: *in vivo* demonstration of a repressor binding site in a methanogen. *Proc. Natl. Acad. Sci. USA* **94**, 1316–1320.

Conover, R. K. and Doolittle, W. F. (1990). Characterization of a gene involved in histidine biosynthesis in *Halobacterium* (*Haloferax*) *volcanii*: isolation and rapid mapping by transformation of an auxotroph with cosmid DNA. *J. Bacteriol.* **172**, 3244–3249.

Danner, S. and Soppa, J. (1996). Characterization of the distal promoter element of halobacteria *in vivo* using saturation mutagenesis and selection. *Mol. Microbiol.* **19**, 1265–1276.

DasSarma, S. (1989). Mechanisms of genetic variability in *Halobacterium halobium*: the purple membrane and gas vesicle mutations. *Can. J. Microbiol.* **35**, 65–72.

DeLong, E. F., Wu, K. Y., Prezelin, B. B. and Jovine, R. V. (1994). High abundance of Archaea in Antarctic marine picoplankton. *Nature* **371**, 695–697.

de Macario, E. C., Guerrini, M., Dugan, C. B. and Macario, A. J. (1996). Integration of foreign DNA in an intergenic region of the archaeon *Methanosarcina mazei* without effect on transcription of adjacent genes. *J. Mol. Biol.* **262**, 12–20.

DiMarco, A. A., Bobik, T. A. and Wolfe, R. S. (1990). Unusual coenzymes of methanogenesis. *Ann. Rev. Biochem.* **59**, 355–394.

Dyall-Smith, M. L. and Doolittle, W. F. (1994). Construction of composite transposons for halophilic Archaea. *Can. J. Microbiol.* **40**, 922–929.

Eisenberg, H. (1995). Life in unusual environments: progress in understanding the structure and function of enzymes from extreme halophilic environments. *Arch. Biochem. Biophys.* **318**, 1–5.

Elferink, M. G., Schleper, C. and Zillig, W. (1996). Transformation of the extremely thermoacidophilic archaeon *Sulfolobus solfataricus* via a self-spreading vector. *FEMS Microbiol. Lett.* **137**, 31–35.

Erauso, G., Marsin, S., Benbouzid-Rollet, N., Baucher, M. F., Barbeyron, T., Zivanovic, Y., Prieur, D. and Forterre, P. (1996). Sequence of plasmid pGT5 from the archaeon *Pyrococcus abyssi*: evidence for rolling-circle replication in a hyperthermophile. *J. Bacteriol.* **178**, 3232–3237.

Ferrando-May, E., Brustmann, B. and Oesterhelt, D. (1993). A C-terminal truncation results in high-level expression of the functional photoreceptor sensory rhodopsin I in the Archaeon *Halobacterium salinarium. Mol. Microbiol.* **9**, 943–953.

Fischer, F., Zillig, W., Stetter, K. O. and Schreiber, G. (1983). Chemolithoautotrophic metabolism of anaerobic extremely thermophilic archaebacteria. *Nature* **301**, 511–513.

Fox, G. F., Magrum, L. J., Balch, W. E., Wolfe, R. S. and Woese, C. R. (1977). Classification of methanogenic bacteria by 16S ribosomal RNA characterization. *Proc. Natl. Acad. Sci. USA* **74**, 4537–4541.

Gambacorta, A., Trincone, A., Nicolaus, B., Lama, L. and De Rosa, M. (1994). Unique features of lipids of Archaea. *System. Appl. Microbiol.* **16**, 518–527.

Gernhardt, P., Possot, O., Foglino, M., Sibold, L. and Klein, A. (1990). Construction of an integration vector for use in the archaebacterium *Methanococcus voltae* and expression of a eubacterial resistance gene. *Mol. Gen. Genet.* **221**, 273–279.

Grogan, D. W. (1989). Phenotypic characterization of the archaebacterial genus *Sulfolobus*: comparison of five wild-type strains. *J. Bacteriol.* **171**, 6710–6719.

Grogan, D. W. (1991). Selectable mutant phenotypes of the extremely thermophilic archaebacterium *Sulfolobus* acidocaldarius. *J. Bacteriol.* **173**, 7725–7727.

Grogan, D. W. (1995). Isolation of *Sulfolobus acidocaldarius* mutants. In *Thermophiles* (F. T. Robb and A. R. Place, eds), pp. 125–131. Cold Spring Harbor Laboratory Press, Plainview, New York.

Grogan, D. W. (1996). Exchange of genetic markers at extremely high temperatures in the archaeon *Sulfolobus acidocaldarius*. *J. Bacteriol.* **178**, 3207–3211.

Grogan, D. W. and Gunsalus, R. P. (1993). *Sulfolobus acidocaldarius* synthesizes UMP via a standard de novo pathway: results of biochemical-genetic study. *J. Bacteriol.* **175**, 1500–1507.

Gropp, F., Gropp, R. and Betlach, M. C. (1995). Effects of upstream deletions on light- and oxygen-regulated bacterio-opsin gene expression in *Halobacterium halobium*. *Mol. Microbiol.* **16**, 357–364.

Halladay, J. T., Jones, J. G., Lin, F., MacDonald, A. B. and DasSarma, S. (1993). The rightward gas vesicle operon in *Halobacterium* plasmid pNRC100: identification of the *gvpA* and *gvpC* gene products by use of antibody probes and genetic analysis of the region downstream of *gvpC*. *J. Bacteriol.* **175**, 684–692.

Halladay, J. T., Ng, W. L. and DasSarma, S. (1992). Genetic transformation of a halophilic archaebacterium with a gas vesicle gene cluster restores its ability to float. *Gene* **119**, 131–136.

Hartl, T., Grossebuter, W., Gorisch, H. and Stezowski, J. J. (1987). Crystalline NAD/NADP-dependent malate dehydrogenase; the enzyme from the thermoacidophilic archaebacterium *Sulfolobus acidocaldarius*. *Biol. Chem. Hoppe Seyler* **368**, 259–267.

Hartmann, R., Sickinger, H. D. and Oesterhelt, D. (1980). Anaerobic growth of halobacteria. *Proc. Natl. Acad. Sci. USA* **77**, 3821–3825.

Hennigan, A. N. and Reeve, J. N. (1994). mRNAs in the methanogenic archaeon *Methanococcus vannelii*: numbers, half-lives and processing. *Mol. Microbiol.* **11**, 655–670.

Henry, S. A., Donahue, T. F. and Culbertson, M. R. (1975). Selection of spontaneous mutants by inositol starvation in yeast. *Mol. Gen. Genet.* **143**, 5–11.

Hershberger, K. L., Barns, S. M., Reysenbach, A. L., Dawson, S. C. and Pace, N. R. (1996). Wide diversity of Crenarchaeota. *Nature* **384**, 420.

Holmes, M. L. and Dyall-Smith, M. L. (1990). A plasmid vector with a selectable marker for halophilic archaebacteria. *J. Bacteriol.* **172**, 756–761.

Holmes, M. L., Nuttall, S. D. and Dyall-Smith, M. L. (1991). Construction and use of halobacterial shuttle vectors and further studies on *Haloferax* DNA gyrase. *J. Bacteriol.* **173**, 3807–3813.

Holmes, M., Pfeifer, F. and Dyall-Smith, M. (1994). Improved shuttle vectors for *Haloferax volcanii* including a dual-resistance plasmid. *Gene* **146**, 117–21.

Huber, G. and Stetter, K. O. (1991). *Sulfolobus metallicus*, sp. nov., a novel strictly chemolithoautotrophic thermophilic Archaeal species of metal-mobilizers. *Sytem. Appl. Microbiol.* **14**, 372–378.

Hudepohl, U., Gropp, F., Horne, M. and Zillig, W. (1991). Heterologous *in vitro* transcription from two archaebacterial promoters. *FEBS Lett.* **285**, 257–259.

Hummel, H. and Bock, A. (1985). Mutations in *Methanobacterium formicicum* conferring resistance to anti-80S ribosome-targeted antibiotics. *Mol. Gen. Genet.* **198**, 529–533.

Hungate, R. E. (1969). A roll tube method for the cultivation of strict anaerobes. In *Methods in Microbiolgy* (J. R. Noriss and D. W. Ribbons, eds), pp. 117–132. Academic Press, New York.

Jain, M. K. and Zeikus, J. G. (1987). Methods for Isolation of auxotrophic mutants of *Methanobacterium ivanovii* and initial characterization of acetate auxotrophs. *Appl. Environ. Microbiol.* **53**, 1387–1390.

Jarrell, K. F., Bayley, D. P., Florian, V. and Klein, A. (1996). Isolation and characterization of insertional mutations in flagellin genes in the archaeon *Methanococcus voltae*. *Mol. Microbiol.* **20**, 657–666.

Jefferson, R. A., Burgess, S. M. and Hirsh, D. (1986). beta-Glucuronidase from *Escherichia coli* as a gene-fusion marker. *Proc. Natl. Acad. Sci. USA* **83**, 8447–8451.

Jenal, U., Rechsteiner, T., Tan, P. Y., Buhlmann, E., Meile, L. and Leisinger, T. (1991). Isoleucyl-tRNA synthetase of *Methanobacterium thermoautotrophicum* Marburg. Cloning of the gene, nucleotide sequence, and localization of a base change conferring resistance to pseudomonic acid. *J. Biol. Chem.* **266**, 10570–10577.

Jolley, K. A., Rapaport, E., Hough, D. W., Danson, M. J., Woods, W. G. and Dyall-Smith, M. L. (1996). Dihydrolipoamide dehydrogenase from the halophilic archaeon *Haloferax volcanii*: homologous overexpression of the cloned gene. *J. Bacteriol.* **178**, 3044–3048.

Jones, W. J., Leigh, J. A., Mayer, F., Woese, C. R. and Wolfe, R. S. (1983a). *Methanococcus jannaschii* sp. nov., an extremely thermophilic methanogen from a submarine hydrothermal vent. *Arch. Microbiol.* **136**, 254–261.

Jones, W. J., Whitman, W. B., Fields, R. D. and Wolfe, R. S. (1983b). Growth and plating efficiency of methanococci on agar media. *Appl. Environ. Microbiol.* **46**, 220–226.

Jurgens, G., Lindstrom, K. and Saano, A. (1997). Novel group within the kingdom Crenarchaeota from boreal forest soil. *Appl. Environ. Microbiol.* **63**, 803–805.

Kamekura, M., Dyall-Smith, M. L., Upasani, V., Ventosa, A. and Kates, M. (1997). Diversity of alkaliphilic halobacteria: proposals for transfer of *Natronobacterium vacuolatum*, *Natronobacterium magadii*, and *Natronobacterium pharaonis* to *Halorubrum*, *Natrialba*, and *Natronomonas* gen. nov., respectively, as *Halorubrum vacuolatum* comb. nov., *Natrialba magadii* comb. nov. and *Natronomonas pharaonis* comb. nov., respectively. *Int. J. Syst. Bacteriol.* **47**, 853–857.

Kandler, O. and Konig, H. (1985). Cell envelopes of archaebacteria. In *Archaebacteria* (C. R. Woese and R. S. Wolfe, eds), pp. 525–544. Academic Press, New York.

Keeling, P. J. and Doolittle, W. F. (1995). Archaea: narrowing the gap between prokaryotes and eukaryotes. *Proc. Natl. Acad. Sci. USA* **92**, 5761–5764.

Keeling, P. J., Charlebois, R. L. and Doolittle, W. F. (1994). Archaebacterial genomes: eubacterial form and eukaryotic content. *Curr. Opin. Genet. Dev.* **4**, 816–822.

Keeling, P. J., Klenk, H. P., Singh, R. K., Feeley, O., Schleper, C., Zillig, W., Doolittle, W. F. and Sensen, C. W. (1996). Complete nucleotide sequence of the *Sulfolobus islandicus* multicopy plasmid pRN1. *Plasmid* **35**, 141–144.

Khorana, H. G. (1988). Bacteriorhodopsin, a membrane protein that uses light to translocate protons. *J. Biol. Chem.* **263**, 7439–7442.

Kiener, A. and Leisinger, T. (1983). Oxygen sensitivity of methanogenic bacteria. *System. Appl. Microbiol.* **4**, 305–312.

Kiener, A., Holliger, C. and Leisinger, T. (1984). Analogue-resistant and auxotrophic mutants of *Methanobacterium thermoautotrophicum*. *Arch. Microbiol.* **139**, 87–90.

Kikuchi, A. and Asai, K. (1984). Reverse gyrase – a topoisomerase which introduces positive superhelical turns into DNA. *Nature* **309**, 677–681.

Kim, W. and Whitman, W. B. (1997) Random insertional mutagenesis of *Methanococcus maripaludis*. 97th General Meeting of the American Society for Microbiology, Miami Beach, Florida, Abstract I–42.

Klein, A. and Horner, K. (1995). Integration vectors for methanococci. In *Methanogens* (K. R. Sowers and H. J. Schreier, eds), pp. 409–411. Cold Spring Harbor Laboratory Press, Plainview, New York.

Knox, M. R. and Harris, J. E. (1988). Isolation and characterization of mutants of mesophilic methanogenic bacteria resistant to analogues of DNA bases and nucleosides. *Arch. Microbiol.* **149**, 557–560.

Kondo, S., Yamagishi, A. and Oshima, T. (1991). Positive selection for uracil auxotrophs of the sulphur-dependent thermophilic archaebacterium *Sulfolobus acidocaldarius* by use of 5-fluoroorotic acid. *J. Bacteriol.* **173**, 7698–7700.

Krebs, M. (1995). Gene replacement in *Halobacterium halobium*. In *Halophiles* (S. DasSarma and E. M. Fleischmann, eds), pp. 205–208. Cold Spring Harbor Laboratory Press, Plainview, New York.

Krebs, M. P. and Khorana, H. G. (1993). Mechanism of light-dependent proton translocation by bacteriorhodopsin. *J. Bacteriol.* **175**, 1555–1560.

Krebs, M. P., Hauss, T., Heyn, M. P., RajBhandary, U. L. and Khorana, H. G. (1991). Expression of the bacterioopsin gene in *Halobacterium halobium* using a multicopy plasmid. *Proc. Natl. Acad. Sci. USA* **88**, 859–863.

Krebs, M. P., Mollaaghababa, R. and Khorana, H. G. (1993a). Gene replacement in *Halobacterium halobium* and expression of bacteriorhodopsin mutants. *Proc. Natl. Acad. Sci. USA* **90**, 1987–1991.

Krebs, M. P., Spudich, E. N., Khorana, H. G. and Spudich, J. L. (1993b). Synthesis of a gene for sensory rhodopsin I and its functional expression in *Halobacterium halobium*. *Proc. Natl. Acad. Sci. USA* **90**, 3486–3490.

Kreisl, P. and Kandler, O. (1986). Chemical structure of the cell wall polymer of *Methanosarcina*. *System. Appl. Microbiol.* **7**, 293–299.

Kushner, D. J. (1985). The Halobacteriaceae. In *Archaebacteria* (C. R. Woese and R. S. Wolfe, eds), pp. 171–214. Academic Press, Inc., New York.

Ladapo, J. and Whitman, W. B. (1990). Method for isolation of auxotrophs in the methanogenic archaebacteria: Role of the acetyl-CoA pathway of autotrophic CO^2 fixation in *Methanococcus maripaludis*. *Proc. Natl. Acad. Sci. USA* **87**, 5598–5602.

Lam, W. L. and Doolittle, W. F. (1989). Shuttle vectors for the archaebacterium *Halobacterium volcanii*. *Proc. Natl. Acad. Sci. USA* **86**, 5478–5482.

Lambowitz, A. M. and Belfort, M. (1993). Introns as mobile genetic elements. *Ann. Rev. Biochem* **62**, 587–622.

Langer, D., Hain, J., Thuriaux, P. and Zillig, W. (1995). Transcription in archaea: similarity to that in eucarya. *Proc. Natl. Acad. Sci. USA* **92**, 5768–5772.

Langworthy, T. A. (1985). Lipids of archaebacteria. In *Archaebacteria* (C. R. Woese and R. S. Wolfe, eds), pp. 459–497. Academic Press, New York.

Leigh, J. A. and Wolfe, R. S. (1983). Carbon dioxide reduction factor and methanopterin, two coenzymes required for CO_2 reduction to methane by extracts of *Methanobacterium*. *J. Biol. Chem.* **258**, 7536–7540.

Liao, H., McKenzie, T. and Hageman, R. (1986). Isolation of a thermostable enzyme variant by cloning and selection in a thermophile. *Proc. Natl. Acad. Sci. USA* **83**, 576–580.

Lindstrom, E. B. and Sehlin, H. M. (1989). High effiency plating of the thermophilic sulphur-dependent Archaebacterium *Sulfolobus acidocaldarius*. *Appl. Environ. Microbiol.* **55**, 3020–3021.

Lunnen, K. D., Morgan, R. D., Timan, C. J., Krzycki, J. A., Reeve, J. N. and Wilson, G. G. (1989). Characterization and cloning of MwoI (GCN7GC), a new type-II restriction-modification system from *Methanobacterium wolfei*. *Gene* **77**, 11–19.

MacGregor, B. J., Moser, D. P., Alm, E. W., Nealson, K. H. and Stahl, D. A. (1997). Crenarchaeota in Lake Michigan sediment. *Appl. Environ. Microbiol.* **63**, 1178–1181.

Maidak, B. L., Olsen, G. J., Larsen, N., Overbeek, R., McCaughey, M. J. and Woese, C. R. (1997). The RDP (Ribosomal Database Project). *Nucl. Acids Res.* **25**, 109–111.

Mankin, A. S. and Garrett, R. A. (1991). Chloramphenicol resistance mutations in the single 23S rRNA gene of the archaeon *Halobacterium halobium*. *J. Bacteriol.* **173**, 3559–3563.

Mankin, A. S., Zyrianova, I. M., Kagramanova, V. K. and Garrett, R. A. (1992). Introducing mutations into the single-copy chromosomal 23S rRNA gene of the archaeon *Halobacterium halobium* by using an rRNA operon-based transformation system. *Proc. Natl. Acad. Sci. USA* **89**, 6535–6539.

Margolin, W., Wang, R. and Kumar, M. (1996). Isolation of an ftsZ homolog from the archaebacterium *Halobacterium salinarium*: implications for the evolution of FtsZ and tubulin. *J. Bacteriol.* **178**, 1320–1327.

Meile, L., Kiener, A. and Leisinger, T. (1983). A plasmid in the archaebacterium *Methanobacterium thermoautotrophicum*. *Mol. Gen. Genet.* **191**, 480–484.

Meile, L., Jenal, U., Studer, D., Jordan, M. and Leisinger, T. (1989). Characterization of ΨM1, a virulent phage of *Methanobacterium thermoautotrophicum* Marburg. *Arch. Microbiol.* **152**, 105–110.

Meile, L., Abendschein, P. and Leisinger, T. (1990). Transduction in the archaebacterium *Methanobacterium thermoautotrophicum* Marburg. *J. Bacteriol.* **172**, 3507–3508.

Meile, L., Stettler, R. and Leisinger, T. (1995). Transduction of *Methanobacterium thermoautotrophicum* Marburg. In *Methanogens* (K. R. Sowers and H. J. Schreier, eds), pp. 425–427. Cold Spring Harbor Laboratory Press, Plainview, New York.

Metcalf, W. W., Zhang, J. K., Apolinario, E., Sowers, K. R. and Wolfe, R. S. (1997). A genetic system for Archaea of the genus *Methanosarcina*: liposome-mediated transformation and construction of shuttle vectors. *Proc. Natl. Acad. Sci. USA* **94**, 2626–2631.

Mevarech, M. and Werczberger, R. (1985). Genetic transfer in *Halobacterium volcanii*. *J. Bacteriol.* **162**, 461–462.

Micheletti, P. A., Sment, K. A. and Konisky, J. (1991). Isolation of a coenzyme M-auxotrophic mutant and transformation by electroporation in *Methanococcus voltae*. *J. Bacteriol.* **173**, 3414–3418.

Miller, T. L. and Wolin, M. J. (1974). A serum bottle modification of the Hungate technique for cultivating obligate anaerobes. *Appl. Microbiol.* **27**, 985–987.

Mukhopadhyay, B. (1993). Coenzyme F420-dependent methylene-H4MPT dehydrogenases of *Methanobacterium thermoautotrophicum* strain Marburg and *Methanosarcina barkeri* and effects of methanogenic substrates on the levels of three catabolic enzymes in *Methanosarcina barkeri*, PhD thesis, University of Iowa, Iowa City, Iowa.

Muller, V., Blaut, M. and Gottschalk, G. (1993). Bioenergetics of methanogenesis. In *Methanogenesis* (J. G. Ferry, ed.), pp. 360–406. Chapman & Hall, New York, New York.

Physarum polycephalum can insert itself and induce point mutations in the nuclear ribosomal DNA of *Saccharomyces cerevisiae*. *Mol. Cell. Biol.* **13**, 1023–1033.

Nagle, D. P., Teal, R. and Eisenbraun, A. (1987). 5-Fluorouracil-resistant strain of *Methanobacterium thermoautotrophicum*. *J. Bacteriol.* **169**, 4119–4123.

Ng, W. L. and DasSarma, S. (1993). Minimal replication origin of the 200-kilobase *Halobacterium* plasmid pNRC100. *J. Bacteriol.* **175**, 4584–4596.

Ni, B. F., Chang, M., Duschl, A., Lanyi, J. and Needleman, R. (1990). An efficient system for the synthesis of bacteriorhodopsin in *Halobacterium halobium*. *Gene* **90**, 169–172.

Nieuwlandt, D. T. and Daniels, C. J. (1990). An expression vector for the archaebacterium *Haloferax volcanii*. *J. Bacteriol.* **172**, 7104–7110.

Nolling, J. and Vos, W. M. D. (1992). Characterization of the archaeal, plasmid-encoded type II restriction-modification system *Mth* TI from *Methanobacterium thermoformicicum* THF: Homology to the bacterial *Ngo*PII system from *Neisseria gonorrhoeae*. *J. Bacteriol.* **174**, 5719–5726.

Nolling, J., Frijlink, M. and Vos, W. M. D. (1991). Isolation and characterization of plasmids from different strains of *Methanobacterium thermoformicicum*. *J. Gen. Microbiol.* **137**, 1981–1986.

Nolling, J., Eeden, F. J. M. v., Eggen, R. I. L. and Vos, W. M. d. (1992). Modular organization of related Archaeal plasmids encoding different restriction-modification systems in *Methanobacterium thermoformicicum*. *Nucl. Acids Res.* **20**, 6501–6507.

Offner, S. and Pfeifer, F. (1995). Complementation studies with the gas vesicle-encoding p-vac region of *Halobacterium salinarium* PHH1 reveal a regulatory role for the p-*gvpDE* genes. *Mol. Microbiol.* **16**, 9–19.

Offner, S., Wanner, G. and Pfeifer, F. (1996). Functional studies of the *gvpACNO* operon of *Halobacterium salinarium* reveal that the GvpC protein shapes gas vesicles. *J. Bacteriol.* **178**, 2071–2078.

Olsen, G. J. (1994). Archaea, Archaea, everywhere. *Nature* **371**, 657–658.

Olsen, G. J. and Woese, C. R. (1996). Lessons from an Archaeal genome: what are we learning from *Methanococcus jannaschii*? *Trends Genet.* **12**, 377–379.

Olsen, G. J., Pace, N. R., Nuell, M., Kaine, B. P., Gupta, R. and Woese, C. R. (1985). Sequence of the 16S rRNA gene from the thermoacidophilic archaebacterium *Sulfolobus solfataricus* and its evolutionary implications. *J. Mol. Evol.* **22**, 301–307.

Orr-Weaver, T. L., Szostak, J. W. and Rothstein, R. J. (1981). Yeast transformation: a model system for the study of recombination. *Proc. Natl. Acad. Sci. USA* **78**, 6354–6358.

Palm, P., Schleper, C., Grampp, B., Yeats, S., McWilliam, P., Reiter, W. D. and Zillig, W. (1991). Complete nucleotide sequence of the virus SSV1 of the archaebacterium *Sulfolobus shibatae*. *Virology* **185**, 242–250.

Patel, G. B., Choquet, C. G., Nash, J. H. E. and Sprott, G. D. (1993). Formation and regeneration of *Methanococcus voltae* protoplasts. *Appl. Environ. Microbiol.* **59**, 27–33.

Patel, G. B., Nash, J. H. E., Agnew, B. J. and Sprott, G. D. (1994). Natural and electroporation-mediated transformation of *Methanococcus voltae* protoplasts. *Appl. Environ. Microbiol.* **60**, 903–907.

Pearl, L. H., Demasi, D., Hemmings, A. M., Sica, F., Mazzarella, L., Raia, C. A., D'Auria, S. and Rossi, M. (1993). Crystallization and preliminary X-ray analysis of an NAD(+)-dependent alcohol dehydrogenase from the extreme thermophilic archaebacterium *Sulfolobus solfataricus*. *J. Mol. Biol.* **229**, 782–784.

Pfeifer, F. and Blaseio, U. (1989). Insertion elements and deletion formation in a halophilic archaebacterium. *J. Bacteriol.* **171**, 5135–5140.

Pfeifer, F. and Ghahraman, P. (1993). Plasmid pHH1 of *Halobacterium salinarium*: characterization of the replicon region, the gas vesicle gene cluster and insertion elements. *Mol. Gen. Genet.* **238**, 193–200.

Pfeifer, F., Weidinger, G. and Goebel, W. (1981). Genetic variability in *Halobacterium halobium*. *J. Bacteriol.* **145**, 375–381.

Pfeifer, F., Griffig, J. and Oesterhelt, D. (1993). The fdx gene encoding the [2Fe–2S] ferredoxin of *Halobacterium salinarium* (*H. halobium*). *Mol. Gen. Genet.* **239**, 66–71.

Possot, O., Gernhardt, P., Klein, A. and Sibold, L. (1988). Analysis of drug resistance in the archaebacterium *Methanococcus voltae* with respect to potential use in genetic engineering. *Appl. Environ. Microbiol.* **54**, 734–740.

Prangishvili, D. A., Vashakidze, R. P., Chelidze, M. G. and Gabriadze, I. (1985). A restriction endonuclease *Sua*I from the thermoacidophilic archaebacterium *Sulfolobus acidocaldarius*. *FEBS Lett.* **192**, 57–60.

Rajagopal, B. S. and LeGall, J. (1994). Pyruvate as a substrate for growth and methanogenesis for *Methanosarcina barkeri*. *Curr. Microbiol.* **28**, 307–311.

Rakhely, G. and Kovacs, K. L. (1996). Plating hyperthermophilic archea on solid surface. *Anal. Biochem.* **243**, 181–183.

Rechsteiner, T., Kiener, A. and Leisinger, T. (1986). Mutants of *Methanobacterium thermoautotrophicum*. *System. Appl. Microbiol.* **7**, 1–4.

Reeve, J. N. (1992). Molecular biology of methanogens. *Ann. Rev. Microbiol.* **46**, 165–191.

Rodriguez-Valera, F. (1995). Cultivation of halophilic Archaea. In *Halophiles* (S. DasSarma and E. M. Fleischmann, eds), pp. 13–16. Cold Spring Harbor Laboratory Press, Plainview, New York.

Rosenshine, I., Tchelet, R. and Mevarech, M. (1989). The mechanism of DNA transfer in the mating system of an archaebacterium. *Science* **245**, 1387–1389.

Rothstein, R. (1991). Targeting, disruption, replacement, and allele rescue: integrative DNA transformation in yeast. *Meth. Enzymol.* **194**, 281–301.

Ruepp, A. and Soppa, J. (1996). Fermentative arginine degradation in *Halobacterium salinarium* (formerly *Halobacterium halobium*): genes, gene products, and transcripts of the *arcRACB* gene cluster. *J. Bacteriol.* **178**, 4942–4947.

Sandbeck, K. A. and Leigh, J. A. (1991). Recovery of an integration shuttle vector from tandem repeats in *Methanococcus maripaludis*. *Appl. Environ. Microbiol.* **57**, 2762–2763.

Sandler, S. J., Satin, L. H., Samra, H. S. and Clark, A. J. (1996). recA-like genes from three archaean species with putative protein products similar to Rad51 and Dmc1 proteins of the yeast *Saccharomyces cerevisiae*. *Nucl. Acids Res.* **24**, 2125–2132.

Santoro, N. and Konisky, J. (1987). Characterization of bromoethanesulfonate-resistant mutants of *Methanococcus voltae*: evidence of a coenzyme M transport system. *J. Bacteriol.* **169**, 660–665.

Sapienza, C., Rose, M. R. and Doolittle, W. F. (1982). High-frequency genomic rearrangements involving archaebacterial repeat sequence elements. *Nature* **299**, 182–185.

Schinzel, R. and Burger, K. J. (1986). A site-specific endonuclease activity in *Halobacterium halobium*. *FEMS Microbiol. Lett.* **37**, 325–329.

Schleper, C. and Zillig, W. (1995). Transfection of *Sulfolobus solfataricus*. In *Thermophiles* (F. T. Robb and A. R. Place, eds), pp. 91–93. Cold Spring Harbor Laboratory Press, Plainview, New York.

Schleper, C., Kubo, K. and Zillig, W. (1992). The particle SSV1 from the extremely thermophilic archaeon *Sulfolobus* is a virus: demonstration of infectivity and of transfection with viral DNA. *Proc. Natl. Acad. Sci. USA* **89**, 7645–7649.

Schleper, C., Roder, R., Singer, T. and Zillig, W. (1994). An insertion element of the extremely thermophilic archaeon *Sulfolobus solfataricus* transposes into the endogenous beta-galactosidase gene. *Mol. Gen. Genet.* **243**, 91–96.

Schleper, C., Holz, I., Janekovic, D., Murphy, J. and Zillig, W. (1995a). A multicopy plasmid of the extremely thermophilic archaeon *Sulfolobus* effects its transfer to recipients by mating. *J. Bacteriol.* **177**, 4417–4426.

Schleper, C., Puehler, G., Holz, I., Gambacorta, A., Janekovic, D., Santarius, U., Klenk, H. P. and Zillig, W. (1995b). *Picrophilus* gen. nov., fam. nov.: a novel aerobic, heterotrophic, thermoacidophilic genus and family comprising archaea capable of growth around pH 0. *J. Bacteriol.* **177**, 7050–7059.

Schmid, K., Thomm, M., Laminet, A., Laue, F. G., Kessler, C., Stetter, K. O. and Schmitt, R. (1984). Three new restriction endonucleases *MaeI*, *MaeII* and *MaeIII* from *Methanococcus aeolicus*. *Nucl. Acids Res.* **12**, 2619–2628.

Schnellen, C. G. T. P. (1947). Onderzoekingen over der methaangisting, PhD thesis, Technical University, Delft, The Netherlands.

Segerer, A. H. and Stetter, K. O. (1991). The Order Sulfolobales. In *The Prokaryotes* (A. Balows, H. G. Truper, M. Dworkin, W. Harder, and K.-H. Schleifer, eds), pp. 684–701. Springer-Verlag, New York.

Segerer, A., Stetter, K. O. and Klink, F. (1985). Two contrary modes of chemolithotrophy in the same archaebacterium. *Nature* 313, 787–789.

Seifert, H. S., Chen, E. Y., So, M. and Heffron, F. (1986). Shuttle mutagenesis: a method of transposon mutagenesis for *Saccharomyces cerevisiae*. *Proc. Natl. Acad. Sci. USA* 83, 735–739.

Shand, R. F. and Betlach, M. C. (1991). Expression of the *bop* gene cluster of *Halobacterium halobium* is induced by low oxygen tension and by light. *J. Bacteriol.* 173, 4692–4699.

Smith, D. R., Doucette-Stamm, L. A., Deloughery, C., Lee, H., Dubois, J., Aldredge, T., Bashirzadeh, R., Blakely, D., Cook, R., Gilbert, K., Harrison, D., Hoang, L., Keagle, P., Lumm, W., Pothier, B., Qiu, D., Spadafora, R., Vicaire, R., Wang, Y., Wierzbowski, J., Gibson, R., Jiwani, N., Prabhakar, S., McDougall, S., Shimer, G., Goyal, A., Pietrokovski, S., Church, G., Daniels, C. J., Mao, J.-I., Rice, P., Nolling, J. and Reeve, J. N. (1997). Complete genome sequence of *Methanobacterium thermoautotrophicum* ΔH: Functional analysis and comparative genomics. *J. Bacteriol.* 179, 7135–7155.

Smith, M. R. and Lequerica, J. L. (1985). *Methanosarcina* mutant unable to produce methane or assimilate carbon from acetate. *J. Bacteriol.* 164, 618–625.

Smith, M. R. and Mah, R. A. (1981). 2-Bromoethanesulfonate: a selective agent for isolating resistant *Methanosarcina* mutants. *Curr. Microbiol.* 6, 321–326.

Sowers, K. R. and Gunsalus, R. P. (1988). Plasmid DNA from the acetotrophic methanogen *Methanosarcina acetivorans*. *J. Bacteriol.* 170, 4979–4982.

Sowers, K. R. and Schreier, H. J. (1995). *Methanogens*. Cold Spring Harbor Laboratory Press, Plainview, New York.

Sowers, K. R., Boone, J. and Gunsalus, R. P. (1993). Disaggregation and growth of *Methanosarcina* spp. as single cells. *Appl. Environ. Microbiol.* 59, 3832–3839.

St Jean, A., Trieselmann, B. A. and Charlebois, R. L. (1994). Physical map and set of overlapping cosmid clones representing the genome of the archaeon *Halobacterium* sp. GRB. *Nucl. Acids Res.* 22, 1476–1483.

Stalker, D. M., Kolter, R. and Helinski, D. R. (1979). Nucleotide sequence of the region of an origin of replication of the antibiotic resistance plasmid R6K. *Proc. Natl. Acad. Sci. USA* 76, 1150–1154.

Stetter, K. O. (1996a). Hyperthermophiles in the history of life. *Ciba Found. Symp.* 202, 1–10; discussion 11–18.

Stetter, K. O. (1996b). Hyperthermophilic prokaryotes. *FEMS Microbiol. Rev.* 18, 149–158.

Tanner, R. S., McInerney, M. J. and Nagle, D. P., Jr. (1989). Formate auxotroph of *Methanobacterium thermoautotrophicum* Marburg. *J. Bacteriol.* 171, 6534–6538.

Tchelet, R. and Mevarech, M. (1994). Interspecies genetic transfer in halophilic archaebacteria. *System. Appl. Microbiol.* 16, 578–581.

Thaur, R. K., Hedderich, R. and Fischer, R. (1993). Reactions and enzymes involved in methanogenesis from CO_2 and H_2. In *Methanogenesis* (J. G. Ferry, ed.), pp. 209–252. Chapman & Hall, New York.

Thomm, M. (1996). Archaeal transcription factors and their role in transcription initiation. *FEMS Microbiol. Rev.* 18, 159–171.

Thomm, M., Frey, G., Bolton, B. J., Laue, F., Kessler, C. and Stetter, K. O. (1988). A restriction enzyme in the archaebacterium *Methanococcus vanielii*. *FEMS Microbiol. Lett.* 52, 229–234.

Tindall, B. J. (1991). The Family *Halobacteriaceae*. In *The Prokaryotes* (A. Balows, H.

G. Truper, M. Dworkin, W. Harder, and K.-H. Schleifer, eds), pp. 768–808. Springer-Verlag, New York.

Tumbula, D. L., Makula, R. A. and Whitman, W. B. (1994). Transformation of *Methanococcus maripaludis* and identification of a Pst I-like restriction system. *FEMS Microbiol. Lett.* **121**, 309–314.

Tumbula, D. L., Bowen, T. L. and Whitman, W. B. (1997). Characterization of pURB500 from the archaeon *Methanococcus maripaludis* and construction of a shuttle vector. *J. Bacteriol.* **179**, 2976–2986.

Wang, X. and Lutkenhaus, J. (1996). FtsZ ring: the eubacterial division apparatus conserved in archaebacteria. *Mol. Microbiol.* **21**, 313–319.

Weidinger, G., Klotz, G. and Goebel, W. (1979). A large plasmid from *Halobacterium halobium* carrying genetic information for gas vacuole formation. *Plasmid* **2**, 377–386.

Woese, C. R. (1977). A comment on methanogenic bacteria and the primitive ecology. *J. Mol. Evol.* **9**, 369–371.

Woese, C. R. (1987). Bacterial evolution. *Microbiol. Rev.* **51**, 221–271.

Woese, C. R., Magrum, L. J. and Fox, G. E. (1978). Archaebacteria. *J. Mol. Evol.* **11**, 245–251.

Woese, C. R., Kandler, O. and Wheelis, M. L. (1990). Towards a natural system of organisms: proposal for the domains Archaea, Bacteria, and Eucarya. *Proc. Natl. Acad. Sci. USA* **87**, 4576–4579.

Woods, W. G. and Dyall-Smith, M. L. (1997). Construction and analysis of a recombination-deficient (*radA*) mutant of *Haloferax volcanii*. *Mol. Microbiol.* **23**, 791–797.

Woods, A. G., Whitman, W. B. and Konisky, J. (1985). A newly-isolated marine methanogen harbors a small cryptic plasmid. *Arch. Microbiol.* **142**, 259–261.

Worrell, V. E. and Nagle, D. P., Jr (1990). Genetic and physiological characterization of the purine salvage pathway in the archaebacterium *Methanobacterium thermoautotrophicum* Marburg. *J. Bacteriol.* **172**, 3328–3334.

Worrell, V. E., Nagle, D. P., Jr, McCarthy, D. and Eisenbraun, A. (1988). Genetic transformation system in the archaebacterium *Methanobacterium thermoautotrophicum* Marburg. *J. Bacteriol.* **170**, 653–656.

Yang, C. F., Kim, J. M., Molinari, E. and DasSarma, S. (1996). Genetic and topological analyses of the bop promoter of *Halobacterium halobium*: stimulation by DNA supercoiling and non-B-DNA structure. *J. Bacteriol.* **178**, 840–845.

Yang, Y.-L., Ladapo, J. A. and Whitman, W. B. (1995). Mutagenesis of *Methanococcus* spp. with ethylmethanesulfonate. In *Methanogens* (K. R. Sowers and H. J. Schreier, eds), pp. 403–404. Cold Spring Harbor Laboratory Press, Plainview, New York.

Zillig, W., Yeats, S., Holz, I., Bock, A., Gropp, F., Rettenberger, M. and Lutz, S. (1985). Plasmid-related anaerobic autotrophy of the novel archaebacterium *Sulfolobus ambivalens*. *Nature* **313**, 789–791.

Zillig, W., Kletzin, A., Schleper, C., Holz, I., Janekovic, D., Hain, J., Lanzendorfer, M. and Kristjansson, J. K. (1994). Screening for *Sulfolobales*, their plasmids and their viruses in Icelandic solfataras. *System. Appl. Microbiol.* **16**, 609–628.

Zillig, W., Prangishvilli, D., Schleper, C., Elferink, M., Holz, I., Albers, S., Janekovic, D. and Gotz, D. (1996). Viruses, plasmids and other genetic elements of thermophilic and hyperthermophilic Archaea. *FEMS Microbiol. Rev.* **18**, 225–236.

Zinder, S. H. (1993). Physiological ecology of methanogens. In *Methanogenesis* (J. G. Ferry, ed), pp. 128–206. Chapman & Hall, New York.

Zusman, T., Rosenshine, I., Boehm, G., Jaenicke, R., Leskiw, B. and Mevarech, M. (1989). Dihydrofolate reductase of the extremely halophilic archaebacterium *Halobacterium volcanii*. The enzyme and its coding gene. *J. Biol. Chem.* **264**, 18878–18883.

PART III

Genetic Approaches to Analyse Specific Complex Phenomena

◆◆

11 Genetic Characterization of the Gastric Pathogen *Helicobacter pylori*

E. Allan, S. Foynes, N. Dorrell and B. W. Wren

Microbial Pathogenicity Research Group, Department of Medical Microbiology, St Bartholomew's and the Royal London School of Medicine and Dentistry, London, UK

◆◆◆

CONTENTS

Helicobacter pylori – The organism and disease
Cloning approaches
Gene expression studies
Mutagenesis
Whole genome approaches

◆◆◆◆◆◆ I. *HELICOBACTER PYLORI* – THE ORGANISM AND DISEASE

A. Historical Review

For over a century the cause of peptic ulcer disease in humans was thought to be induced by psychological stress and to be prevalent in hyperacid producers. The normal human stomach was considered to be sterile and the notion that a bacterium could survive in this hostile environment and cause disease was inconceivable. However in 1982, after an accidental period of prolonged incubation of agar plates, Robin Warren isolated a slow-growing bacterium from gastric tissues (Warren and Marshall, 1984). This Gram-negative, microaerophilic, spiral bacterium was initially named *Campylobacter pyloridis*. Subsequent phylogenetic analysis placed the bacterium in its own genus with the name *Helicobacter pylori*.

The discovery and isolation of *H. pylori* in pure culture from gastric biopsies has revolutionized our view of gastric disease and ensuing therapies. We now know that *H. pylori* infects the human gastric mucosa, triggering chronic gastritis, peptic ulcer disease, gastric lymphoma and possibly gastric adenocarcinoma. Persistent *H. pylori* infection is one of the commonest infections world-wide. Its prevalence in the Western world is estimated at 50% (Graham *et al.*, 1991a) and in developing

Genetic Characterization of *Helicobacter pylori*

countries this may be as high as 80% (Graham *et al.*, 1991b). While most infected individuals are a symptomatic, a significant number develop serious gastrointestinal disease. The development of gastric disease depends on multiple factors, including the strain of *H. pylori*, host genetic predisposition and dietary intake.

This chapter aims to give an overview of the genetic approaches that can be used to study this unique pathogen and to speculate upon how the availability of the whole genome sequence could be exploited for functional studies. Detailed practical protocols on the isolation and characterization of *H. pylori* have been published recently (Clayton and Mobley, 1997).

B. Unusual Features of *H. pylori*

Helicobacter pylori is a remarkable microorganism because of its ability to readily colonize a major proportion of the human population and to persist successfully for long periods in a hostile environment. The organism has adapted to survive in the mucous layer overlying the gastric epithelium. There is a wide pH gradient in the stomach, ranging from pH2 in the gastric lumen to near neutral in the mucosal vessels. The acidic nature of the stomach also results in the generation of free radicals of nitrogen and oxygen. Counteracting the damaging effects of free-radical damage is likely to be an important factor in the survival mechanisms of *H. pylori* in the stomach. As no other bacteria are known to persist in the human stomach, *H. pylori* appears to be unique in its ability to survive in this inhospitable ecological niche.

The ability to survive exposure to the gastric environment is a prerequisite for successful colonization of the gastric epithelium. *H. pylori* has a strong affinity for gastric epithelial cells. Cell lines such as adult gastric adenocarcinoma (AGS) and KATO-III cells are frequently used to quantitate the binding efficacy of *H. pylori* (Nilius *et al.*, 1994). Once bound, *H. pylori* interacts with the host immune system in such a way as to permit long-term survival (Blaser, 1995). Another less obvious yet highly significant feature of *H. pylori* is its high degree of interstrain genome diversity. Macrodiversity – the variability in gene order – in addition to microdiversity – the variability in nucleic acid sequence – is a characteristic of the *H. pylori* genome (Jiang *et al.*, 1996). *H. pylori* is highly competent for the uptake of exogenous DNA and the identification of at least one pathogenicity island and numerous insertion sequences suggests that horizontal gene transfer readily occurs (Tomb *et al.*, 1997). In contrast to most bacteria, the natural population structure appears to be non-clonal (Go *et al.*, 1996), although the selective advantage that this diversity provides is not clear.

C. Putative Virulence Determinants

The pathogenicity of *H. pylori* is undoubtedly complex, involving the interaction of numerous determinants, many of which remain to be identified. Despite the importance and prevalence of *H. pylori* infection, information on pathogenesis is limited.

The production of a potent multi-subunit urease is considered an essential virulence determinant. The ammonia generated by this enzyme's activity is thought to protect the organism by buffering gastric acid (Eaton *et al.*, 1991; Eaton and Krakowka, 1994).

Helicobacter pylori produces a 39 kDa cytotoxin termed VacA, which induces vacuolation in a variety of mammalian cell types and produces gastric epithelial cell lesions and ulcers in mice (Ghiara *et al.*, 1995). Although the *vacA* gene is present in almost all strains, only half express an active toxin. VacA-producing strains occur more frequently in patients with peptic ulcer disease than in those patients with gastritis alone. Furthermore, *vacA* exhibits marked variation in signal sequence and midgene coding regions. *H. pylori* strains of *vacA* signal sequence type s1a are more likely to be associated with enhanced gastric inflammation and duodenal ulceration, whereas *vacA* s2-type strains are associated with less inflammation and lower ulcer prevalence (Atherton *et al.*, 1996). Another marker for virulence in strains, associated with disease outcome (peptic ulceration), is a 120–140 kDa immunodominant protein, termed CagA (cytotoxin-associated gene A). The function of CagA is unknown. However, the gene is at the limits of a pathogenicity island (PAI), termed the *cag* PAI (Censini *et al.*, 1996). The *cag* PAI encompasses other genes, some of which encode proteins with similarities to several prokaryotic secretory pathways, including the type IV secretion systems involved in the export of potential virulence factors. Mutations in several genes on the *cag* PAI results in a marked reduction in the ability of *H. pylori* to induce IL-8 secretion (a proinflammatory cytokine, believed to play a major role in the pathogenesis of gastritis and gastroduodenal ulceration) from KATO-III gastric epithelial cells (Censini *et al.*, 1996).

Other potential determinants of pathogenicity include motility (Eaton *et al.*, 1992; Jenks *et al.*, 1997), adhesion (Jones *et al.*, 1997; Ilver *et al.*, 1998), the abilities to attract neutrophils and to interfere with host cell signal transduction systems (Evans *et al.*, 1995; Segal *et al.*, 1997). All of these properties, plus its ability to thrive in a restricted ecological niche, are encoded on a relatively small genome of exactly 1 667 867 base pairs (at least for strain ATCC 26695) (Tomb *et al.*, 1997).

D. *Helicobacter pylori* – Unanswered Questions

The mechanisms by which the organism causes disease and is transmitted are still poorly understood and the current options for treatment or prevention are far from optimal. To cause disease *H. pylori* has to: (i) survive the hostile environment of the stomach; (ii) penetrate the mucous layer; (iii) adhere to and interact with gastric epithelial cells; (iv) cause damage to host cells; (v) avoid and subvert the immune system; and (vi) be transmitted to new hosts. Each of these stages of infection poses important questions which remain to be answered. In particular, very little is known about the regulation of virulence determinants such as urease, VacA, CagA and other gene products encoded by the *cag* PAI.

E. Which Strain to Study?

The choice of strain is critical to obtain meaningful biological data. Repeated culturing reduces pathogenicity and ideally a fresh CagA+/VacA+ clinical isolate should be used.

Table 1 outlines the advantages and disadvantages of the most commonly used laboratory strains. Routinely, we culture 20 aliquots of new strains and store them in 15% v/v glycerol and 10% serum at –80°C. Prior to any functional studies we ensure that the strains are motile on semi-solid media. Currently, we favour studies using the SSI or Sydney strain which is freely available (Lee *et al.*, 1997). In addition to readily colonizing mice, this strain is also suitable for the gnotobiotic piglet model for *H. pylori* infection (unpublished).

Table I. A comparison of laboratory *H. pylori* strains

Strain	Growth in mice	Growth in gnotobiotic piglets	Genetic use	Strain stability	Genome sequence
NCTC 11638	+	+	++	+	No
N6	–	++	+++	++	No
ATCC 26695	?	+++	+	++	Yes
SS1	+++	+++	+++	+++	No

◆◆◆◆◆◆ II. CLONING APPROACHES

A. Gene Libraries

Studies to date on the isolation of virulence genes from *H. pylori* have frequently relied on the use of gene libraries. A number of the studies described below have used phage libraries such as λEMBL3 (Clayton *et al.*, 1989), λZAP II (Tummuru *et al.*, 1993) or λgt11 (Macchia *et al.*, 1993). Compared with some bacteria, for example *C. jejuni*, *H. pylori* DNA is stable when cloned in *Escherichia coli*. This is demonstrated by the stability of an ordered library of *H. pylori* strain NCTC 11638 DNA cloned into the cosmid Lorist6 (Bukanov and Berg, 1994). This cosmid library has enabled a high resolution physical genetic map of the organism to be constructed. Drazek *et al.* (1995) have used the library to map putative haemolysin genes and Ilver *et al.*, (1998) have used it to co-locate *vacA* and *babAs* (Lewis b binding antigens) on the *H. pylori* NCTC 11638 genome.

B. Screening Libraries with Antiserum

The first *H. pylori* gene cloned and expressed in *E. coli* was the urease gene which was identified by screening a λEMBL3 library with whole cell antiserum raised in rabbits (Clayton *et al.*, 1989). In principle, antiserum against outer membrane proteins should give an increased possibility of identifying proteins located on the bacterial surface. However, *H. pylori* is prone to lysis or is "autolytic" (Phadnis *et al.*, 1996) and intracellular proteins are readily isolated from culture supernatants. Such proteins have initially been incorrectly interpreted as surface proteins (O'Toole *et al.*, 1991; Doig *et al.*, 1992). Library screening has also been carried out using serum from *H. pylori*-infected patients. This approach identified the *cagA* (Tummuru *et al.*, 1993) and *hsp60* (Macchia *et al.*, 1993) genes.

C. Functional Complementation

These studies require the host cell to be deficient for an identifiable phenotypic characteristic, then by introducing cloned genes from *H. pylori*, a phenotypic change is observed. This method was used to isolate the antioxidant superoxide dismutase (SOD) gene from *H. pylori* (Spiegelhalder *et al.*, 1993). In this study, a SOD-deficient *E. coli* mutant strain, used as a recipient of a *H. pylori* cosmid library, was subcultured under oxidative stress conditions. Surviving clones contained an active SOD enzyme with sufficient activity for complementation in *E. coli*.

D. Direct Cloning

A method similar to functional complementation was used to identify a haemolytic gene in *H. pylori* (Drazek *et al.*, 1995). In this study, the haemolytic gene of *H. pylori* was identified by its ability to confer haemolytic activity on a non-haemolytic *E. coli*.

E. Probes Based on Protein Sequence Information

Several putative *H. pylori* virulence genes have been identified following purification of the gene products. The purified protein is sequenced and probes based on either N-terminal or internal amino acid sequences can be used as probes to isolate the gene. Screening of genomic libraries with N-terminal-based probes were used to isolate the flagellin gene, *flaA* (Leying *et al.*, 1992), the flagellar-hook protein, *flgE*, (O'Toole *et al.*, 1994) and the catalase gene, *katA* (Odenbreit *et al.*, 1996b). A probe based on internal amino acid sequence of the cytotoxin was used to isolate the *vacA*, gene (Cover *et al.*, 1994).

F. PCR-based Methods

Polymerase chain reaction with degenerate oligonucleotide primers (PCRDOP) allows the isolation of conserved genes from target DNA without having an assay for their function (Wren *et al.*, 1992). The high degree of amino acid sequence conservation identified within some protein families from numerous genera allows the design of degenerate primers to amplify these genes. PCRDOP is advantageous in that only small gene fragments are isolated, thus avoiding problems associated with over-expression of toxic gene products. This method is also convenient and rapid. To date PCRDOP has been used successfully to isolate a number of genes from *H. pylori* including *recA* (Thompson and Blaser, 1995), *fliI* (Jenks et al., 1997) and *cheY* (Foynes and Wren, unpublished).

G. Post-genomic Era

The entire annotated genome sequence of *H. pylori* strain ACTC 26695 was recently published by The Institute for Genomic Research (TIGR) (Tomb *et al.*, 1997). Taking advantage of this available sequence information (http://www.tigr.org), it is now possible to design PCR primers to amplify any gene from the entire complement of the organism. The PCR products can be conveniently cloned into a suitable vector and then used for subsequent expression or functional analysis (see sections III, IV and V below).

◆◆◆◆◆◆ III. GENE EXPRESSION STUDIES

A. Production of Recombinant Proteins

To produce recombinant antigens for vaccination or for determination of patients' immune response by ELISA, several *H. pylori* genes, including *ureA*, *ureB*, *hspA*, *hspB* and *clpB*, have been cloned in the pMAL expression vector, and expressed in *E. coli* as translational fusions with maltose-binding protein (MBP) (Ferrero *et al.*, 1994; Suerbaum *et al.*, 1994; Allan, 1997). Following purification of the fusion protein by amylose affinity chromatography, the recombinant *H. pylori* protein may be cleaved from MBP by virtue of a Factor Xa protease cleavage site, located just 5′ to the polylinker insertion site in the pMAL vector. The yield of fusion protein obtained using this system appears to be inversely proportional to the molecular mass: in the studies of Suerbaum *et al.* (1994), the yield of a MBP–HspB fusion (100 kDa) was fivefold lower than that of MBP–HspA (55 kDa). Similarly, Ferrero *et al.* (1994) reported a twofold greater yield of MBP–UreA (68 kDa) than of MBP–UreB (103 kDa).

B. Reporter Gene Technology

To study differential expression of *cagA*, Karita *et al.* (1996) constructed a transcriptional fusion with the promoterless *xylE* gene of *Pseudomonas putida*. The *xylE* gene fusions were introduced into the chromosome of wild-type *H. pylori* by allelic exchange and the activity of the *xylE* gene product, catechol 2,3-dioxygenase, was determined in crude cell extracts, providing quantitative analysis of gene expression (Karita *et al.*, 1996).

A promoterless chloramphenicol acetyltransferase (*cat*) gene has also been used as a reporter gene for studies of *H. pylori* flagellin gene expression. Josenhans *et al.* (1995a) showed that the expression of the flagellin genes was growth-phase-dependent, by monitoring expression of the *cat* gene product.

The autofluorescent Green Fluorescent Protein (GFP), originally from the jellyfish *Aequorea victoria*, has also been used as a reporter for gene expression in *H. pylori*. Josenhans *et al.* (1997) constructed chromosomal transcriptional gene fusions of the *flaA* and *flaB* flagellin genes with the promoterless *gfp* gene as a reporter, allowing expression of these genes to be monitored by fluorescence. These workers have also constructed a mini-Tn3-*gfp* transposon to be used for random shuttle mutagenesis (Josenhans *et al.*, 1997).

◆◆◆◆◆◆ IV. MUTAGENESIS

To elucidate putative virulence mechanisms, genetic mutants of *H. pylori* must be obtained and analysed in *in vitro* assays and in appropriate animal models.

A. Selection for Naturally Occurring Variants

Eaton *et al.* (1992) showed that non-motile variants of *H. pylori*, which occurred with low frequency, were poor colonizers of gnotobiotic piglets, indicating that motility is a probable colonization factor of *H. pylori* (Eaton *et al.*, 1992).

B. Random Chemical Mutagenesis

Studies investigating the role of urease in colonization also used undefined mutants of *H. pylori* (Eaton *et al.*, 1991). Using N-methyl-N-nitro-N-nitrosoguanidine as mutagen, these workers demonstrated that urease-negative mutants, obtained at very low frequency (less than 1/1000), were unable to colonize gnotobiotic piglets, in contrast to the parental strain which was fully virulent. Another mutagen, ethyl methanesulphonate (EMS) has also been used for random chemical mutagenesis of *H. pylori* (Segal *et al.*, 1992).

Although the studies described above showed the likely importance of motility and urease production in virulence of *H. pylori*, it was possible that the inability to colonize was the result of a mutation in a second undefined gene. In order to eliminate the role of putative secondary factors, it was necessary to construct defined isogenic mutants in both of these genes using precise methods.

C. Construction of Defined *H. pylori* Mutants

1. Shuttle mutagenesis

Transposable elements are widely used genetic tools for random mutagenesis of bacterial genomes (see Chapter 4). A transposon that is functional in *H. pylori*, has not yet been identified and defined mutants have to date been constructed by shuttle mutagenesis. This involves introducing a mutation into a *H. pylori* gene cloned in *E. coli* and then the transfer of this mutated gene to the *H. pylori* genome by homologous recombination-mediated allelic exchange. Plasmids with *E. coli* replication origins (eg. pUC and pBluescript vectors) are commonly used for delivery of mutated genes as these are unable to replicate in *H. pylori*.

2. Insertional inactivation of *H. pylori* genes cloned in *E. coli*

Where the cloned *H. pylori* gene contains a unique restriction site, the gene may be inactivated by insertion of an antibiotic-resistance gene (O'Toole *et al.*, 1994; Tummuru *et al.*, 1994; Josenhans *et al.*, 1995b; Bauerfeind *et al.*, 1996). A kanamycin-resistance gene originally from *Campylobacter coli* (*aph3'*-III; Labigne-Roussel *et al.*, 1988) is expressed well in *H. pylori* and is commonly used as a selectable marker. Alternatively, a gene conferring resistance to chloramphenicol (*cat*$_{GC}$) has been used (Haas *et al.*, 1993a). (See Chapter 2, Table 1 for genearl discussion of resistance markers.) Defined mutants in the flagellin genes, *flaA* and *flaB*, constructed by insertional inactivation, were used to investigate the role of these genes in colonization (Suerbaum *et al.*, 1993). The expression of both flagellin genes was shown to be required for normal colonization in the gnotobiotic piglet model (Eaton *et al.*, 1996).

3. Transposon shuttle mutagenesis

Transposable elements which function in *E. coli* have been used to generate random mutations in *H. pylori* genes cloned in *E. coli* (see also Chapter 4, VIII). Labigne (1997) adapted a Tn3-based transposon for shuttle mutagenesis of *H. pylori* because it preferentially inserts into AT-rich DNA sequences (the *H. pylori* genome has a GC content of 39%; Tomb *et al.*, 1997). This transposable element, which carries the *E. coli* kanamycin-resistance gene *aph3'*-III, has been used to generate isogenic urease-negative mutants of *H. pylori* (Ferrero *et al.*, 1992). Failure of these defined mutants to colonize the gnotobiotic piglet confirmed the essential role of the urease enzyme (Eaton and Krakowka, 1994).

Other transposable elements which have been used for shuttle mutagenesis of *H. pylori* genes are derivatives of Tn*Max* (Haas *et al.*, 1993b; see also Chapter 4, VIII and Chapter 5, IV). These transposons, consisting of a series of Tn*1721*-based mini-transposons, carry a gene conferring resistance to chloramphenicol and have been used to generate isogenic *H. pylori* mutants in the flagellin (Haas *et al.*, 1993a), cytotoxin (Schmitt and Haas, 1994) and catalase (Odenbreit *et al.*, 1996b) genes.

Another Tn*Max* mini-transposon, developed by Odenbreit *et al.* (1996a), was used to detect and simultaneously mutate exported *H. pylori* proteins, which are likely to be involved in colonization and/or virulence. Tn*Max9* has been used for insertion mutagenesis of a *H. pylori* gene library in *E. coli*. In addition to a chloramphenicol-resistance gene, this mini-transposon carries an unexpressed open reading frame encoding a β-lactamase gene without the promoter and leader sequence. Extracytoplasmic β-lactamase (resulting in ampicillin resistance) is produced only by those clones in which Tn*Max9* has inserted downstream of a *H. pylori* sequence specifying a signal for export. To ensure expression of cloned genes in *E. coli*, the *H. pylori* gene library was constructed in a vector containing a weak *E. coli* constitutive promoter upstream of the multiple cloning site. This method identified mutant strains impaired in motility, in competence for natural transformation and in adherence to gastric epithelial cell lines (Odenbreit *et al.*, 1996a).

4. PCR-based shuttle mutagenesis

An alternative method for shuttle mutagenesis of *H. pylori* genes is inverse PCR mutagenesis (IPCRM), developed in our laboratory (Wren *et al.*, 1994; Dorrell *et al.*, 1996). This method allows a precise deletion to be introduced into the cloned *H. pylori* gene and simultaneously creates a unique restriction site for insertion of an antibiotic-resistance gene. IPCRM is an adaptation of inverse PCR where plasmid DNA with cloned inserts is used as circular templates for PCR (Figure 1). Primers are designed in opposite orientations at a gap that defines the deletion in the target DNA. Primer design may also ensure a frameshift mutation and allow stop codons to be introduced. This method has been used to construct isogenic mutants in *hpaA* (Jones *et al.*, 1997), *fliI* (Jenks *et al.*, 1997) and *clpB* (Allan *et al.*, 1998). The major advantage of this technique is that small motifs known to be essential for protein function can be specifically deleted.

D. Allelic Exchange Mutagenesis

The mutated *H. pylori* gene, carried on a plasmid unable to replicate in *H. pylori* (e.g. pUC and pBluescript), is introduced into wild-type bacteria.

Figure 1. Diagrammatic representation of tagged inverse PCR mutagenesis. Solid boxes represent cloned DNA. "U" represents unique restriction site. "T" represents unique identifying DNA tag.

Usually a double crossover homologous recombination event occurs resulting in elimination of the vector sequences and replacement of the wild-type gene with the mutated allele. In our experience, single crossovers, resulting in integration of the vector and mutated allele, occur very infrequently. For a more detailed description of allelic exchange see Chapter 10, II, III and IV. McGowan *et al.* (1997), following repeated failures to isolate viable mutants in the gene encoding an F_1F_0-ATPase, performed PCR on chromosomal DNA from bacteria incubated overnight with the construct. This PCR indicated that integration of the mutated gene had occurred on the *H. pylori* chromosome suggesting that mutagenesis of this gene is lethal to the organism.

E. Introduction of the Mutated Gene into Wild-type *H. pylori*

As most strains of *H. pylori* are naturally competent and transformable, recombinant plasmid DNA isolated by standard techniques (we use Promega Wizard™ kits) can be introduced into wild-type bacteria by natural transformation with reasonable efficiency (Ge and Taylor, 1997). Methods for natural transformation of both agar and broth cultures of *H. pylori* have been described (Haas *et al.*, 1993a; Ge and Taylor, 1997). Several mutated *H. pylori* genes carried on plasmids, including *flaA*, *flgE*, *hpaA* and *clpB*, were introduced into wild-type bacteria by natural transformation (Haas *et al.*, 1993a; O'Toole *et al.*, 1994, Jones *et al.*, 1997; Allan *et al.*, 1998).

Electroporation (Chapter 1, II.D.5) has also been used to introduce mutated genes into wild-type *H. pylori* and several detailed protocols are available (Ferrero *et al.*, 1992; Thompson and Blaser, 1995; Ge and Taylor, 1997). Ge and Taylor (1997) have quoted efficiencies of electrotransformation of $1.0–2.0 \times 10^3$ transformants per microgram of plasmid DNA, dependent on the strain of *H. pylori* used. Electroporation may be more efficient than natural transformation: McGowan and co-workers (1997) found that electroporation was fivefold more efficient than natural broth transformation in producing *H. pylori vacA* mutants.

◆◆◆◆◆◆ V. WHOLE GENOME APPROACHES

A. The *H. pylori* Genome Sequence – New Opportunities

Scientists at The Institute for Genomic Research (TIGR) (http://www.tigr.org) have sequenced, annotated and published the entire genome sequence of *H. pylori* strain ACTC 26695, just 15 years after the organism was first cultured (Warren and Marshall, 1984). Initial computer analysis suggests the presence of 1590 predicted coding sequences, of which nearly 70% can be matched to genes encoding proteins of known function. The availability of a complete genetic data set heralds a new era in *H. pylori* research as it will provide a framework for global studies of virulence and other aspects of the organism's biology.

Examples where these data will provide key insights include the following:

- **Outer membrane proteins (OMPs) and DNA repeat sequences**. The OMPs identified provide a tractable subset of the total genome, comprising most of the proteins known to be involved in virulence (e.g. those required for adherence to gastric epithelial cells and evasion of the immune system). Of particular interest is the presence of tandem repeat sequences upstream of some OMPs. In other mucosal pathogens, increasing or decreasing the numbers of repeats by slipped-strand mispairing and recombination affects transcription of the downstream genes (High *et al.*, 1993). In this way, minor reversible mutations rapidly change the antigen profile of the pathogen, leading to evasion of the host immune system (High *et al.*, 1993). Repeat DNA sequences upstream of OMPs and other repeat sequences could act as "hotspots" for potential virulence genes (Hood *et al.*, 1996).
- **Metabolism**. Strictly aerobic and anaerobic enzymes are both missing, implying that the only ecological niche for *H. pylori* is microaerophilic. Free-living *H. pylori* in the environment appears unlikely.
- **Regulatory proteins**. Compared to the *E. coli* genome sequence (Blattner *et al.*, 1997), *H. pylori* has tenfold less identified regulatory sequences, calculated as a percentage of the whole genome. This remarkable economy of regulatory elements in *H. pylori* may reflect the very limited range of environments in which it survives, suggesting a highly evolved interrelationship between man and microbe.
- **ORFans**. ORFans refer to predicted open reading frames (ORFs) that have no known homologues or any clues as to function. Almost one-third of the predicted coding sequences are ORFans, but this figure is likely to reduce significantly as more bacterial genomes are sequenced. These ORFans may encode proteins unique to *H. pylori*. In this respect, they are perhaps the most intriguing genes/gene products to study and may provide selective targets for antibiotic therapy.

B. Global Approaches to the Analysis of Gene Function

The acquisition and analysis of sequence data is not an end in itself; instead it is a starting point for generating hypotheses that can be tested in the laboratory. Homology provides clues, but does not prove the gene function.

Two complementary approaches have emerged to study function. *Functional genomics* emphasizes the roles of DNA and RNA in the natural progression from information (DNA) to function (protein), while with *proteomics* the emphasis is on the proteins themselves. Knowledge of the entire genome sequence allows an integrated approach to the functional genomics and proteomics of *H. pylori* at the mutational, transcriptional and protein expression levels (see Figure 2.)

The release of the *H. pylori* genome sequence has coincided with important technological advances in four areas:

1. Bioinformatics
2. High density hybridization technology
3. Mutagenesis
4. Protein chemistry and mass spectrometry.

When combined, these advances will liberate scientific understanding from the piecemeal study of individual genes or operons, towards a comprehensive analysis of the entire gene and protein complement of the bacterial cell (see Figure 2). The following provides an outline of what is now possible.

C. Bioinformatics (*in silico* Analysis of Sequence Data)

The past few years have seen vast improvements in the algorithms used to analyse important sequence data. Furthermore, an increasing range of bioinformatics software has been developed and released into the public domain by way of the Internet. Careful and intelligent use of this software can afford important new insights into protein structure and function and allow the generation of testable hypotheses.

The published annotated form of the genome sequence of *H. pylori* by scientists at TIGR is not definitive. The authors used a narrow set of programs in their analysis and doubtless several re-analyses of the data will be undertaken. Already one such analysis is available at the PEDANT web site: (http://pedant.mips.biochem.mpg.de/frishman/pedant.html). In addition to the benefits of such "static" analysis, an on-going dynamic analysis is needed, constantly re-evaluating the *H. pylori* sequence data in the light of newly published sequences.

D. Differential Gene Expression (DGE) and High Density Hybridization Arrays

In terms of genome analysis, DGE refers to the studies of the transcriptional activity of all the organism's genes. This is possible by extracting the mRNA expressed under a range of environmental conditions and hybridizing these sequences to a high-density gridded array of the DNA content of an organism. The availability of ever-cheaper oligonucleotides, 96-well PCR technology, and complete genome sequence data, make

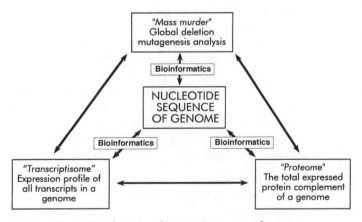

Figure 2. Complementary functional genomics approaches.

possible the highly attractive option of using gridded libraries of PCR products, constituting a defined and complete set of ORFs and inter-genic regions. The Affymetrix Biochip™ is an example of the emerging technology of nucleic acid arrays which extends this principle (Chee *et al.*, 1996; Editorial, 1996). Biochip technology makes it possible to perform thousands of hybridizations in parallel, so that, for example, the effect of a given stimulus on transcription, such as low pH or the interaction with gastric epithelial cells, can be assayed simultaneously for all genes in a genome. A "*H. pylori* Biochip" could also be used for differential genomics by comparing the gene complements of various clinical isolates.

E. Global Gene Deletion Analysis (the "Mass Murder" Approach)

Traditional mutagenesis methods using phage transduction and transposons have proven a fruitless approach to the genetic characterization of *H. pylori*. By contrast, as outlined in section IV above, the construction of defined mutants by allele replacement has proven to be reliable and relatively simple (Ferrero *et al.*, 1992; Jenks *et al.*, 1997; Jones *et al.*, 1997). Thus not only is *H. pylori* one of the first bacterial genomes to be sequenced, but it is also possible to study gene function by gene deletion/disruption analysis. The availability of the entire genomic sequence of *H. pylori* means that the large-scale systematic construction of defined mutants is now possible. In conjunction with global gene deletion analysis is the potential to label each mutant with a unique DNA signature tag (Hensel *et al.*, 1995). The original use of DNA signature tags used randomly tagged transposons (Hensel *et al.*, 1995). More recently, we have developed this technology by coupling the incorporation of tags with allele replacement (see section IV.D), referred to as STAR (signature tagged allele replacement). Thus the systematic and unbiased construction of individually tagged *H. pylori* mutants means that hundreds of mutants can be analysed simultaneously for phenotypic features, such as the ability to survive acid shock. Although *in vitro* screens are useful, they cannot be expected to identify all virulence genes of a pathogen because they do not reflect the complex environment that a pathogen encounters within the host. The greatest potential for the wholesale tagging of mutants is for *in vivo* studies. This will allow a drastic reduction in the number of animal experiments required for the assessment of virulence. Furthermore, the need for onerous and repetitive filter-based radioactive hybridizations can be avoided by quantitating the survival of mutants using a specifically designed Affymetrix Biochip containing all the different sequence tags used (Shoemaker *et al.*, 1996).

F. Proteome Analysis (the Entire Complement of Proteins Expressed by a Cell)

The protein products of many of the newly identified *H. pylori* putative ORFs are unknown. Recent improvements in high-sensitivity biological mass spectrometry have provided a powerful adjunct to traditional 2-D gel electrophoresis (Pennington *et al.*, 1997). Proteins cut out of a 2-D gel can now be peptide-mass-fingerprinted, and constituent peptides can be sequenced by mass spectrometry. New software takes data from mass spectrometry and uses it to find the best match in a sequence database, allowing one to go from a spot on a PAGE gel to protein identification in a matter of hours. Thus the entire complement of proteins expressed by a cell (the proteome) can be defined. This kind of approach has already been used to provide insights into the function of an anatomical subset of the proteome such as the cell envelope from *Salmonella typhimurium* (Qi *et al.*, 1996). Proteome studies are made even more powerful when applied to an organism whose genome has been sequenced. Synergistic interactions between the two approaches maximize information return, so that, for example, ORFan gene products can be shown to be real proteins.

G. Practical Implications for Genomic Studies on *H. pylori*

A multi-dimensional analysis of *H. pylori*, looking at sequences, mutants, transcripts and proteins, will result in a quantum leap in our understanding of the biology and pathogenesis of *H. pylori*. In determining the activity of large sets of genes, proteins and the interactions between them, an important step towards constructing a functional model of the entire organism will be taken. This basic information will provide the framework for future research and for highlighting potential drug targets or vaccine components. In the long term this information will strengthen our ability to combat the two major gastrointestinal diseases, namely peptic ulceration and gastric cancer.

It is ironic that *H. pylori*, one of the most recently identified pathogens, promises to be one of the most well characterized. The knowledge garnered by combining traditional genetic methods with those from genomics approaches will not only enhance our understanding of *H. pylori*, but also that of other bacteria.

References

Allan, E. (1997). Identification and characterisation of potential pathogenicity determinants of *Helicobacter pylori*. PhD Thesis, University of London, London.

Allan, E., Mullany, P. and Tabaqchali, S. (1998). Construction and characterisation of a *Helicobacter pylori clpB* mutant and its role in the stress response. *J. Bacteriol.* **180**, 426–429.

Atherton, J. C., Peek, R. M. Jr, Tham, K. T., Cover, T. L. and Blaser, M. J. (1996). Clinical and pathological importance of heterogeneity in *vacA*, the vacuolating cytotoxin gene of *Helicobacter pylori*. *Gastroenterology* **112**, 92–99.

Bauerfeind, P., Garner, R. M. and Mobley, H. L. T. (1996). Allelic exchange mutagenesis of *nixA* in *Helicobacter pylori* results in reduced nickel transport and urease activity. *Infect. Immun.* **64**, 2877–2880.

Blaser, M. J. (1995). The role of *Helicobacter pylori* in gastritis and its progression to peptic ulcer disease. *Aliment. Pharmacol. Ther.* **9**, 27–30.

Blattner, F. R., Plunket III, G., Bloch, C. A., Perna, N. T., Burland, V., Riley, M.,

Collado-Vides, J., Glasner, J. D., Rode, C.K., Mayhew, G.F., Gregor, J., Davis, N. W., Kirkpatrick, H. A., Goeden, M. A., Rose, D. J., Mau, B. and Shao, Y. (1997). The complete sequence of *Escherichia coli* K12. *Science* **277**, 1453–1462.

Bukanov, N. O. and Berg, D. E. (1994). Ordered cosmid library and high-resolution physical-genetic map of *Helicobacter pylori* strain NCTC 11638. *Mol. Microbiol.* **11**, 509–523.

Censini, S., Lange, C, Xiang, Z., Crabtree, J.E., Ghiara, P, Borodovsky, M, Rappuoli, R. and Covacci, A. (1996). Cag, a pathogenicity island of *Helicobacter pylori*, encodes type I-specific and disease-associated virulence factors. *Proc. Natl. Acad. Sci. USA*, **93**, 14648–14653.

Chee, M., Yang, R., Hubbell, E., Berno, A., Huang, X. C., Stern, D., Winkler, J., Lockhart, D. J., Morris, M. S. and Fodor, S. P. A. (1996). Accessing genetic information with high-density DNA arrays. *Science* **274**, 610–614.

Clayton, C. L., and Mobley, H. L. T. (1997). *Methods in Molecular Medicine,* Helicobacter pylori *Protocols*. Humana Press, Totowa, New Jersey, USA.

Clayton, C. L., Wren, B. W., Mullany, P., Topping, A. and Tabaqchali, S. (1989). Molecular cloning and expression of *Campylobacter pylori* species-specific antigens in *Escherichia coli* K-12. *Infect. Immun.* **57**, 623–629.

Cover, T. L., Tummuru, M. K. R., Cao, P., Thompson, S. A. and Blaser, M. J. (1994). Divergence of genetic sequences for the vacuolating cytotoxin among *Helicobacter pylori* strains. *J. Biol. Chem.* **269**, 10566–10573.

Doig, P., Austin, J. W., Kostrzynska, M. and Trust, T. J. (1992). Production of a conserved adhesion by the human gastroduodenal pathogen *Helicobacter pylori*. *J. Bacteriol.* **174**, 2539–2547.

Dorrell, N., Gyselman, V. G., Foynes, S., Li, S-R. and Wren, B. W. (1996). Improved efficiency of inverse PCR mutagenesis (IPCRM). *Biotechniques* **21**, 604–608.

Drazek, S. E., Dubois, A., Holmes, R. K., Kersulyte, D., Akopyants, N. S., Berg, D. E. and Warren, R. L. (1995). Cloning and characterisation of hemolytic genes from *Helicobacter pylori*. *Infect. Immun.* **63**, 4345–4349.

Eaton, K. A. and Krakowka, S. (1994). Effect of gastric pH on urease-dependent colonization of gnotobiotic piglets by *Helicobacter pylori*. *Infect. Immun.* **62**, 3604–3607.

Eaton, K. A., Brooks, C. L., Morgan, D. R. and Krakowka, S. (1991). Essential role of urease in pathogenesis of gastritis induced by *Helicobacter pylori* in gnotobiotic piglets. *Infect. Immun.* **59**, 2470–2475.

Eaton, K. A., Morgan, D. R. and Krakowka, S. (1992). Motility as a factor in the colonisation of gnotobiotic piglets by *Helicobacter pylori*. *J. Med. Microbiol.* **37**, 123–127.

Eaton, K. A., Suerbaum, S., Josenhans, C. and Krakowka, S. (1996). Colonization of gnotobiotic piglets by *Helicobacter pylori* deficient in two flagellin genes. *Infect. Immun.* **64**, 2445–2448.

Editorial. (1996). To affinity and beyond. *Nat. Genet.* **14**, 367–370 and 441–447.

Evans, D. J., Evans, D. G., Takemura, T., Nakano, H., Lampert, H. C., Graham, D. Y., Granger, D. N. and Kvietys, P. R. (1995). Characterisation of a *Helicobacter pylori* neutrophil-activating protein. *Infect. Immun.* **63**, 2213–2220.

Ferrero, R. L., Cussac, V., Courcoux, P. and Labigne, A. (1992). Construction of isogenic urease-negative mutants of *Helicobacter pylori* by allelic exchange. *J. Bacteriol.* **174**, 4212–4217.

Ferrero, R. L., Thiberge, J-M., Huerre, M. and Labigne, A. (1994). Recombinant antigens prepared from the urease subunits of *Helicobacter* spp.: evidence of protection in a mouse model of gastric infection. *Infect. Immun.* **62**, 4981–4989.

Ge, Z. and Taylor, D. E. (1997). *H. pylori* DNA transformation by natural competence and electroporation. In *Methods in Molecular Medicine*, Helicobacter pylori

Protocols (C. L. Clayton and H. L. T. Mobley, eds), pp. 145–152. Humana Press, Totowa, New Jersey, USA.

Gershon, D. (1997). Bioinformatics in a post-genomic age. *Nature* **389**, 417–418.

Ghiara, P., Marchetti, M., Blaser, M. J., Tummuru, M. K., Cover, T. L., Segal, E. D., Tompkins, L. S. and Rappuoli, R. (1995). Role of the *Helicobacter pylori* virulence factors vacuolating cytotoxin, CagA, and urease in a mouse model of disease. *Infect. Immun.* **63**, 4154–4160.

Go, M. F., Kapur, V., Graham, D. Y. and Musser, J. M. (1996). Population genetic analysis of *Helicobacter pylori* by multilocus enzyme electrophoresis: extensive allelic diversity and recombinational population structure. *J. Bacteriol.* **178**, 3934–3938.

Graham, D. Y., Malaty, H. M., Evans, D. G., Evans, D. J., Klein, P. D. and Adam, E. (1991a). Epidemiology of *Helicobacter pylori* in an asymptomatic population in the United States. Effect of age, race, and socio-economic status. *Gastroenterology* **100**, 1495–1501.

Graham, D. Y., Adam, E., Reedy, G. T., Agarwal, J. P., Agarwal, R., Evans, D. J., Malaty, H. M. and Evans, D. G. (1991b). Sero-epidemiology of *Helicobacter pylori* infection in India. Comparison of developing and developed countries. *Dig. Dis. Sci.* **36**, 1084–1088.

Haas, R., Meyer, T. F. and van Putten, J. P. M. (1993a). Aflagellated mutants of *Helicobacter pylori* generated by genetic transformation of naturally competent strains using transposon shuttle mutagenesis. *Mol. Microbiol.* **8**, 753–760.

Haas, R., Kahrs, A. F., Facius, D., Allmeier, H., Schmitt, R. and Meyer, T. F. (1993b). Tn*Max* – a versatile mini-transposon for the analysis of cloned genes and shuttle mutagenesis. *Gene* **130**, 23–31.

Hensel, M., Shea, J. E., Gleeson, C., Jones, M. D., Dalton, E. and Holden, D. W. (1995). Simultaneous identification of bacterial virulence genes by negative selection. *Science* **269**, 400–403.

High, N. J., Deadman, M. E. and Moxon, E. R. (1993). The role of the repetitive DNA motif (CAAT) in the variable expression of *Haemophilus influenzae* LPS epitope αGal(1–4)βGal, *Mol. Microbiol.* **9**, 1275–1282.

Hood, D. W., Deadman, M. E., Jennings, M. P., Bisercic, M., Leischmann, R. C., Venter, J. C. and Moxon, E. R. (1996). DNA repeats identify novel virulence genes in *Haemophilus influenzae*. *Proc. Natl. Acad. Sci. USA*, **93**, 11121–11125.

Ilver, D., Arnquist, A., Ögren, J., Frick, I-M., Kersulyte, D., Incecik, E. T., Berg, D. E., Covacci, A., Engstrand, L. and Borén, T. (1998). *Helicobacter pylori* adhesin binding fucosylated histo-blood group antigens revealed by retagging. *Science* **279**, 373–377.

Jenks, P. J., Foynes, S., Ward, S. J., Constantinidou, C., Penn, C. W. and Wren, B. W. (1997). A flagellar-specific ATPase (FliI) is necessary for flagellar export in *Helicobacter pylori*. *FEMS Microbiol. Lett.* **152**, 205–211.

Jiang, Q., Hiratsuka, K. and Taylor, D. E. (1996). Variability of gene order in different *Helicobacter pylori* strains contributes to genome diversity. *Mol. Microbiol.* **20**, 833–842.

Jones, A. C., Logan, R. P. H., Foynes, S., Cockayne, A., Wren, B. W. and Penn, C. W. (1997). A flagellar sheath protein of *Helicobacter pylori* is identical with HpaA, a putative N-acetylneuraminyllactose-binding haemagglutinin, but is not an adhesin for AGS cells. *J. Bacteriol.* **179**, 5643–5647.

Josenhans, C., Labigne, A. and Suerbaum, S. (1995a). Reporter gene analyses show that expression of both *H. pylori* flagellins is dependent on the growth phase. [Abstract] *Gut* **37**, (Suppl. 1), A62.

Josenhans, C., Labigne, A. and Suerbaum S. (1995b). Comparative ultrastructural and functional studies of *Helicobacter pylori* and *Helicobacter mustelae* flagellin mutants: both flagellin subunits, FlaA and FlaB, are necessary for full motility in *Helicobacter* species. *J. Bacteriol.* **177**, 3010–3020.

Josenhans, C., Friedrich, S. and Suerbaum S. (1997). Gene fusions with GFP (green fluorescent protein gene) as a tool for reporter gene and marker studies in *Helicobacter pylori*. [Abstract] *Gut* **41** (Suppl. 1), A14.

Karita, M., Tummuru, M. K. R., Wirth, H-P. and Blaser, M. (1996). Effect of growth phase and acid shock on *Helicobacter pylori cagA* expression. *Infect. Immun.* **64**, 4501–4507.

Labigne, A. (1997). Random Mutagenesis of the *H. pylori* genome. In *Methods in Molecular Medicine*, Helicobacter pylori *Protocols* (C. L. Clayton and H. L. T. Mobley eds), pp. 153–163. Humana Press, Totowa, New Jersey, USA.

Labigne-Roussel, A. (1987). Gene disruption and replacement as a feasible approach for mutagenesis of *Campylobacter jejuni*. *J. Bacteriol.* **170**, 1704–1708.

Lee, A., O'Rourke, J., Corazon de Ungria, M., Robertson, B., Daskalopoulos, G. and Dixon, M.F. (1997). A standardised mouse model of *Helicobacter pylori* infection: introducing the Sydney strain. *Gastroenterology* **112**, 1386–1397.

Leying, H., Suerbaum, S., Geis, G. and Hass, R. (1992). Cloning and genetic characterisation of a *Helicobacter pylori* flagellin gene. *Mol. Microbiol.* **6**, 2863–2874.

Macchia, G., Massone, A., Burroni, D., Covacci, A., Censini, S. and Rappuoli, R. (1993). The Hsp60 protein of *Helicobacter pylori*: structure and immune response in patients with gastroduodenal diseases. *Mol. Microbiol.* **9**, 645–652.

McGowan, C. C., Cover, T. L. and Blaser, M. J. (1997). Analysis of F_1F_0-ATPase from *Helicobacter pylori*. *Infect. Immun.* **65**, 2640–2647.

Nilius, M., Bode, G., Buchler, M. and Malfertheiner, P. (1994). Adhesion of *Helicobacter pylori* and *Escherichia coli* to human and bovine surface mucus cells in *vitro*. *Eur. J. Clin. Invest.* **7**, 454–459.

Odenbreit, S., Till, M. and Haas, R. (1996a). Optimized BlaM-transposon shuttle mutagenesis of *Helicobacter pylori* allows the identification of novel genetic loci involved in bacterial virulence. *Mol. Microbiol.* **20**, 361–373.

Odenbreit, S., Wieland, B. and Haas, R. (1996b). Cloning and genetic characterization of *Helicobacter pylori* catalase and construction of a catalase-deficient mutant strain. *J. Bacteriol.* **178**, 6960–6967.

O'Toole, P., Logan, S. M., Kostrzynska, M., Wadstorm, T. and Trust, T. J. (1991). Isolation, biochemical characterisation and molecular analysis of a species specific protein antigen produced by the gastric pathogen *Helicobacter pylori*. *J. Bacteriol.* **173**, 505–513.

O'Toole P. W., Kostrzynska, M. and Trust, T. J. (1994). Non-motile mutants of *Helicobacter pylori* and *Helicobacter mustelae* defective in flagellar hook production. *Mol. Microbiol.* **14**, 691–703.

Pennington, S. R., Wilkins, M. R., Hochstrasser, D. F. and Dunn, M. J. (1997). Proteome analysis: from protein characterisation to biological function. *Trends Cell Biol.* **7**, 168–173.

Phadnis, S. H., Parlow, M. H., Levy, M., Ilver, D., Caulkins, C. M., Conners, J. B. and Dunn, B. E. (1996). Surface localisation of *Helicobacter pylori* urease and heat shock protein homolog requires bacterial autolysis. *Infect. Immun.* **64**, 905–912.

Qi, S. Y., Moir, A. and O'Connor, C. D. (1996). Proteome of *Salmonella typhimurium* SL1344: identification of novel abundant cell envelope proteins and assignment to a 2-D reference map. *J. Bacteriol.* **178**, 5032–5038.

Schmitt, W. and Haas, R. (1994). Genetic analysis of the *Helicobacter pylori* vacuolating cytotoxin: structural similarities with the IgA protease type of exported protein. *Mol. Microbiol.* **12**, 307–319.

Segal, E. D., Shon, J. and Tompkins, L. S. (1992). Characterization of *Helicobacter pylori* urease mutants. *Infect. Immun.* **60**, 1883–1889.

Segal, E. D., Lange, C., Covacci, A., Tompkins, L. S. and Falkow, S. (1997). Induction of host signal transduction pathways by *Helicobacter pylori*. *Proc. Natl. Acad. Sci. USA*, **14**, 7595–7599.

Shoemaker, D. D., Lashkari, D. A., Morris, D., Mittmann, M. and Davis, R. W. (1996). Quantitative phenotypic analysis of yeast deletion mutants using a highly parallel molecular bar-coding strategy. *Nat. Genet.* **14**, 450–456.

Spiegelhalder, C., Gerstenecker, B., Kersten, A., Schiltz, E. and Kist, M. (1993). Purification of *Helicobacter pylori* superoxide dismutase and cloning and sequencing of the gene. *Infect. Immun.* **62**, 5315–5325.

Suerbaum, S., Josenhans, C. and Labigne, A. (1993). Cloning and genetic characterization of the *Helicobacter pylori* and *Helicobacter mustelae flaB* flagellin genes and construction of *H. pylori flaA*-and *flaB*-negative mutants by electroporation-mediated allelic exchange. *J. Bacteriol.* **175**, 3278–3288.

Suerbaum, S., Thiberge, J-M., Kansau, I., Ferrero, R. L. and Labigne, A. (1994). *Helicobacter pylori hspA–hspB* heat-shock gene cluster: nucleotide sequence, expression, putative function and immunogenicity. *Mol. Microbiol.* **14**, 959–974.

Thompson, S. A. and Blaser, M. J. (1995). Isolation of the *Helicobacter pylori recA* gene and involvement of the *recA* region in resistance to low pH. *Infect. Immun.* **63**, 2185–2193.

Tomb, J. F., White, O., Kerlavage, A. R., Clayton, R. A., Sutton, G. G., Fleischmann, R. D., Ketchum, K. A., Klenk, H. P., Gill, S., Dougherty, B. A., Nelson, K., Quackenbush, J., Zhou, L., Kirkness, E. F., Peterson, S., Loftus, B., Richardson, D., Dodson, R., Khalak, H. G., Glodek, A., McKenney, K., Fitzegerald, L. M., Lee, N., Adams, M. D., Hickey, E. K., Berg, D. E., Gocayne, J. D., Utterback, T. R., Peterson, J. D., Kelley, J. M., Cotton, M. D., Weidman, J. M., Fujii, C., Bowman, C., Watthey, L., Wallin, E., Hayes, W. S., Borodovsky, M., Karp, P. D., Smith, H. O., Fraser, C. M. and Venter, J. C. (1997). The complete genome sequence of the gastric pathogen *Helicobacter pylori*. *Nature* **388**, 539–547.

Tummuru, M. K. R., Cover, T. L. and Blaser, M. J. (1993). Cloning and expression of a high-molecular-mass major antigen of *Helicobacter pylori*: Evidence of linkage to cytotoxin production. *Infect. Immun.* **61**, 1799–1809.

Tummuru, M. K. R., Cover, T. L. and Blaser, M. J. (1994). Mutation of the cytotoxin-associated *cagA* gene does not affect the vacuolating cytotoxin activity of *Helicobacter pylori*. *Infect. Immun.* **62**, 2609–2613.

Warren, J. R. and Marshall B. J. (1984). Unidentified curved bacilli in the stomach of patients with gastritis and peptic ulceration. *Lancet* **i**, 1310–1314.

Wren, B. W., Colby, S. M., Cubberley, R. R. and Pallen, M. J. (1992). Degenerate PCR primers for the amplification of fragments of genes encoding response regulators from a range of pathogenic bacteria. *FEMS Microbiol. Lett.* **78**, 287–291.

Wren, B. W., Henderson, J. and Ketley, J. M. (1994). A PCR-based strategy for the rapid construction of defined bacterial deletion mutants. *Biotechniques* **16**, 994–996.

12 Virulence Determinants in the Bacterial Phytopathogen *Erwinia*

Nicholas R. Thomson, Joanna D. Thomas and George P. C. Salmond
Department of Biochemistry, University of Cambridge, Cambridge, UK

◆◆

CONTENTS

◆◆◆◆◆◆ I. INTRODUCTION

In this chapter we summarize the approaches taken in the genetic analysis of the Gram-negative bacterial phytopathogen, *Erwinia*. The main interest in bacteria of this genus is due to their importance as the aetiological agents of various rots and wilt diseases of diverse crop plants. However, in addition to this agricultural importance, some erwinias produce antibiotics (Parker *et al.*, 1982; Axelrood *et al.*, 1988; Ishimaru *et al.*, 1988) and some strains are used for the industrial manufacture of antileukaemic L-asparaginases (Robert-Baudouy, 1991). Finally, the ability to secrete multiple proteins is a common trait of several *Erwinia* species and so the genetic tractability of bacteria of this genus has played a significant role in enabling the fundamental study of protein targeting in bacteria. Overall, therefore, the erwinias are of agricultural, industrial and fundamental biological interest. Nevertheless, it is the phytopathological aspects of erwinias that have attracted most interest and has driven considerable research activity.

The molecular biological analysis of pathogenesis and virulence necessitates access to genetic systems that are readily applicable to the organism under study. It is not surprising, therefore, that the taxonomic relatedness of *Erwinia* to *Escherichia coli* encouraged attempts by several workers to transfer a battery of genetic tools from the latter to the former. To a large extent this approach has worked very well indeed – though, it must be said, in a totally strain-dependent fashion. When researchers

METHODS IN MICROBIOLOGY, VOLUME 29
0580-9517 $30.00

have attempted to transfer some of these "*E. coli* type" genetic tools to other bacterial plant pathogens such as *Pseudomonas, Ralstonia, Agrobacterium* and *Xanthomonas* they have been less successful, presumably largely because of the taxonomic gulf when moving out of the *Enterobacteriaceae*. Nevertheless, strains of the other plant pathogens have also been made genetically amenable, usually by exploiting broad host range vectors for transposon mutagenesis and for cloning technology.

This chapter will concentrate on the genetic analysis of the soft rotting erwinias, to highlight examples of the spectrum of techniques that can be applied to an investigation of a bacterial pathogen – which also has a variety of other interesting phenotypes worthy of molecular biological analysis! But, where appropriate, some examples of technique applications in other phytopathogens will also be discussed. So, this chapter is not intended as a comprehensive review of the molecular genetics of bacterial phytopathogenesis but is simply an illustration and discussion of multiple approaches that can be taken when trying to "tool up" an organism for detailed dissection of one or more biological processes of interest.

◆◆◆◆◆◆ II. STRATEGIES TAKEN TO IDENTIFY THE MAJOR VIRULENCE DETERMINANTS OF *ERWINIA* AND OTHER PHYTOPATHOGENS

Two basic strategies for defining virulence determinants in *Erwinia* have been attempted. Firstly, the classical "black box" approach (Daniels *et al.*, 1984), where mutants, generated at random and *en masse*, are screened for their ability to cause disease *in planta*. The advantage of this approach is that, in principle, it should be possible to identify all the rare or non-obvious genes involved in virulence, requiring no prior knowledge of the process.

The alternative strategy is a directed approach which focuses on phenotypes thought likely to affect virulence. The obvious candidates for virulence factors produced by *Erwinia* were the array of extracellular enzymes (pectinases, cellulases (Cel) and proteases (Prt)) which, by their very nature, were expected to play a role in virulence. Consequently, much research has focused on the biosynthesis, structure, regulation and secretion of the plant cell wall degrading enzymes. Directed approaches have also explored the role of the cell surface in the plant–pathogen interaction. A third directed approach has been to identify genes which are induced in the presence of plant cell extracts, on the assumption that some of these are likely to affect virulence. All of these strategies will be discussed in the following sections, detailing some of the genetic "tools" and strategies that have contributed to the advancement of our knowledge of plant–microbe interactions, focusing on the erwinias.

A. A "Black Box" Approach to Virulence Gene Identification

The black box approach has been applied to a number of plant pathogens including *Agrobacterium tumefaciens* (Thomashow *et al.*, 1987), *Pseudomonas syringae* (Anderson and Mills, 1985; Lindgren *et al.*, 1986; Malik *et al.*, 1987), *Pseudomonas solanacearum* (Boucher *et al.*, 1985) and *Xanthomonas campestris* (Dow *et al.*, 1987; Shaw *et al.*, 1988). In addition, reduced virulence mutants of *Erwinia amylovora* (Eam; Bauer and Beer, 1987; Vanneste *et al.*, 1990), *Erwinia carotovora* subsp. *atroseptica* (Eca; Hinton *et al.*, 1989b), *Erwinia carotovora* subsp. *carotovora* (Ecc; Hianda *et al.*, 1987; Pirhonen *et al.*, 1991), and *Erwinia stewartii* (McCammon *et al.*, 1985) have also been isolated. The black box approach requires only a method for the generation of random mutants and an appropriate host plant assay with which to screen for mutants affected in virulence. In most instances, because bacterial virulence determinants are both numerous and complex, mutants isolated are rarely completely non-pathogenic and so we shall refer to them as reduced virulence (Rvi) mutants.

The generation of random Rvi mutants can be achieved by two principal techniques, either by chemical mutagenesis or by gene disruption using transposons (Tns). Both approaches to mutagenesis have been taken by various workers in the genetic analysis of *Erwinia* soft rot disease and indeed for the study of other phytopathogens such as *Xanthomonas* as discussed in the following sections.

1. Chemical mutagenesis

Chemical mutagenesis has been used on numerous occasions to identify virulence determinants of *Erwinia* spp. (Chatterjee and Star, 1977; Andro *et al.*, 1984; Hinton *et al.*, 1985; Hugovieux-Cotte-Pattat *et al.*, 1986; Murata *et al.*, 1990). Chemical mutagens, such as ethyl methane sulphonic acid (EMS), have the advantage over Tn insertions of being able to generate subtle random mutations resulting from single base changes. These point mutations can sometimes give rise to conditional mutations, including temperature-sensitive ones. However, there are several drawbacks to using chemical mutagens. The major disadvantage is that the mutated gene cannot be easily isolated and cloned because, unlike Tn mutagenesis, the gene is not "tagged" with a selectable drug-resistance marker. Chemically-induced mutations may also exhibit "leaky" mutant phenotypes where the point mutation has not entirely abolished gene activity. In addition, it is very difficult to determine whether the mutant phenotype is the consequence of single or multiple mutational events. Consequently, in the study of *Erwinia* and other phytopathogens, Tn mutagenesis has been the favoured technique.

2. Transposon mutagenesis

Transposable elements encoding drug resistance are vital "tools" for bacterial molecular genetic research (Chapter 4) and information derived by their use has formed the basis of numerous aspects of research into *Erwinia* spp. pathogenicity (Chatterjee *et al.*, 1983; Zink *et al.*, 1984; Salmond *et al.*, 1986; Hinton *et al.*, 1987, 1989b). There are various Tns available encoding different antibiotic-resistance markers, including Tn*10* (tetracycline), Tn*3* (ampicillin; Ap) and Tn7 (trimethoprim, streptomycin (Sm) and spectino-mycin (Sp); Kleckner, 1981). However, arguably the most widely used is Tn5 (kanamycin; Kn). Tn5 is a 5700 bp linear element composed of two insertion sequences (IS) that border the region encoding antibiotic resistance (Mazodier *et al.*, 1985). Both of the two IS elements are highly related in their sequence. In Tn5, only one of the IS elements is transcriptionally active and it encodes two independent transcripts expressed from the same open reading frame (ORF). The first transcript encodes a transposase enzyme which catalyses the transposition of Tn5, the other transcript encodes a protein that regulates this transposition event.

A key feature that makes Tns particularly useful for genetic analysis is their ability to insert randomly into the host chromosome, thereby causing gross interruptions to any target gene, although some Tns (to varying degrees) show site preferences ("hot spots"; Kleckner *et al.*, 1979). Reports of Tn5 hot spots have been made previously in Eca, Ecc and *Escherichia coli* (Shaw and Berg, 1979; Hinton *et al.*, 1987; 1989b). For example, in Eca, a disproportionate number of auxotrophic Rvi mutants, generated using Tn5, required either uracil or tryptophan for growth (Hinton *et al.*, 1989b).

As mentioned earlier Tns, by virtue of their antibiotic resistance marker, "tag" the gene into which they insert, thereby aiding subsequent subcloning and sequencing of this disrupted gene. These gene tags are also important tools for strain construction when Tns are used in conjunction with generalized transducing phage. The phage, propagated on a particular Tn mutant strain, can be used to shuttle the Tn insertion into another strain. In this way, by transducing the Tn-linked mutation into a "clean" wild type genetic background the linkage between the insertion mutation and the phenotype can be confirmed. In addition, this strategy can be used to construct defined bacterial strains carrying multiple mutations and, if one of the Tns has gene fusion capabilities, the effect of one mutation on the expression of another gene can also be determined.

Unlike with chemical mutagens, it is possible to determine the exact number of Tns that have inserted into the bacterial chromosome. This is achieved by designing DNA probes to the Tn sequence and using them in Southern blot analysis. Alternatively, when transducing the Tn insertion mutation into a clean genetic background using generalized transducing phage (as discussed above), the co-inheritance of the Tn drug-resistance marker and the mutant phenotype can also be scored. A single Tn insertion should give a co-inheritance score of 100% for the phenotype and the drug resistance. This is reduced to 50% if there are two Tns present, and so on.

(a) Mu-based plasmid delivery vectors

Bacteriophage Mu represents another form of Tn. Mu is a temperate bacteriophage which is able to integrate, at random, into the chromosome of its host generating genetic lesions (hence Mu–Mutator). Once inserted into the chromosome Mu replicates by a process involving transposition and it is this ability to integrate randomly into DNA that has made Mu a very useful tool for genetic manipulation.

The successful adsorption of a bacteriophage to a bacterial cell requires an outer membrane receptor. For Mu, this receptor is located in the lipopolysaccharide core region (Sandulache et al., 1985). The exact structure of the receptor recognized by Mu is determined by the orientation of an invertible "G" segment on the Mu genome. The G segment carries two sets of genes, one of which is transcribed depending on the orientation of the G segment (either G(+) or G(−)), which encode the complementary phage tail receptor binding site (Sandulache et al., 1985). The host range of Mu is also altered as a consequence of the orientation of the G segment. One of the major drawbacks in using bacteriophage Mu for molecular genetic research is its restricted host range. Originally it was thought to infect only strains of E. coli, Citrobacter and Shigella (Howe and Bade, 1975), but several studies have since shown a wider host range and derivatives of Mu have been used to study virulence determinants in specific strains of Erwinia chrysanthemi and Eam. Various mutants have been isolated in these Erwinia strains using Mu d1 phage (d–defective for phage proliferation but can actively transpose; Casadaban and Cohen, 1979) such as the Out⁻ secretion mutants (Hugovieux-Cotte-Pattat and Robert-Baudouy, 1985; Ji et al., 1987). Mu d1 insertions are selectable by virtue of an ampicillin-resistance gene. This phage carries a promoterless lacZYA operon such that, if inserted in the correct orientation within the gene, the expression of lacZ will be driven by the promoter of the inactivated gene. By measuring the β-galactosidase activity, using the chromogenic substrate 5-bromo-4-chloro-indoyl-β-D-galatopyranoside (X-gal), the regulation of the inactivated target gene can be studied. The major problem with the use of Mu d1 as a mutagen is that it enters the lytic cycle if exposed to temperatures exceeding 37°C, due to the inactivation of the cts repressor (Baker et al., 1983). However, the development of Mu dX, a derivative of Mu d1 with the Tn Tn9 inserted in the Mu B gene (Baker et al., 1983), has overcome this problem of thermo-induction and has also been used successfully to isolate insertion mutants of phytopathogens such as Eam where mutants affected in pathogenicity and the ability to provoke a plant hypersensitive response were isolated (Vanneste et al., 1990).

The general principles and procedures behind the use of Mu d phages are as follows. Mu d derivatives are unable to replicate (see Casadaban and Cohen, 1979) and so lysates of these phages are prepared from an E. coli strain dilysogenic for Mu d and a thermoinducible helper prophage which provides the functions for the lytic cycle in trans. The culture of the strain to be mutated is grown to early stationary phase and infected for 20 min, at 30°C, with the Mu lysate at a multiplicity of infection of 0.4 giving approximately 10^{-3} antibiotic resistance Mu d particles per plaque-forming helper phage (see Baker et al., 1983). The infected culture is incubated for a further hour to allow expression of the antibiotic resistance marker on the Mu d Tn, after which transductants are selected on plates containing the appropriate antibiotics. All mutants isolated in this way should be further screened for the presence of a co-integrated helper phage. This is achieved by incubating the cells at a temperature of 42°C for 20 min followed by a further period of 1 h incubation at 37°C. If a helper prophage has co-integrated into the chromosome, then thermal induction will induce this phage to enter the lytic cycle, thereby resulting in cell lysis and the release of Mu phage particles.

The drawbacks associated with using Mu as a mutagen centre around the instability of the Mu insertions because, once inserted, Mu remains able to transpose to a second site (albeit at a lower frequency for Mu *dX*; Baker *et al.*, 1983) making phenotypic analysis potentially problematic. In addition, although this phage is useful for the construction of gene fusions for the study of regulation (Hugovieux-Cotte-Pattat and Robert-Baudouy, 1985), the large size of the Mu DNA (~35 kb) inserted into a gene makes it difficult to clone the DNA flanking the insert. Consequently, derivatives unable to replicate and undergo secondary transposition have been developed, called mini-Mu bacteriophages (Castilho *et al.*, 1984), which have large deletions of Mu DNA reducing their size to between 9 and 22 kb. The advantage of these mini-Mu Tns is that they form stable insertions and because of their reduced size they are more useful for mapping and cloning.

One method of overcoming the limited host range of Mu has been to co-integrate the bacteriophage on to a self-transmissible broad host range plasmid to serve as a delivery vehicle (Boucher *et al.*, 1977). Derivatives of the Inc-P plasmid RP4 have been used for just such a purpose, and they are able to transfer, by conjugal mobilization, into most Gram-negative bacteria (Datta *et al.*, 1971; Olsen and Shipley, 1973). Transfer of the RP4::Mu plasmid is achieved in a patch mating of a suitable *E. coli* donor and the recipient cells, allowing transfer of a single strand of the plasmid DNA via the sex pilus, encoded by the plasmid *tra* genes. The donor and recipient cells can then be counter-selected by the use of antibiotic resistance or by the omission of a nutritional supplement in selective media. Once conjugally transferred into the new host, Mu is able to undergo zygotic induction leading to its random insertion into the host genome. A similar system has been employed successfully to generate Rvi mutants of *E. stewartii* (McCammon *et al.*, 1985).

Other strategies for introducing Mu into Mu-resistant strains include the construction of Mu–P1 hybrids. These phages share a homologous region which determines host specificity. Recombination of these regions between the two phages results in the generation of Mu phage hybrids which can infect Mu-resistant, P1-sensitive strains (Toussaint *et al.*, 1978).

(b) Suicide vectors

Tns, other than Mu, cannot replicate autonomously and so, for molecular genetic use, are inserted into a vector for their replication and transmission into the recipient cell. Several delivery systems have been used for the genetic analysis of *Erwinia* spp. Widely used delivery systems are those based on bacteriophage Mu and λ.

(i) Mu-based Tn delivery systems

Suicide vectors based on bacteriophage Mu exploit the property of Mu to interfere with the replication of IncP plasmids in certain genetic backgrounds. It is thought that restriction by the host and at least one of the Mu gene products are responsible for the failure of these plasmids to become established (Boucher *et al.*, 1977; van Vliet *et al.*, 1978). Thus an RP4::Mu plasmid can act as a suicide vector (van Vliet *et al.*, 1978). Beringer *et al.* (1978) constructed such a RP4::Mu suicide plasmid carrying Tn5, pJB4JI, which has been used extensively in the erwinias (Gantotti *et al.*, 1981; Chatterjee *et al.*, 1983; Zink *et al.*, 1984).

Plasmid pJB4JI is stably propagated in *E. coli*. It confers gentamycin (Gm) resistance and carries Tn5 inserted into the prophage Mu. The RP4-based plasmid can be transferred to many Gram-negative bacteria by conjugal mobilization, as described above. After transmission into the recipient, pJB4JI is unstable and is lost from the cell. Donor and recipient cells are counter-selected using a drug susceptibility and/or auxotrophy. Due to the unstable nature of the vector in the transconjugant *Erwinia* cells, selecting for the antibiotic marker carried on Tn5 will select for survivors of transposition events, where Tn5 has inserted into the host chromosome. The resultant transconjugants can be identified as being Gm^s (due to loss of plasmid pJB4JI) and Kn^R (the drug resistance marker carried on Tn5).

Unfortunately the instability of the RP4::Mu-based vectors is not universal to all *Erwinia* strains. Various studies have reported the stable maintenance of these plasmids in some Ecc and Eca (Zink *et al.*, 1984), Eam (Vanneste *et al.*, 1990) and *E. stewartii* (McCammon *et al.*, 1985) isolates. In addition, Mcade *et al.* (1982) report that 25% of the putative Tn-insertion mutants recovered after conjugation possessed, in addition to Tn5, Mu-derived sequences inserted into their chromosomes, making the process of phenotypic analysis arduous. Observations of this nature highlight the need for alternative Tn delivery systems for the study of some of the erwinias. One such alternative is based on another bacteriophage, lambda (λ).

(ii) Lambda (λ)-based delivery systems

Bacteriophage λ is a powerful genetic tool developed originally for work with *E. coli* K12. As with Mu, its use in other Gram-negative bacteria as a delivery system for Tns or cosmids was limited by its host range. This restricted host range is partly the result of λ requiring a single receptor protein, the product of the *lamB* gene, for its adsorption. The LamB protein is a component of the *E. coli* outer membrane and functions as part of an uptake system for maltose and maltodextrins, in addition to being parasitized as a receptor by several coliphages. All the information required for the targeting and the insertion of the LamB protein in the outer membrane is encoded by the *lamB* structural gene. This has been exploited in order to extend the host range of bacteriophage λ by simply cloning the *lamB* gene into various multicopy plasmids which can be delivered into new hosts (Clement *et al.*, 1982; De Vries *et al.*, 1984; Harkki and Palva, 1985).

Plasmids carrying *lamB* have been used to make λ-sensitive derivatives of a variety of bacteria including strains of *A. tumefaciens, Rhizobium meliloti* and *P. aeruginosa* (Ludwig, 1987), as well as numerous strains of *Erwinia* in which the "power" of λ-mediated Tn mutagenesis has been amply demonstrated (Salmond *et al.*, 1986; Hinton *et al.*, 1987, 1989b; Steinberger and Beer, 1988; Ellard *et al.*, 1989; Pirhonen *et al.*, 1991; Mulholland *et al.*, 1993).

The introduction of *lamB*+ plasmids into *Erwinia* is achieved by either conjugation or direct transformation, using electroporation. Conjugal transfer can be performed in a tri-parental mating. Many of the LamB encoding plasmids have a mobilization site (Mob; commonly from RP4) which forms the origin of transfer permitting the transfer of the plasmid into the recipient. A "helper" strain, containing a transfer-proficient (Tra+) plasmid, is also required to encode the production of the sex pilus. The Tra+ plasmid is transferred into the helper strain containing the Mob plasmid before transferring the Mob plasmid into the recipient strain. Electroporation is the most efficient system with which to transform *Erwinia* directly (see Mulholland and Salmond, 1995).

The transfer of LamB+ plasmids to the erwinias has not been without its problems: λ-mediated Tn mutagenesis in the erwinias is strongly strain dependent (Ellard *et al.*, 1989; Hinton *et al.*, 1989b). Strain-dependent difficulties were experienced with features such as plasmid host range and with phage absorption. However, problems in transferring the relatively narrow host range plasmid, pHCP2 (based on pBR322; Clement *et al.*, 1982), to some *Erwinia* strains was somewhat overcome by the use of pTROY9 (De Vries *et al.*, 1984). The latter vector is based on a broad host range plasmid selected for its high constitutive expression of *lamB*. Where plasmid transfer was successful, but Tn mutagenesis failed, it was thought to be the result of either the inability of the LamB to be expressed and/or inserted into the outer membrane or because the receptor binding site had been occluded by extracellular polysaccharides thereby preventing absorption of the bacteriophage (Salmond *et al.*, 1986). In some cases the incorporation of a sodium chloride wash prior to phage absorption alleviates this problem by removing the extracellular polysaccharide coating the cell (see Ellard *et al.*, 1989).

To make proper use of this technology it is desirable that λ is unable to replicate and form plaques in the host but is still able to inject the DNA contained within its phage head. Fortuitously, for the study of Ecc and Eca, all of the most studied strains are able to express the LamB protein and can be infected by λ, but cannot support its replication.

The use of λ as a delivery system offers several advantages over the use of plasmid-based systems, such as pJB4JI, in *Erwinia*. Mutagenesis with λ::Tn5 is a very simple technique, involving phage infection (approximately 30 min) followed by a short period for transposition and expression of the mutant phenotype (1–2 h), after which transductants are selected on complex media supplemented with the appropriate antibiotics. Plasmid-mediated Tn mutagenesis is more laborious and time consuming, requiring a clear ability to counter-select against the *E. coli* donor by either auxotrophy, carbon source utilization or an additional antibiotic resistance. But perhaps the most compelling reason for using lambda as a genetic tool is the range of Tns that are available on this vector. λ-Based Tns include those for tagging genes and making *in vivo* deletions (Tn*5*, Tn*10*; Way *et al.*, 1984), Tns for studying export, secretion and membrane topology of proteins (Tn*phoA* and Tn*blaM*; Manoil and Beckwith, 1985; Hinton and Salmond, 1987; Tadayyon and Broome-Smith, 1992). In addition are λ-based Tns for assaying transcription or translation of the target gene (Tn*10LacZ*; Way *et al.*, 1984).

If λ::Tn*10* is used then mutants which are isolated can be used to generate stable deletion derivatives by eduction around the insertion site (Bochner *et al.*, 1980). Bacterial cultures to be cured of Tn*10* are plated on to media containing fusaric acid and chlortetracycline, which are toxic to cells carrying Tn*10*. Colonies able to survive under these conditions have usually lost part or all of the Tn plus some flanking sequences. This can be an extremely useful method of strain construction, especially if problems are encountered with the secondary transposition of the inserted Tn or there are a restricted number of antibiotic markers available for use (see Mulholland and Salmond, 1995).

Although gene fusion Tns such as Tn*phoA* and Tn*blaM* are more commonly used in directed genetic approaches to study gene regulation and protein targeting, there is, in principle, no reason why these Tns cannot be used in a black box approach. Tn*phoA*, a derivative of Tn*5*, carries a truncated version of the *E. coli phoA* gene encoding alkaline phosphatase (Manoil and Beckwith, 1985). If the Tn inserts in the correct orientation and reading frame a hybrid protein is formed. The PhoA hybrid protein is only catalytically active if it is exported to the periplasm requiring a *sec*-dependent export signal to be provided by the gene into which Tn*phoA* has inserted. Alkaline phosphatase can be assayed by the breakdown of the chromogenic substrate 5-bromo-4-chloro-3-indoyl phosphate (XP; Manoil and Beckwith, 1985). Use of Tn*phoA* offers an enrichment for mutants affected in genes encoding a subset of proteins, as transductants appearing blue on XP plates must have a Tn*phoA* insertion in an exported, secreted or transmembrane protein (see Hinton and Salmond, 1987). There are also a number of Tn*phoA* derivatives which allow switching of the fusion type by homologous recombination (Wilmes-Riesenberg and Wanner, 1992). An existing Tn*phoA* fusion may be exchanged to allow the production of translational or transcriptional fusions or for transposition-defective derivatives, lacking the transposase, and forming stable insertions.

Tn*blaM* carries a truncated β-lactamase gene in addition to a Sp-resistance gene. As with Tn*phoA*, if this Tn inserts in the correct orientation and reading frame, gene fusions encoding a hybrid BlaM protein are generated. This system can be used to isolate translational fusions to both cytoplasmic and exported proteins. Cytoplasmic protein fusion-containing strains are identified as being able to grow on media containing ampicillin, but only with heavy inocula. Strains with fusions in genes encoding exported proteins, however, can grow as individual colonies on ampicillin-containing media (Tadayyon and Broome-Smith, 1992).

Transductants in which Tn*blaM* has inserted into sites from which it transposes at a high frequency can also be isolated ("high hoppers"). Random insertion mutants which are Sp resistant but ampicillin sensitive are patched out on to media containing increasing concentrations of ampicillin (10 µg ml^{-1}–1 mg ml^{-1}). This selects for secondary transposition events where high hoppers can be identified as generating a number of individual colonies within this patch, resulting from the secondary transposition of Tn*blaM* into a suitably-expressed gene leading to a productive BlaM fusion (see Mulholland and Salmond, 1995). This allows the isolation of a variety of mutants with translational fusions without the need for further Tn mutagenesis. To confirm the phenotype, the insertions can be transduced out into the progenitor strain, in the case of Ecc with ΦKP (Toth *et al.*, 1993). The advantage of this technique is that it circumvents the need for a high level of efficiency in the initial λ- mediated transduction.

It is also important to note that the availability of λ-sensitive strains of *Erwinia* has been useful because the presence of LamB allows direct cosmid complementation of any mutant and so enables the rapid cloning of corresponding genes by complementation (Murata *et al.*, 1990; Mulholland *et al.*, 1993).

3. *In planta* screening protocols looking for reduced virulence mutants

Success of a classical "black box" approach to identifying pathogenicity determinants is dependent on the availability of simple tests with which mutagenized bacterial colonies can be screened *en masse* for reduced virulence in their host plant. Hinton *et al.* (1989b) devised such a screen for blackleg caused by Eca using micropropagated potato plants (*Solanum tuberosum*). Micropropagated plants were grown aseptically (under controlled conditions: $20°C$ and a 16 h light regime) until they reached the three leaf node stage, after approximately two weeks. Nodal segments, with the leaves, were cut and further propagated allowing for root formation (approximately 1 week). Plants were then transferred into peat pots and grown until they were between 5 and 10 cm tall (stem approximately 2 mm thick).

Cells from individual bacterial colonies, recovered after either Tn or chemical mutagenesis, were taken from an agar plate using a sterile tooth pick. The tooth pick was then used to inoculate the micropropagated plants by stabbing the cells into a pre-cut 1 mm deep, 2 mm long, slit cut in the stem of the potato plants just below the leaf node. Paraffin wax was used to seal the hole to prevent desiccation. The plants were then re-incubated, as before, for a period of 5–7 days allowing the symptoms of blackleg to develop fully. The extent of the symptoms caused by the inoculated mutants was then scored on a numerical stem rot index. A score of "0" indicated that there was no visible reaction, "1" a slight browning around the inoculation site, "2" a slight blackening around this site, "3" the appearance of a small area of black rot spreading from the inoculation site, "4" the appearance of a medium area of black rot spreading from the inoculation site and "5" a large area of black rot accompanied by complete stem collapse.

Various other workers have used similar types of screening regime; for example the screening of Ecc on axenic tobacco seedlings by Pirhonen *et al.* (1991) or the screening of *Xanthomonas* on turnip seedlings by Daniels *et al.* (1984). The drawback with screening campaigns of this nature, apart from the inherent subjectivity of the scoring systems, is that they lack sufficient sensitivity for mutants with very subtle reductions in their ability to cause disease (rot index score of 4). Problems can also be encountered with the reproducibility of the results. Hinton *et al.* (1989b) observed that more reliable assay results could be obtained by re-testing any strains identified in the initial screen as being Rvi mutants, on more mature potato plants. The latter tests were thought to be more reproducible assays for blackleg because the high water content and undifferentiated nature of the parenchymous tissue of the micropropagated plants made them highly susceptible to infection by *Erwinia*.

In searches for virulence determinants of a number of *Erwinia* species potato tuber assays have proven to be important assays in both the black box approach and the directed approach (Keen *et al.*, 1984; Collmer *et al.*, 1985; Hinton *et al.*, 1989b; Pirhonen *et al.*, 1991; Jones *et al.*, 1993; Mulholland *et al.*, 1993). This assay represents a facile screen for the ability to macerate plant tissue. Potato tubers are surface sterilized by washing in 1% NaOCl, rinsed with distilled water, and air dried. Dilutions of bacterial suspension are injected into a hole drilled using a 200 μl micropipette tip inserted 15 mm into the intact tuber. Inoculation sites are sealed with paraffin wax and the tubers wrapped in alternate layers of moist tissue and "cling film" to create a moist and anaerobic environment (Collmer *et al.*, 1985), optimal for potato tuber tissue maceration (Maher and Kelman, 1983). Tubers are then incubated at 25°C for up to six days. The extent of rot is determined by gently scraping out the rotted tissue and weighing it. It is also important to plate out and enumerate the infecting bacterial cells to determine their ability to grow *in planta* (Payne *et al.*, 1987) to identify any growth rate mutants.

Possible problems associated with this assay centre on its relatively crude nature. Hinton *et al.* (1989b) found that high standard deviation values were recorded when testing Eca Rvi mutants, despite the use of multiple replicates. In addition, there is a significant variation in the susceptibility of various potato cultivars to soft rot disease (Lapwood and Read, 1986). It is also important to note that there is an element of bias introduced with all of the plant assays where direct injection is used. In this type of crude inoculation assay mutations affecting factors important in the early stages of infection may not be detected. However, there is a range of other assays which do address these questions.

Plant leaf assays can measure the invasiveness of the bacteria by simply spotting the inocula directly on to the leaf surface. One such assay is used to assay virulence in Echr where leaves, aseptically removed from Saintpaulia plants are placed in Petri dishes on moist filter paper (to maintain humidity). Bacterial cultures can then be applied to the leaf surface or end of the petiole and incubated in a sealed Petri dish for between 2 and 5 days. The extent of rot is then scored, on a rot index, in a similar manner to above (Expert and Toussaint, 1985). Daniels *et al.* (1984) also used a similar type of assay looking for Rvi equivalent mutants of *Xanthomonas* on turnip leaves.

Virulence Determinants in Erwinia

4. Reduced virulence mutants (Rvi) isolated by random mutagenesis

Reduced virulence mutants have been isolated from a number of plant pathogens using the black box approach followed by plant tests. Mutants of Ecc and Eca fall into three broad categories: (i) mutants affected in motility; (ii) those affected in the synthesis and/or secretion of extracellular enzymes; and (iii) those affected in growth, either in rate of growth or requiring nutritional supplements.

Pirohnen *et al.* (1991) generated 6200 Kn^R transductants using λ::Tn5. Of these, 298 were found to be Rvi mutants on axenic tobacco seedlings (*Nicotiana tabacum* L. "*Samsum*"). The majority of these (268) were motility mutants; most being completely non-motile (Mot⁻), others exhibiting intermediate levels of motility. Motility mutants were also isolated for Eca. Mulholland *et al.* (1993) screened 1384 λ::Tn5 transductants on micro-propagated potato plants. The Mot⁻ mutants were examined under the electron microscope and found to lack flagella (Fla⁻). Further examination of the Eca Mot⁻ mutants, using a series of Eca bacteriophage, revealed that they exhibited a pleiotropic phenotype showing a differing pattern of resistance when compared to the wild type. This suggested that these mutants were altered in their cell surface. Indeed these mutants were also more sensitive to detergents and surface active agents than the progenitor strain.

The role of motility in plant pathogens is unclear. Initially it had only been shown to be important for virulence of Eam, *P. syringae* and *P. phaseolicola* only when the bacteria are externally applied to the leaf or stem. (Panopoulos and Schroth, 1974; Bayot and Ries, 1986; Hattermann and Ries, 1989). However, Pirhonen *et al.* (1991) have shown motility in Ecc is also important for the virulence of directly injected bacteria.

The second category of mutant isolated was composed of Rvi mutants affected in the synthesis and/or secretion of exoenzymes. It had always been anticipated that this class of mutant would exist because of the very nature of the array of exoenzymes the erwinias and other phytopathogens produce. Pirhonen *et al.* (1991) identified three classes of enzyme mutant: Out⁻, Exp⁻ and Pnl⁻.

Out⁻ mutants had been previously reported in Ecc and Echr; they synthesize polygalacturonases (Peh), pectate lyases (Pel) and Cel normally but are unable to secrete them properly and so the enzymes accumulate in the periplasmic space (Chatterjee *et al.*, 1985b; Andro *et al.*, 1984). The synthesis and secretion of Prt, however, is unaffected in Out⁻ mutants. This is now known to be because Peh, Pel and Cel are secreted by a common pathway (the Type II pathway) with a periplasmic intermediate and Prt is secreted via an alternative route (the Type I pathway).

Mutants resembling Out⁻ mutants were also isolated in Eca (Hinton *et al.*, 1989b). Eca Pep⁻ mutants (pectolytic enzyme production) are similar to Out⁻ mutants in that they are defective for secretion of Pel and Peh. However, they are distinct from Out⁻ mutants in that the synthesis of these enzymes is also reduced.

Exp⁻ mutants produce significantly reduced levels of Cel and Prt and are reduced for the production of Pel and Peh, the residual Pel and Peh activity being located in the periplasm. These highly pleiotropic mutants appeared to be global regulatory mutants, impaired in the synthesis and secretion of multiple exoenzymes (see section II.B.2(b)(iii)).

Single enzyme mutants have also been isolated. Pnl⁻ mutants produce wild-type levels of all the other enzymes but are defective for the production of pectin lyase (Pirhonen et al., 1991).

The last category of Ecc and Eca Rvi mutants that has been isolated by the black box approach are those which have specific nutritional requirements or those with a reduced growth rate. Hinton et al. (1989b) screened 3245 λ::Tn5 transductants on potato seedlings. Nine of these were auxotrophs which were also Rvi. However, not all of the auxotrophs were Rvi. Interestingly, some mutants, with the same nutritional requirement, displayed different abilities to cause disease in planta. In addition, some of the auxotrophs, which were found to be Rvi when tested on seedlings, were unimpaired in virulence when tested in potato tubers. One of the more interesting auxotrophic mutants was a cysB mutant. The cysB gene product is a positive regulator known to play a key role in sulphate assimilation and had been previously shown to affect exoenzyme production under sulphate limiting conditions (Hinton et al., 1989b).

Growth rate mutants have been isolated in Xanthomonas. Daniels et al. (1984) isolated mutants which grew normally in culture or on agar plates but had a reduced ability to grow in planta. Similar mutants have also been described in other plant pathogens such as P. syringae pv. phaseolicola (Lindgren et al., 1986).

Other Rvi mutants that have been isolated include three classes of Eam mutants. Using Mu dX, Vanneste et al. (1990) isolated mutants which were avirulent on apple root calli and, when tested on tobacco leaves, these mutants were unable to elicit a hypersensitive response (HR). The mutants were termed hypersensitive response and pathogenicity mutants (Hrp). Similar mutants have been isolated in many plant pathogens including P. solanacearum (Boucher et al., 1987) and P. syringae pv. tomato (Cuppels, 1986).

Vanneste et al. (1990) also isolated mutants that were avirulent but remained able to provoke a HR, termed disease-specific mutants (Dsp), and a third class of mutants the majority of which were impaired for extracellular polysaccharide (EPS) production. Random EPS mutants have also been isolated for E. stewartii (McCammon et al., 1985).

In addition to being EPS⁻, one of the class three mutants was also unable to grow on media deficient in iron because it lacked a functional siderophore, implying that the iron uptake system of Eam is a virulence factor (Vanneste et al., 1990). This has also been found to be true for a number of other phytopathogens (see section II.B.4). Clearly a wide variety of Rvi mutants have been isolated using the black box approach. A more detailed account and the genetic explanation for some of these mutants will be given in the following sections.

B. A Directed Approach to Virulence Gene Identification

1. The biosynthesis and role of exoenzymes in virulence

The plant cell-wall-degrading enzymes of the soft-rot erwinias were, by their very nature, predicted to play a role in virulence. This was confirmed by the isolation of a set of Rvi⁻ mutants in which the loss of virulence was associated with the organism's inability to produce individual plant cell-wall-degrading enzymes (Andro *et al.*, 1984). These enzymes are generally called exoenzymes because the majority are secreted to the exterior of the cell. The activities and structures of the exoenzymes were initially characterized biochemically. However, the era of genetic technology brought about a more detailed genetic investigation of their biosynthesis and regulation. This section will briefly describe the array of exoenzymes employed by the soft-rot erwinias to macerate plant tissue and the important biochemical methods used in identifying their enzymatic activities. We will then concentrate on the genetic approaches taken to study the biosynthesis of the exoenzymes. In contrast, section II.B.2 concerns the regulation of the exoenzymes. A number of standard genetic techniques have been utilized in the characterization of the exoenzymes. However, these will not be discussed in detail.

(a) The exoenzyme repertoire

The soft-rot erwinias employ an arsenal of exoenzymes which act synergistically to break down pectin to yield the substrates of the pectinolysis pathway (Figure 1). This repertoire includes several endo- and exo-Pel isozymes, pectin lyase (Pnl), pectin acetyl esterase (Pae Y), pectin methylesterases (Pme), and endo- and exo-Peh (Table 1). The nomenclature

---

Figure 1. A diagramatic illustration of the degradation of pectin by *Erwinia chrysanthemi*. The figure describes the modes of action of pectinolytic exoenzymes Pme (pectin methylesterase A), Pnl (pectin lyase), Pel (pectate lyases – PelA, PelB, PelC, PelD, PelE, PelI, PelL, PelX (the location of PelX is currently unclear: see Table 1) and PelZ) and Peh (polygalacturonase). Pectin is first demethoxylated, by Pme (encoded by *pemA*) to form PGA (polygalacturonate). The activity of Pel and Peh is low on pectin, because of the methoxyl groups, and so the main role of these enzymes is to cleave PGA, produced by Pme, generating saturated and unsaturated oligogalacturonides which can then be taken into the cell via a sugar transport system. Unlike Pel and Peh, Pnl is able to cleave pectin efficiently. The degradative products of Pnl activity are thought to be demethoxylated by PemB (pectin methylesterase B) before being catabolized by Ogl(*). The intracellular catabolism of pectin is mediated by the enzymes: Ogl (oligogalacturonate lyase), KduI (5-keto-4-deoxyuronate isomerase), KduD (2-keto-3-deoxygluconate oxidoreductase), KdgK (2-keto-3-deoxygluconate kinase) and KdgA (2-keto-3-deoxy-6-phosphogluconate aldolase). Also shown in the figure are the three types of secretion system used to transport exoenzymes into the extracellular milieu. In addition to the exoenzymes depicted, pectin acetyl esterase (Pae Y; not shown) is important for the removal of acetyl groups from sugar beet pectin prior to enzymatic degradation by Pel (see section II.B.1(a)). Echr, a representation of the *Erwinia chrysanthemi* cell; IM, inner membrane; OM, outer membrane; Cyto, cytoplasm; Peri, periplasm; Prt, proteases; Cel, cellulases. The unfilled arrow represents a sugar uptake system. Types I, II and III represent simplistic representations of the protein secretion systems described in section II.B.3. Harpins are thought to be transported from the cytoplasm to the extracellular milieu in a single step (see section II.B.3(b)(i) for further details).

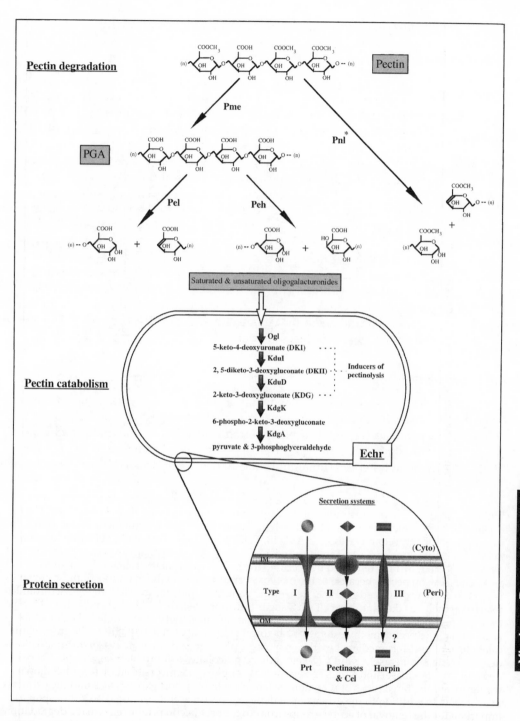

Table I. (a). Plant cell-wall-degrading pectinases and cellulases produced by *Erwinia chrysanthemi* (Echr)

Strain	Enzyme	Activity[1]	Class[2]	Location[3]	pI	pH opt.	T opt. (°C)	Gen-Bank	References
3937	PelA	endo-Pel	Ib	E	4.6	8.5	55	M77808	Favey et al. (1992); Tardy et al. (1997)
	PelB	endo-Pel	Ia	E	7.9	9.3	60	X67475	Hugouvieux-Cotte-Pattat and Robert-Baudouy (1992); Tardy et al. (1997)
	PelC	endo-Pel	Ia	E	8.2	9.2	60	None	Hugouvieux-Cotte-Pattat and Robert-Baudouy (1992); Tardy et al. (1997)
	PelD	endo-Pel	Ib	E	9.8	8.8	50	None	Kotoujansky et al. (1985); Tardy et al. (1997)
	PelE	endo-Pel	Ib	E	10.3	~8	50	M33584	Reverchon et al., 1989; Tardy et al. (1997)
	PelI	endo-Pel (P)	III	E	~9	9.2	ND	Y13340	Shevchik (1997a)
	PelL	endo-Pel (P)	IV	E	9.5	8–9	ND	X81136	Lojowska et al. (1995)
	PelZ	endo-Pel (P)	V	?	8.8	ND	ND	X97119	Pissavin et al. (1996)
	KdgC	none[4]	II	?	ND	ND	ND	X62073	Condemine and Robert-Baudouy (1991)
	EGZ (CelZ)	endo-Cel	A	E	4.5	7	ND	Y00540	Boyer et al. (1987a); Guiseppi et al. (1988)
	EGY (CelY)	endo-Cel	D	?	8.2	5.7	ND	M74044	Boyer et al. (1987a); Guiseppi et al. (1991)
	PemA	Pme		E	9.64	ND	ND	L07644	Laurent et al. (1993)
	PemB	Pme		OM[5]	9.5	7.5	40	X84665	Shevchik (1996)
	PaeY	Pae		E	9	8	ND	Y09828	Shevchik and Hugouvieux-Cotte-Pattat (1997)
EC16	PelA	endo-Pel	Ib	E	4.2–4.6	8.5–9.5	ND	M14509	Barras et al. (1987); Tamaki et al. (1988)
	PelB	endo-Pel	Ia	E	8.8	8.5–9.5	ND	M14510	Barras et al. (1987); Keen and Tamaki 1986
	PelC	endo-Pel	Ia	E	9	8.5–9.5	ND	M19411	Barras et al. (1987); Tamaki et al. (1988)
	PelE	endo-Pel	Ib	E	9.8–10	8.5–9.5	ND	M14509	Barras et al. (1987); Keen and Tamaki (1986)
	PelL	endo-Pel (P)	IV	E	alkaline	8.5–9	ND	L42248	Alfano et al. (1995)
	PelX	exo-Pel (P)	IV	?	8.6	7.5–8	ND	M62739	Brooks et al. (1990)
	PehX	exo-Peh		E	8.3	ND	ND	M31308	He and Collmer (1990)
B374	PelA	endo-Pel		E	ND	ND	ND	None	van Gijsegem (1989)
	PelD	endo-Pel	Ib	E	ND	ND	ND	X17284	van Gijsegem (1989)
	PelE	endo-Pel	Ib	E	ND	ND	ND	X17284	van Gijsegem (1989)
	Pme	Pme		E	9.9	ND	ND	Y00549	Plastow (1988)
3665	EGZ (CelZ)	endo-Cel		E	4.6	7	ND	None	Boyer et al. (1984b); Boyer et al. (1987a)
	EGY (CelY)	endo-Cel		?	8.2	5.5–6	ND	None	Boyer et al. (1987b)

Table I. (b). Plant cell-wall-degrading pectinases and cellulases produced by *Erwinia carotovora* subsp. *carotovora* (Ecc)

Strain	Enzyme	Activity[1]	Class[2]	Location[3]	pI	pH opt.	T opt. (°C)	Gen-Bank	References
SCRI193	PelA	endo-Pel	II	P	7.2	ND	ND	None	Hinton et al. (1989a)
	PelB	endo-Pel	II	P	8.2	ND	ND	X16397	Hinton et al. (1989a)
	PelC	endo-Pel	Ia	E	10.3	ND	ND	X16398	Hinton et al. (1989a)
	PelD	endo-Pel	Ia	E	11	ND	ND	None	Hinton et al. (1989a)
	CelV	endo-Cel	A[g]	E	4.4–4.6	7	42	X76000	Cooper and Salmond (1993)
	Peh	endo-Peh		E	ND	ND	ND	X52944	Hinton et al. (1990)
EC153	Pel153	endo-Pel	II	P	8.8	ND	ND	J03673	Trollinger et al. (1989)
Ecc71	Pel-1	endo-Pel	Ia	E	10	ND	ND	L32171	Willis et al. (1987); Chatterjee et al. (1995a)
	Pel-2	endo-Pel		E	9.7	ND	ND	None	Willis et al. (1987)
	Pel-3	endo-Pel	VI	E	7.96–9.2	ND	ND	L32172	Willis et al. (1987); Liu et al. (1994a)
	Pel-4	endo-Pel		P	8.	ND	ND	None	Willis et al. (1987)
	Peh-1	endo-Peh		E	9.96	5.5	ND	L32172	Willis et al. (1987); Liu et al. (1994a)
	Peh-2	exo-Peh		?	ND	ND	ND	None	Willis et al. (1987)
	PnlA	Pnl		?[*]	9.92	ND	ND	M59909	McEvoy et al. (1990); Chatterjee et al. (1991)
SCC3193	PelA	NC		E	alkaline	ND	ND	None	Heikinheimo et al. (1995)
	PelB	endo-Pel[7]	VI	E	alkaline	9.5	ND	X79232	Heikinheimo et al. (1995)
	PelC	NC		E	alkaline	ND	ND	None	Heikinheimo et al. (1995)
	PelD	NC		E	alkaline	ND	ND	None	Heikinheimo et al. (1995)
	CelV1	Cel	A[8]	E	ND	ND	ND	X79241	Mäe et al. (1995)
	CelS	Cel	Novel	E	5.5	6.8	45–55	M32399	Saarilahti et al. (1990a)
	PehA	endo-Peh		E	10.19	5.5	35–45	X51701	Saarilahti et al. (1990b)
TG3	Pel	endo-Pel		?	ND	ND	ND	None	Hu et al. (1987)

Virulence Determinants in Erwinia

Table I. (c). Plant cell-wall-degrading pectinases produced by *Erwinia carotovora* subsp. *atroseptica* (Eca)

Strain	Enzyme	Activity[1]	Class[2]	Location[3]	pI	pH opt.	T opt. (°C)	Gen-Bank	References
EC	PelA	endo-Pel	Ia	E	ND	8.5	40	M18859	Lei *et al.* (1985a); Lei *et al.* (1988)
	PelB	endo-Pel	Ia	E	ND	8.3	40	M17264	Lei *et al.* (1985a); Lei *et al.* (1987)
	PelC	endo-Pel		E	ND	ND	ND	None	Lei *et al.* (1985a)
	Peh	endo-Peh		E	ND	5.5	37	M87433	Lei *et al.* (1985b); Lei *et al.* (1992)
Er	PLI (PAL-I)	endo-Pel		E	10.7	10	50	None	Suguira *et al.* (1984); Ito *et al.* (1988)
	PLII	endo-Pel	Ia	E	10.1	10	60	S51475	Suguira *et al.* (1984); Yoshida *et al.* (1992)
	PLIII	endo-Pel	Ia	E	ND	10	60	D10064	Yoshida *et al.* (1991a, 1991b)
	Pnl	Pnl	?[5]	ND	ND	ND	ND	M65057	Nishida *et al.* (1990); Ohnishi *et al.* (1991)
C18	Pel1	endo-Pel	Ia	E	>10	8.5	40	X81847	Bartling *et al.* (1995)
	Pel2	endo-Pel	Ia	E	>10	8	50	X81847	Bartling *et al.* (1995)
	Pel3	endo-Pel	Ia	E	>10	8.5	40	X81847	Bartling *et al.* (1995)

Plant cell-wall-degrading pectinases and cellulases produced by the soft-rot *Erwinia* spp. Echr, Ecc and Eca. This table describes the erwinia pectinase and cellulase activities for which the corresponding structural genes have been sequenced (GenBank accession numbers provided where available) and/or cloned. Enzymatic activities for which the structural genes have not been cloned are not included. The biochemical characteristics of each protein are described, including pI value, and pH and temperature (T) optimum (opt.) requirements for enzymatic activity (ND – not determined). Values were calculated under a variety of conditions either in the cognate host, or by *in trans* expression in *E. coli*. All pI values were determined by IEF, except for Ecc strain 71 PnlA for which the pI was calculated from the amino acid sequence. The characteristics described for the Echr 3937 Pels are according to a recent re-characterization by Tardy *et al.* (1997). These are very similar to the original values determined by Kotoujansky *et al.* (1985) and Bertheau *et al.* (1984). The extent to which each enzyme contributes to pathogenicity varies and is not described.

[1] P, plant-inducible; endo-Pel, endo-pectate lyase; exo-Pel, exo-pectate lyase; endo-Cel, endo-glucanase; Pme, pectin methylesterase; Pae, pectin acetyl esterase; endo-Peh, endo-polygalacturonase; exo-Peh, exo-polygalacturonasc; NC, activity not characterized.

[2] Class refers only to Pels and Cels. Pels have been classified according to dendogram analysis using the AllAll program (data not shown). Only Pels with a GenBank accession number have been classified, with the exception of Echr 3937 PelB and PelC and Ecc SCRI193 PelA and PelD, which have previously been assigned to the class shown. Cels are characterized based on their amino acid sequence homology, as described by Henrissat *et al.* (1989), with the exception of Echr 3665 EGZ and EGY which have not been sequenced. Ecc SCC3193 CelS falls into a novel class which has not previously been described.

[3] E, extracellular; P, periplasmic; OM, outer membrane; ?, cellular localization has not been described.

[4] Although KdgC shows 36% amino acid homology to Ecc PelB, no pectate lyase activity could be demonstrated for this protein. KdgC does not possess a *sec*-dependent export signal, suggesting that KdgC resides in the cytoplasm.

[5] PemB has a characteristic *N*-terminal lipoprotein signal sequence and is localized in the outer membrane. It does not appear to be exposed to the extracellular medium, and therefore is probably involved in the degradation of methylated oligogalacturonides present in the periplasm, rather than in a direct interaction with extracellular pectin.

[6] The Pnl enzymes produced by the strains Ecc71 and Eca Er do not possess a *sec*-dependent export signal, suggesting that these proteins reside in the periplasm.

[7] Ecc SCC3193 PelB, unlike any other Pel, has a preference for partially methyl esterified pectin.

[8] Class only refers to the catalytic domains of CelV and CelV1. The cellulose binding domains of these proteins show greater homology with those from many different Cel families.

is somewhat confusing since genes have been cloned from several different strains and the exact type and number of exoenzymes varies between individual strains. For example, there are generally five major Pel isozymes in Echr (PelABCDE) and four in Ecc (PelABCD). However, a spontaneous deletion in the Echr EC16 chromosome removed the 5' region of the *pelD* gene (Tamaki *et al.*, 1988). The systematic construction of deletion strains lacking the known *pel* genes led to the revelation of a second set of so-called minor Pels in Echr (Beulieu *et al.*, 1993; Kelemu and Collmer, 1993). These include the endo-pectate lyases PelI, PelL, and PelZ and an exo-Pel, PelX (Brookes *et al.*, 1990; Alfano *et al.*, 1995; Lojkowska *et al.*, 1995; Pissavin *et al.*, 1996; Shevchik *et al.*, 1997a). In addition to the pectinases, Echr and Ecc each produce a major extracellular and a minor periplasmic Cel isozyme which degrade cellulose. These are termed EGZ (*celZ*) and EGY (*celY*) respectively in Echr, and CelV/CelV1 (*celV/celV1*) and CelS (*celS*) respectively in Ecc. Some *Erwinia* spp. also produce a set of extracellular metalloproteases. However, these enzymes may not play a significant role in the virulence of *Erwinia* and will not be discussed further here (Dahler *et al.*, 1990). Table 1 describes the plant cell-wall-macerating enzymes of the soft-rot erwinias and some of their biochemical characteristics. Computer generated amino acid pile-ups and phylogenetic trees have been utilized to compare the known amino acid sequences of the enzymes within the Pel and Cel families. These methods have allowed each enzyme to be assigned to a class or family of related proteins as described in Table 1.

(b) Transcriptional organization

Genes encoding virulence determinants of microbial pathogens are often located in a single region of the chromosome, termed a "pathogenicity island". However, gene mapping and cloning revealed that the exoenzyme genes are located in small clusters at several different chromosomal loci. The exoenzyme genes within each cluster are organized as individual transcriptional units which are transcribed in a non-operonic fashion. Each cluster contains genes expressing different exoenzyme activities. For example, *pelAED*, *paeY* and *pemA* are all located within the same cluster on the Echr 3937 chromosome. The *pelBC* gene cluster appears to have arisen by a gene duplication event. Perhaps the scattered nature of the exoenzyme genes may be due to the fact that they are essentially catabolic genes which have been "hi-jacked" to aid survival in the plant host.

(c) A biochemical characterization of the exoenzymes

Three main methods have been used to detect exoenzyme activity: enzyme detection media; spectrophotometric enzyme assays; and activity staining of isoelectric focusing (IEF) gels. These methods have been used for screening both growing cells and crude and purified exoenzyme extracts. Standard methods such as osmotic shock, lysozyme treatment and sphaeroplasting were used to obtain crude periplasmic and cytoplasmic fractions, and the fidelity of cell fractionation was routinely observed via assay of the cytoplasmic and periplasmic markers β-galactosidase (Miller, 1972) and β-lactamase (O'Callaghan *et al.*, 1972; Sykes and Matthew, 1979) respectively. The protein concentration of exoenzyme extracts was determined using standard methods. Viscosimetric assays were also used to determine Pel or Cel activity (Boyer *et al.*, 1984b, 1987b; Zink and Chatterjee, 1985).

(i) Enzyme detection media

Growth media supplemented with enzyme substrates have been used to identify the enzymatic properties of microorganisms for many years. In most cases enzymatic activity was detected directly or after substrate staining with an appropriate detection solution. Plates were inoculated with either a single colony using a sterile toothpick, or by spotting a small volume of an overnight culture on to the plate. Alternatively, purified enzyme extracts, cultures, or cell fractionation extracts were incubated in wells by the "cup-plate" method. Plates were incubated at a temperature conducive to enzyme production in the *E. coli* host or the wild-type *Erwinia* strain. These methods were usually qualitative. However, occasionally exoenzyme detection media were used in a semi-quantitative manner, usually by the incorporation of a standard curve (Walker *et al.*, 1994; Bortolli-German *et al.*, 1995; Shevchik *et al.*, 1996).

Pectinase activity was routinely observed by growing strains on L- or YC-agar containing PGA for 16–24 h, after which plates were flooded with 1M $CaCl_2$ for 5–12 h to reveal pronounced depressions or "pitting" caused by the degradation of PGA to unsaturated and saturated oligogalacturonates (Keen *et al.*, 1984). In a similar method, plates were flooded with 7.5% (w/v) copper acetate, which forms a blue precipitate with PGA, for 1–2 h. Pel[+] colonies were identified by the presence of a double, cream halo around the colony on a translucent blue background (Andro *et al.*, 1984; Bertheau *et al.*, 1984). The addition of EDTA to the growth media (to chelate the Ca^{2+} ions essential for Pel activity) allowed the sole detection of Peh activity as a single clear halo on a translucent blue background (Andro *et al.*, 1984; Bertheau *et al.*, 1984). Pnl and Pme activity were detected by growing cells on L-agar for 4 h, then overlaying an agarose–pectin mixture and re-incubating for a further 15 h. The plates were flooded with 0.05% (w/v) Ruthenium red to reveal dark haloes around positive colonies (McEvoy *et al.*, 1990; Shevchik *et al.*, 1996). By contrast, clear haloes represented the further breakdown of PGA to oligogalacturonates by Pel and Peh activities. To detect Cel activity strains were grown on agar containing carboxymethylcellulose (CMC). Plates were developed by staining with 0.2% (w/v) Congo red for 15 min, followed by bleaching with 1M NaCl (Teather and Wood, 1982). Cel activity was observed as a clear halo on a red background. Occasionally the xylanase activity of the Cels was demonstrated by replacing CMC with 0.5% xylan, or by staining with Remazol brilliant blue xylan (Biely *et al.*, 1985; Cooper and Salmond, 1993).

(ii) Spectrophotometric enzyme assays

Spectrophotometric enzyme detection assays were used to quantify the exoenzyme activities of either purified enzyme fractions or cellular fractions. Pel activity was assayed by following the degradation of PGA to unsaturated oligosaccharides, which results in a change of absorbance at OD_{235nm} (Moran *et al.*, 1968). Pnl was assayed as for Pel, but in the presence of 1 mM EDTA in the reaction medium to inhibit Pel activity (McEvoy *et al.*, 1990; Lojkowska *et al.*, 1995). Peh activity was quantified by one of two spectrophotometric methods based on the liberation of reducing groups from a reaction mixture containing PGA: the *p*-hydroxybenzoic acid hydrazide method of Lever (1972), and the arsenomolybdate method of Nelson (1944). Pme is the first enzymatic activity in the hierarchy of pectin degradation and converts pectin to pectate (PGA; Figure 1.). PGA is the sole substrate for Pel activity. Hence, Pme activity is a prerequisite for Pel activity when pectin is provided as the sole enzymatic substrate. For this reason Pme activity was often assayed indirectly by measuring Pel activity (McMillan *et al.*, 1993; Shevchik *et al.*, 1996). Finally, endo-Cel activity was monitored by one of two different methods. In the first method the amount of liberated reducing sugars released from CMC as D-glucose equivalents was determined (Nelson, 1944; Park and Johnson; 1949; Somoygi, 1952). By using crystalline cellulose as a substrate this method was modified to assay for exo-Cel activity (Boyer *et al.*, 1984b; Saarilahti *et al.*, 1990a). The second method utilized the substrate Ostazin brilliant red cellulose (Biely *et al.*, 1985). Some of the Cel assays were also adapted to microtitre well plate assays (Boyer *et al.*, 1987b; Py *et al.*, 1991a; Barras *et al.*, 1992; Bortolli-German *et al.*, 1995). For example, the microcrystalline cellulose-binding study, using Avicel (PH1012), of Py *et al.* (1991a).

(iii) Activity staining of isoelectric focusing gels

Although the enzyme detection methods described above provided an insight into the types of enzymatic activity in each *Erwinia* strain they did not reveal whether each type of enzymatic activity was due to a single enzyme or multiple isozymes. In this respect, IEF was perhaps one of the most useful biochemical techniques employed in the *Erwinia* field. IEF allows the separation of proteins based on their isoelectric point (pI). Proteins are loaded on a polyacrylamide gel, containing carrier ampholytes in an appropriate pH range, and focused to equilibrium between the cathode and anode. This technique was modified to allow the visualization of enzymatic activities by activity staining. A charged substrate, such as polygacturonate (PGA) or CMC was either included in the IEF acrylamide gel and stained directly (Kotoujansky *et al.*, 1985), or alternatively and more commonly, a second substrate-containing acrylamide gel was layered over the IEF gel and incubated overnight to allow protein transfer by diffusion (Bertheau *et al.*, 1984). In each case the enzymatic activity was visualized by staining with an appropriate detection solution in a similar manner to the development of enzyme detection plates as described in section II.B.1(c)(i). This technique was critical in confirming the existence of multiple isozymes with similar enzymatic activities and has been routinely used to characterize the expression products of cloned *Erwinia* exoenzyme genes (Bertheau *et al.*, 1984; Keen *et al.*, 1984; Kotoujansky *et al.*, 1985; Ried and Collmer, 1985, 1988: Roeder and Collmer, 1985; Payne *et al.*, 1987; Reverchon and Robert-Baudouy, 1987b; Willis *et al.*, 1987; Tamaki *et al.*, 1988; Hinton *et al.*, 1989a; He and Collmer, 1990; Cooper and Salmond, 1993; Alfano *et al.*, 1995; Heikinheimo *et al.*, 1995; Lojkowsky *et al.*, 1995; Pissavin *et al.*, 1996; Shevchik *et al.*, 1997a).

(iv) Protein purification

Some of the exoenzymes were purified to various degrees of homogeneity by standard methods, such as ion-exchange chromatography, well before any genetic analysis occurred. This was simplified by their extracellular nature, which allowed purification directly from the *Erwinia* culture supernatant. The purified proteins were also characterized biochemically to reveal several distinct properties including molecular weight, pI, the optimum pH and temperature ranges, and the substrate specificities conducive to maximal activity. However, the cloning of individual exoenzyme genes facilitated the over-expression of each gene product and, in some cases, its purification to near homogeneity. In most cases the biochemical properties of each cloned exoenzyme gene product were redefined. Purified exoenzyme extracts were often tested *in planta* to determine their ability to cause disease (section II.B.1(d)(v)). The availability of purified proteins allowed antibodies to be generated towards the exoenzymes. Recently, crystallography analysis demonstrated that PelC and PelE consist of an unusual backbone of parallel β-strands coiled in a large right-hand helix (Yoder *et al.*, 1993a, 1993b; Lietzke *et al.*, 1994). However, despite similarities in their global structures, PelC and PelE display differences in the localization of charges at their surface and variation in the lateral chains. A comparison of the structurally conserved amino acids has led to the identification of two putative distinct active sites (Heffron *et al.*, 1995; Henrissat *et al.*, 1995).

(d) The genetical approach to cloning and mutational analysis of the exoenzyme genes

Biochemical methods helped to identify the types and numbers of exoenzymes present in each *Erwinia* strain. However, many of the isozymes appeared to have similar molecular weights and pI values. Hence, some of the isozymes could not be purified to homogeneity from *Erwinia* culture extracts. It was also not clear whether all the activities observed by IEF activity staining corresponded to individual exoenzymes or their degradation products. Therefore, the focus of research turned to recombinant DNA technology to allow the cloning and expression of individual exoenzyme genes in *E. coli*. However, the existence of multiple isozymes with varying activities made it difficult to use the standard plate methods to screen (i) mutations in individual exoenzyme genes and (ii) cosmid library complementation of mutant exoenzyme phenotypes. In addition, the extracellular enzymes were often retained in the periplasm when expressed in *E. coli*. Therefore, the structural exoenzyme genes of *Erwinia* were cloned by a number of different genetic approaches. Since the genetic analysis of these enzymes is an amalgamation of the information generated in several different strains we will highlight only key examples of the techniques used.

(i) Chromosomal mapping

Several workers localized the exoenzyme genes on the *Erwinia* chromosome with respect to multiple auxotrophic markers or existing exoenzyme genes. For example, Reverchon and Robert-Baudouy (1987b) localized the *pelADE* gene cluster close to the *pro-l* marker on the Echr B374 genetic map by chromosomal mobilization. *pel::lacZ* Kn^R chromosomal fusions were transduced, using the generalized transducing phage φEC2, into a polyauxotrophic strain of Echr. An RP4::mini-Mu plasmid allowed chromosomal mobilization from the wild-type Echr strain into these newly generated strains. Co-transfer of the Kn^s phenotype with each polyauxotrophic marker was calculated. Aymeric *et al.* (1988) localized the *celZ* gene between *ura* and *pan*, and the *celY* gene between *xyl* and *met* on the linkage map of Echr 3937 in a similar manner.

(ii) Cloning of the major exoenzymes

Genomic libraries

Genomic DNA libraries, generally screened in an *E. coli* host, were used to clone the majority of the exoenzyme structural genes. Kotoujansky *et al.* (1985) utilized the λL47–1 bacteriophage vector to clone some of the exoenzyme genes of Echr 3937. This method had two advantages. Firstly, it allowed expression of the cloned genes in *E. coli* by transcription from the λ pL promoter. Secondly, it avoided the problem of the exoenzymes remaining intracellular when expressed in *E. coli*, since lysis at the end of the phage lytic cycle released the exoenzymes to the exterior of the cell. The authors developed modified exoenzyme detection plates to reveal Pel or Cel activity around individual phage plaques without killing the phage. These plates consisted of a bottom layer of media supplemented with either PGA or CMC (the substrates for Pel and Cel respectively), separated by a nylon membrane from a top layer of L top agar containing the phage genomic library and an indicator *E. coli* strain. After incubation at 37°C for 24 h the top layer was removed, and the bottom layer was flooded with saturated copper acetate, or 0.1% Congo red followed by bleaching with 1M NaCl, to reveal Pel or Cel activity respectively. Positive clones were then purified from the top agar layer.

Other workers discovered that when exoenzyme genes were expressed in *E. coli* a small proportion of the exoenzyme activity was in fact detectable in the culture media simply due to cell lysis which occurs naturally within a bacterial colony. This phenomenon allowed cosmid libraries carried in *E. coli* to be screened on exoenzyme detection media. Several different cosmid and plasmid vectors have been used to clone the

Erwinia exoenzyme genes in this manner (Barras *et al.*, 1984; Keen *et al.*, 1984; Lei *et al.*, 1985a, 1985b; Zink and Chatterjee, 1985; Boyer *et al.*, 1987b; Hu *et al.*, 1987; Willis *et al.*, 1987; Plastow, 1988; Saarilahti *et al.*, 1990ab; Karlsson *et al.*, 1991; Cooper and Salmond, 1993; Mäe *et al.*, 1995; Shevchik *et al.*, 1996).

Early experiments suggested that *recA* was required for induction of Pnl synthesis in Ecc strain 71 (Ecc71) (Zink *et al.*, 1985), yet attempts to clone the *pnl* gene from cosmid libraries screened in either a RecA$^+$ or RecA *E. coli* background failed. However, it was noticed that Ecc71 produced higher levels of Pnl activity than Ecc strain SCRI193. McEvoy *et al.* (1990) used this observation to their advantage to clone the Ecc71 *pnlA* gene. An Ecc71 pSF6 cosmid library was mobilized into strain SCRI193 and transconjugants were screened for hyperproduction of Pnl. The presence of the *pnlA* gene on hyperproducing clones was confirmed by their ability to restore Pnl production in a *pnlA*::Tn5 mutant of Ecc71. The *pnlA* gene was subsequently sequenced and characterized by Chatterjee *et al.* (1991).

RP4::mini-Mu plasmids

RP4::mini-Mu plasmids were used for *in vivo* cloning of some of the *pel* and *cel* genes of Echr strain B374 (van Gijsegem *et al.*, 1985). These plasmids carry a mini-Mu with a functional transposase. This feature permits them to integrate into their host chromosome by replicon fusion and, by subsequent mini-Mu mediated excision, generate R-prime plasmids carrying random pieces of the host chromosome (van Gijsegem and Toussaint, 1982). *E. coli* recipients which were lysogenic for λ were originally used to facilitate detection of the enzymatic activities as with λL47–1. However, this was found not to be necessary, and B374/RP4::mini-Mu donor strains were mated with different polyauxotrophic non-lysogenic derivatives of *E. coli*. From each mating, *E. coli* transconjugants were selected which had lost one or other of their requirements and these were replica-plated on to pectic and cellulolytic enzyme detection media. Mapping in this manner allowed the exoenzyme genes in each RP4-prime plasmid to be localized on the B374 chromosome by co-inheritance. Several R-prime plasmid-containing strains exhibiting exoenzyme activity were then characterized biochemically by IEF.

Antibodies

The availability of antibodies raised against some of the *Erwinia* exoenzymes aided the cloning of their respective genes. Immunoscreening was particularly useful for identifying genes which were not expressed sufficiently in *E. coli* to allow detection by the standard exoenzyme plate methods. For example, the Echr EC16 *pehX* gene was isolated from a pUC19 genomic library by immunoscreening *E. coli* transformants with an antibody raised against the exoPeh protein (He and Collmer, 1990). A Pel-deficient strain of Echr EC16 was used as the source of the library to reduce interference from the Pels in identification of *pehX⁺* colonies. Transformants were screened in the following manner: colonies were picked to LB plates with the appropriate antibiotic selection and incubated at 37°C overnight. The bacteria were transferred to a nitrocellulose membrane and lysed. *peh⁺* clones were identified by a standard immunoblot procedure using anti-exoPeh antibodies. Clones encoding the Ecc Pels and endo-Peh were isolated in a similar manner (Lei *et al.*, 1985a, 1985b).

Immunoblots have also been used to identify exoenzyme protein homologues in different *Erwinia* strains. For example, *Erwinia* total culture extracts were analysed by immunoblotting with anti-PemB antibodies after separation by SDS-PAGE (Shevchik *et al.*, 1996). Anti-PemB antibodies were pre-incubated with a lysate of *E. coli* cells overproducing PemA to avoid any cross-reaction with PemA. The authors demonstrated the presence of PemB in all of the Echr strains, but none of the Ecc or Eca strains, tested.

Antibodies have also been used to distinguish between different isozymes with the same enzymatic activity. For example, biochemical analysis revealed the presence of two celluloytic activities in Echr strains (Bertheau *et al.*, 1984; Boyer *et al.*, 1984b). The screening of different genomic libraries in various Echr strains revealed the presence of two genes, *celZ* and *celY*, encoding EGZ and EGY respectively, which displayed a degree of protein homology (Barras *et al.*, 1984; Kotoujansky *et al.*, 1985; van Gijsegem *et al.*, 1985; Boyer *et al.*, 1987a, 1987b; Guiseppi *et al.*, 1988, 1991). Boyer *et al.* (1987b) showed that the Echr 3937 *celZ* gene expressed a protein capable of cross-reacting with an Echr 3665 EGZ-antibody (which had previously been shown not to cross-react with EGY) using Ouchterlony double immunodiffusion tests. A cross-reaction was characterized by the formation of a precipitin line between an agar plate well containing the antibody and a well containing a purified protein sample. This result was further supported by the inability of a DNA fragment carrying the Echr 3665 *celY* gene to hybridize to the Echr 3937 *celZ* gene (Boyer *et al.*, 1987a).

(iii) Cloning of the minor Pels

As alluded to earlier, the construction of deletion strains of Echr lacking the known *pel* genes revealed a set of minor Pels, some of which appeared to be plant inducible (Beaulieu *et al.*, 1993; Kelemu and Collmer, 1993). Several of these were subsequently cloned, including genes for an exo-pel, PelX, and the endo-Pels PelL, PelZ and PelI (Brooks *et al.*, 1990; Alfano *et al.*, 1995; Lojkowska *et al.*, 1995; Pissavin *et al.*, 1996; Shevchik *et al.*, 1997b). *pelL* of Echr 3937 was cloned by screening a genomic library of an Echr PelABCDE⁻ strain on various media for pectinolytic activity (Lojkowska *et al.*, 1995). However, when a genomic library of an Echr EC16 PelABCEX⁻ PehX⁻ strain was screened in this manner clones with a stable Pel⁺ phenotype could not be isolated (Alfano *et al.*, 1995). This suggested that some of the minor *pel* genes were toxic when expressed in *E. coli*. Therefore, Alfano *et al.* (1995) isolated the Echr EC16 *pelL* gene by Tn5tac1 mutagenesis of a

PelABCEX⁻ PehX⁻ strain. Tn5tac1 carries $lacI^q$ and the P_{tac} promoter which permit isopropyl-β-D-thio-galactoside (IPTG)-dependent expression of downstream genes. Tn5tac1 can therefore be used to define and isolate genes which would normally be toxic to the host strain. By contrast, *pelI* and *paeY* of Echr 3937 were identified by selecting for Mu-*lacZ* insertions which generated PGA-inducible *lacZ* transcriptional fusions (Hugouvieux-Cotte-Pattat and Robert-Baudouy, 1989; Shevchik *et al.*, 1997a; Shevchik and Hugouvieux-Cotte-Pattat, 1997). Finally, the *pelZ* gene of Echr 3937, formerly termed *pecZ*, was originally thought to be involved in the regulation of *pelB* and *pelC* transcription (Yankosky *et al.*, 1989; Hugouvieux-Cotte-Pattat and Robert-Baudouy, 1992). However, a genetic and biochemical characterization of the *pecZ* locus demonstrated that the "regulatory" effect of *pecZ* was due to the titration of regulatory DNA-binding proteins by a novel Pel isozyme, now termed PelZ (Pissavin *et al.*, 1996). Despite their low activity, the secondary Pels appear to have important roles in either plant-tissue infection or host specificity (Lojkowska *et al.*, 1995; Pissavin *et al.*, 1996; Shevchik *et al.*, 1997a).

(iv) Mutational analysis

Attempts to isolate chromosomal mutations in the exoenzyme genes using random approaches generally proved negative simply because of the existence of multiple isozymes which masked any phenotypic analysis on exoenzyme detection media. Therefore, several alternative strategies were devised to allow the generation of specific chromosomal mutations, including marker exchange, Tn mutagenesis and mini-Mu mutagenesis. In some cases a combination of specific mutation events have been used to generate strains devoid of several exoenzyme activities (Ried and Collmer, 1988; He and Collmer, 1990; Kelemu and Collmer, 1993; Alfano *et al.*, 1995). Mutations constructed in this manner were subsequently used to investigate the regulation of the exoenzymes (section II.B.2) and/or their contribution towards pathogenicity *in planta* (section II.B.1(d)(v)). Mutations have occasionally been constructed in plasmid-encoded exoenzyme genes and studied in an *E. coli* host background. For example, Mäe *et al.* (1995) used plasmid-encoded *lacZ* transcriptional fusions to study the regulation of the Ecc71 *celV1* and *celS* genes. However, it is important to interpret the data generated using these types of mutations in heterologous backgrounds with caution. Finally, site-directed mutagenesis has been used to study the importance of individual amino acid residues in some of the exoenzyme genes (Py *et al.*, 1991a; Bortoli-German *et al.*, 1995).

Marker exchange mutagenesis

Marker exchange mutagenesis has proven to be a useful tool for manipulating genes which do not confer an easily screened phenotype but for which a cloned copy is available (Ruvken and Ausebel, 1981). In the insertion mutagenesis method the cloned gene is inactivated by the insertion of a gene cassette encoding a selectable phenotypic marker such as antibiotic resistance. Effective marker exchange requires the delivery of the mutated gene on an unstable replicon or "suicide vector", and recombinational exchange of the mutated allele with the wild-type chromosomal copy. The suicide vector is usually transferred into the recipient strain by plasmid mobilization from an *E. coli* donor strain. Allelic exchange is carried out by growth of the recipient strain under conditions that simultaneously promote plasmid loss and select for the phenotypic marker.

Several different suicide vectors have been used to generate chromosomal insertion mutations in *Erwinia*, the most common of which are pBR322 based. However, in each case plasmid loss was promoted by one of three different methods: (i) sucrose sensitivity encoded by the *sacB* gene (for example pKNG101). The *sacB* gene is derived from *Bacillus subtilis* and encodes levansucrase, an enzyme that accumulates in the periplasm of Gram-negative bacteria, resulting in lethal synthesis of levan in the presence of 5% sucrose (Gay *et al.*, 1985). (ii) Phosphate limitation. This was the most common method used in *Erwinia*. For example pBR322 was quickly lost from Echr strains grown in the presence of high concentrations of phosphate (Roeder and Collmer, 1985). (iii) Non-selection of the vector. For example, Roeder and Collmer (1985) also noticed that the loss of pBR322 from Echr strains could be promoted simply by growth in the absence of ampicillin.

Antibiotic resistance has often been used as a phenotypic marker for insertion either by direct insertion into the gene of interest (Ried and Collmer, 1985; Payne *et al.*, 1987) or via the interposon Ω (Boccara *et al.*, 1994; Mäe *et al.*, 1995; Heikinheimo *et al.*, 1995). However, phenotypic markers which generate either transcriptional and/or translational fusions when inserted downstream of the promoter of the gene of interest have also been used. For example, promoterless versions of the *uidA* gene which encodes β-glucuronidase activity (Bardonnet and Blanco, 1992; Hugouvieux-Cotte-Pattat *et al.*, 1992; Lojkowska *et al.*, 1995; Pissavin *et al.*, 1996; Shevchik *et al.*, 1996, 1997a).

Marker exchange eviction mutagenesis was used to generate chromosomal deletions in the exoenzyme genes in the following manner: (i) An *nptI-sacB-sacR* insertion cassette which encodes Kn^R and sucrose sensitivity was cloned into the gene(s) of interest in pBR322 in an *E. coli* host. (ii) The resulting construct was then conjugally mobilized into the recipient strain. (iii) The deletion mutation was introduced into the chromosome by growing the recipient strain under phosphate-limiting conditions, to promote plasmid loss, and subsequently selecting for Kn^R recombinants. (iv) A second pBR322 derivative was then generated by excising the majority of the *nptI-sacB-sacR* insertion cassette (Chapter 4, VI) from the derivative generated in (i). (v) This unmarked deletion derivative was then mobilized into the strain obtained in (iii). (vi) The marked deletion was evicted by a second recombinational exchange with the unmarked deletion derivative by selecting for sucrose tolerance. The resulting Kn^s mutants had an unmarked deletion within the chromosomal gene(s) of interest. Ried and Collmer (1988) and Boccara *et al.* (1994) used this method to generate deletion mutations in the *pelB* and *pelC* genes, and *celY*, respectively.

Tn mutagenesis

Directed insertional inactivation has also been carried out by Tn mutagenesis of plasmid-encoded versions of the exoenzyme genes. For example, He and Collmer (1990) used λ::Tn*phoA* to mutagenize a plasmid-encoded version of the Echr EC16 *pehX* gene in an *E. coli* CC118 host by the method of Manoil and Beckwith (1985). The mutagenized constructs were subsequently mobilized into Echr and used to marker exchange the chromosomal *pehX* gene as described by Roeder and Collmer (1985). Willis *et al.* (1987) generated a chromosomal Tn5 mutant of the Ecc71 *peh-1* gene in a similar manner. A cloned version of the *peh-1* gene was mutagenized with λ467 as described by Zink *et al.* (1985). The mutated construct was transformed into Ecc71 and subsequently marker exchanged into the chromosome.

Mini-Mu mutagenesis

The mini-Mu bacteriophage derivative Mu *dl*1724, which contains the *lacZ* reporter gene (Castilho *et al.*, 1984), has been used to generate *lacZ* transcriptional fusions in many of the exoenzyme genes in Echr 3937, including *celZ* and *celY* (Aymeric *et al.*, 1988) and *pelZ* of Echr 3937 (Pissavin *et al.*, 1996). Allelic exchange was carried out by either Mu-mediated transduction followed by non-selection of the vector to promote plasmid loss, or electroporation followed by selection on low phosphate media. By contrast, McEvoy *et al.* (1990) used Mu *dl*1734 to construct plasmid-encoded Ecc 71 *pnlA::lacZ* fusions in an *E. coli* host. Individual plasmid-encoded *pnlA::lacZ* fusions were transferred into a *pnlA*::Tn5 strain of Ecc71 by triparental matings, and subsequently screened for Pnl activity by the Pnl agar plate method. Hu *et al.* (1987) generated an Ecc TG3 *pel::lacZ* fusion in a similar manner. However, this mutation was only characterized in an *E. coli* host.

In vitro site-directed mutagenesis

Py *et al.* (1993) used the *in vitro* mutagenesis method of Sayers *et al.* (1988) to generate site-directed point mutations in a plasmid-encoded version of the Echr *celZ* gene. This identified two residues involved in the cellulolytic activity of EGZ. In a similar experiment Bortoli-German *et al.* (1995) used informational suppression to investigate the structural, functional and evolutionary aspects of the Echr EGZ. Amber mutations were constructed in 16 codons of a cloned version of the *celZ* gene. Each construct was subsequently introduced into 13 *E. coli* strains, each carrying a different suppresser tRNA that inserts an amino acid at the mutated position. The *in vivo* stability of each mutated version of CelZ was determined by immunoblots with an anti-EGZ antibody, and the Cel activity present in the periplasm of *E. coli* host cells was tested by a semi-quantitative plate method using CMC as a substrate. This method identified several amino acid residues which were important for the catalytic activity and/or *in vivo* stability of EGZ.

The *in planta* tests previously described in section II.A.3 were used to determine the contribution of each exoenzyme to the pathogenicity of *Erwinia*. In the early stages of research, culture supernatant extracts of *Erwinia* (Zink and Chatterjee, 1985), and periplasmic fractions of *E. coli* expressing cloned exoenzyme genes (Willis *et al.*, 1987; Tamaki *et al.*, 1988; Lei *et al.*, 1988), were tested for their ability to macerate either whole potato tubers, or potato or cucumber slices. Among the proteins shown to macerate plant tissue were Peh1 from Ecc71, PelA from the Eca strain EC, and various Pels from Echr EC16, including the wild-type PelA, PelB, PelC and PelE proteins, a recombinant PelE–PelA protein, and a PelA::LacZ fusion protein (Willis *et al.*, 1987; Lei *et al.*, 1988; Tamaki *et al.*, 1988).

An alternative strategy was to use *E. coli* and Echr strains expressing cloned exoenzyme genes to inoculate plant host tissue (Keen *et al.*, 1984; Lei *et al.*, 1985b; Keen and Tamaki, 1986; Payne *et al.*, 1987; Hu *et al.*, 1987). Various plant hosts and initial inoculum densities were used. However, it was not possible to select for the maintenance of the plasmid constructs *in planta* and the exoenzymes were generally retained in the periplasm in *E. coli*. Despite these facts tissue maceration was often observed. For example, two independent studies demonstrated that cells of *E. coli* or Echr EC16 carrying *pelE* clones macerated potato tuber tissue to a greater degree than those containing *pelB* clones (Keen and Tamaki, 1986; Payne *et al.*, 1987). Payne *et al.* (1987) also showed that under anaerobic conditions inoculations of less than 100 bacteria per site produced significant maceration. However, even though *E. coli* and Echr were capable of multiplying substantially in aerobic conditions tissue maceration was not observed under such conditions.

Although these experiments provided an insight into the macerating capacities of the individual exoenzymes, they did not reflect the true contributions of each exoenzyme towards the ability of the *Erwinia* spp. to cause soft-rot. For example, cloned genes were often expressed from high copy number vectors and strong promoters, potentially resulting in exoenzyme expression levels non-representative of the true levels *in planta*. For this reason, *Erwinia* strains containing directed chromosomal mutations in the exoenzyme genes were tested for their ability to cause disease *in planta* (Roeder and Collmer, 1985; Payne *et al.*, 1987; He and Collmer, 1990; Boccara *et al.*, 1994; Walker *et al.*, 1994; Mäe *et al.*, 1995; Lojkowska *et al.*, 1995; Shevchik and Hugouvieux-Cotte-Pattat, 1997). In general *Saintpaulia ionantha* plants or whole potato tubers were used as the host plant.

Echr 3937 strains with an interposon Ω in *celZ*, a deletion in *celY*, or both were as virulent as the wild-type strain when used to inoculate three-month-old *S. ionantha* plants at high cell inoculum density (Boccara *et al.*, 1994). However, a slight delay in the appearance of symptoms was observed for the *celY* strain. This suggested that EGY might be involved in the early stages of infection. However, the authors postulated that the Cels of Echr might be necessary for saprophytic degradation of dead tissues in which the cellulose has already been attacked by fungi. Interestingly, the CelV and CelV1 mutants of Ecc strains SCRI193 and SCC3193 exhibited a reduced virulence in whole potato tubers and tobacco seedlings, suggesting that, although not absolutely required for pathogenicity, the Ecc Cels enhanced the ability of the parental strains to macerate plant tissue (Walker *et al.*, 1994; Mäe *et al.*, 1995).

The individual Pel isozymes also appear to contribute to different extents to the maceration of plant tissue. A *pelC::kn^R* marker exchange mutant of Echr EC16 was shown to macerate potato tuber tissue to the same degree as the wild-type strain when inoculated at a high cell density (Roeder and Collmer, 1985). Likewise, a *pelB* mutant of Ecc SCC3193 was shown to be as virulent as the wild-type strain on axenic tobacco seedlings. However, the authors suggested that this might be a reflection of the unusual substrate specificity of PelB which, unlike most endo-Pels, can degrade methyl-esterified pectin. By contrast, some of the Pels were shown to be essential for full pathogenicity. A *pelE::kn^R* marker exchange mutant of Echr EC16 caused significantly less tissue maceration than the wild-type strain on intact potato tuber (Payne *et al.*, 1987). Finally, results from study of PelL and PaeY marker exchange mutants of Echr 3937 demonstrated that PelL and PaeY are essential for virulence *in planta* (Lojkowska *et al.*, 1995; Shevchik and Hugouvieux-Cotte-Pattat, 1997).

For years workers have been puzzled as to why *Erwinia* produces so many isozymes with similar enzymatic activities. Recently, Bartling *et al.* (1995) demonstrated that purified recombinant forms of the Pels of Eca could act synergistically to degrade 68% esterified pectin, against which the activity of the individual Pel isozymes was lower, by up to 64%. Indeed, the characterization of the individual exoenzyme genes *in planta* has revealed that some of the exoenzymes are necessary for the early stages of soft-rot (Shevchik and Hugouvieux-Cotte-Pattat, 1997), while others appear to play a role in host specificity (Ried and Collmer, 1988; Beaulieu *et al.*, 1993; Pissavin *et al.*, 1996; Shevchik *et al.*, 1997a). Therefore, it is likely that each member of the soft-rot erwinias has evolved a specific repertoire of exoenzymes which facilitate both the survival of the organism on certain host plants, and the degradation of the different plant tissues which the organism encounters during the progression of the disease.

2. The regulation of phytopathogenicity

The majority of studies which have looked at the regulation of pathogenicity in the erwinias have followed a directed research approach. The overriding strategy taken is similar to the black box approach in that large numbers of random chemical or Tn mutants are generated, but it relies on direct screening methods to identify mutations which cause deregulation of a phenotype known to be important for virulence, principally the production of Pels. From these studies it is clear that the regulation of exoenzymes in the erwinias is extremely complex, being affected by both intrinsic and environmental factors. This section will look at the identification and characterization of KdgR, a repressor of pectinolysis in Echr, and discuss how a directed genetic approach was used to dissect the regulatory hierarchy that governs virulence in Echr. This section will then consider other mechanisms of regulation known to affect phytopathogenicity of the erwinias, including the roles of the regulatory proteins PecS/M, PecT and RsmA, in addition to the roles of bacterial pheromones and environmental factors. For a more comprehensive review of the regulatory proteins governing pectinolysis and virulence of *Erwinia* spp. see Hugouvieux-Cotte-Pattat *et al.* (1996).

(a) Dedicated regulators of phytopathogenicity

(i) KdgR

The ability to break down pectin in plant cell walls is not only important for virulence as the catabolic products of pectin are also used as a source of carbon and energy. The catabolism of pectin begins with the elaboration of the extracellular enzymes Pel, Peh, Pme and Pnl. These enzymes act in concert to demethoxylate and cleave the pectin chains into saturated and unsaturated oligogalacturonides. Oligogalacturonides are then broken down intracellularly by the gene products of *ogl, kduI, kduD, kdgK* and *kdgA* resulting in the formation of pyruvate and 3-phosphoglyceraldehyde which are directed into primary metabolism (Condemine *et al.*, 1986; Hugouvieux-Cotte-Pattat and Robert-Baudouy, 1987; Figure 1).

Some of the earliest observations of the regulation of pectin catabolism showed that the expression of the exoenzyme genes increased towards the end of the exponential growth phase and that the exoenzyme genes and all of the genes within the pectin catabolic pathway were inducible in the presence of PGA or galacturonate (Collmer and Bateman, 1981; Hugouvieux-Cotte-Pattat and Robert-Baudouy, 1987). However, analysis of mutants blocked at various steps within this pathway revealed that the true intracellular inducers of pectin catabolism were in fact 2-keto-3-deoxygluconate (KDG), 5-keto-4-deoxyuronate (DKI) and 2,5-diketo-3-deoxygluconate (DKII) (Condemine *et al.*, 1986; Chatterjee *et al.*, 1985; Hugouvieux-Cotte-Pattat and Robert-Baudouy, 1987).

To investigate the regulation of pectinolysis in Echr, Condemine and Robert-Baudouy (1987) attempted to isolate Echr mutants for which the expression of genes within the pectin catabolic pathway was deregulated in the absence of inducer. A LacZ⁻ Echr strain was constructed with a *lacZ* transcriptional fusion (made using Mu *d*A*plac*) to a pectinolysis regulon gene (*kduD*) with a low basal level of expression. Because of the low expression of the *kduD::lacZ* fusion, this strain was unable to grow on minimal salts media supplemented with lactose as sole carbon source, under non-inducing conditions. Therefore, by performing a second round of Tn mutagenesis and selecting transductants on minimal salts media with lactose it was possible to isolate mutants with increased *kduD::lacZ* expression. Tn5 was used for the second round of mutagenesis. One of the isolates recovered after mutagenesis grew strongly on the selective media and exhibited a constitutive level of expression of the *kduD::lacZ* fusion. Pel activity was also increased in this mutant. IEF revealed that PelA,

PelD and PelE were constitutively expressed whereas PelB and PelC were only slightly affected, if at all, by this Tn5 insertion.

Using the Echr generalized transducing phage φEC2, the Tn5-tagged insertion was transduced into other Echr strains which had *lacZ* transcriptional fusions to other genes within the pectin catabolic pathway, including *ogl, kduI, kdgT, kdgA* and *kdgK*. All of these genes were also constitutively expressed throughout the growth cycle in a mutant carrying this insertion. Thus it was clear that this mutation defined a novel genetic locus, denoted *kdgR*, which negatively regulated all the genes involved in the intracellular steps of pectin catabolism and at least some of the exoenzymes, *pelA, pelD* and *pelE* (Condemine and Robert-Baudouy, 1987). More recent studies have shown that KdgR also regulates the enzyme genes *pelB, pelC, pemA, pemB* and *paeY*. In addition to a gene for exoenzyme secretion, *outT* (Hugouvieux-Cotte-Pattat and Robert-Baudouy, 1889; Condemine *et al.*, 1992; Nasser *et al.*, 1994; Condemine and Robert-Baudouy, 1995; Shevchik *et al.*, 1996; Shevchik and Hugouvieux-Cotte-Pattat, 1997).

KdgR mutants were also isolated by Hugouvieux-Cotte-Pattat *et al.* (1986) using chemical mutagenesis and by isolating spontaneous KdgR⁻ mutants able to grow on minimal media supplemented with KDG as sole carbon source. KDG can only be taken into the cell in sufficient quantities to support growth when *kdgT*, encoding the KDG transport system, is over-expressed. Consequently, KdgR⁻ mutants which constitutively express *kdgT* can grow on this medium (Hugouvieux-Cotte-Pattat *et al.*, 1986).

The *kdgR* gene was cloned using RP4::mini-Mu plasmids which were fortuitously isolated while attempting to complement an Echr Ogl⁻ mutant (Reverchon and Robert-Baudouy, 1987a). On the assumption that some bacterial genes involved in the same catabolic pathway are clustered, plasmids complementing the Ogl⁻ mutants were also tested for the ability to complement a range of other Echr mutants, including a KdgR⁻ mutant. All of the Ogl⁻ complementing plasmids also complemented the KdgR⁻ mutant, suggesting that these two genes were closely linked (Reverchon and Robert-Baudouy, 1987a).

Problems were encountered during the subcloning of the *kdgR* gene because plasmids carrying *kdgR*, although stable in *E. coli*, are unstable in Echr (Reverchon *et al.*, 1991; Nasser *et al.*, 1992). This meant that subcloning of the *kdgR* gene could not proceed by complementation of its respective null mutant. To overcome this problem, a *kdgR*::Mu *d* mutant was generated and the disrupted gene cloned, by virtue of its antibiotic resistance, and used as a probe in a Southern blot to identify subclones carrying *kdgR*. The *kdgR* gene was localized to a 1.9 kb fragment which was sequenced. As predicted by the high level of linkage between *ogl* and *kdgR*, sequence analysis revealed that *kdgR* was contiguous with the *ogl* gene and was transcribed in the same direction. The predicted KdgR protein shared a significant level of homology with GylR, an activator of the glycerol operon from *Streptomyces* and IclR a repressor of the acetate operon from *E. coli* and *Salmonella* (Reverchon *et al.*, 1991). There was a notable region of homology between all three proteins which corresponded to the helix-turn-helix DNA binding motifs of GylR and IclR (Reverchon *et al.*, 1991). Therefore, it was likely that KdgR was also a DNA-binding protein. Consistent with this, comparison of the upstream regions of all of the genes known to be regulated by KdgR led to the identification of a highly conserved 25 bp motif called the KdgR-box. The KdgR-box is composed of a 13 bp and a 12 bp segment separated by between three and five bases A(A/T)AAAA(A/T)GAAA(C/T)N$_{3-5}$TGTTTCATT(A/T)T(A/T)T (Reverchon *et al.*, 1989; Condemine and Robert-Baudouy, 1991; Nasser *et al.*, 1994). The underlined bases were totally conserved in all of KdgR-boxes known and are thought to be essential for KdgR-binding. KdgR-boxes have also been identified upstream of the Ecc *pelC* and *pehA* genes (Hinton *et al.*, 1989a; Saarilahti *et al.*, 1990b).

For most of the genes within the KdgR regulon the KdgR-box is situated either adjacent to, or overlapping with, the transcriptional promoter regions, indicating that KdgR represses transcription by competing with RNA polymerase for binding or by disrupting RNA elongation (Nasser *et al.*, 1994).

To confirm that the proposed KdgR-box was the operator site for KdgR mediated regulation, a 25 bp oligonucleotide, based on the putative KdgR-box of *kdgT* was designed. The KdgR-box was then cloned into a plasmid carrying the *E. coli lacZ* gene, such that it replaced the *lac* operator region (Reverchon *et al.*, 1991). The expression of the *lacZ* gene was assayed, in an *E. coli* background, in the presence and absence of kdgR cloned on a compatible plasmid. In the absence of *kdgR*, the *lacZ* gene was highly expressed, as determined by the level of β-galactosidase activity. However, in the presence of *kdgR*, expressed *in trans*, the *lacZ* gene activity was repressed. This repression was abolished in the presence of galacturonate. Although galacturonate is not a true inducer of pectinolysis itself, *E. coli*, as with Echr, is able to catabolize galacturonate to KDG. Therefore, it was apparent that the 25 bp KdgR-box was sufficient for KdgR-mediated regulation *in vivo*. The expression of a plasmid borne *pelE::uidA* fusion was also analysed in *E. coli*. In this instance the expression of the *pelE* gene was measured by assaying β-glucuronidase activity. As with the *lacZ kdgR*-operator fusion plasmid, *pelE::uidA* expression was repressed by *kdgR in trans* and this repression was relieved by the exogenous addition of galacturonate (Reverchon *et al.*, 1991).

The direct interaction of the purified KdgR protein and the KdgR-box was shown by Nasser *et al.* (1992). Purification of the KdgR protein was difficult because the *kdgR* ORF possessed a poorly-conserved promoter and lacked a discernible ribosome binding motif and so could only be weakly expressed (Reverchon *et al.*, 1991). To overcome this the *kdgR* ORF was cloned into an expression vector with a strong hybrid promoter which was inducible by IPTG (Reverchon *et al.*, 1991; Nasser *et al.*, 1992). However, low yield and instability of the purified protein further hampered investigations. A second attempt to over-express KdgR used the bacteriophage T7 RNA polymerase expression system which yielded up to 10 mg of purified protein per litre of culture (Nasser *et al.*, 1992, 1994). The purified KdgR protein had an approximate molecular mass of 34 kDa, consistent with that predicted from the sequence, but during purification

KdgR migrated with an apparent molecular mass of 62 kDa – thus implying that the intracellular form of KdgR was a dimer (Nasser et al., 1992). This was later confirmed (Nasser et al., 1994).

To demonstrate the binding of KdgR to its operator region a 25 bp KdgR-box oligonucleotide (Reverchon et al., 1991) was used as a target in gel retardation experiments (Nasser et al., 1992). If a protein can interact and form a complex with a DNA target then the DNA/protein complex will have a slower electrophoretic mobility than the "free" unbound DNA. When purified, the KdgR protein was mixed with the 25 bp KdgR-box oligonucleotide and run on a non-denaturing polyacrylamide gel then, depending on the concentration of KdgR repressor, two bands could be observed. The higher band represented the retarded KdgR/operator band and the lower represented the more mobile free-DNA band, thereby confirming that KdgR interacts directly with the KdgR-box and that no other sequences outside of the 25 bp motif are required for efficient binding by KdgR (Nasser et al., 1992). When KDG was added in increasing concentrations to the KdgR and operator reaction mix it caused the protein/DNA complex to dissociate. The same effect was not observed for galacturonate, confirming KDG to be the true inducer which acts by direct interaction with KdgR causing KdgR to dissociate from its operator perhaps through conformational changes of the repressor protein, thereby allowing transcription to occur (Nasser et al., 1991, 1992).

Gel retardation experiments were also conducted on 13 of the operator regions of genes known to be regulated by KdgR, highlighting some interesting aspects of KdgR-mediated regulation (Nasser et al., 1994). Of the operator regions tested, KdgR was only able to bind to eight in vitro. For pelB, outT and the joint kduI–kdgF operator regions the number of retarded bands corresponded directly with the number of KdgR-boxes identified from the sequence. However, using a 275 bp subclone of the kdgT operator region as the target DNA, two retarded complexes were observed indicating the presence of a second, as yet unidentified, binding site within the kdgT operator region. To determine what additional sequences were involved in KdgR-binding to this region DNase I footprinting was performed. The results revealed that, in the presence of high concentrations of repressor, an additional sequence contiguous with the identified KdgR-box was protected. The sequence of this secondary binding site was equivalent to a half KdgR-box motif (TTGTTTGCAAGC). A similar half KdgR-box was also identified in the pelE operator region.

KdgR-binding was not observed for pem, kdgC, kduD, kdgA or outC and so in order to determine which specific purine bases within the KdgR-box were important for the efficient binding of KdgR, Nasser et al. (1992) used the technique of dimethyl sulphate (DMS) interference. The principle behind this technique is to randomly modify individual purine bases. By this method, when a base important for the protein–DNA interaction is modified, it stops the regulatory protein binding (or reduces the efficiency of binding). By gel retardation experiments the operators which are not bound by the protein can be separated from those that it retains the ability to bind. The sequence of these operators can then be determined and the position of the important bases ascertained (Siebenlist and Gilbert,

1980). The KdgR-box oligonucleotide was ^{32}P-end labelled and then treated with DMS to partially methylate the purine bases within the KdgR-box. The methylated DNA and the purified KdgR protein were mixed and run on a polyacrylamide gel as for gel retardation experiments. Any of the DMS treated KdgR-boxes that had been methylated on residues which were important for KdgR-binding were unable to interact with KdgR and so migrated more quickly than the complexed DNA. The DNA was extracted from both the free and retarded bands and purified. Treatment with piperidine cleaved the extracted DNA at the methylated purine bases and so when the cleaved DNA was size fractionated on a sequencing gel, bands were seen to be missing from the complexed DNA sample and present in the free DNA sample. These bands revealed the positions of 14 purine bases within the KdgR-box that are required for the efficient binding of KdgR (Nasser *et al.*, 1992). Included in these 14 bases were 6 adenine residues that were conserved on both DNA strands of all known KdgR-boxes, including those which did not bind KdgR *in vitro*. As a consequence of this it was necessary to redefine the KdgR-box. Nasser *et al.* (1994) conducted a missing contact assay on the operator regions of *pelE* and *ogl*. The missing contact assay is similar in principle to the DMS interference assay (Brunelle and Schleif, 1987). Operator fragments are partially depurinated or depyrimidated at a ratio of slightly less than 1 base per DNA fragment. As with DMS interference, operators are then complexed with the regulator protein and those that are still capable of binding can then be separated from those that are not by polyacrylamide gel electrophoresis. Electrophoresed DNA is purified from the two classes of operator, complexed and free, and run on a sequencing polyacrylamide gel. The sequence obtained after missing contact assays for the operator regions of *pelE* and *ogl* showed that the original consensus for the KdgR-box (Reverchon *et al.*, 1989) could be shortened to two 8 bp motifs separated by a single base: AAT(G/A)AAA(C/T)N N(T/C)(G/A)TTT(C/T)A (Nasser *et al.*, 1992). Interestingly, this consensus is well conserved in the operators to which KdgR is able to bind *in vitro* but is degenerate in those it cannot, leading to the proposal that there were two classes of operator to which KdgR is able to bind.

The differential expression of the pectinolysis genes may, therefore, be influenced by several factors such as the varying affinity of KdgR for different operators. However, other factors were also predicted to affect the repression KdgR mediates on genes within its regulon, including the position of the operator relative to the promoter and the relative number of KdgR operators (Nasser *et al.*, 1992, 1994).

One other aspect of KdgR-mediated regulation highlighted in this study (and the study of Reverchon *et al.*, 1989) was that, unlike the genes involved in the intracellular stages of pectin catabolism, the *pel* and *pem* genes were still inducible by pectin catabolic intermediates in a KdgR⁻ mutant. In addition, KdgR does not influence the growth phase induction of Pel and Cel production (Hugouvieux-Cotte-Pattat *et al.*, 1992). This indicated that there must be other regulatory systems which also govern exoenzyme production in the erwinias. Some of these will be considered below.

In an attempt to define other genetic loci which regulated pectinolysis, researchers employed similar techniques to those used in the identification and characterization of KdgR. Reverchon *et al.* (1994) isolated bacteriophage Mu *d* insertion mutants of Echr, denoted PecS⁻ mutants, which produced elevated levels of Pel and Cel in addition to a blue pigment. This pigment, thought to be indigoidine, is cryptic in the wild-type strain and is proposed to protect the invading bacteria against the plant host defences (Reverchon *et al.*, 1994).

The *pecS* locus is composed of two genes, *pecS* and *pecM*, which are divergently transcribed. The first ORF, *pecS*, was predicted to encode a cytoplasmic protein which shared a significant level of homology with a family of novel regulatory proteins exemplified by MarR. MarR is a repressor of the *marROAB* operon mediating multiple antibiotic resistance (MAR) in *E. coli*. Mutations in *marR* or *marO* (the *mar* operator region) led to a constitutive MAR phenotype giving resistance to a number of structurally unrelated compounds (reviewed by Miller and Sulavik, 1996). Although MarR has no discernible DNA-binding domain, it does bind DNA in a sequence-specific manner (Martin and Rosner, 1995). MarR also acts as a sensor, whereby salicylate prevents MarR binding to *marO* thereby relieving repression on the *marROAB* operon and inducing MAR (Cohen *et al.*, 1993). PecS also shares a significant level of sequence similarity with Rap and Hor, from *Serratia marcescens* and Ecc respectively. Both Rap and Hor are global regulatory proteins that control pigment, antibiotic and exoenzyme production (Thomson *et al.*, 1997).

The second Echr ORF identified at this locus, *pecM*, was predicted to encode an integral membrane protein with ten possible transmembrane domains. Database searches revealed that PecM shared a significant level of sequence identity with an ORF of unknown function from *E. coli* (Reverchon *et al.*, 1994).

Both PecS⁻ and PecM⁻ mutants exhibit the same phenotype indicating that the scope of the regulation of these two genes is the same, although a PecS⁻ mutant exhibited a higher level of derepression for Pel activity than did a PecM⁻ mutant. The same was also true for the derepression of a *celZ::lacZ* transcriptional fusion. Transcriptional fusions were also used to demonstrate that PecS and PecM exert a significant regulatory effect on some of the exoenzyme genes (*pemA, pelB, pelC, pelD, pelE* and *pelL*) but, unlike KdgR, do not regulate the expression of the genes involved in the intracellular catabolism of pectin (*ogl, kduD, kdgT* and *kdgK*; Reverchon *et al.*, 1994; Lojkowska *et al.*, 1995). The expression of the *outC* secretion gene was also derepressed in a PecS⁻ mutant (Condemine and Robert-Baudouy, 1995).

Based on the confirmed subcellular location of PecS and PecM, and in an attempt to explain the common phenotype of the PecS⁻ and PecM⁻ mutants, an interesting model was proposed (Reverchon *et al.*, 1994). It was suggested that PecS and PecM define a novel two-component regulatory system for Pel, Cel and pigment production in Echr. In this model the transmembrane protein PecM would act as the signal transducer and PecS as the cytoplasmic response regulator. However, it should be noted that neither of these proteins displays any homology to previously characterized two-component regulatory proteins.

In Echr Pel and Cel production is induced during the latter stages of the growth phase (Hugouvieux-Cotte-Pattat *et al.*, 1986, 1992; Aymeric *et al.*, 1988). No other extrinsic factors are known to influence Cel production in Echr. In contrast a range of environmental factors regulate Pel production (see section II.B.2(c)). To identify possible signals PecM might respond to, PecS⁻ and PecM⁻ mutants were tested for Pel and Cel production under various growth conditions (Reverchon *et al.*, 1994). However, there was no significant difference detected between the PecS⁻ or the PecM⁻ mutants and the progenitor strain for growth phase regulation of Pel and Cel, nor could any extrinsic factors such as temperature, oxygen limitation or nitrogen starvation, which are known to affect Pel expression, be attributed to PecS- or PecM-mediated regulation (Reverchon *et al.*, 1994; Praillet *et al.*, 1997a). It was also shown that neither PecS nor PecM mutations had an effect on the expression of KdgR and vice versa.

Protein expression studies showed that the intracellular form of PecS is dimeric, with the total cellular complement of PecS being 50 dimers per cell. Like MarR, the PecS protein lacks a clearly identifiable DNA-binding domain and yet has been shown to bind specifically to DNA (Praillet *et al.*, 1996). However, the binding affinity of the purified PecS protein for its operator regions was found to be very weak and did not strictly correlate with the observed *in vivo* repression PecS exerts on genes within its regulon (Praillet *et al.*, 1996). This was thought to be consistent with the idea that the DNA-binding activity of PecS is potentiated by PecM or some as yet unidentified factor.

Gel retardation experiments using the operator regions of *outC*, *celZ* and *pecS* itself, revealed that several different protein–DNA complexes were formed with increasing PecS concentrations. Under saturating PecS concentrations large nucleoprotein complexes were formed, composed of multiple PecS dimers. These nucleoprotein complexes were shown by DNase I footprinting experiments to protect a region of DNA of approximately 150 bp. At sub-saturating PecS concentrations two operator regions were apparent for *outC* and *pecS* and a single protected site for *celZ* (Praillet *et al.*, 1996, 1997a). None of these operators exhibited any of the common characteristics of regulatory regions, such as inverted repeats, and when directly compared to each other were only related by a high AT content (~60–70%). These features were thought to suggest that PecS might act in an analogous manner to the histone-like proteins (H–NS) which are global regulators controlling expression of numerous genes (for a review see Atlung and Ingmer, 1997). Despite sharing some similarities with H–NS proteins, PecS also exhibits several differences (see Praillet *et al.*, 1996).

The PecS binding sites identified so far either overlap or are downstream of the putative promoter regions, suggesting that PecS, like KdgR, acts by competing for binding with RNA polymerase or interfering with RNA chain elongation.

Praillet *et al.* (1997a) also showed that PecS, in addition to repressing the expression of exoenzyme genes, also represses the expression of *pecM* and negatively autoregulates itself, such that the expression of the *pecS* and *pecM* genes was increased by 10- and 100-fold in a PecS⁻ mutant, respectively. The expression of *pecS* was also derepressed in a PecM⁻ mutant, but to a lesser extent. This was consistent with the idea that PecS and PecM form a novel two-component system, where in the absence of an inducing signal PecS is unmodified by PecM and is able to bind DNA, albeit at a low affinity. When PecM is activated, it confers on PecS a higher DNA-binding affinity allowing PecS to fully repress the transcription of genes within its regulon (Praillet *et al.*, 1997a). Recent biochemical evidence, using PecS purified from *E. coli*, Echr and a Echr PecM⁻ strain, appears to support the theory that the activity of PecS is potentiated by PecM (Praillet *et al.*, 1997b).

(iii) PecT

PecT⁻ mutants were also isolated as random Tn mutants exhibiting a hyper Pel phenotype (Surgey *et al.*, 1996). The *pecT* gene encodes a 316 amino acid protein which shares sequence homology with members of the LysR family of transcriptional regulatory proteins which characteristically have a helix-turn-helix DNA-binding motif in the N-terminus (for a review see Schell, 1993). A chromosomaly located *pecT::cm^R* insertion was constructed by allelic exchange to determine the spectrum of regulation exerted by PecT. Transcriptional gene fusions to the various genes involved in pectinolysis, exoenzyme secretion and cellulase production were analysed in this *pecT::cm^R* mutant, and revealed that PecT only regulates the expression of the pectinolytic genes, with the *pelC, pelD, pelE* and *pelL* genes being induced by between two and fivefold and the *pelB* gene showing a twofold reduction in expression.

Double mutants of *pecT* and either *kdgR* or *pecS* revealed that the expression of *pecT* was independent of KdgR or PecS. In addition, the Pel activity of the resultant double mutants, and a triple PecT⁻, KdgR⁻ and PecS⁻ mutant, was much higher than the Pel activity exhibited by the respective isogenic mutants. The effect of each mutation of Pel activity was additive. This confirmed that all three of these proteins belong to independent regulatory systems. By assaying the Pel activity under inducing and non-inducing conditions Surgey *et al.* (1996) also showed that, unlike the PecT⁻ PecS⁻ and KdgR⁻PecS⁻ double mutants, the Pel activity of a PecT⁻ KdgR⁻ mutant was similar in the presence or absence of inducer, exhibiting only a small induction by the addition of PGA. Thus it was likely that the activity of PecT was also responsible for the induction of the pectinolytic genes by pectin catabolic intermediates.

Castillo and Reverchon (1997) have recently identified an additional Echr exoenzyme mutant, *pec-1*, with strongly reduced Pel activity. This mutant was found to have an insertion in one of two *cis*-acting regulatory regions, denoted R_1 and R_2, located upstream of *pecT* and separated from each other by 150 bp. PecT is able to bind to these operator regions in Echr and in doing so negatively autoregulates its own expression. These results conflict with previous findings which indicated that PecT positively autoregulates its own expression. However, experiments showing positive autoregulation were performed in *E. coli* with *pecT* on a multicopy plasmid which may explain the disparity between the findings. Thus, in a *pec-1* mutant PecT can no longer autorepress its own transcription and so is over-expressed leading to a "super-repression" of the genes within the PecT regulon. The requirement for both the R_1 and R_2 regulatory regions for PecT autorepression is predicted to indicate that PecT regulates its own expression by binding simultaneously to both regions generating a distortion in the DNA and preventing transcription (Castillo and Reverchon, 1997).

(b) Global regulators of phytopathogenicity

(i) CRP

It is well established that Pel synthesis in the erwinias is subject to cyclic AMP (cAMP)-controlled catabolite repression when grown in the presence of glucose (Hubbard *et al.*, 1978). Similarly Pel production is also repressed by a phenomenon known as "self-catabolite repression", where Pel production is repressed by high concentrations of catabolic intermediates generated during pectin degradation (Tsuyumu, 1979; Collmer and Bateman, 1981). In addition, mutants unable to produce cAMP are also defective for the production of Pel (Mount *et al.*, 1979) suggesting that catabolite repression could also play a significant role in *Erwinia* spp. pathogenicity.

Because the consensus binding sites of the cAMP receptor protein (CRP) had been located upstream of some of the pectinolysis genes and, in the case of *pelE*, that modification of this site led to a reduction of *pelE* expression (Hugovieux-Cotte-Pattat *et al.*, 1986; Ji *et al.*, 1987; Gold *et al.*, 1992), Reverchon *et al.* (1997) attempted to define the specific role of cAMP-CRP in the control of exoenzyme production. The Echr *crp* gene was cloned by the heterogeneric cross-complementation of an *E. coli* CRP⁻ mutant.

After the construction of a Echr CRP⁻ mutant, by allelic exchange, it was apparent that cAMP-CRP was not involved in the regulation of Cel but severely affected Pel production. Transcriptional fusions to various pectinolytic genes revealed that the *crp* mutation caused a reduction in the expression of *pemA, pelC, pelC, pelD, pelE, ogl, kduI* and *kduT*. In contrast, the expression of *kduD* and *kdgK* was unaffected and the expression of *pelA* was elevated in the absence of cAMP-CRP (Reverchon *et al.*, 1997). As anticipated from the importance of the Pels in the pathogenicity of *Erwinia* spp., *in planta* virulence assays showed that the tissue macerating ability of Echr was markedly reduced by the CRP⁻ mutation.

The direct interaction of the purified CRP protein with the regulatory regions of the pectin catabolic genes was shown by Nasser *et al.* (1997). Gel retardation assays showed that cAMP-CRP interacted specifically with the 5′ regulatory regions of *pelA, pelB, pelC, pelC, pelD, pelE, ogl, kduI* and *kduT*. Consistent with the *in vivo* transcriptional studies, no protein complexes were observed between cAMP-CRP and the *kdgK* or *kduD* upstream regulatory regions (Reverchon *et al.*, 1997). However, the results of the gel retardation studies for *pelA* did not agree with the *in vivo* trancriptional data. No complexes could be observed, perhaps indicating that cAMP-CRP regulates the expression of *pelA* via another unidentified intermediary.

DNase I protection assays confirmed that cAMP-CRP interacts directly with DNA. All the operator regions tested exhibited a protected region of approximately 35 bp, the sequence of which essentially agreed with the *E. coli* CRP binding consensus (TGTGAN$_6$TCACA; Nasser *et al.*, 1997). By comparing the locations of the KdgR and CRP binding sites Nasser *et al.* (1997) noticed that, although the operator sites of *pelB*, *pelD*, *pelE* and *kdul* were distinct, those for *pelC* partially overlapped and the KdgR and cAMP-CRP binding sites for *ogl* and *kdgT* were superimposed.

Gel retardation assays using sub-saturating concentrations of both KdgR and cAMP-CRP with, for example, the operator region of *pelD*, produced three complexes: KdgR–DNA, cAMP-CRP–DNA and cAMP-CRP–KdgR–DNA. At saturating concentrations of KdgR and cAMP-CRP only the ternary complex could be observed. Thus for the majority of the operator regions to which both cAMP-CRP and KdgR bind, both proteins can bind concomitantly, without interference. This was also shown to be true for the *pelC* operator, despite the fact that the cAMP-CRP and KdgR operators partially overlapped (Nasser *et al.*, 1997). In contrast the binding of either cAMP-CRP or KdgR to the operators of *ogl* and *kdgT*, under all of the conditions tested, was mutually exclusive. However, because KdgR was able to form a protein–DNA complex of a considerably higher stability with these operators, when compared to cAMP-CRP, it bound preferentially. Consequently, the only band observed in Gel retardation experiments with the *ogl* or *kdgT* operators, in the presence of both cAMP-CRP and KdgR, was the KdgR–DNA complex.

Nasser *et al.* (1997) have proposed a model which may relate to the wider picture of how Pel production is governed in *Erwinia*. From the *in vivo* transcription data it was evident that KdgR-mediated repression under non-inducing conditions predominates over the activation by cAMP-CRP, because of the relative stabilities of the protein–DNA complexes. However, *in vivo* transcription data also revealed that the expression of the majority of pectinolytic genes under non-inducing conditions was reduced in a CRP$^-$ mutant, when compared to the wild-type strain. This indicated that under non-inducing conditions cAMP-CRP-mediated activation still occurred, albeit at a low level. The physiological significance of this might be that in the absence of inducer cAMP-CRP facilitates the production of a basal level of Pel production which, under favourable conditions, could lead to the rapid induction of all the pectinolysis genes. Consistent with the idea that the low basal expression of several of the Pel genes would ensure the rapid induction of pectinolysis, the sequence of the upstream region of the *ogl* gene revealed the existence of two possible transcriptional promoters, one of which had the potential to be independent of KdgR-mediated repression. Thus, when Echr encounters pectin, the presence of basal levels of Pel and Ogl would lead to the rapid production of pectin catabolic products which in turn induce the genes encoding pectinolysis.

A CRP-like protein from *X. campestris* has also been implicated in the regulation of phytopathogenicity. The CLP protein was found to regulate the production of xantham gum, pigment and exoenzymes (De Crecy-Lagard *et al.*, 1990).

(ii) RsmA

The original RsmA⁻ mutants (regulator of secondary metabolites) were isolated in Ecc after mini-Tn5 mutagenesis. The RsmA⁻ mutation dramatically upregulated the production of Pel, Peh, Cel and Prt, where the basal level of enzyme activity displayed by the mutant was higher than the induced enzyme activity of the wild type in the presence of plant cell extracts (Chatterjee *et al.*, 1995b). This elevated exoenzyme phenotype of the RsmA⁻ mutant dramatically increased tissue maceration making the mutants "hypervirulent" when compared to the wild-type Ecc strain. Northern blot analysis revealed that the RsmA⁻ mutation affected enzyme production by reducing the level of mRNA transcripts. It was not clear at what level this occurred, or indeed whether this was a direct effect mediated by the putative *rsmA* gene product.

The nucleotide sequence of the cloned *rsmA* gene predicted that it encoded a product of only 61 amino acids. Database homology searches indicated that the putative *rsmA* gene shared 95% sequence identity with the *CsrA* gene (carbon storage regulator) from *E. coli* (Cui *et al.*, 1995). The CsrA protein is proposed to have an RNA-binding motif (also present in RsmA) and to regulate gene expression by affecting the rate of mRNA transcript decay (Liu *et al.*, 1995; Liu and Romeo, 1997). Therefore by analogy, RsmA is proposed to exert its effect in a similar manner (Cui *et al.*, 1995).

Southern hybridization analysis using an *rsmA* gene probe revealed that *rsmA* is virtually ubiquitous in the erwinias and present in all of the other *Enterobacteriaceae* tested, including: *Enterobacter aerogenes*, *E.coli*, *Salmonella typhimurium*, *S. marcescens*, *Shigella flexneri* and *Yersinia pseudotuberculosis* (Cui *et al.*, 1995; Mukherjee *et al.*, 1996).

The *rsmA* gene also had a dramatic effect on gene expression when cloned into a multicopy plasmid. When this plasmid was introduced into an array of different bacterial strains it suppressed phenotypes such as exoenzyme production, pathogenicity *in planta*; motility (by affecting flagella formation), EPS production, pigment production, antibiotic production, the production of a bacterial pheromone and the elicitation of the hypersensitive response (HR; Cui *et al.*, 1995; Mukherjee *et al.*, 1996).

More recent work has shown that the ability of RsmA⁻ mutants to elicit a HR is directly related to the increased expression of the gene *hrpN*, the expression of which is virtually undetectable in the wild-type RsmA⁺ Ecc strain (Cui *et al.*, 1996; Mukherjee *et al.*, 1997; see section II.B.3(b)(i).

(iii) Bacterial pheromones

It is well documented that exoenzyme production in the erwinias is induced during the late exponential growth phase (Boyer *et al.*, 1984a;

Hugouvieux-Cotte-Pattat *et al.*, 1986, 1992; Aymeric *et al.*, 1988). The molecular mechanism for growth phase or cell-density-dependent gene regulation in some of the *Erwinia* spp. has been shown to centre on the production of a small diffusible molecule (commonly denoted a bacterial pheromone or autoinducer).

Several research groups have isolated Ecc mutants that displayed a global down regulation in the production of exoenzymes. These mutants were isolated via different approaches. Aep⁻ mutants (activation of extracellular protein production) were isolated as Tn mutants that on plate assays displayed a reduction in exoenzyme production (Murata *et al.*, 1991). Exp⁻ mutants were isolated by following a black box approach as detailed in section II.A (Pirhonen *et al.*, 1991). Rex⁻ and Car⁻ mutants were discovered following research into production of the antibiotic carbapenem by Ecc (Jones *et al.*, 1993). It had been previously shown that mutants unable to produce carbapenem (Car⁻) fell into two classes: class I mutants secreted a low molecular weight, diffusible factor which restored antibiotic production to the class II Car⁻ mutants. This factor was shown to be *N*-(3-oxohexanoyl)-L-homoserine lactone (OHHL; Bainton *et al.*, 1992a, 1992b). Subsequently the class II, OHHL non-producing mutants, were also found to exhibit a reduced Pel, Cel and Prt phenotype and so were phenotypically Rex⁻ (regulation of exoenzymes).

The genes responsible for the Exp⁻ and Rex⁻ phenotypes of these mutants, *expI* and *carI* respectively, were found to encode proteins which shared significant levels of sequence identity with LuxI encoded by the *lux* regulon from *Photobacterium fischeri* (Pirhonen *et al.*, 1993; Swift *et al.*, 1993).

The *lux* regulon can be divided into two operons, one composed of *luxR* and the other *luxICDABEG*, separated by an operator region. The genes *luxCDABEG* are responsible for providing the substrate and encoding the enzyme luciferase for the production of light. The product of the *luxI* gene is proposed to be the autoinducer synthase, required for the production of endogenous OHHL. LuxR is the transcriptional activator, thought to be activated by an interaction with OHHL. When LuxR has been activated by OHHL it is able to activate the expression of the *luxICDABEG* genes, thereby inducing the expression of *lux* (Devine *et al.*, 1988; Stevens *et al.*, 1994; for a review see Meighen, 1991). This entire regulatory system impinges on the specific threshold concentration of OHHL at which the LuxR protein is activated. In physiological terms, because of the proposed low basal level of *luxI* expression, only when a bacterial culture reaches high cell densities (such as those reached in culture at late exponential–early stationary phase) will the OHHL threshold be reached and light be produced.

Cross-complementation with the *luxI* gene *in trans*, or the exogenous addition of OHHL, restored exoenzyme production to the Ecc ExpI⁻ and CarI⁻ (Rex⁻) mutants. The exogenous addition of OHHL also restored the ability of these mutants to cause disease *in planta* (Jones *et al.*, 1993; Pirhonen *et al.*, 1993). Consistent with the idea that exoenzyme production in Ecc is regulated by an analogous system, LuxR homologues have also been identified in Ecc (Pirhonen *et al.*, 1993). Therefore exoenzyme production in Ecc is regulated by a system analogous to the *lux* system (bio-luminescence) in *P. fischeri*.

To demonstrate that the production of OHHL by *expI* was responsible for the growth phase production of exoenzymes in Ecc, Pirhonen *et al.* (1993) constructed an Ecc strain with a *pehA::lacZ* chromosomal fusion. The β-galactosidase activity of the ExpI+ strain was induced at the onset of early stationary phase. No such induction of the expression of the *pehA::lacZ* fusion was observed in the ExpI- double mutant. These results were also confirmed by Northern blot analysis, measuring the level of mRNA transcripts of a number of exoenzyme genes through the growth cycle (Jones *et al.*, 1993; Pirhonen *et al.*, 1993). This mode of OHHL-dependent regulation of virulence factors in *Erwinia* may ensure that genes encoding virulence determinants are only expressed when Ecc cells reach a high enough cell density to mount a successful attack on the plant host.

(c) Environmental factors controlling virulence in the erwinias

Like the diversity of genetic regulatory mechanisms that govern exoenzyme production in the erwinias, physiological studies have also shown that Pel and Cel production are subject to a variety of other environmental influences. Studies looking at the extrinsic factors affecting Pel production have monitored total Pel activity using liquid enzyme assays. In addition, to establish the role of the individual *pel* genes in *Erwinia* pathogenicity, gene fusions have been used to analyse the expression of individual *pel* genes under various growth conditions. As mentioned previously, growth phase has a major influence on Pel production. In Echr, the total Pel activity and the corresponding expression of the *pel* genes, are at a relatively low basal level during the first few hours of growth. When the culture enters late exponential phase there is an induction of the *pel* genes which can result in an increase in the total Pel activity by 40-fold (Hugouvieux-Cotte-Pattat *et al.*, 1992). The explanation for this low basal level of Pel production and the apparent induction of Pel synthesis by pectin catabolic intermediates is thought to be due to the activity of the regulator KdgR (see section II.B.2(a)(i)). However, because KdgR does not affect the growth phase induction of Pels (Hugouvieux-Cotte-Pattat *et al.*, 1992), other regulatory mechanisms were thought to be involved. In Ecc at least, the growth phase control of exoenzyme production is mediated by OHHL (see section II.B.2(b)(iii)). However, it is not yet clear whether a similar system is also responsible for the growth phase induction of Pel production in Echr.

In addition to the induction of Pel synthesis by pectin catabolic intermediates, other growth factors also influence Pel synthesis. Plant cell extracts act in synergy with inducers such as PGA, augmenting Pel production (Hugouvieux-Cotte-Pattat *et al.*, 1992). Plant cell extracts, when added alone, have no effect on Pel production. Similar effects on Pel production were also observed when $CaCl_2$ was added to the growth medium in combination with PGA. In Ecc Ca^{2+} ions were also found to stimulate Pel production. However, Ca^{2+} ions inhibit the production of Peh (Collmer and Keen, 1986; Flego *et al.*, 1997). The biological

Virulence Determinants in Erwinia

389

significance of calcium modulation of Pel and Peh synthesis may be explained by the following model (Flego *et al.*, 1997). Calcium in plants is concentrated in the cell wall. Assuming that the Pel isozymes are important for different stages of the infection process, and are induced accordingly (Pagel and Heitefuss, 1990; Saarilahti *et al.*, 1992), then PehA would be required in the early stages of infection, but would be less important during the latter stages. As a consequence of Peh activity calcium would be released from the plant cell wall resulting in inhibition of *peh* transcription. However, the released Ca^{2+}ions would induce the synthesis of the Pel isozymes which would continue the infection process.

Many other extrinsic factors also affect Pel production including: catabolite repression (see section II.B.2(b)(i)), anaerobiosis, nitrogen starvation, growth temperature, osmolarity, iron availability and the presence of DNA-damaging agents such as Mitomycin C (Hugovieux-Cotte-Pattat *et al.*, 1992; Sauvage and Expert, 1994; Liu *et al.*, 1994b; for a review see Hugovieux-Cotte-Pattat *et al.*, 1996). Cel production in *Erwinia* is also subject to growth phase regulation. However, no other growth condition tested was found to significantly affect Cel production (Boyer *et al.*, 1984a; Aymeric *et al.*, 1988).

3. Secretion systems in the erwinias

The standard "black box" genetical approach initially used to investigate virulence in *Erwinia* produced several classes of mutants which displayed a reduced virulence *in planta*. However, not all of these classes appeared to be due to mutations in either the structural exoenzyme genes or the complex exoenzyme regulatory gene network. Of particular interest was a class of Echr and Ecc mutants (Out⁻) which produced normal levels of exoenzymes but retained pectinases and Cel in the periplasm. These mutants still targeted Prt to the extracellular medium, suggesting that the secretion of the exoenzymes is facilitated by at least two different secretory pathways. In fact *Erwinia* possesses at least three genetically distinct, yet highly conserved, secretory pathways (Figure 1) for the secretion of Prt (Type I), pectinases and Cel (Type II or General Secretory Pathway (GSP)) and Harpin (Type III). The Type I secretion system of Echr has been extensively studied and is a good example of progress due to a "tractable system" (Wandersman *et al.*, 1986, 1987; Delepelaire and Wandersman, 1989, 1990, 1991; Létoffé *et al.*, 1989, 1990; Dahler *et al.*, 1990; Ghigo and Wandersman, 1995). However, Prt may not play a significant role in the virulence of *Erwinia* (Dahler *et al.*, 1990). Therefore an understanding of the mechanism of Prt targeting may not contribute much towards an elucidation of soft-rot disease mechanisms. For this reason this pathway will not be discussed any further (for a recent review see Binet *et al.*, 1997). However, the role of the Type II, and to a lesser degree, the Type III secretory pathway in the virulence of *Erwinia* clearly defines secretion as an important virulence determinant in this organism. Therefore, it is essential that we understand the mechanisms by which proteins are secreted from the *Erwinia* cell, in order to fully understand the organism's ability to cause disease *in planta*. Hence, a large number of genetic techniques have been used to further our understanding of the Type II and Type III secretory systems of *Erwinia*.

(a) Secretion of Pel, Cel and Peh via the Type II (GSP) pathway

In Ecc and Echr, the existence of a common machinery facilitating the secretion of pectinases and Cel was first implied by the isolation of pleiotropic, so-called Out⁻ mutants. Chemical mutagenesis using EMS, and insertion mutagenesis using Tn and phage Mu derivatives, were used to generate mutants of Ecc and Echr which were defective in the secretion of Pel, Peh and Cel but unaltered for Prt secretion (Andro *et al.*, 1984; Thurn and Chatterjee, 1985; Salmond, *et al.*, 1986; Ji *et al.*, 1987; Hinton and Salmond, 1987; Murata *et al.*, 1990; He *et al.*, 1991a). Such mutants synthesized normal levels of Pel, Peh and Cel but accumulated these enzymes in the periplasm. This observation suggested that the secretion process can be divided into two distinct steps: one which exports the exoenzymes across the inner membrane; and the second which secretes the exoenzymes across the outer membrane. The Out⁻ mutants appeared to be blocked in this second step. It was assumed that the first step in exoenzyme transport occurred in an analogous manner to *sec*-dependent protein export in *E. coli* (Puglsey, 1993; for a review see Blaauwen and Driessen, 1996), since when Echr PelE was expressed *in trans* in *E. coli* it was exported to the periplasm (He *et al.*, 1991a). A direct dependence on *sec*-mediated export for the transport of exoenzymes across the Echr inner membrane was demonstrated by He *et al.* (1991b). Secretion kinetics of ³⁵S-labelled Pre-PelE indicated that the Pre-PelE was rapidly processed by the removal of the *N*-terminal signal sequence, and rapidly secreted to the bacterial medium.

(i) Isolation of the out gene clusters

The Echr and Ecc *out* gene clusters were identified by cosmid complementation of Out⁻ mutants with genomic DNA cosmid libraries (Ji *et al.*, 1989; Murata *et al.*, 1990; Lindeberg and Collmer, 1992; Reeves *et al.*, 1993). Cosmids which restored the ability of existing Out⁻ mutants to secrete Pel and Cel were characterized by exoenzyme detection plates, and subsequently restriction mapped. The subsequent subcloning of cosmids, sequence analysis and database homology searches revealed a common cluster of 15 *out* genes (*outC-O, outS, outT and outB*) in Echr and Ecc which were homologous to genes involved in the highly conserved Type II or GSP, as exemplified by the *pul* genes of *Klebsiella oxytoca* (Pugsley, 1993).

The *outC-O* genes are organized in at least two transcriptional units. Evidence for independent transcription of Echr *outC-M* and *outO* is presented by the presence of an inverted repeat between *outM* and *outO* (indicating a possible site for rho-independent termination) and a consensus site for transcription initiation (Lindeberg and Collmer, 1992). Very few intergenic regions exist within the *out* clusters. Additionally, many of the *out* genes either overlap or are separated by a minimal number of base pairs, indicative of translational coupling. Therefore, it is probable that the *outC-M* genes are transcribed from a single promoter upstream of *outC*. However, *in vitro* T7-expression data suggest that Ecc OutG might be expressed from its own promoter within the proximal upstream region of *outG* (Reeves *et al.*, 1993). Tn*phoA* insertions in the upstream region of Echr *outC* displayed a wild-type phenotype for exoenzyme secretion, suggesting that no open reading frames exist in either direction for at least 700 bp upstream of *outC* (Lindeberg and Collmer, 1992). However, a putative regulatory gene, *outT*, lies further upstream of *outC* and in the same orientation (Condemine *et al.*, 1992). In contrast, in Ecc two large open reading frames reside in the position of *outT* in the same orientation as *outC* (S. Wharam, personal communication). These encode an exo-Peh and a Pel, which are probably only expressed *in planta*. *outS* and *outB* are translated in a divergent fashion at the left end of the *out* clusters (Condemine *et al.*, 1992; G. P. C. Salmond, personal communication).

The Echr Out secretory pathway has been successfully reconstituted in *E. coli* (He *et al.*, 1991a). Secretion of a plasmid-encoded form of PelE was successfully demonstrated when the Echr *out* cluster was provided *in trans* on a cosmid. This suggested that no other Echr proteins are necessary for the secretion of PelE in *E. coli*. By contrast, it has not been possible to reconstitute the Ecc Out secretory pathway in *E. coli* by similar methods (Reeves *et al.*, 1993). It has been predicted that the Out apparatus forms a (possibly transient) structure which spans the inner and outer membranes to allow transport of proteins to the exterior of the cell. Individual roles have been assigned to some of the Out proteins and these will be discussed in the following sections.

(ii) Visualizing the Out proteins

The T7 gene (g)*10* promoter expression system devised by Tabor and Richardson (1985) was used to identify the proteins encoded by the Ecc SCRI193 *out* genes (Reeves *et al.*, 1993). This expression system uses an *E. coli* K38(pGP1–2) host which contains the T7 RNA polymerase gene under the control of the λ pL promoter. Exclusive expression of the *out* genes was obtained by thermal induction (allowing expression of T7 RNA polymerase) in the presence of rifampicin (to inhibit host cell gene expression). In this manner some of the Out products were tentatively assigned to specific *out* genes. In particular, a doublet migrating at *ca.*70 kDa was predicted to represent the pre- and pro-forms of OutD, suggesting that OutD traverses the inner membrane in a *sec*-dependent manner.

(iii) Proteolytic processing of OutG, -H, -I and -J by the OutO peptidase

The OutG, -H, -I and -J proteins are termed "pseudopilins" based on their sequence homology at the N-terminus to the Type IV NMePhe pilin subunits, exemplified by PilA of *Pseudomonas aeruginosa* (Johnson *et al.*, 1986). The pilin subunits are expressed as precursor proteins. A short, usually 6–8 amino acid, N-terminal sequence is cleaved and subsequently methylated as the pilin subunits cross the cytoplasmic membrane. During this process methylation of the *N*-terminal phenylalanine residue of the mature pilin subunit occurs. OutO shows homology to the bifunctional peptidase and methylase enzymes (exemplified by PilD/XcpA in *P. aeruginosa* (Bally *et al.*, 1991)) which carry out this processing event. Therefore, it was proposed that OutO might process the pseudopilins in an analogous manner, facilitating the formation of a pilus-type structure which might span the periplasmic space. However, despite the fact that pseudopilin multimerization has been demonstrated in two Type II systems, there is conflicting evidence as to the involvement of a pilus-type structure (Pugsley, 1996; Lu *et al.*, 1997). Moreover, such a structure has not been identified in any of the known Type II systems.

The T7 gene (g)*10* promoter expression system allowed the overexpression of the Ecc Out proteins in an *E. coli* host (Reeves *et al.*, 1993). This system functioned in Ecc. Hence it was possible to investigate the OutO-dependent processing of OutG in the cognate background (Reeves *et al.*, 1994). Cultures of the wild-type Ecc strain SCRI193 and several Out⁻ mutants, expressing OutG were pulsed with [^{35}S]-methionine, chased over a period of time, and visualized by autoradiography. In the wild-type Ecc, pre-OutG was processed incompletely to a mature form, consistent with DNA sequence predictions. Furthermore, pre-OutG was processed in all Out⁻ mutants investigated, except RJP249 (OutO⁻), clearly demonstrating that pre-OutG processing was OutO-dependent. Interestingly, an OutO–BlaM fusion lacking the final three *C*-terminal residues could complement RJP249. This implied that these residues were not essential for the peptidase function of OutO.

Since the other pseudopilins were not efficiently expressed in the T7 expression system, BlaM fusion proteins generated during BlaM topology mapping studies (see section II.B.3(a)(iv)) were exploited to demonstrate OutO peptidase processing of OutH-J. Total cell samples of Ecc strains expressing *blaM* fusions were analysed by SDS-PAGE and immunoblotted using an anti-BlaM antisera. The pre-sequences of the BlaM hybrid proteins were all recognized and processed in the wild-type Ecc SCRI193 background, but not in RJP249. Processing was also observed in Echr strain 3937 and Eca strain SCRI1043, suggesting that the *out* cluster is conserved in several *Erwinia* species.

(iv) Cellular localization

Hydropathy profiling

In order to gain a greater understanding of the function of the Out proteins, it was important to determine their precise cellular location. The method of Kyte and Doolittle (1982) was used to determine the hydropathy profiles of the deduced Ecc *out* gene products and, revealed that all, except *outE*, contain one or more highly hydrophobic segment. Surprisingly, most of the Out proteins were predicted to reside in the inner membrane; an intriguing phenomenon considering that the Out proteins appear to direct protein transport across the outer membrane. Nine proteins (OutC, -D, -G, -H, -I, -J, -K, -M and -N) have *N*-terminal

393

domains predicted to act as inner membrane anchors or cleavable export signals. OutL has an internal putative transmembrane region, and OutF and OutO contain multiple hydrophobic regions, and are therefore predicted to be polytopic integral cytoplasmic membrane proteins. OutS and OutD appeared to be the only Out proteins which were located in the outer membrane. OutE and OutB appeared to be mainly cytoplasmic.

BlaM topology probe technology

Genetical approaches, generally using alkaline phosphatase (PhoA) or β-lactamase (BlaM) have been used to determine the topology of a number of cytoplasmic membrane proteins (for reviews see Broome-Smith *et al.*, 1990; Manoil *et al.*, 1990). However, unlike PhoA, the BlaM topology probe enables the convenient selection of in-frame translational fusions to both cytoplasmic and periplasmic domains. The BlaM technology relies on the principle that only BlaM (which encodes ampicillin resistance (Ap^R)) fusions to extracytoplasmic proteins confer sufficient Ap^R to protect individual cells from Ap. Therefore the cellular location of an in-frame plasmid-encoded BlaM fusion can be determined by the ability of host cells harbouring the fusion to grow as patches on medium containing ampicillin (i.e. Ap^R). The extent of Ap^R can be quantified by determining the minimum inhibitory concentration (MIC) of Ap required to prevent growth of the host.

A detailed topological analysis of OutC, -F, -G, -H, -I, -J, -K, -L, -N and -O was carried out using the BlaM topology probe in both *E. coli* and Ecc (Reeves *et al.*, 1994; Thomas *et al.*, 1997; G. P. C. Salmond personal communication). Progressively 3′ truncated forms of each *out* gene were generated using a nested deletion kit and ligated to a promoterless *blaM* gene. Fusions were characterized as either Ap^R or Ap^S and their MIC values were determined. The stability of individual BlaM fusion proteins was assessed by immunoblot detection of total protein extracts using an anti-BlaM antibody. The results confirmed the computer-predicted locations of the Out proteins investigated. Additionally, OutF was shown to be a polytopic inner membrane protein with three transmembrane domains which connect a large *N*-terminal cytoplasmic domain, a smaller periplasmic domain, and a large cytoplasmic loop; OutO was shown to be a polytopic inner membrane protein with at least seven transmembrane domains, and a hydrophilic cytoplasmic loop towards the *N*-terminus of the protein. BlaM topology mapping also revealed that the OutO catalytic domain responsible for processing the OutG–J pseudopilins lies within the cytoplasmic loop of OutO (Reeves *et al.*, 1994). Hence, this process must occur on the cytoplasmic face of the inner membrane.

The OutD/OutS interaction

The OutD proteins are members of a highly conserved family of outer membrane pore-forming proteins which facilitate the transport of a diverse array of macromolecules to the exterior of the cell. The *N*-termini of the OutD homologues is conserved to a lesser degree than the *C*-termini, suggesting a potential role in species specificity. A number of elegant experiments utilizing co-immunoprecipitation techniques and co-sedimentation in sucrose density gradients have been used to further investigate the multimerization and cellular localization of Echr OutD. Co-immunoprecipitation experiments demonstrated that OutD is capable of forming mixed multimers with pIV, its homologous counterpart in f1 phage, probably through their conserved C-termini (Kazmierczak et al., 1994). OutD has also been shown to form homomultimers in vivo (Shevchik et al., 1997b). The localization of OutD to the outer membrane has been shown to be dependent on OutS, which appears to stabilize OutD by interacting with the C-terminus of OutD (Shevchik et al., 1997b). Electron microscopy has recently revealed that purified YscC, the OutD homologue in the Type III secretion system of Y. enterocolitica, can form a ring-shaped structure of ca. 20 nm with an apparent central pore (Koster et al., 1997). Therefore, it is possible that the OutD proteins adopt a similar conformation in the outer membrane of Erwinia.

(v) Localized mutagenesis of the Ecc out cluster

Although generally an advantage, one possible drawback of chemical and Tn mutagenesis is their random nature; large numbers of mutations must be screened to obtain a small proportion within any gene(s) of interest. However, chemical mutagenesis is the most common approach to generating single base mutations in bacterial genomes. It is important to note that the secretion of exoenzymes by Erwinia does not confer a selective advantage for growth, therefore the hunt for mutants relies solely on screening exoenzyme halo size on detection media – a laborious process. Therefore, localized mutagenesis, utilizing the strain-specific generalized transducing phage φKP (Toth et al., 1993) and Tn5, was exploited to generate single base alterations in the Ecc SCRI193 out cluster (Housby et al., 1998).

A Tn5-out linked derivative of the wild-type Ecc strain was generated by Tn5 mutagenesis followed by φKP co-transduction of the out cluster, and Tn5 (phenotypically Kn^R), into a clean genetic background. A φKP lysate was made on this derivative strain. The lysate was mutagenized using hydroxylamine, and the mutagenized lysate used to transduce the wild-type strain. To investigate the possibility of generating temperature-sensitive (ts) mutations (which might indicate an alteration in protein conformation) Kn^R transductants were screened for an Out^− phenotype on exoenzyme detection plates at 26, 30 and 33°C. The percentage of co-transduction frequency was calculated to determine whether each mutation was truly linked to Tn5, and cell fractionation and quantitative spectrophotometric exoenzyme assays were carried out to determine whether each hydroxylamine generated mutant was a "true" secretion mutant. Cosmid complementation allowed each mutant to be assigned a complementation group. The mutated alleles were amplified by polymerase chain reaction (PCR), cloned and sequenced (at least twice) to confirm the exact location of the lesions responsible for the Out^− phenotype.

This localized mutagenesis approach generated 19 novel Out^− mutants, including two ts mutants, defective in the OutE ATPase and OutL, which only secreted exoenzymes at the lower, permissive temperature of 26°C (Housby *et al.*, 1998). A Pro[159] →Leu transition was found to be responsible for the OutL[ts] phenotype. This residue, based on BlaM topology mapping, resides in the cytoplasmic domain of OutL. The Arg[166]→His transition responsible for the OutE[ts] phenotype was found to lie within the N-terminal third of OutE. Sandkvist *et al.* (1995) showed that in the Type II system of *Vibrio cholerae* the N-terminal cytoplasmic domain of the OutL homologue, EpsL, potentially interacts with the N-terminal region of the OutE homologue, EpsE, in order to maintain the association of EpsE with the cytoplasmic face of the inner membrane. Therefore, by analogy, one might predict that the Ecc OutL protein plays a similar role to EpsL. Several lines of evidence suggest that some OutE homologues exist as dimers (Turner *et al.*, 1993; Possot and Pugsley, 1994; Turner *et al.*, 1997). In fact, the minimal dimerization domain of XcpR of *P. aeruginosa* has recently been localized to within an 85 amino acid region in its N-terminal domain (Turner *et al.*, 1997). Therefore, it is possible that the N-terminal region of OutE is also necessary for dimerization. The exact role of OutE in the secretion process is still unclear. OutE might act to energize the actual secretion process or, more probably, OutE might energize the assembly of the secretion apparatus. However, whatever the specific roles of OutL and OutE might be, it is clear that interactions on the cytoplasmic face of the inner membrane (besides pseudopilin processing and *sec*-dependent export) are essential for secretion across the outer membrane of Ecc.

(vi) The species-specificity paradox

Sequence comparisons revealed that the *out* systems of Echr and Ecc are the most related of the GSP systems, the amino acid identity of individual proteins ranging from 47 to 83%. However, despite this high degree of similarity several differences were noted: (i) Echr does not have an OutN homologue, and Ecc does not have an OutT homologue; (ii) unlike the Echr system, the Ecc system can not be reconstituted in *E. coli* by the presence of the *out* genes *in trans*; (iii) despite a high degree of homology between the Pel and Cel isozymes of Echr and Ecc (e.g. Echr PelC and Ecc PelC exhibit 80% amino acid identity and are likely to display the same three-dimensional structure), neither system is capable of secreting the exoenzymes of the other when provided *in trans* (Py *et al.*, 1991b). These observations suggest that subtle differences exist which fine-tune the secretion systems to the individual requirements of each host. Somehow these differences result in the ability of each system to selectively secrete functionally diverse cognate host proteins, yet prevent the secretion of closely related non-cognate proteins. This phenomenon has been termed the *species specificity paradox* and invites several questions. How are the proteins targeted through the Out apparatus? Is a secretion motif involved, and if so is it specific to the individual host or secreted proteins, and what form does it take? Finally, which Out proteins are involved in recognizing this secretion motif? These questions have been approached in a number of ways.

Investigations into the nature of the secretion motif have proved difficult. Somehow, an array of biochemically and structurally diverse proteins are targeted through the same exit. A comparison of the *Erwinia* exoenzymes which are secreted in a Type II-dependent manner failed to reveal any putative conserved secretion motifs. However, it was noted that the Echr CelZ must adopt its final tertiary conformation prior to secretion across the outer membrane (Bortoli-German *et al.*, 1994). This suggested that the secretion motif might take the form of a three-dimensional "patch" signal consisting of several regions of the secreted protein which only come into close proximity with one another upon folding of the protein into its final conformation. Hybrid-fusion technology (using translational fusions to BlaM) and amino acid substitution experiments have leant further evidence towards this theory. It was shown that three regions of the Echr endoglucanase EGZ must be present for secretion to occur: EGZ catalytic domain, cellulose-binding domain (CBD), and Ser/Thr linker region (Py *et al.*, 1991a, 1993). However, in Ecc PehA the extreme C-terminus appears to be essential for both secretion and protein stability (Palomäki and Saarilahti, 1995, 1997). The combined evidence from these studies, and similar results in other Type II secretory systems, suggest that the secretion motif does take the form of a three-dimensional patch signal.

If the exoenzymes are targeted through the Out pathway via a three-dimensional patch signal this would imply that folding of the exoenzymes into their final conformation in the periplasm is an essential prerequisite for exoenzyme secretion, and hence virulence in *Erwinia*. Periplasmic disulphide bond formation in *E. coli* has been extensively studied and requires several oxidoreductases, termed Dsb proteins (for review see Missiakis and Raina, 1997). Many of the *Erwinia* exoenzymes contain cysteine residues which could potentially form disulphide bonds. As a result of this observation several Echr and Ecc Dsb⁻ mutants have been constructed by marker exchange. The secretion of the cellulase EGZ in Echr requires intact copies of both *dsbA* and *dsbC* for wild-type levels of secretion and enzyme activity (Shevchik *et al.*, 1994, 1995). A number of experiments involving exoenzyme spectrophotometric assays, *pelC::lacZ* and *celV::lacZ* transcriptional fusion data, and protein analysis by SDS-PAGE and immunoblot detection with anti-CelV antisera have been used to investigate the requirement of periplasmic protein folding in the secretion of exoenzymes in Ecc (Vincent-Sealey, 1997). When a plasmid-encoded form of Ecc SCRI193 CelV was expressed in an *E. coli* DsbA⁻ strain it remained in the periplasm but retained cellulase activity. Indeed, the absence of any cysteine residues in CelV suggested that disulphide bond formation is not a prerequisite for the correct folding of CelV. In contrast, Ecc PelC does contain cysteine residues, and is inactive when expressed in an *E. coli* DsbA⁻ mutant. Similarly, in an Ecc SCRI193 DsbA⁻ mutant PelC remains in an inactive non-secretable form. However, the same mutant produces two to three times the amount of CelV (which backs up in the cytoplasm and periplasm), suggesting that CelV might be regulated by a negative feedback loop.

Based on the assumption that the recognition of the exoenzymes must be an important early step in the secretion process, Lindeberg *et al.* (1996) devised a method to allow the systematic exchange of each Echr and Ecc Out component, to determine the "gatekeepers" of the Out systems. Non-polar mutations were constructed in the Echr *outC-M* operon carried on the plasmid pCPP2006. *E. coli* cells harbouring each mutated version of pCPP2006 were assessed for their ability to secrete a plasmid-encoded Echr PelE in the absence or presence of the corresponding Ecc *out* gene *in trans*. Four major points could be concluded from this study. (i) Each *out* gene is required for secretion of Echr PelE from *E. coli*, with the exception of *outH*. However, it must be noted that an OutH mutant of Ecc displays a secretion-incompetent phenotype (Thomas, 1997). (ii) Each mutation can be complemented by its homologue from Ecc SCRI193, except *outC* and *outD*. Hence OutC and OutD have been termed the "gatekeepers" of the Out pathway. (iii) The addition of Ecc *outC-D in trans* in Echr does not confer secretion of Pel1 in this organism, suggesting that *outC* and/or *outD* interact with other proteinaceous components of the Out apparatus. Perhaps Ecc OutD requires Ecc OutS for its correct localization to the outer membrane. (iv) Pel1 secretion can be conferred on the Echr *out* system by the presence of the whole Ecc *out* cluster (*outS, outB, outC-O*).

Of all the components of the Out apparatus, OutD appeared to be the most likely candidate for the recognition of the exoenzyme secretion motif based on several criteria. (i) A "gatekeeper" role determining species specificity has been demonstrated (Lindeberg *et al.*, 1996). (ii) The OutD homologues display a reduced level of homology in the N-terminal, puta-tive periplasmic domain, which could facilitate species-specific recogni-tion of the exoenzymes. (iii) OutD and its homologues form multimers in the outer membrane (Kazmierczak *et al.*, 1994; Koster *et al.*, 1997; Shevchik *et al.*, 1997b), suggesting that they might act as gated pores, which could open in response to some sort of trigger.

A recent study by Shevchik *et al.* (1997b) investigated the ability of OutD to interact with the secreted proteins. Co-expression of *pelB*, *outD* and *outS* in *E. coli* caused a bacteriolytic effect, suggesting that when OutD is correctly inserted into the outer membrane (by OutS), an interaction between PelB and OutD locks OutD into an "open" conformation, causing cell leakage. By comparing the location of PelB in *E. coli* cells in the absence and presence of OutD (assessed by flotation in a sucrose density gradient), Shevchik *et al.* (1997b) demonstrated a direct interaction between PelB and OutD *in vivo*. This interaction was reproduced *in vitro* with purified proteins by performing ligand blotting experiments (Shevchik *et al.*, 1997b). However, despite the high degree of homology between Echr PelB and Ecc PelC (76%) a similar interaction between Echr OutD and Ecc PelC was not detected. Therefore, the interaction between OutD and the exoenzymes might hold the key to the species-specific nature of the Out apparatus.

Few studies have been conducted to investigate the regulation of the *out* clusters in *Erwinia*. Several of the Echr pectinases are inducible by PGA degradation products. This regulation is mediated, in part, by the KdgR repressor protein which binds to a consensus sequence (KdgR box) located upstream of the pectinase genes (see section II.B.2(a)(i)). Candidate KdgR boxes have been identified upstream of the Echr *outT* and *outC–E* genes, and the Ecc *outC* gene (Condemine *et al.*, 1992; Reeves *et al.*, 1993). Therefore, although some of the *out* genes are constitutively expressed, it appeared likely that others might display differential expression in the presence and absence of PGA, in a KdgR-like manner (Condemine *et al.*, 1992). In fact, regulation of the *outC* operon expression by kdgR requires the *outT* gene product, suggesting that OutT could be a transcriptional activator of *outC* (Condemine and Robert-Baudouy, 1995). Lindeberg and Collmer (1992) generated *out*::Tn5*gusA1* fusions in plasmid-encoded versions of the Echr and Ecc *outC–M* regions. The mutated *outC–M* regions were then marker exchanged into the respective wild-type strain and each strain assayed for Pel and β-glucuronidase activity. The *out* clusters of both Echr and Ecc were shown to be significantly induced, albeit weakly, by PGA. The expression of the *out* clusters was also shown to be strongly growth-phase dependent (Lindeberg and Collmer, 1992). However, the molecular basis for this is currently unclear. Recent evidence suggests that expression of the *xcp* gene-encoded Type II secretion apparatus of *Pseudomonas aeruginosa* is regulated in response to small signalling molecules that are synthesized in a cell-density-dependent fashion (Chapon-Hervé *et al.*, 1997). It has been known for some time that these bacterial pheromones are major global regulators of exoenzyme levels in *Erwinia* (see section II.B.2(b)(iii)). Therefore, it is feasible that the Echr and Ecc *out* clusters might be regulated in a similar fashion. Co-ordinate global synthesis of the exoenzymes and their extracellular targeting machinery in this manner would ensure efficient secretion of *de novo* synthesized exoenzymes.

(b) The Type III secretory systems of the erwinias

(i) The Hrp system

The hypersensitive response (HR)

When bacterial phytopathogens infiltrate non-host (or resistant) plant tissue in large numbers an incompatible reaction, often correlated with the HR, occurs. The HR is characterized by a rapid, localized plant cell death, which restricts bacterial movement and further bacterial growth. The role of the HR in the virulence of *Erwinia* was first implied by the isolation of MudIIPR13 chromosomal insertion mutants of Eam strain CFBP1430 which displayed a reduced virulence in pear and apple seedlings (host), and were unable to induce a HR in tobacco (non-host) (Barny *et al.*, 1990). Further characterization of these and similar Eam mutants revealed the existence of a large (~25 kB) chromosomally encoded *hrp* gene cluster

Virulence Determinants in Erwinia

consisting of eight transcriptional units (Steinberger and Beer, 1988; Barny *et al.*, 1990; Vanneste *et al.*, 1990; Walters *et al.*, 1990; Bauer and Beer, 1991; Wei *et al.*, 1992a; Kim *et al.*, 1997). The Eam *hrp* genes are physically and functionally conserved among several bacterial phytopathogens which elicit a HR in non-host plants (Laby and Beer, 1992). However, comparative analysis of the genetic organization, DNA sequence and regulatory systems of these *hrp* clusters suggested that the *Pseudomonas syringae hrp* system is the most closely related system to that of Eam (Alfano and Collmer, 1996).

hrp Genes have also been identified in *Erwinia* spp. which do not normally elicit an observable HR. PelABCE⁻ mutants of Echr strain EC16 were found to display an HR phenotype on various plants (Bauer *et al.*, 1994). The Echr *hrp* genes were identified by hybridization of an Echr EC16 genomic DNA cosmid library to Eam *hrp* sequence DNA probes, and *hrp*::Tn10mini-Kn marker exchange mutants were generated in both the wild-type and PelABCE⁻ strains (Bauer *et al.*, 1994). *pelABCE⁻hrp⁻* strains did not elicit a HR. However, *pelABCE⁺ hrp⁻* strains produced and secreted normal levels of exoenzymes but showed a reduced virulence on witloof chicory. Similarly, RsmA⁻ mutants of Ecc were found to have a reduced virulence in chicory leaves and elicited a HR in tobacco leaves (Cui *et al.*, 1996; Mukherjee *et al.*, 1996). RsmA is a global repressor of exoenzymes in Ecc (Cui *et al.*, 1995; Chatterjee *et al.*, 1995b). Therefore, it has been suggested that under normal growth conditions the HR of Ecc and Echr is masked by the arsenal of pectinolytic enzymes produced by Ecc. The degree of importance the HR plays in the pathogenicity of Ecc and Echr is currently unclear. Recently, the *wts* region of *Erwinia stewartii*, which encodes products required for a water-soaking phenotype, has also been shown to constitute a functional *hrp* system (Frederick, 1991; Coplin *et al.*, 1992; Ahmed *et al.*, 1996).

The *Erwinia* harpins

Genetical and biochemical studies have elucidated the roles of some of the *Erwinia hrp* genes. These include regulatory proteins, components of Type III secretory systems, and elicitors of the HR (termed harpins). In general, harpins are glycine-rich, hydrophilic proteinaceous elicitors of the HR which lack cysteine residues. The HrpN$_{Eam}$ protein of Eam strain Ea321 was isolated from a centrifuged and filter-sterilized preparation of sonicated cells of *E. coli* containing a *hrp* cosmid, and was the first Harpin molecule to be identified in phytopathogenic bacteria (Wei *et al.*, 1992a). This preparation showed a strong HR in tobacco leaves. However, the use of metabolic inhibitors demonstrated that the HR elicited by HrpN$_{Eam}$ requires active plant metabolism (He *et al.*, 1994). The *hrpN$_{Eam}$* gene has been identified in two Eam strains (Ea321 and CFBP1430) (Wei *et al.*, 1992a; Barny, 1995). Although *hrpN$_{Eam}$⁻* marker exchange mutants of these strains failed to produce HrpN$_{Eam}$, they displayed different phenotypes. The Ea321 derivatives were non-pathogenic in immature pear fruits and did not elicit a HR in tobacco, whereas the CFB1430 derivatives remained pathogenic, but were reduced in virulence on pear and apple seedlings, and elicited a variable HR on tobacco. These observations suggested that the ability of harpins to induce a HR is strain specific.

The $hrpN_{Echr}$ gene was identified by hybridization of an Echr EC16 genomic DNA cosmid library with the $hrpN_{Eam}$ gene (Bauer et al., 1995). Purified $HrpN_{Echr}$ induced an HR in leaves of several non-host plants, whereas $hrpN_{Echr}$::Tn5-gusA1 mutants did not. These mutants also displayed a reduced ability to incite infection in witloof chicory leaves. The Ecc $hrpN_{Ecc}$ was isolated from an $RsmA^-$ mutant of strain Ecc71 which showed reduced virulence in chicory leaves and elicited a HR in tobacco leaves (Cui et al., 1996; Mukherjee et al., 1996). These observations suggest that $HrpN_{Eam}$ is the only example pv of an *Erwinia* harpin which has an apparent role in virulence.

The Hrp secretory apparatus

Harpins do not have N-terminal signal sequences suggesting that they are secreted to the exterior of the cell in a *sec*-independent manner (Figure 1). The *hrpA* and *hrpC* operons of Eam have been shown to encode components of a Type III secretory pathway that secretes harpin (Kim et al., 1997). Some phytopathogenic bacteria also secrete avirulence (Avr) proteins via the Hrp Type III pathway (Gopalan et al., 1996; Scofield et al., 1996; Tang et al., 1996; van den Ackerveken et al., 1996; for a review concerning avirulence see Vivian and Gibbon, 1997). It has recently been shown that the Eam DspA protein, which displays homology with AvrE of *P. aeruginosa*, is secreted in a Hrp-dependent manner (Gaudriault et al., 1997). The Eam *dsp* region is located only 4 kb away from the *hrp* cluster, and *dsp* mutants are non-pathogenic but can still elicit an HR on tobacco (Barny et al., 1990; Tharaud et al., 1994). Therefore, DspA is the first example of a pathogenicity factor being secreted via the Hrp secretion pathway. It is possible that other proteins, besides Harpin and DspA, are also secreted in a Hrp-dependent fashion in Eam.

Nine of the Eam *hrp* genes involved in secretion are also conserved in the Type III secretion systems of animal pathogens, hence the designation *hrc* for **h**ypersensitive **r**esponse **c**onserved (Bogdanove et al., 1996b). The Hrc proteins include one outer membrane protein with significant homology to the Type II secretory pathway outer membrane pore proteins, exemplified by OutD in Echr (Shevchik et al., 1997b), one outer membrane lipoprotein, five inner membrane proteins, and two cytoplasmic proteins (including a putative ATPase). The Eam HrpJ protein is one of only two Hrp proteins which has sequence homology to secreted Yop proteins in *Yersinia* spp. (Bogdanove et al., 1996a; Leach and White, 1996).

Harpins are thought to be transported from the cytoplasm to the extracellular milieu in a single step (Figure 1). Indeed, none of the proteins currently known to be secreted in a Type III-dependent manner have been shown to accumulate naturally in the periplasm prior to transport across the outer membrane. However, altered localization of HrpZ in *Pseudomonas syringae* pv. *syringae* has recently demonstrated that Hrp secretion in this organism can be viewed as two genetically distinguishable steps, each encoded by different *hrp* operons: the first, *sec*-independent export across the inner membrane, requires *hrcN* and *hrcR–V*, and the second, secretion across the outer membrane, involves *hrcC* (Charkowski *et al.*, 1997). Interestingly, a novel *sec*-independent periplasmic protein translocation pathway has recently been identified in *E. coli* (Santini *et al.*, 1998). Proteins are directed through this pathway via an *N*-terminal signal sequence which contains a twin-arginine motif. Perhaps the Harpins traverse the inner membrane in an analogous manner. Although, recent evidence suggests that the *Y. enterocolitica* Yop proteins are targeted through the Type III Ysc pathway via a mRNA signal rather than a peptide sequence (Anderson and Schneewind, 1997).

Regulation of the *hrp* genes

Expression of the Eam *hrp* genes in response to environmental stimuli has been intensively investigated using *hrp*::Tn5–*gusA1* gene fusions (Wei *et al.*, 1992b). The *hrp* genes are actively transcribed in plants and minimal media, but are repressed in complex media. However, expression of the *hrp* loci was delayed and severely reduced in pear (host) compared with tobacco leaves (non-host). Expression of the Eam *hrp* genes is affected by various carbon and nitrogen sources, ammonia, nicotinic acid, temperature and pH. These observations allowed the development of an "inducing medium" which facilitated over-expression of some of the *hrp* gene products (Wei *et al.*, 1992b).

Species-specific differences appear to exist in the regulation mechanisms controlling *hrp* gene expression in the Group 1 and Group 2 strains. The *hrp* clusters of Group 2 strains are activated by members of the AraC family (Genin *et al.*, 1992; Oku *et al.*, 1995; Wengenik and Bonas, 1996). However, the Eam and *Pseudomonas syringae* pv. *syringae* Pss61 *hrp* clusters (Group 1) are regulated by the *hrpS* and *hrpL* gene products, which are members of a two-component regulator family (Sneath *et al.*, 1990; Lonetto *et al.*, 1994; Xiao *et al.*, 1994; Wei and Beer, 1995). In fact, transcriptional fusions demonstrated that HrpL also regulates the avirulence-associated *dsp* region (Gaudriault *et al.*, 1997). No functional cross-complementation is observed between the *hrpL* genes of Eam and Pss61 suggesting that subtle differences between the regulatory systems exist (Wei and Beer, 1995). *hrp* consensus sequences, so-called "*hrp*-boxes", are located in the promoter regions of a number of Eam genes, suggesting these are regulated by HrpS/HrpL in a similar manner to the *hrp* and *avr* genes in Pss61 (Bogdanove *et al.*, 1996a; Kim *et al.*, 1997).

(ii) Mot⁻ mutants of *Erwinia carotovora* subsp. *atroseptica* (Eca)

All of the *hrc* genes, with the exception of *hrcC*, also have homologues in flagellum biogenesis. A class of Rvi⁻ mutants of Eca was isolated which showed a reduced virulence *in planta* (Mulholland *et al.*, 1993). These mutants were also non-motile and hence termed Mot⁻ mutants. The corresponding *mopA* and *mopC–E* genes (<u>mo</u>tility and <u>p</u>athogenicity) were subsequently sequenced and found to encode protein homologues of Type III secretory systems and components of flagella biogenesis. Therefore, it is likely that flagella biogenesis, and as a result motility, in Eca also requires the secretion of proteins via a Type-III-like system.

4. Is there a role for the bacterial cell surface in pathogenicity?

The role of the bacterial cell surface in the virulence of phytopathogens has been approached using several strategies. Expert and Toussaint (1985) used bacteriocins to select for spontaneous cell surface mutants of Echr. The toxicity of bacteriocins relies on the adsorption of the toxin to specific receptors on the outer membrane (OM; see Konisky, 1982). Thus wild-type Echr cells which survive treatment with these toxins are likely to have altered cell surfaces which prevent bacteriocin adsorption. Spontaneous bacteriocin-resistant mutants were isolated by treating actively growing cells with a number of different bacteriocins produced by various *Erwinia* strains. The treated cells were incubated at 30°C for 15 min and then plated out on to nutrient agar plates. After treatment with the bacteriocins there were very few surviving colonies. The OM integrity of several surviving, putative OM mutants was partially assessed by testing the sensitivity of the mutants to detergents and chelating agents, as well as their susceptibility to bacteriophage Mu.

Pathogenicity studies were performed on these bacteriocin-resistant mutants to determine if any of the putative spontaneous OM alterations had affected pathogenicity. None of the mutants were affected for the production or secretion of pectinases and Cel, as shown on exoenzyme assay plates and by IEF. In addition, *Saintpaulia* leaf assays revealed that these mutants retained the ability to macerate plant tissue. However, when tested on whole plants they were no longer able to cause soft-rot because they provoked a plant HR which limited the spread of the infection (Expert and Toussaint, 1985).

OM profiles of the bacteriocin-resistant mutants showed that the majority of them lacked one or more OM polypeptides that were specifically induced under iron-limiting conditions. These proteins were thought to be involved in iron uptake, possibly as a receptor for a siderophore. Congruent with this, Enard *et al.* (1988) showed that the production of these polypeptides is directly associated with that of a catechol-like siderophore, chrysobactin. Moreover these three polypeptides form the OM receptors for several siderophores produced by Echr (for a review see Expert *et al.*, 1996). Enard *et al.* (1988) have also shown that Echr mutants affected in iron assimilation fail to induce a systemic infection producing, at best, only localized symptoms. Consequently, this implicated iron uptake capability as a virulence factor in Echr.

An alternative strategy using the Echr temperate bacteriophage ΦEC2 (Resibois *et al.*, 1984) was used to isolate spontaneous cell surface mutants (Schoonejans *et al.*, 1987). As with bacteriocins, bacteriophages also require a specific OM receptor and so cells surviving lysis may also have spontaneous cell surface alterations affecting phage adsorption. High titre lysates were spotted, undiluted, on to soft agar lawns seeded with the ΦEC2-sensitive wild-type Echr strain. After incubation at 30°C for 24 h putative ΦEC2-resistant (ΦEC2^R) single colonies were isolated from the phage plaque. The lipopolysaccharide (LPS) profiles of these ΦEC2^R mutants indicated that the majority of them had lost the putative O-antigenic side chain, thought to make up at least part of the OM receptor of ΦEC2 (Schoonejans *et al.*, 1987). In addition to lacking the O-antigen, two of the mutants tested also exhibited alteration in the LPS core region (shown by their failure to adsorb bacteriophage Mu; see Sandulache *et al.*, 1985). Analysis of the OM proteins failed to identify any qualitative changes under both high iron or iron-limiting conditions.

Virulence tests showed that these ΦEC2^R mutants were unaffected in the production and secretion of exoenzyme virulence factors and, when tested *in planta*, only two of the mutants were unable to elicit normal soft-rot symptoms. Both of the mutants affected in virulence had alterations in their LPS core and so it was proposed that those mutants with the most severe alterations to their LPS were likely to be more susceptible to plant antimicrobial agents, such as phytoalexins, or that these alterations might expose a structure which could be recognized by, and thereby provoke, the plant host defences.

These results show that both the cell surface of Echr and the chrysobactin siderophore have roles as virulence factors. However, attempts to identify similar siderophores acting in the pathogenicity of Ecc and Eca have been fruitless. In addition, unlike the phage-resistant mutants of Echr, T4-resistant Ecc mutants were unaffected in virulence when tested on tobacco, even though they had alterations in their LPS (Pirhonen *et al.*, 1988). However, in contrast to the T4-resistant studies, recent work has shown that multiply phage resistant mutants of Eca exhibit reduced virulence on potato in stem inoculation tests (Thorpe *et al.*, submitted). In addition, some of the latter mutants were very pleiotropic showing: reduced exoenzyme production; the virtual absence of flagella; an altered LPS profile; and a heightened sensitivity to surface active agents. The explanation for this pleiotropic phenotype is unclear at present.

LPS has also been shown to be important for the pathogenicity of other bacterial phytopathogens. Titarenko *et al.* (1997) used Tn5 to generate mutants of *Ralstonia solanacearum* that showed an increased sensitivity to plant-derived antimicrobial compounds. Sensitive mutants were found to carry Tn5 insertions within the *rfaF* gene encoding a heptosyltransferase enzyme which is involved in LPS biosynthesis. RfaF mutants were sensitive to a number of plant antimicrobial compounds and, when tested on tobacco, failed to cause any observable disease symptoms, with the number of recoverable infecting bacteria falling rapidly post-inoculation. Complementation of the RfaF$^-$ mutant with the *rfaF$^-$* gene *in trans* partially restored the ability of the mutants to grow and infect tobacco plants. Similar LPS-defective mutants have also been isolated from *X. campestris* (Dow *et al.*, 1995). As with *R. solanacearum*, a decrease in viability *in planta* could also be related to a reduced pathogenicity in *X. campestris* RfaX$^-$ and RfaY$^-$ mutants. The explanation for the avirulent phenotypes of these LPS mutants was again thought to reflect the role of LPS as a permeability barrier protecting the invading pathogen from host defences. However, it is also possible that many of these cell surface mutations, especially the spontaneous OM mutants, are highly pleiotropic and that other possible effects are responsible for their pathogenicity phenotype in plants.

Bacteriophages Mu and EC2 were also used to try to identify the cell surface alterations. The receptor binding site for Mu lies in the inner core of LPS (Sandulache et al., 1985) and at least part of the receptor of ΦEC2 is thought to be composed of the O-antigen-like structure of the LPS molecule (Schoonejans et al., 1987). Thus alterations in the LPS layer will affect the ability of these bacteriophages to adsorb to the Echr cells. The majority of the bacteriocin-resistant mutants remained sensitive to Mu but were resistant to the infection by ΦEC2 implying that these mutants possessed alterations in the LPS layer but not in the core region. It was also noted that mutants isolated as being resistant to ΦEC2 were also insensitive to the bacteriocins.

5. Plant-inducible genes

Considerable understanding of the mechanisms of phytopathogenicity have come from constructing mutations in genes thought likely to affect pathogenicity and from the direct selection of mutants affected in virulence (see previous sections). These approaches, although fundamental to understanding the plant–pathogen interaction, are unlikely to identify all of the genes that are required to mount a successful infection. Some less obvious virulence determinants may not be discovered because a mutation in the gene concerned may be undetectable (or too subtle to detect) using the mass screening protocols commonly employed. Consequently, several researchers have developed methods for identifying genes, the expression of which is specifically induced in planta. This follows the rationale that some of these plant-inducible genes may be important for survival in planta and so may define novel pathogenicity determinants.

Osbourn et al. (1987) constructed a broad host range promoter probe plasmid for use in X. campestris p.v. campestris. Plasmid pIJ3100 carried a promoterless chloramphenicol (Cm) acetyltransferase gene (CAT) cloned alongside a multicloning site, such that when chromosomal DNA was "shotgun" cloned into this multicloning site any DNA exhibiting promoter activity could direct the expression of the CAT gene and thereby confers Cm[R] on the host bacterium.

A pool of the random genomic inserts cloned into pIJ3100 was introduced into Xanthomonas by conjugation. To identify any of the transconjugants that were Cm[R] in planta, individual colonies containing the recombinant pIJ3100 plasmids were inoculated into turnip seedlings that had been previously treated with Cm. The wild-type strain transformed with plasmid pIJ3100 acted as the negative control and was unable to survive in the antibiotic-treated plants. Of 1100 transconjugants tested 19 were found to be Cm[R] in planta after rigorous repeat testing. Of these 19 Cm[R] strains, 5 were found to be Cm[R] in planta and in vitro, i.e. the CAT gene was expressed constitutively. However, 14 strains harbouring the recombinant pIJ3100 plasmids were only Cm[R] in planta and were thought to define plant-inducible promoters (Osbourn et al., 1987). To isolate the genes associated with these promoters a chromosomal library was screened by Southern blot analysis using the plant-inducible promoter as a DNA probe. Unfortunately, subsequent molecular characterization of these plant-inducible regions has failed to identify any plant-inducible genes within the original *Xanthomonas* genomic DNA fragments that were identified *in planta*, although a different ORF involved in pathogenicity was fortuitously isolated during this investigation (Osbourn *et al.*, 1987, 1990).

An alternative approach to isolating Echr plant-inducible genes was taken by Beaulieu and Van Gijsegem (1990) using bacteriophage Mu *d*IIPR3. Like plasmid pIJ3100, Mu *d*IIPR3 carries a promoterless antibiotic-resistance gene, conferring Kn^R, which is expressed only if Mu *d*IIPR3 inserts in the correct orientation and 3' of an active promoter sequence. Echr was mutagenized with Mu *d*IIPR3. After mutagenesis, transductants were pooled and grown in the presence of *S. ionantha* plant cell extract for 6 h before Kn was added to the growth media. Therefore, only those transductants in which the Kn resistance gene was expressed could survive. The use of plant cell extract, in place of inoculating transductants directly into plants, ensured that mutants in which Mu *d*IIPR3 had inserted into genes essential for survival *in planta*, and therefore presumably virulence, were not lost. To screen out mutants in which Kn^R was constitutively expressed, recovered transductants were grown on media supplemented with Kn in the presence and absence of plant cell extracts. Mutants which were only Kn^R in the presence of plant cell extracts were retained and were tested for virulence using potato tuber and whole plant virulence assays (Beaulieu and Van Gijsegem, 1990).

The advantage of the Mu *d*IIPR3 system is that, in a single step, mutants can be isolated that have insertions in plant-inducible genes which, unlike the plasmid-based system, are directly tagged and can be further characterized for possible alterations in virulence. In addition, the antibiotic selection for plant-inducible genes used by both systems dramatically reduces the number of mutants that must be screened, especially when compared to methods using, for example, *lacZ* or *uidA* gene fusions to follow gene activity. However, there is also a disadvantage associated with this selection procedure, whereby genes which are expressed for only short periods during growth may not protect the cell from the antibiotic selection and so would be unrecoverable.

Plant-inducible gene mutants isolated using Mu *d*IIPR3 fell into three categories: those that were severely attenuated for virulence *in planta*; those that were mildly affected for virulence; and those that were unaffected for virulence. Within these three groups further phenotypic analysis revealed that one of the severely attenuated strains exhibited a reduced PelA phenotype. PelA although only weakly expressed *in vitro* has been previously shown to be important for virulence *in planta* (Boccara *et al.*, 1988). Other plant-inducible mutants were affected in iron assimilation and so could not grow on media in the presence of an iron chelator. In addition, one of the mildy attenuated strains was unable to catabolize galacturonate, suggesting that the ability to use this carbon source *in planta* is important for a successful infection (Beaulieu and Van Gijsegem, 1990).

In an attempt to determine if the reduced virulence phenotypes of these mutants were host-specific, *in planta* assays were performed using a number of plant hosts. The majority of the plant-inducible mutants showed a reduced virulence phenotype on all of the plants tested. However, one mutant that was virulent on *S. ionantha* plants appeared to be hypervirulent on pea plantlets (*Pisum sativum* L.) and Witloof chicory leaves. In addition, the attenuation of virulence of two mutants was found to be plant specific. This was exemplified by one mutant that was fully virulent on pea plantlets and in potato tubers but exhibited a reduced virulence phenotype on *S. ionantha* plants and Witloof chicory leaves (Beaulieu and Van Gijsegem, 1992). The genetic explanation for all of these mutants has yet to be addressed, although limited progress has been made on identifying a plant-derived inducer from *S. ionantha* cell extract (see Bourson *et al.*, 1993). More recent studies have shown that there are many genes induced *in planta* which are important for the pathogenicity of *Erwinia* spp. These are exemplified by the genes encoding the secondary pectate lyases (discussed in section II.B.1(d)(iii)).

◆◆◆◆◆◆ III. FUTURE PROSPECTS

As discussed in the previous sections, it is clear that the spectrum of genetic and molecular biological tools that has been applied to the erwinias is now as comprehensive as those used for the analysis of the molecular biologist's "workhorse", *E. coli.* Further applications of this impressive array of techniques will continue to improve our understanding of molecular phytopathology, protein secretion, global gene regulation systems, antibiotic production and various other interesting traits shown by the erwinias and other plant pathogens. However, in the past few years there has been an increased effort in the application of several new techniques to the study of bacterial pathogenesis in various bacterial pathogens.

These "new" approaches include the use of genomics for the analysis of predicted functions of gene products, study of gene organization and regulation, pathogenicity island analysis, and the investigation of gene evolution and horizontal transfer. At the time of writing, no *Erwinia* genome sequence has been completed. However, the determination of an *Erwinia* genome sequence would undoubtedly produce some exciting information. Why is this plant pathogen in the *Enterobacteriaceae* beside mostly human and animal pathogens? Is it simply an *E. coli* with pathogenicity islands of genes for plant virulence factors?

Proteomic analysis has yet to be applied to *Erwinia* and, as above, there can be little doubt of the wealth of information that such an analysis will produce. Similarly, the direct application of new versions of *in vivo* expression technology (IVET) and signature-tagged mutagenesis (STM) (Chapter 4, XIV) to uncover every aspect (during pathogenesis and survival outside of plants) of *Erwinia* physiology will open up new research projects. It was shown some time ago that the spread of *Erwinia in planta* could be followed by exploiting the *lux* genes from *Vibrio* in a biolumi-nescence assay of real time migration of the pathogen. In addition to bio-luminescence, more recently some workers have turned to the green fluorescent protein (GFP) as a tag for gene expression *in vivo*, in various bacterial pathogens, and so this protein fusion approach will also be attractive in *Erwinia* and in other plant pathogens.

In summary, the current catalogue of genetic tools available for the study of *Erwinia* is already very good. The superimposition of the "new" approaches, mentioned above, on this existing list of techniques can only lead us to deeper understanding of the diversity of physiological processes that we currently see in such interesting bacteria. Or, to put it another way, as with the use of *lux* and GFP, the future looks very bright (and exciting) indeed!

Acknowledgements

This work was supported by generous funding from the BBSRC to George P. C. Salmond.

References

Ahmed, M., Majerczak, D. R., and Coplin, D. L. (1996). Harpin is not necessary for the pathogenicity of *Erwinia stewartii* on maize (Abstr.). In *Abstr. Book 8th Int. Congr. Mol. Plant–Microbe Interact.*, pp. G-11.

Alfano, J. R. and Collmer, A. (1996). Bacterial pathogens in plants: life up against the wall. *Plant Cell* **8**, 1683–1698.

Alfano, J. R., Hyun Ham, J., and Collmer, A. (1995). Use of Tn5tac1 to clone a *pel* gene encoding a highly alkaline, asparagine-rich pectate lyase isozyme from an *Erwinia chrysanthemi* EC16 mutant with deletions affecting the major pectate lyase isozymes. *J. Bacteriol.* **177**, 4553–4556.

Anderson, D. M. and Mills, D. (1985). The use of transposon mutagenesis in the isolation of nutritional and virulence mutants in two pathovars of *Pseudomonas syringae*. *Phytopathology* **75**, 104–108.

Anderson, D. M. and Schneewind, O. (1997) A mRNA signal for the type III secre-tion of YOP proteins by *Yersinia enterocolitica*. *Science* **278**, 1140–1143.

Andro, T., Chambost, J. P., Kotoujansky, A., Cattaneo, J., Bertheau, Y., Barras, F., van Gijsegem, F. and Coleno, A. (1984). Mutants of *Ech* defective in secretion of pectinase and cellulase. *J. Bacteriol*, **160**, 1199–1203.

Atlung, T. and Ingmer, H. (1997). H-NS: a modulator of environmentally regu-lated gene expression. *Mol. Microbiol.* **24**, 7–17.

Axelrood, P. E., Rella, M. and Schroth, M. N. (1988) Role of antibiosis in competi-tion of *Erwinia* strains in potato infection Courts. *Appl. Environ. Microbiol.* **54**, 1222–1229.

Aymeric, J. L., Guiseppi, A., Pascal, M. C. and Chippaux, M. (1988). Mapping and regulation of the *cel* genes in *Erwinia chrysanthemi*. *Mol. Gen. Genet.* **211**, 95–101.

Bainton, N. J., Bycroft, B. W., Chhabra, S. R., Stead, P., Gledhill, L., Hill, P., Rees, C. E. D., Winson, M. K., Salmond G. P. C., Stewart G. S. A. B. and Williams, P. (1992a). A general role for the *lux* autoinducer in bacterial cell signalling: control antibiotic biosynthesis in *Erwinia*. *Gene* **116**, 87–91.

Bainton, N. J., Stead, P., Chhabra, S. R., Bycroft, B. W., Salmond, G. P. C., Stewart,

G. S. A. B. and Williams, P. (1992b). *N*-(3-oxohexanoyl)-L-homoserine lactone regulates carbapenem antibiotic production in *Erwinia carotovora*. *J. Biochem.* **288**, 997–1004.

Baker, T. A., Howe, M. M. and Gross, C. A. (1983). Mu *d*X, a derivative of Mu *d*1 (lac Ap[1]) which makes stable LacZ fusions at high temperature. *J. Bacteriol.* **156**, 970–974.

Bally, M., Bal, G., Badere, A. and Lazdunski, A. (1991). Protein secretion in *Pseudomonas aeruginosa*: The *xcpA* gene encodes an integral inner membrane protein homologous to *Klebsiella pneumoniae* secretion function protein PulO. *J. Bacteriol.* **173**, 479–486.

Bardonnet, N. and Blanco, C. (1992). *uidA* antibiotic resistance cassettes for insertion mutagenesis, gene fusion and genetic constructions. *FEMS Microbiol. Lett.* **93**, 243–248.

Barny, M. A. (1995). *Erwinia amylovora hrpN* mutants, blocked in harpin synthesis, express a reduced virulence on host plants and elicit variable hypersensitive reactions on tobacco. *Eur. J. Plant Pathol.* **101**, 333–340.

Barny, M. A., Guinebretiere, M. H., Marcais, B., Coissac, E., Paulin, J. P. and Laurent, J. (1990). Cloning of a large gene cluster involved in *Erwinia amylovora* CFBP1430 virulence. *Mol. Microbiol.* **4**, 777–786.

Barras, F., Boyer, M-H., Chambost, J-P. and Chippaux, M. (1984). Construction of a genomic library of *Erwinia chrysanthemi* and molecular cloning of cellulase gene. *Mol. Gen. Genet.* **197**, 513–514.

Barras, F., Thurn, K. K. and Chatterjee, A. K. (1987). Resolution of four pectate lyase structural genes of *Erwinia chrysanthemi* (EC16) and characterisation of the enzymes in *Escherichia coli*. *Mol. Gen. Genet.* **209**, 319–325.

Barras, F., Bortoli-German, I., Bauzan, M., Rouvier, J, Gey, C., Heyraud, A. and Henrissat, B. (1992). Stereochemistry of hydrolysis reaction catalysed by endoglucanase Z from *Erwinia chrysanthemi*. *FEBS Lett.* **300**, 145–148.

Bartling, S., Wegener, C. and Olsen, O. (1995). Synergism between *Erwinia* pectate lyase isoenzymes that depolymerise. *Microbiology* **141**, 873–881.

Bauer, D. W. and Beer, S. V. (1987). Cloning of a gene from *Erwinia amylovora* involved in induction of hypersensitivity and pathogenicity. In *Proc. 6th Int. Conf. Plant Path. Bact.* (E. L. Civerolo, A. Collmer, R. E. Davis, and A. G. Gillespie, eds), pp. 425–429. Martinus Nijhoff, Dordrecht.

Bauer, D. W. and Beer, S. V. (1991). Further characterisation of an *hrp* gene cluster of *Erwinia amylovora*. *Mol. Plant–Microbe Interact.* **4**, 493–499.

Bauer, D. W., Bogdanove, A. J., Beer, S. V. and Collmer, A. (1994). *Erwinia chrysanthemi hrp* genes and their involvement in soft rot pathogenesis and elicitation of the hypersensitive response. *Mol. Plant–Microbe Interact.* **7**, 573–581.

Bauer, D. W., Wei, Z. M., Beer, S. V. and Collmer, A. (1995). *Erwinia chrysanthemi* Harpin$_{Ech}$ that contributes to soft-rot pathogenesis. *Mol. Plant–Microbe Interact.* **8**, 484–491.

Bayot, R. G. and Ries, S. M. (1986). Role of motility in apple blossom infection by *Erwinia amylovora* and studies of fire blight control with attractant and repellent compounds. *Phytopathology* **76**, 441–445.

Beaulieu, C. and van Gijsegem, F. (1990). Identification of plant inducible genes in *Erwinia chrysanthemi* 3937. *J. Bacteriol.* **172**, 1569–1575.

Beaulieu, C. and van Gijsegem, F. (1992). Pathogeneic behaviour of several mini-Mu-induced mutants of *Erwinia chrysanthemi* on different plants. *Mol. Plant–Microbe Interact.* **5**, 340–346.

Beaulieu, C., Boccara, M. and van Gijsegem, F. (1993). Pathogenic behaviour of pectinase-defective *Erwinia chrysanthemi* mutants on different plants. *Mol. Plant–Microbe Interact.* **6**, 197–202.

Virulence Determinants in *Erwinia*

409

Beringer, J. E., Beynon, J. L., Buchanan-Wollaston, A. V. and Johnston, A. W. B. (1978). Transfer of drug-resistance transposon Tn5 into *Rhizobium*. *Nature* **276**, 633–634.

Bertheau, Y., Madgidihervan, E., Kotoujansky, A., Nguyenthe, C., Andro, T. and Coleno, A. (1984). Detection of depolymerase isoenzymes after electrophoresis or electrofocusing, or in titration curves. *J. Anal. Biochem.* **139**, 383–389.

Biely, P., Mislovicova, D. and Toman, R. (1985). Soluble chromogenic substrates for the assay of endo-1,4-beta-xylanases and endo-1,4-beta-glucanases. *J. Anal. Biochem.* **144**, 42–146.

Binet, R. and Wandersman, C. (1995). Protein secretion by hybrid ABC-transporters: specific functions of the membrane ATPase and the membrane fusion protein. *EMBO J.* **14**, 2298–2306.

Binet, R., Letoffe, S., Ghigo, J. M., Delepelaire, P. and Wandersman, C. (1997). Protein secretion by Gram-negative bacterial ABC exporters – a review. *Gene* **192**, 7–11.

Blaauwen, T. and Driessen, A. J. M. (1996). Sec-dependent preprotein translocation in bacteria. *Arch. Microbiol.* **165** 1–8.

Boccara, M., Diolez, A., Rouve, M. and Kotoujansky, A. (1988). The role of individual pectate lyases of *Ech* strain 3937 in pathogenicity on *Saintpaulia* plants. *Physiol. Mol. Plant Path.* **33**, 95–104.

Boccara, M., Aymeric, J. L. and Camus, C. (1994). Role of endoglucanases in *Erwinia chrysanthemi* 3937 virulence on *Saintpaulia ionantha*. *J. Bacteriol.* **176**, 1524–1526.

Bochner, B. R., Huang, H. C., Schieven, G. L. and Ames, B. (1980). Positive selection for loss of tetracycline resistance. *J. Bacteriol.* **143**, 926–933.

Bogdanove, A. J., Wei, Z. M., Zhao, L. P. and Beer, S. V. (1996a). *Erwinia amylovora* secretes Harpin via a Type III pathway and contains a homologue of *yopN* of *Yersinia* spp. *J. Bacteriol.* **178**, 1720–1730.

Bogdanove, A. J., Beer, S. V., Bonas, U., Boucher, C. A., Collmer, A., Coplin, D. L., Cornelis, G. R., Huang, H. C., Hutcheson, S. W., Panopoulos, N. J. and van Gijsegem, F. (1996b). Unified nomenclature for broadly conserved *hrp* genes of phytopathogenic bacteria. *Mol. Microbiol.* **20**, 681–683.

Bortoli-German, I., Brun, E., Py, B., Chippaux, M. and Barras, F. (1994). Periplasmic disulphide bond formation is essential for cellulase secretion by the plant pathogen *Erwinia chrysanthemi*. *Mol. Microbiol.* **11**, 545–553.

Bortoli-German, I., Haiech, J., Chippaux, M. and Barras, F. (1995). Informational suppression to investigate structural functional and evolutionary aspects of the *Erwinia chrysanthemi* cellulase EGZ. *J. Mol. Biol.* **246**, 82–94.

Boucher, C., Bergeron, B., deBertalmio, M. B. and Denarie, J. (1977). Introduction of bacteriophage Mu into *Pseudomonas solanacearum* and *Rhizobium meliloti* using the R factor RP4. *J. Gen. Microbiol.* **98**, 253–263.

Boucher, C. A., Barberis, P. A., Trigalet, A. P. and Demery, D. (1985). Transposon mutagenesis of *Pseudomonas solanacearum*: isolation of Tn5-induced avirulent mutants. *J. Gen. Microbiol.* **131**, 2449–2457.

Boucher, C., van Gijsegem, F., Barberis, P., Arlat, M. and Zischek, C. (1987). *Pseudomonas solanacearum* genes controlling both pathogenicity on tomato and hypersensitivity on tobacco are clustered. *J. Bacteriol.* **169**, 5626–5632.

Bourson, C., Favey, S., Reverchon, S. and Robert-Baudouy, J. (1993). Regulation of the expression of a *pelA::uidA* fusion in *Erwinia chrysanthemi* and demonstration of the synergistic action of plant extract with polygalacturonate on pectate lyase synthesis. *J. Gen. Microbiol.* **139**, 1–9.

Boyer, M. H., Chambost, J. P., Magnan, M. and Cattaneo, J. (1984a). Carboxymethyl-cellulase from *Erwinia chrysanthemi*. I Production and regulation of extracellular carboxymethyl-cellulase. *J. Biotechnol.* **1**, 229–239.

Boyer, M-H., Chambost, J. P., Magnan, M. and Cattanéo, J. (1984b). Carboxymethyl-cellulase from *Erwinia chrysanthemi*. II. Purification and partial characterisation of an endo-β-1,4-glucanase. *J. Biotechnol.* **1**, 241–252.

Boyer, M-H., Cami, B., Kotoujansky, A., Chambost, J-P., Frixon, C. and Cattanéo, J. (1987a). Isolation of the gene encoding the major endoglucanase of *Erwinia chrysanthemi*. Homology between *cel* genes of two strains of *Erwinia chrysanthemi*. *FEMS Microbiol. Lett.* **41**, 351–356.

Boyer, M-H., Cami, B., Chambost, J-P., Magnan, M. and Cattanéo, J. (1987b). Characterisation of a new endoglucanase from *Erwinia chrysanthemi*. *Eur. J. Biochem.* **162**, 311–316.

Brooks, A. D., He, S-Y., Gold, S., Keen, N. T., Collmer, A. and Hutcheson, S. W. (1990). Molecular cloning of the structural gene for exopolygalacturonate lyase from *Erwinia chrysanthemi* EC16 and characterisation of the enzyme product. *J. Bacteriol.* **172**, 6950–6058.

Broome-Smith, J. K., Tadayyon, M., and Zhang, Y. (1990). β-Lactamase as a probe of membrane protein assembly and protein export. *Mol. Microbiol.* **4**, 1637–1644.

Brunelle, A. and Schleif, R. (1987). Missing contact probing of DNA– protein interactions. *Proc. Natl. Acad. Sci. USA.* **84**, 6673–6676.

Casadaban, M. J. and Cohen, S. N. (1979). Lactose genes fused to exogenous promoters in one step using a Mu-*lac* bacteriophage: *In vivo* probe for transcriptional control sequences. *Proc. Natl. Acad. Sci. USA* **76**, 4530–4533.

Castilho, B. A., Olfson, P. and Casadaban, M. J. (1984). Plasmid insertion mutagenesis and *lac* gene fusion with mini-Mu bacteriophage transposons. *J. Bacteriol.* **158**, 488–495.

Castillo, A. and Reverchon, S. (1997). Characterisation of the pecT control region from *Erwinia chrysanthemi* 3937. *J. Bacteriol.* **179**, 4909–4918.

Chambost, J. P., Boyer, M. H., Cami, B., Barras, E. and Cattaneo, J. (1985). *Erwinia* cellulases. In *Proc. 6th Int. Conf. Plant Path. Bact.* (E.L. Civerolo, A. Collmer, A.G. Gillespie, and R. E. Davis, eds), pp. 150–159. Martinus Nijhoff, Dordrecht.

Chapon-Herve, V., Akrim, M., Latifi, A., Williams, P., Lazdunski, A. and Bally, M. (1997) Regulation of the xcp secretion pathway by multiple quorum-sensing modulons in *Pseudomonas aeruginosa*. *Mol. Microbiol.* **246**, 1169–1178.

Charkowski, A. O., Huang, C. and Collmer, A. (1997). Altered localisation of HrpZ in *Pseudomonas syringae* pv. *syringae hrp* mutants suggest that different components of the Type III secretion pathway control protein translocation across the inner and outer membrane of gram-negative bacteria. *J. Bacteriol.* **179**, 3866–3874.

Chatterjee, A. K. and Starr, M. P. (1977). Donor strains of soft-rot bacterium *Erwinia chrysanthemi* and conjugational transfer of the pectolytic capacity. *J. Bacteriol.* **132**, 862–869.

Chatterjee, A. K., Thurn, K. K. and Feese, D. A. (1983). Tn5-induced mutations in the enterobacterial phytopathogen *Erwinia chrysanthemi*. *Appl. Environ. Microbiol.* **45**, 644–650.

Chatterjee, A. K., Thurn, K. K. and Tyrell, D. J. (1985). Isolation and characterisation of Tn5 insertion mutants of *Erwinia chrysanthemi* that are deficient in polygalacturonate catabolic enzymes oligogalacturonate lyase and 3-deoxy-D-glycero-2,5-hexodiulosonate dehydrogenase. *J. Bacteriol.* **162**, 708–714.

Chatterjee, A., Mcevoy, J.L., Chambost, J.P., Blasco, F. and Chatterjee, A.K. (1991). Nucleotide sequence and molecular characterisation of *pnlA*, the structural gene for damage-inducible pectin lyase of *Erwinia carotovora* subsp. *carotovora* 71. *J. Bacteriol.* **173**, 1765–1769.

Chatterjee, A., Liu, Y. and Chatterjee, A.K. (1995a). Nucleotide sequence of a pectate lyase structural gene, *pel1* of *Erwinia carotovora* subsp. *carotovora* Strain 71 and structural relationship of *pel1* with other *pel* genes of *Erwinia* species. *Mol. Plant–Microbe Interact.* **8**, 92–95.

Chatterjee, A., Cui, Y. Y., Liu, Y., Dumenyo, C. K. and Chatterjee, A. K. (1995b). Inactivation of *rsmA* leads to overproduction of extracellular pectinases, cellulases, and proteases in *Erwinia carotovora* subsp. *carotovora* in the absence of the starvation cell density-sensing signal, N-(3-oxohexanoyl)-L-homoserine lactone. *Environ. Microbiol.* **61**, 1959–1967.

Clement, J. M., Perrin, D. and Hedgpeth, J. (1982). Analysis of λ receptor and β-lactamase synthesis and export using cloned genes in a minicell system. *Mol. Genet.* **185**, 302–310.

Cohen, S. P., Levy, S. B., Foulds, J. and Rosner, J. L. (1993). Salicylate induction of antibiotic resistance in *Escherichia coli*: activation of the *mar*-independent pathway. *J. Bacteriol.* **175**, 7856–7862.

Collmer, A. and Bateman, D. F. (1981) Impaired induction and self-catabolite repression of extracellular pectate lyase in *Erwinia chrysanthemi* mutants deficient in oligogalacturonide lyase. *Proc. Natl. Acad. Sci. USA* **78**, 3920–3924.

Collmer, A. and Keen, N. T. (1986) The role of pectic enzymes in plant pathogenesis. *Ann. Rev. Phytopathol.* **24**, 383–409.

Collmer, A., Schoedel, C., Roeder, D. L., Ried, J. L. and Rissler, J. F. (1985). Molecular cloning in *Escherichia coli* of *Erwinia chrysanthemi* genes encoding multiple forms of pectate lyase. *J. Bacteriol.* **161**, 913–920.

Condemine, G. and Robert-Baudouy, J. (1987). Tn5 insertion in *kdgR*, a regulatory gene of the polygalacturonate pathway in *Erwinia chrysanthemi*. *FEMS Microbiol. Lett.* **42**, 39–46.

Condemine, G. and Robert-Baudouy, J. (1991). Analysis of an *Erwinia chrysanthemi* gene cluster involved in pectin degradation. *Mol. Microbiol.* **5**, 2191–2202.

Condemine, G. and Robert-Baudouy, J. (1995). Synthesis and secretion of *Erwinia chrysanthemi* virulence factors are co-regulated. *Mol. Plant–Microbe Interact.* **8**, 632–636.

Condemine, G., Hugouvieux-Cotte-Pattat, N. and Robert-Baudouy, J. (1986). Isolation of *Erwinia chrysanthemi kduD* mutants altered in pectin degradation. *J. Bacteriol.* **165**, 937–941.

Condemine, G., Dorel, C., Hugouvieux-Cotte-Pattat, N. and Robert-Baudouy, J. (1992). Some of the *out* genes involved in secretion of pectate lyases in *Erwinia chrysanthemi* are regulated by *kdgR*. *Mol. Microbiol.* **6**, 3199–3211.

Cooper, V. J. C. and Salmond, G. P. C. (1993). Molecular analysis of the major cellulase (CelB) of *Erwinia carotovora*: evidence for an evolutionary "mix-and-match" of enzyme domains. *Mol. Gen. Genet.* **241**, 341–350.

Coplin, D. L., Frederick, R. D., Majerczak, D. R. and Tuttle, L. D. (1992). Characterisation of a gene cluster that specifies pathogenicity in *Erwinia stewartii*. *Mol. Plant-Microbe Interact.* **5**, 81–88.

Cui, Y., Chatterjee, A., Liu, Y., Dumenyo, C. K. and Chatterjee, A. K. (1995). Identification of a global repressor gene, *rsmA*, of *Erwinia carotovora* subsp. *carotovora* that controls extracellular enzymes, N-(3-oxanoyl)-L-homoserine lactone, and pathogenicity in soft-rotting *Erwinia* spp. *J. Bacteriol.* **177**, 5108–5115.

Cui, Y., Madi, L., Mukherjee, A., Dumenyo, C. K. and Chatterjee, A. K. (1996). The RsmA(-) mutants of *Erwinia carotovora* subsp. *carotovora* strain Ecc71 overexpress *hrpN*(Ecc) and elicit a hypersensitive reaction-like response in tobacco leaves. *Mol. Plant–Microbe Interact.* **9**, 565–573.

Cuppels, D. A. (1986). Generation and characterisation of Tn5 insertion mutations in *Pseudomonas syringae* pv. *tomato*. *Appl. Environ. Microbiol.* **51**, 323–327.

Dahler, G. S., Barras, F. and Keen, N. T. (1990). Cloning of genes encoding metalloproteases form *Erwinia chrysanthemi* EC16. *J. Bacteriol.* **172**, 5803–5813.

Daniels, M. J., Barber, C. E., Turner, P. C., Cleary, W. G. and Sawczyc, M. K. (1984). Isolation of mutants of *Xanthomonas campestris* pv *campestris* showing altered pathogenicity. *J. Gen. Microbiol.* **103**, 2447–2455.

Datta, N., Hedges, R. W., Shaw, E. J., Sykes, R. B. and Richmond, M. H. (1971). Properties of an R factor from *Pseudomonas aeruginosa. J. Bacteriol.* **108**, 1244–1249.

De Crecy-Lagard, V., Glaser, P., Lejeune, P., Sismeiro, O., Barber, C. E., Daniels, M.J. and Danchin, A. (1990). A *Xanthomonas campestris* pv. *campestris* protein similar to catabolite activation factor is involved in regulation of phytopathogenicity. *J. Bacteriol.* **172**, 5877–5883.

Delepelaire, P. and Wandersman, C. (1989). Protease secretion by *Erwinia chrysanthemi*. Proteases B and C are synthesised and secreted as zymogens without a signal peptide. *J. Biol. Chem.* **264**, 9083–9089.

Delepelaire, P. and Wandersman, C. (1990). Protein secretion in Gram-negative bacteria – the extracellular metalloprotease-B from *Erwinia chrysanthemi* contains a C-terminal secretion signal analogous to that of *Escherichia coli* alpha hemolysin. *J. Biol. Chem.* **265**, 17118–17125.

Delepelaire, P. and Wandersman, C. (1991). Characterisation, localisation and transmembrane organisation of the three proteins PrtD, PrtE and PrtF necessary for protease secretion by the Gram-negative bacterium *Erwinia chrysanthemi. Mol. Microbiol.* **5**, 2427–2434.

Devine, J. H., Countryman, C. and Baldwin, T. O (1988). Nucleotide sequence of the *luxR* and *luxI* genes and structure of the primary regulatory region of the *lux* regulon of *Vibrio fischeri. Biochemistry.* **27**, 837–842.

De Vries, G. E., Raymond, C. K. and Ludwig, R. A. (1984). Extension of bacteriophage λ host range: Selection, cloning and characterisation of a constitutive λ receptor gene. *Proc. Natl. Acad. Sci. USA* **81**, 6080–6084.

Dow, J. M., Scofield, G., Trafford, K., Turner, P. C. and Daniels, M. J. (1987). A gene cluster in *Xanthomonas campestris* pv. *campestris* required for pathogenicity controls the excretion of polygalacturonate lyase and other enzymes. *Physiol. Mol. Plant. Pathol.* **31**, 261–271.

Dow, J. M., Osbourne, A. E., Wilson, T. J. G. and Daniels, M. J. (1995). A locus determining pathogenicity of *Xanthomonas campestris* is involved in lipopolysaccharide biosynthesis. *Mol. Plant–Microbe Interact.* **8**, 768–777.

Ellard, F. M., Cabelo, A. and Salmond, G. P. C. (1989). Bacteriophage Lambda-mediated transposon mutagenesis of phytopathogenic and epiphytic *Erwinia* species is strain dependant. *Mol. Gen. Genet.* **218**, 491–498.

Enard, C., Diolez, A. and Expert, D. (1988), Systemic virulence of *Erwinia chrysanthemi* 3937 requires a functional iron assimilation system. *J. Bacteriol.* **170**, 2419–2426.

Expert, D. and Toussaint, A. (1985). Bacteriocin-resistant mutants of *Erwinia chrysanthemi*: possible involvement of iron acquisition in phytopathogenicity. *J. Bacteriol.* **163**, 221–227.

Expert, D., Enard, C. and Masclaux, C. (1996). The role of host iron in plant–pathogen interactions. *T.I.M.* **4**, 232–237.

Favey, S., Bourson, C., Bertheau, Y., Kotoujansky, A. and Boccara, M. (1992). Purification of the acidic pectate lyase and nucleotide-sequence of the corresponding gene (*pelA*) of *Erwinia chrysanthemi* strain 3937. *J. Gen. Microbiol.* **138**, 499–508.

Flego, D., Pirhonen, M., Saarilahti, H., Palva, T. K. and Palva, E. T (1997). Control of virulence gene expression by plant calcium in the phytopathogen *Erwinia carotovora. Mol. Microbiol.* **25**, 831–838.

Frederick, R. D. (1991). Characterisation of *wts* genes from *Erwinia stewartii* and

Virulence Determinants in Erwinia

their homology with the *hrp* gene cluster from *Pseudomonas syringae phaseolicola*. *Phytopathology* **81**, 1187 (Abstr.).

Gantotti, B. V., Kindle, K. K. and Beer, S. V. (1981). Transfer of the drug resistance transposon Tn5 to *Erwinia herbicola* and the induction of insertion mutations. *Curr. Microbiol.* **6**, 377–381.

Gaudriault, S., Malandrin, L., Paulin, J. P. and Barny, M. A. (1997). DspA, an essential pathogenicity factor of *Erwinia amylovora* showing homology with AvrE of *Pseudomonas syringae*, is secreted via the Hrp–secretion pathway in a DspB-dependent way. *Mol. Microbiol.* **26**, 1057–1069.

Gay, P., Lecoq, D., Steinmetz, M., Berkelman, T. and Kado, C. I. (1985). Positive selection procedure for entrapment of insertion sequence elements in Gram-negative bacteria. *J. Bacteriol.* **164**, 918–921.

Genin, S., Gough, C. L., Zischek, C. and Boucher, C. A. (1992). Evidence that the *hrpB* gene encodes a positive regulator of pathogenicity genes from *Pseudomonas solanacearum*. *Mol. Microbiol.* **6**, 3065–3076.

Ghigo, J. M. and Wandersman, C. (1992a). Cloning and nucleotide sequence and characterisation of the gene encoding the *Erwinia chrysanthemi* B374 PrtA metal-loprotease: a third metalloprotease secreted via a C-terminal secretion signal. *Mol. Gen. Genet.* **236**, 135–144.

Ghigo, J. M. and Wandersman, C. (1992b). A fourth metalloprotease gene in *Erwinia chrysanthemi*. *Res. Microbiol.*. **143**, 857–867.

Ghigo, J. M. and Wandersman, C. (1994). A carboxyl-terminal 4-amino acid motif is required for secretion of the metalloprotease PrtG through the *Erwinia chrysanthemi* protease secretion pathway. *J. Biol. Chem.* **269**, 8979–8985.

Gold, S., Nishio, S., Tsuyumu and Keen, N. T. (1992). Analysis of the *pelE* pro-moter in *Erwinia chrysanthemi* ECI6. *Mol. Plant–Microbe Interact.* **5**, 170–178.

Gopalan, S., Bauer, D. W., Alfano, J. R., Loniello, A. O., He, S. Y. and Collmer, A. (1996). Expression of the *Pseudomonas syringae* avirulence protein AvrB in plant cells alleviates its dependence on the hypersensitive response and pathogenic-ity (Hrp) secretion system in eliciting genotype-specific hypersensitive cell death. *Plant Cell* **8**, 1095–1105.

Guiseppi, A., Cami, B., Aymeric, J. L., Ball, G. and Creuzet, N. (1998). Homology between endoglucanase Z of *Erwinia chrysanthemi* and endoglucanases of *Bacillus subtillis* and alkalophilic *Bacillus*. *Mol. Microbiol.* **2**, 159–164.

Guiseppi, A., Aymeric, J. L., Cami, B., Barras, F. and Creuzet, N. (1991). Sequence analysis of the cellulase-encoding *celY* gene of *Erwinia chrysanthemi*: a possible case of interspecies gene transfer. *Gene* **106**, 109–114.

Handa, A. K., Bressan, R. A., Korty, A. G., Jayaswal, R. K. and Charles, D. L. (1987). In (E. L. Civerolo, A. Collmer, A. G. Gillespie, and R. E. Davis, eds, *Proc. 6th Int. Conf. Plant Path. Bact., Isolation and characterisation of pectolytic nonpathogenic mutants of Erwinia carotovora subsp. carotovora (Ecc)*. pp. 212–217. Martinus Nijhoff, Dordrecht.

Harkki, A. and Palva, E. T. (1985). A *lamB* expression plasmid for extending the host range of λ to other enterobacteria. *FEMS Microbiol. Lett.* **27**, 183–187.

Hattermann, D. R. and Ries, S. M. (1989). Motility of *Pseudomonas syringae* pv. *glycinea* and its role in infection. *Phytopathogenicity.* **79**, 284–289.

He, S. Y. and Collmer. A. (1990). Molecular cloning, nucleotide sequence, and marker-exchange mutagenesis of the exo-poly-α-D-galacturonosidase-encoding *pehX* gene of *Erwinia chrysanthemi* EC16. *J. Bacteriol.* **172**, 4988–4995.

He, S. Y., Lindeberg, M., Chatterjee, A. K. and Collmer, A. (1991a). Cloned *Erwinia chrysanthemi out* genes enable *Escherichia coli* to selectively secrete a diverse family of heterologous proteins to its milieu. *Proc. Natl. Acad. Sci. USA* **88**, 1079–1083.

He, S. Y., Schoedel, C., Chatterjee, A. K. and Collmer, A. (1991b). Extracellular secretion of pectate lyase by the *Erwinia chrysanthemi* out pathway is dependent upon Sec-mediated export across the inner membrane. *J. Bacteriol.* **173**, 4310–4317.

Heffron, S., Henrissat, B., Yoder, M. D., Lietzke, S. and Jurnak, F. (1995). Structure-based multiple alignment of extracellular pectate lyase sequences. *Mol. Plant–Microbe Interact.* **8**, 331–334.

Heikinheimo, R., Flego, D., Pirhonen, M., Karlsson, M. B., Eriksson, A., Mäe, A., Koiv, V. and Palva, E. T. (1995). Characterisation of a novel pectate lyase from *Erwinia carotovora* subsp. *carotovora*. *Mol. Plant–Microbe Interact.* **8**, 207–217.

Henrissat, B., Claeyssens, M., Tomme, P., Lemesle, L. and Mornon, J. P. (1989). Cellulase families revealed by hydrophobic cluster analysis. *Gene* **81**, 83–95.

Henrissat, B., Heffron, S. E., Yoder, M. D., Lietzke, S. E. and Jurnak, F. (1995). Functional implications of structure-based sequence alignment of proteins in the extracellular pectate lyase superfamily. *Plant. Physiol.* **107**, 963–976.

Hinton, J. C. D. and Salmond, G. P. C. (1987). Use of Tn*phoA* to enrich for extracellular enzyme mutants of *Erwinia carotovora* subsp. *carotovora*. *Mol. Microbiol.* **1**, 381–386.

Hinton, J. C. D., Perombelon, M. C. M. and Salmond, G. P. C. (1985). Efficient transformation of *Erwinia carotovora* subsp. *carotovora* and *Erwinia carotovora* subsp. *atroseptica*. *J. Bacteriol.* **161**, 786–788.

Hinton, J. C. D., Perombelon, M. C. M. and Salmond, G. P. C. (1987) Cloning of the *cysB* gene of *Erwinia carotovora*, and the identification of its product. *Mol. Gen. Genet.* **207**, 466–470.

Hinton, J. C. D., Sidebotham, J. M., Gill, D. R. and Salmond, G. P. C. (1989a). Extracellular and periplasmic isoenzymes of pectate lyase from *Ecc* belong to different gene families. *Mol. Microbiol.* **3**, 1785–1795.

Hinton, J. C. D., Sidebotham, J. M., Hyman, L. J., Perombelon, M. C. M. and Salmond, G. P. C. (1989b). Isolation and characterisation of transposon-induced mutants of *Erwinia carotovora* subsp. *atroseptica* exhibiting reduced virulence. *Mol. Gen. Genet.* **217**, 141–148.

Hinton, J. C. D., Gill, D. R., Lalo, D., Plastow, G. S. and Salmond, G. P. C. (1990). Sequence of the *peh* gene of *Erwinia carotovora*: homology between *Erwinia* and plant enzymes. *Mol. Microbiol.* **4**, 1029–1036.

Housby, J. N., Thomas, J. D., Wharam, S. D., Reeves, P. J. and Salmond, G. P. C. (1998). Conditional mutations in OutE and OutL block exoenzyme secretion across the *Erwinia carotovora* outer membrane. *FEMS Microbiol. Lett.* **165**, 91–102.

Howe, M. M. and Bade, E. G. (1975). Molecular biology of bacteriophage Mu. *Nature.* **190**, 624–632.

Hu, N-T., Wang, Y. M., Lee, W. Y., Yang, S. M. and Tseng, Y. H. (1987). Characterisation of a genomic clone of *Erwinia carotovora* subsp. *carotovora* TG3 encoding a pectolytic enzyme of apparent molecular weight 78 kDa. *Mol. Gen. Genet.* **210**, 294–298.

Hubbard, J. P., Williams, J. D., Niles, R. M. and Mount, M. S. (1978). The relation between glucose repression and endopolygalacturonate transeliminase and adenosine 3′, 5′ cyclic monophosphate levels in *Erwinia carotovora*. *Phytopathology* **68**, 95–99.

Hugovieux-Cotte-Pattat, N. and Robert-Baudouy, J. (1985). Isolation of *kdgK-lac* and *kdgA-lac* gene fusions in the phytopathogenic bacterium *Erwinia chrysanthemi*. *J. Gen. Microbiol.* **131**, 1205–1211.

Hugovieux-Cotte-Pattat, N. and Robert-Baudouy, J. (1987). Hexuronate catabolism in *Erwinia chrysanthemi*. *J. Bacteriol.* **169**, 1223–1231.

Hugouvieux-Cotte-Pattat, N. and Robert-Baudouy, J. (1989). Isolation of *Erwinia chrysanthemi* mutants altered in pectinolytic enzyme production. *Mol. Microbiol.* **3**, 1587–1597.

Hugouvieux-Cotte-Pattat, N. and Robert-Baudouy, J. (1992). Analysis of the regulation of the *pelBC* genes in *Erwinia chrysanthemi* 3937. *Mol. Microbiol.* **6**, 2363–2376.

Hugouvieux-Cotte-Pattat, N., Reverchon, S., Condemine, G. and Robert-Baudouy, J. (1986). Regulatory mutations affecting the synthesis of pectate lyase in *Erwinia chrysanthemi*. *J. Gen. Microbiol.* **132**, 2099–2106.

Hugouvieux-Cotte-Pattat, N., Dominguez, H. and Robert-Baudouy, J. (1992). Environmental conditions affect transcription of the pectinase genes of *Erwinia chrysanthemi* 3937. *J. Bacteriol.* **174**, 7807–7818.

Hugouvieux-Cotte-Pattat, N., Condemine, G., Nasser, W. and Reverchon, S. (1996). Regulation of pectinolysis in *Erwinia chrysanthemi*. *Ann. Rev. Microbiol.* **50**, 213–257.

Ishimaru, C. A., Klos, E. J. and Brubaker, R. R. (1988). Multiple antibiotic production by *Erwinia herbicola*. *Phytopathology*. **78**, 746–750.

Ito, K., Kobayashi, R., Nikaido, N. and Izaki, K. (1988). DNA structure of pectate lyase I gene cloned from *Erwinia carotovora*. *Agric. Biol. Chem.* **52**, 479–487.

Ji, J., Hugouvieux-Cotte-Pattat, N. and Robert-Baudouy, J. (1987). Use of Mu-*lac* insertions to study the secretion of pectate lyases by *Erwinia chrysanthemi*. *J. Gen. Microbiol.* **133**, 793–802.

Ji, J., Hugouvieux-Cotte-Pattat, N. and Robert-Baudouy, J. (1989). Molecular cloning of the *outJ* gene involved in pectate lyase secretion by *Erwinia chrysanthemi*. *Mol. Microbiol.* **3**, 285–293.

Johnson, K., Parker, M. L. and Lory, S. (1986). Nucleotide sequence and transcriptional initiation site of two *Pseudomonas aeruginosa* pilin genes. *J. Biol. Chem.* **261**, 15703–15708.

Jones, S. E., Yu, B., Bainton, N. J., Birdsall, M., Bycroft, B. W., Chhabra, S. R., Cox A. J. R., Golby, P., Reeves. P. J., Stephens, S. Winson, M. K., Salmond, G. P. C. and Williams, P. (1993). The *lux* autoinducer regulates the production of exoenzyme virulence determinants in *Erwinia carotovora* and *Pseudomonas aeruginosa*. *EMBO J.* **12**, 2477–2482.

Karlsson, M-B., Pirhonen, M., Saarilahti, H. T. and Palva, E. T. (1991). Molecular cloning of *ompRS*, a regulatory locus controlling production of outer membrane proteins in *Erwinia carotovora*. subsp. *carotovora*. *Mol. Gen. Genet.* **226**, 353–360.

Kazmierczak, B. I., Mielke, D. L., Russel, M. and Model, P. (1994). pIV, a filamentous phage protein that mediates phage export across the bacterial cell envelope, forms a multimer. *J. Mol. Biol.* **238**, 187–198.

Keen, N. T. and Tamaki, S. J. (1986). Structure of two pectate lyase genes from *Erwinia chrysanthemi* EC16 and their high-level expression in *Escherichia coli*. *J. Bacteriol.* **168**, 595–606.

Keen, N. T., Dahlbeck, D., Staskawicz, B. and Belser, W. (1984). Molecular-cloning of pectate lyase genes from *Erwinia chrysanthemi* and their expression in *Escherichia coli*. *J. Bacteriol.* **159**, 825–831.

Kelemu, S. and Collmer, A. (1993). *Erwinia chrysanthemi* EC16 produces a second set of plant-inducible pectate lyase isoenzymes. *Appl. Environ. Microbiol.* **59**, 1756–1761.

Kim, J. F., Wei, Z. M. and Beer, S. V. (1997). *hrpA* and *hrpC* operons of *Erwinia amylovora* encode components of a type III pathway that secretes harpin. *J. Bacteriol.* **179**, 1690–1697.

Kleckner, N. (1981). Transposable elements in prokaryotes. *Annu. Rev. Genet.* **15**, 341–404.

Kleckner, N., Steele, D. A., Reichardt, K. and Botstein, D. (1979). Specificity of insertion by the translocatable tetracycline-resistance element Tn*10*. *Genetics*. **92**, 1023–1040.

Konisky, J. (1982). Colicins and other bacteriocins with established modes of action. *Annu. Rev. Microbiol.* **36**, 125–144.

Koster, M., Bitter, W., de Cock, H., Allaoui, A., Cornelis, G. R. and Tommassen, J. (1997). The outer membrane component, YscC, of the Yop secretion machinery of *Yersinia enterocolitica* forms a ring-shaped multimeric complex. *Mol. Microbiol.* **26**, 789–797.

Kotoujansky, A., Diolez, A., Boccara, M., Bertheau, Y., Andro, T. and Coleno, A. (1985). Molecular cloning of *Erwinia chrysanthemi* pectinase and cellulase structural genes. *EMBO J.* **4**, 781–785.

Kyte, J. and Doolittle, R. F. (1992). A simple method for displaying the hydrophobic character of a protein. *J. Mol. Biol.* **157**, 105–132.

Laby, R. J. and Beer, S. V. (1992). Hybridisation and functional complementation of the *hrp* gene cluster from *Erwinia amylovora* strain Ea321 with DNA of other bacteria. *Mol. Plant–Microbe Interact.* **5**, 412–419.

Lapwood, D. H. and Read, P. J. (1986). The susceptibility of different potato cultivars to blackleg caused by *Erwinia carotovora* subspecies *atroseptica*. *Annu. Appl. Biol.* **109**, 555–560.

Laurent, F., Kotoujansky, A., Labesse, G. and Bertheau, Y. (1993). Characterisation and overexpression of the *pem* gene encoding pectin methylesterase of *Erwinia chrysanthemi* strain 3937. *Gene*. **131**, 17–25.

Leach, J. E. and White, F. F. (1996). Genes and proteins involved in aggressiveness and avirulence of *Xanthomonas oryzae pv. oryzae* to rice. In *Biology of Plant–Microbe Interactions* (G. Stacey, B. Mullion, and P.M. Gresshoff, eds.), pp. 191–196. *Int. Soc. Mol. Plant-Microbe Interact.*, St Paul.

Lei, S-P. Lin, H. C., Heffernan, L. and Wilcox, G. (1985a). Cloning of the pectate lyase genes from *Erwinia carotovora* and their expression in *Escherichia coli*. *Gene* **35**, 63–70.

Lei, S-P., Lin, H. C., Heffernan, L. and Wilcox, G. (1985b). Evidence that polygalacturonase is a virulence determinant in *Erwinia carotovora*. *J. Bacteriol.* **164**, 831–835.

Lei, S-P., Lin, H. C., Wang, S. S., Callaway, J. and Wilcox, G. (1987). Characterisation of the *Erwinia carotovora pelB* gene and its product pectate lyase. *J. Bacteriol.* **169**, 4379–4383.

Lei, S-P., Lin, H. C., Wang, S. S. and Wilcox, G. (1988). Characterisation of the *Erwinia carotovora pelA* gene and its product pectate lyase A. *Gene* **62**, 159–164.

Lei, S-P., Lin, H. C., Wang, S. S., Higaki, P. and Wilcox, G. (1992). Characterisation of the *Erwinia carotovora peh* gene and its product polygalacturonase. *Gene* **117**, 119–124.

Létoffé, S., Delepelaire, P. and Wandersman, C. (1989). Characterisation of a protein inhibitor of extracellular proteases produced by *Erwinia chrysanthemi*. *Mol. Microbiol.* **3**, 79–86.

Létoffé, S., Delepelaire, P. and Wandersman, C. (1990). Protease secretion by *Erwinia chrysanthemi*: the specific secretion functions are analogous to those of *Escherichia coli* α-haemolysin. *EMBO J.* **9**, 1375–1382.

Lever, M. (1972). A new reaction for colorimetric determination of carbohydrates. *Analyt. Biochem.* **47**, 273–279.

Lietzke, S. E., Yoder, M. D., Keen, N. T. and Jurnak, F. (1994). The three dimensional structure of pectate lyase E, a plant virulence factor from *Erwinia chrysanthemi*. *Plant Physiol.* **106**, 849–862.

Lindeberg, M. and Collmer, A. (1992). Analysis of eight *out* genes in a cluster required for pectate enzyme secretion by *Erwinia chrysanthemi*: sequence comparison with secretion genes from other Gram-negative bacteria. *J. Bacteriol.* **174**, 7385–7397.

Lindeberg, M., Salmond, G. P. C. and Collmer, A. (1996). Complementation of deletion mutations in a cloned functional cluster of *Erwinia chrysanthemi out* genes with *Erwinia carotovora out* homologues reveals OutC and OutD as candidate gatekeepers of species-specific secretion of proteins via the type II pathway. *Mol. Microbiol.* **20**, 175–190.

Lindgren, P. B., Peet, R. C. and Panopoulos, N. J. (1986). Gene cluster of *Pseudomonas syringae* pv. *"phaseolicola"* controls pathogenicity on bean plants and hypersensitivity on non-host plants. *J. Bacteriol.* **168**, 512–522.

Liu, M. Y. and Romeo, T. (1997). The global regulator CsrA of *Escherichia coli* is a specific mRNA-binding protein. *J. Bacteriol.* **179**, 4639–4642.

Liu, M. Y., Yang, H. and Romeo, T. (1995). The product of the pleiotropic *Escherichia coli* gene *csrA* modulates glycogen biosynthesis via effects on mRNA stability. *J. Bacteriol.* **177**, 2663–2672.

Liu, Y., Chatterjee, A. and Chatterjee, A. K. (1994a). Nucleotide sequence and expression of a novel pectate lyase (*pel-3*) gene and a closely linked endo-polygalacturonase (*peh-1*) gene of *Erwinia carotovora* subsp. *carotovora* strain 71. *Appl. Environ. Microbiol.* **60**, 2545–2552.

Liu, Y., Chatterjee, A. and Chatterjee, A. K. (1994b). Nucleotide sequence, organisation and expression of *rdgA* and *rdgB* genes that regulate pectin lyase production in the plant pathogenic bacterium *Erwinia carotovora* subspecies *carotovora* in response to DNA-damaging agents. *Mol. Microbiol.* **14**, 999–1010.

Lojkowska, E., Masclaux, C., Boccara, M., Robert-Baudouy, J. and Hugouvieux-Cotte-Pattat, N. (1995). Characterisation of the *pelL* gene encoding a novel pectate lyase of *Erwinia chrysanthemi* 3937. *Mol. Microbiol.* **16**, 1183–1195.

Lonetto, M. A., Brown, K. L., Rudd, K. E. and Buttner, M. J. (1994). Analysis of the *Streptomyces coelicolor sigE* gene reveals the existence of a subfamily of eubacterial RNA polymerase F factors involved in regulation of extracytoplasmic functions. *Proc. Natl. Acad. Sci. USA* **91**, 7573–7577.

Lu, H.-M., Motley, S. T. and Lory, S. (1997). Interactions of the components of the general secretion pathway: role of *Pseudomonas aeruginosa* type IV pilin subunits in complex formation and extracellular protein secretion. *Mol. Microbiol.* **25**, 247–259.

Ludwig, R. A. (1987). Gene tandem-mediated selection of coliphage λ-receptive *Agrobacterium, Pseudomonas* and *Rhizobium* strains. *Proc. Natl. Acad. Sci. USA* **84**, 3334–3338.

Mäe, A., Heikinheimo, R. and Palva, E. T. (1995). Structure and regulation of the *Erwinia carotovora* subspecies *carotovora* SCC3193 cellulase gene *celV1* and the role of cellulase in phytopathogencity. *Mol. Gen. Genet.* **247**, 17–26.

Maher, E. A. and Kelman, A. (1983). Oxygen status of potato-tuber tissue in relation to maceration by pectic enzymes of *Erwinia carotovora. Phytopathology* **73**, 536–539.

Malik, A. N., Vivian, A. and Taylor, J. D. (1987). Isolation and partial characterisation of three classes of mutant in *Pseudomonas syringae* pathovar *pisi* with altered behaviour towards their host, *Pisum sativum. J. Gen. Microbiol.* **133**, 2393–2399.

Manoil, C. and Beckwith, J. (1985). TnphoA: a transposon probc for protein export signals. *Proc. Natl. Acad. Sci. USA* **82**, 8129–8133.

Manoil, C., Mekalanos, J. J. and Beckwith, J. (1990). Alkaline phosphatase fusions: sensors of subcellular location. *J. Bacteriol.* **172**, 515–518.

Martin, R. G. and Rosner, J. L. (1995). Binding of purified multiple antibiotic-resistance repressor protein (MarR) to *mar* operator sequences. *Proc. Natl. Acad. Sci. USA* **92**, 5456–5460.

Mazodier, P., Cossart, P., Giraud, E. and Gasser, F (1985). Completion of the nucleotide-sequence of the central region of Tn5 confirms the presence of 3 resistance genes. *Nucl. Acid. Res.* **13**, 195–205.

McCammon, S. L., Coplin, D. L. and Rowan, R. G. (1985). Isolation of virulent mutants of *Erwinia stewartii* using bacteriophage Mu PF7701. *J. Gen. Microbiol.* **131**, 2993–3000.

McEvoy, J. L., Murata, H. and Chatterjee, A. K. (1990). Molecular cloning and characterisation of an *Erwinia carotovora* subsp. *carotovora* pectin lyase gene that responds to DNA-damaging agents. *J. Bacteriol.* **172**, 3284–3289.

McMillan, G. P., Johnston, D. J., Morel, J. B. and Perombelon, M. C. M. (1993). A pH-independent assay for pectin methyl esterase for use in column chromato-graphy. *Anal. Biochem.* **209**, 377–379.

Meade, H. M., Long, S. R., Ruvkun, G. B., Brown, S. E. and Ausubel, F. M. (1982). Physical and genetic characterisation of symbiotic and auxotrophic mutants of *Rhizobium meliloti* induced by transposon Tn5 mutagenesis. *J. Bacteriol.* **149**, 114–122.

Meighen, E. A. (1991). Molecular biology of bacterial luminescence. *Microbiol. Rev.* **55**, 123–142.

Miller, J. H. (1972). *Experiments in Molecular Genetics*. Cold Spring Harbour Laboratory Press, New York.

Miller, P. F. and Sulavik, M.C. (1996). Overlaps and parallels in the regulation of intrinsic multiple antibiotic resistance in *Escherichia coli Mol. Microbiol.* **21**, 441–448.

Missiakas, D. and Raina, S. (1997) Protein folding in the bacterial periplasm. *J. Bacteriol.* **179**, 2465–2471.

Moran, F., Nasuno, S. and Starr, M. P. (1968). Extracellular and intracellular poly-galacturonic acid trans-eliminases of *Erwinia carotovora*. *Arch. Biochem. Biophys.* **123**, 298–306.

Mount, M. S., Berman, P. M., Mortlock, R. P. and Hubbard, J. P. (1979). Regulation of endo-polygalacturonate trans-eliminase in an adenosine 3′, 5′-cyclic monophos-phate deficient mutant of *Erwinia carotovora*. *Phytopathology* **69**, 117–120.

Mukherjee, A., Cui, Y. Y., Liu, Y., Dumenyo, C. K. and Chatterjee, A. K. (1996). Global regulation in *Erwinia* species by *Erwinia carotovora* RsmA, a homologue of *Escherichia coli* CsrA – repression of secondary metabolites, pathogenicity and hypersensitive reaction. *Microbiology* **142**, 427–434.

Mukherjee, A., Cui, Y. Y., Liu, Y. and Chatterjee, A. K. (1997). Molecular charac-terisation and expression of the *Erwinia carotovora hrpN* (Ecc) gene, which encodes an elicitor of the hypersensitive reaction. *Mol. Plant–Microbe Interact.* **10**, 462–471.

Mulholland, V. and Salmond, G. P. C. (1995). Use of coliphage λ and other bacte-riophages for molecular genetic analysis of *Erwinia* and related Gram-negative bacteria. In *Microbial Gene Techniques*, Vol. 6 (K.W. Adolph, ed.) Academic Press, London.

Mulholland, V., Hinton, J. C., Sidebotham, J., Toth, I. K., Hyman, L. J., Pérombelon, M. C. M., Reeves, P. J. and Salmond, G. P. C. (1993). A pleiotropic reduced virulence (Rvi⁻) mutant of *Erwinia carotovora* subspecies *atroseptica* defective in flagella assembly proteins that are conserved in plant and animal pathogens. *Mol. Microbiol.* **9**, 343–356.

Murata, H., Fons, M., Chatterjee, A., Collmer, A. and Chatterjee, A. K. (1990). Characterisation of transposition insertion Out⁻ mutants of *Erwinia carotovora* subsp. *carotovora* defective in enzyme export and of a DNA segment that com-plements *out* mutations in *Erwinia carotovora* subsp. *carotovora*, *Erwinia carotovora* subsp. *atroseptica* and *Erwinia chrysanthemi*. *J. Bacteriol.* **172**, 2970–2978.

Murata, H., McEvoy, J. L., Chatterjee, A., Collmer, A. and Chatterjee, A. K. (1991). Molecular cloning of an *aepA* gene that activates production of extracellular pectolytic, cellulolytic and proteolytic enzymes in of *Erwinia carotovora* subsp. *carotovora*. *Mol. Plant–Microbe Interact.* **4**, 239–246.

Nasser, W., Condemine, G., Plantier, R., Anker, D. and Robert-Baudouy, J. (1991). Inducing properties of analogs of 2-keto-3-deoxygluconate on the expression of pectinase genes of *Erwinia chrysanthemi*. *FEMS Microbiol. Lett.* **65**, 73–78.

Nasser, W., Reverchon, S. and Robert-Baudouy, J. (1992). Purification and functional characterisation of the KdgR protein, a major repressor of pectinolysis genes of *Erwinia chrysanthemi*. *Mol. Microbiol.* **6**, 257–265.

Nasser, W., Reverchon, S., Condemine, G. and Robert-Baudouy, J. (1994). Specific interactions of *Erwinia chrysanthemi* KdgR repressor with different operators of genes involved in pectinolysis. *J. Mol. Biol.* **236**, 427–440.

Nasser, W., Robert-Baudouy, J. and Reverchon, S. (1997). Antagonistic effect of CRP and KdgR in the transcription control of the *Erwinia chrysanthemi* pectinolysis genes. *Mol. Microbiol.* **26**, 1071–1082.

Nelson, N. (1944). A photometric adaptation of the Somogyi method for the determination of glucose. *J. Biol. Chem.* **153**, 375–380.

Nishida, T., Suzuki, T., Ito, K., Kamio, Y. and Izaki, K. (1990). Cloning and expression of pectin lyase gene from *Erwinia carotovora* in *Escherichia coli*. *Biochem. Biophys. Res. Commun.* **168**, 801–808.

O'Callaghan, C. H., Morris, A., Kirby, S. M. and Shingler, A. H. (1972). Novel method for detection of β-lactamase by using a chromogenic cephalosporin substrate. *Antimicrob. Agents Chemother.* **1**, 283–288.

Ohnishi, H., Hishida, T., Yoshida, A., Kamio, Y. and Izaki, K. (1991). Nucleotide sequence of pnl gene from *Erwinia carotovora* ER. *Biochem. Biophys. Res. Commun.* **176**, 321–327.

Oku, T., Alvarez, A. M. and Kado, C. I. (1995). Conservation of the hypersensitivity–pathogenicity regulatory gene *hrpX* of *Xanthomonas campestris* and *X. oryzae*. *DNA Sequence* **5**, 245–249.

Olsen, R. H. and Shipley, P. (1973). Host range and properties of the *Pseudomonas aeruginosa* R-factor R1822. *J. Bacteriol.* **113**, 772–780.

Osbourn, A. E., Barker, C. E. and Daniels, M. J. (1987). Identification of plant-induced genes of the bacterial pathogen *Xanthomonas campestris* pathovar *campestris* using a promoter probe plasmid. *EMBO J.* **6**, 23–28.

Osbourn, A. E., Clarke, B. R. and Daniels, M. J. (1990). Identification and DNA sequence of a pathogenicity gene of *Xanthomonas campestris* pathovar *campestris*. *Mol. Plant–Microbe Interact.* **3**, 280–285.

Pagel, W. and Heitefuss, R. (1990). Enzyme activities in soft rot pathogenesis of potato tubers: effects of calcium, pH and degree of pectin esterification on the activities of polygalacturonase and pectate lyase. *Physiol. Mol. Plant Pathol.* **37**, 9–25.

Palomäki, T. and Saarilahti, H. T. (1995). The extreme C-terminus is required for secretion of both the native polygalacturonase (PehA) and PehA–Bla hybrid proteins in *Erwinia carotovora* subsp. *carotovora*. *Mol. Microbiol.* **17**, 449–459.

Palomäki, T. and Saarilahti, H. T. (1997). Isolation and characterisation of new C-terminal substitution mutations affecting secretion of polygalacturonase in *Erwinia carotovora* subsp. *carotovora*. *FEBS Lett.* **243**, 400–407.

Panopoulos, N. J. and Schroth, M. N. (1974). Role of flagella motility in the invasion of bean leaves by *phaseolicola*. *Phytopathology* **64**, 1389–1397.

Park, J. T. and Johnson, M. S. (1949). A submicrodetermination of glucose. *J. Biol. Chem.* **181**, 149–151.

Parker, W. L., Rathnum, M. L., Wells, J. S., Trejo, W. H., Principe, P. A. and Sykes,

R. B. (1982) SQ27860, a simple carbapenem produced by species of *Serratia* and *Erwinia*. *J. Antibiot*. **35**, 653–660.

Payne, J. H., Schoedel, C., Keen, N. T. and Collmer, A (1987). Multiplication and virulence in plant tissues of *Escherichia coli* clones producing pectate lyase isozymes Plb and Ple at high levels and of an *Erwinia chrysanthemi* mutant deficient in Ple. *Appl. Environ. Microbiol*. **53**, 2315–2320.

Pirhonen, M., Heino, P., Helander, I., Harju, P. and Palva, E. T. (1988). Bacteriophage-T4 resistant mutants of the plant pathogen *Erwinia carolovora*. *Microb. Pathogen*. **4**, 359–367.

Pirhonen, M., Saarilahti, H., Karlsson, M. B. and Palva, E. T. (1991). Identification of pathogenicity determinants of *Ecc* by transposon mutagenesis. *Mol. Plant–Microbe Interact*. **4**, 276–283.

Pirhonen, M., Felgo, D, Heikinheimo, R. and Palva, E. T. (1993). A small diffusible molecule is responsible for the global control of virulence and exoenzyme production in *Erwinia carotovora*. *EMBO J*. **12**, 2467–2476.

Pissavin, C., Robert-Baudouy, J. and Hugouvieux-Cotte-Pattat, N. (1996). Regulation of *pelZ*, a gene of the *pelB–pelC* cluster encoding a new pectate lyase of *Erwinia chrysanthemi* 3937. *J. Bacteriol*. **178**, 7187–7196.

Plastow, G. S. (1988). Molecular cloning and nucleotide sequence of the pectin methyl esterase gene of *Erwinia chrysanthemi* B374. *Mol. Microbiol*. **2**, 247–254.

Possot, O. and Pugsley, A. P. (1994). Molecular characterisation of PulE, a protein required for pullulanase secretion. *Mol. Microbiol*. **12**, 287–299.

Praillet, T., Nasser, W., Robert-Baudouy, J. and Reverchon, S. (1996). Purification and functional characterisation of PecS, a regulator of virulence-factor synthesis in *Erwinia chrysanthemi*. *Mol. Microbiol*. **20**, 391–402.

Praillet, T., Reverchon, S. and Nasser, W. (1997a). Mutual control of the PecS/PecM couple, two proteins regulating virulence-factor synthesis in *Erwinia chrysanthemi*. *Mol. Microbiol*. **24**, 803–814.

Praillet, T., Reverchon, S., Robert-Baudouy, J. and Nasser, W. (1997b). The PecM protein is necessary for the DNA-binding capacity of the PecS repressor, one of the regulators of virulence-factor synthesis in *Erwinia chrysanthemi*. *FEMS Lett*. **154**, 265–270.

Pugsley, A. P. (1993). The complete general secretory pathway in gram-negative bacteria. *Microbiol. Rev*. **57**, 50–108.

Pugsley, A. P. (1996). Multimers of the precursor of a type IV pilin-like component of the general secretory pathway are unrelated to pili. *Mol. Microbiol*. **13**, 973–985.

Py, B., Bortoli-German, I., Haiech, J., Chippaux, M. and Barras, F. (1991a). Cellulase EGZ of *Erwinia chrysanthemi*: structural organisation and importance of His98 and Glu133 residues for catalysis. *Protein Engineering* **4**, 325–333.

Py, B., Salmond, G. P. C., Chippaux, M. and Barras, F. (1991b). Secretion of cellulase in *Erwinia chrysanthemi* and *Erwinia carotovora* is species specific. *FEMS Microbiol. Lett*. **79**, 315–322.

Py, B., Chippaux, M. and Barras, F. (1993). Mutagenesis of cellulase EGZ for studying the general protein secretory pathway in *Erwinia chrysanthemi*. *Mol. Microbiol*. **7**, 785–793.

Reeves, P. J., Douglas, P., Mulholland, V., Stevens, S., Walker, D. and Salmond, G.P.C. (1993). Molecular cloning and characterisation of 13 *out* genes from *Erwinia carotovora* subsp. *carotovora*: genes encoding members of a general secretion pathway (GSP) widespread in Gram-negative bacteria. *Mol. Microbiol*. **8**, 443–456.

Reeves, P. J., Douglas, P. and Salmond, G. P. C. (1994). Beta-lactamase topology probe analysis of the OutO NMePhe peptidase, and six other out protein components of the *Erwinia carotovora* general secretion pathway apparatus. *Mol. Microbiol.* **12**, 445–457.

Resibois, A., Colet, M., Faelen, M., Schoonejans, E. and Toussaint, A. (1984). φEC2, a new generalised transducing phage of *Erwinia chrysanthemi*. *Virology.* **137**, 102–112.

Reverchon, S. and Robert-Baudouy, J. (1987a). Molecular cloning of an *Erwinia chrysanthemi* oligogalacturonate lyase gene involved in pectin degradation. *Gene* **55**, 125–133.

Reverchon, S. and Robert-Baudouy, J. (1987b). Regulation of expression of pectate lyase genes *pelA, pelD* and *pelE* in *Erwinia chrysanthemi*. *J. Bacteriol.* **169**, 2417–2423.

Reverchon, S., Huang, Y., Bourson, C. and Robert-Baudouy, J. (1989). Nucleotide-sequences of the *Erwinia chrysanthemi ogl* and *pelE* genes negatively regulated by the *kdgR* gene product. *Gene* **85**, 125–134.

Reverchon, S., Nasser, W. and Robert-Baudouy, J. (1991). Characterisation of *kdgR*, a gene of *Erwinia chrysanthemi* that regulates pectin degradation. *Mol. Microbiol.* **5**, 2203–2216.

Reverchon, S., Nasser, W. and Robert-Baudouy, J. (1994). PecS – a locus controlling pectinase, cellulase and blue pigment production in *Erwinia chrysanthemi*. *Mol. Microbiol.* **11**, 1127–1139.

Reverchon, S., Expert, D., Robert-Baudouy, J. and Nasser, W. (1997). The cyclic AMP receptor protein is the main activator of pectinolysis genes in *Erwinia chrysanthemi*. *J. Bacteriol.* **179**, 3500–3508.

Ried, J. L. and Collmer, A. (1985). Activity stain for rapid characterisation of pectic enzymes in isoelectric-focusing and sodium dodecyl sulfate-polyacrylamide gels. *Appl. Environ. Microbiol.* **50**, 615–622.

Ried, J. L. and Collmer, A. (1988). Construction and characterisation of an *Erwinia chrysanthemi* mutant with directed deletions in all of the pectate lyase structural genes. *Mol. Plant–Microbe Interact.* **1**, 32–38.

Robert-Baudouy, J. (1991). Molecular biology of *Erwinia* – from soft-rot to antileukemics. *TIBTECH* **9**, 325–329.

Roeder, D. L. and Collmer, A. (1985). Marker-exchange mutagenesis of pectate lyase isozyme gene in *Erwinia chrysanthemi*. *J. Bacteriol.* **164**, 51–56.

Ruvken, G. B. and Ausebel, F. M. (1981). A general method for site-directed mutagenesis in prokaryotes. *Nature* **289**, 85–89.

Saarilahti, H. T., Henirssat, B. and Palva, E. T. (1990a). CelS: a novel endoglucanase identified from *Erwinia carotovora* subsp. *carotovora*. *Gene* **90**, 9–14.

Saarilahti, H. T., Heino, P., Pakkanen, R., Kalkkinen, N., Palva, I., and Palva, E.T. (1990b). Structural analysis of the *pehA* gene and characterisation of its protein product, endopolygalacturonase, of *Erwinia carotovora* subspecies *carotovora*. *Mol. Microbiol.* **4**, 1037–1044.

Saarilahti, H. T., Pirhonen, M., Karlsson, M. B., Flego, D. and Palva, E.T. (1992). Expression of *pehA–bla* gene fusions in *Erwinia carotovora* subsp. *carotovora* and isolation of regulatory mutants affecting polygalacturonase production. *Mol. Gen. Genet.* **234**, 81–88.

Salmond, G. P. C., Hinton, J. C. D., Gill, D. R. and Pérombelon, M. C. M. (1986). Transposon mutagenesis of *Erwinia* using phage λ vectors. *Mol. Gen. Genet.* **203**, 524–528.

Sandkvist, M., Bagdasarian, M., Howard, S. P. and Dirita, V. J. (1995). Interaction between the autokinasc EpsE and EpsL in the cytoplasmic membrane is required for extracellular secretion in *Vibrio cholerae*. *EMBO J.* **14**, 1664–1673.

Sandulache, R., Prehm, P., Expert, D., Toussaint, A. and Kamp, D. (1985). The cell wall receptor for bacteriophage Mu G(–) in *Erwinia* and *Escherichia coli*. *FEMS Microbiol Lett.* **28**, 307–310.

Santini, C.-L., Ize, B., Chanal, A., Müller, M., Giordano, G. and Wu, L-F. (1998). A novel Sec-independent periplasmic protein translocation pathway in *Escherichia coli. EMBO J.* **17**, 101–112.

Sauvage, C. and Expert, D. (1994) Differential regulation by iron of *Erwinia chrysanthemi* pectate lyases: pathogenicity of iron transport regulatory (*cbr*) mutants. *Mol. Plant–Microbe. Interact.* **7**, 71–77.

Sayers, J. R., Schmidt, W. and Eckstein, F. (1988). 5′-3′ exonucleases in phosphorothioate-based oligonucleotide-directed mutagenesis. *Nucl. Acids Res.* **16**, 791–802.

Schell, M. (1993). Molecular biology of the LysR family of transcriptional regulators. *Annu. Rev. Microbiol.* **47**, 597–626.

Schoonejans, E., Expert, D. and Toussaint, A. (1987). Characterisation and virulence properties of *Erwinia chrysanthemi* lipopolysaccharide-defective, φEC2-resistant mutants. *J. Bacteriol.* **169**, 4011–4017.

Scofield, S. R., Tobias, C. M., Rathjen, J. P., Chang, J. H., Lavelle, D. T., Michelmore, R. W. and Staskawicz, B. J. (1996). Molecular bases of gene-for-gene specificity in bacterial speck disease of tomato. *Science* **274**, 2063–2065.

Shaw, J. J., Settles, C. G. and Kado, C. I. (1988). Transposon Tn*4431* mutagenesis of *Xathomonas campestris*: characterisation of a non-pathogenic mutant and cloning of a locus for pathogenicity. *Mol. Plant–Microbe Interact.* **1**, 39–45.

Shaw, K. J. and Berg, C. M. (1979). *Escherichia coli* K12 auxotrophs induced by insertion of the transposable element Tn5. *Mol. Gen. Genet.* **92**, 741–747.

Shevchik, V. E. and Hugouvieux-Cotte-Pattat, N. (1997). Identification of a bacterial pectin acetyl esterase in *Erwinia chrysanthemi. Mol. Microbiol.* **24**, 1285–1301.

Shevchik, V. E., Condemine, G. and Robert-Baudouy, J. (1994). Characterisation of DsbC, a periplasmic protein of *Erwinia chrysanthemi* and *Escherichia coli* with disulphide-isomerase activity. *EMBO J.* **13**, 2007–2014.

Shevchik, V. E., Bortoli-German, I., Robert-Baudouy, J., Robinet, S., Barras, F. and Condemine, G. (1995). Differential effect of DsbA and DsbC mutations on extracellular enzyme secretion in *Erwinia chrysanthemi. Mol. Microbiol.* **16**, 745–753.

Shevchik, V. E., Condemine, G., Hugouvieux-Cotte-Pattat, N. and Robert-Baudouy, J. (1996). Characterisation of pectin methylesterase B, an outer membrane lipoprotein of *Erwinia chrysanthemi* 3937. *Mol. Microbiol.* **19**, 455–466.

Shevchik, V. E., Robert-Baudouy, J. and Hugouvieux-Cotte-Pattat, N. (1997a). Pectate lyase PelI of *Erwinia chrysanthemi* 3937 belongs to a new family. *J. Bacteriol.* **179**, 7321–7330.

Shevchik, V. E., Robert-Baudouy, J. and Condemine, G. (1997b). Specific interaction between OutD, an *Erwinia chrysanthemi* outer membrane protein of the general secretory pathway, and secreted proteins. *EMBO J.* **16**, 3007–3016.

Sienbenlist, U. and Gilbert, W. (1980). Contacts between *Escherichia coli* RNA polymerase and an early promoter of phage T7. *Proc. Natl. Acad. Sci. USA* **77**, 122–126.

Sneath, B. J., Howson, J. M. and Beer, S. V. (1990). A pathogenicity gene from *Erwinia amylovora* encodes a predicted protein product homologous to a family of prokaryotic response regulators. *Phytopathology* **80**, 1038.

Somoygi, M. (1952). Notes on sugar determination. *J. Biol. Chem.* **195**, 19–23.

Steinberger, E. M. and Beer, S. V. (1988). Creation and complementation of pathogenicity mutants of *Erwinia amylovora. Mol. Plant–Microbe Interact.* **1**, 135–144.

Stevens, A. M., Dolan, K. M. and Greenberg, E. P. (1994). Synergistic binding of the *Vibrio fischeri* LuxR transcriptional activator domain and RNA polymerase to the *lux* promoter region. *Proc. Natl. Acad. Sci. USA* **91**, 12619–12623.

Suguira, J., Uasuda, M., Kamimiya, S., Izaki, K. and Takahashi, H. (1984). Purification and properties of two pectate lyases produced by *Erwinia carotovora. J. Gen. Appl. Microbiol.* **30**, 167–175.

Surgey, N., Robert-Baudouy, J. and Condemine, G. (1996). The *Erwinia chrysanthemi* PecT gene regulates pectinase gene expression. *J. Bacteriol.* **178**, 1593–1599.

Swift, S., Winson, M. K., Chan, P. F., Bainton, N. J., Birdsall, M., Reeves, P. J., Rees, C. E. D., Chhabra, S. R. and Hill, P. J. (1993). A novel strategy for the isolation of *luxI* homologues: Evidence for the widespread distribution of a LuxR : LuxI superfamily in enteric bacteria. *Mol. Microbiol.* **10**, 511–520.

Sykes, R. B. and Matthew, M. (1979). Detection assay and immunology of β-lactamase. In *Beta-lactamases* (M. T. Hamilton-Miller and J. T. Smith, eds), pp. 17–19. Academic Press, New York.

Tabor, S. and Richardson, C. C. (1985). A bacteriophage T7 RNA polymerase/promoter system for controlled exclusive expression of specific genes. *Proc. Natl. Acad. Sci. USA* **82**, 1074–1078.

Tadayyon, M. and Broome-Smith, J. K. (1992). Tn*blaM* – a transposon for directly tagging bacterial genes encoding cell-envelope and secreted proteins. *Gene* **111**, 21–26.

Tamaki, S. J., Gold, S., Robeson, M., Manulis, S. and Keen, N. T. (1988). Structure and organisation of the *pel* genes from *Erwinia chrysanthemi* EC16. *J. Bacteriol.* **170**, 3468–3478.

Tang, X., Frederick, R. D., Zhou, J. M., Halterman, D. A., Jia, Y. L. and Martin, G. B. (1996). Initiation of plant disease resistance by physical interaction of AvrPto and Pto kinase. *Science* **274**, 2060–2063.

Tardy, F., Nasser, W., Robert-Baudouy, J. and Hugouvieux-Cotte-Pattat, N. (1997). Comparative analysis of the five major *Erwinia chrysanthemi* pectate lyases: Enzyme characteristics and potential inhibitors. *J. Bacteriol.* **179**, 2503–2511.

Teather, R. M. and Wood, P. J. (1982). Use of Congo red-polysaccharide interactions in enumeration and characterisation of cellulolytic bacteria from bovine remen. *Appl. Environ. Microbiol.* **43**, 777–780.

Tharaud, M., Menggad, M., Paulin, J. P. and Laurent, J. (1994). Virulence, growth, and surface characteristics of *Erwinia amylovora* mutants with altered pathogenicity. *Microbiology* **140**, 659–669.

Thomas, J. D. (1997). The general secretory pathway (GSP) of *Erwinia carotovora* subsp. *carotovora* (Ecc). PhD Thesis, Warwick University.

Thomas, J. D., Reeves, P. J. and Salmond, G. P. C. (1997). The general secretion pathway of *Erwinia carotovora* subspecies *carotovora*: Analysis of the membrane topology of OutC and OutF. *Microbiology* **143**, 713–720.

Thomashow, M. F., Karlinsey, J. E., Marks, J. R. and Hurlbert, R. E. (1987). Identification of a new virulence locus in *Agrobacterium tumefaciens* that affects polysaccharide composition and plant cell attachment. *J. Bacteriol.* **169**, 3209–3216.

Thomson, N. R., Cox, A., Bycroft, B. W., Stewart, G. S. A. B., Williams, P. and Salmond, G. P. C. (1997). The Rap and Hor proteins of *Erwinia, Serratia* and *Yersinia:* a novel subgroup in a growing superfamily of proteins regulating diverse physiological processes in bacterial pathogens. *Mol. Microbiol.* **26**, 531–544.

Thorpe, C., Toth, I., Bentley, S. and Salmond, G. P. C. (submitted). Mutation in a gene required for both lipopolysaccharide and enterobacterial common antigen biosynthesis affects virulence in the plant pathogen, *Erwinia carotovora* subspecies *atroseptica*. *Mol. Microbiol.*

Thurn, K. K. and Chatterjee, A. K. (1985). Single-site chromosomal Tn5 insertions affect the export of pectolytic and cellulolytic enzymes in *Erwinia chrysanthemi* EC16. *Appl. Environ. Microbiol.* **50**, 894–898.

Titarenko, E., Lopez-Solanilla, E., Garcia-Olmedo, F. and Rodriguez-Palenzuela, P.

(1997). Mutants of *Ralstonia (Pseudomonas) solanacearum* sensitive to antimicrobial peptides are altered in their lipopolysaccharide structure and are avirulent in tobacco. *J. Bacteriol.* **179**, 6699–6704.

Toth, I., Perombelom, M. C. M. and Salmond, G. P. C. (1993). Bacteriophage φKP mediated generalised transduction in *Erwinia carotovora* subsp. *carotovora*. *J. Gen. Microbiol.* **139**, 2705–2709.

Toussaint, A., Lefebvre, N., Scott, J., Cowan, J. A., De Bruijn, F. and Bukhari, A. I. (1978). Relationships between temperate phages Mu and P1. *Virology* **89**, 146–161.

Trollinger, D., Berry, S., Belser, W. and Keen, N. T. (1989). Cloning and characterization of a pectate lyase gene from *Erwinia carotovora* EC153. *Mol. Plant–Microbe Interact.* **2**, 17–25.

Tsuyumu, S. (1979). "Self-catabolite repression" of pectate lyase in *Erwinia carotovora*. *J. Bacteriol.* **137**, 1035–1036.

Turner, L. R., Lara, J. C., Nunn, D. N. and Lory, S. (1993). Mutations in the consensus ATP-binding sites of XcpR and PilB eliminate extracellular protein secretion and pilus biogenesis in *Pseudomonas aeruginosa*. *J. Bacteriol.* **175**, 4962–4969.

Turner, L. R., Olson, J. W. and Lory, S. (1997). The XcpR protein of *Pseudomonas aeruginosa* dimerises via its N-terminus. *Mol. Microbiol.* **26**, 877–887.

van den Ackerveken, G., Marois, E. and Bonas, U. (1996). Recognition of the bacterial avirulence protein AvrBs3 occurs inside the host plant cell. *Cell.* **87**, 1307–1316.

van Gijsegem, F. (1989). Relationship between the *pel* genes of the *pelADE* cluster in *Erwinia chrysanthemi* strain B374. *Mol. Microbiol.* **3**, 1415–1424.

van Gijsegem, F. and Toussaint, A. (1982). Chromosome transfer and R-prime formation by an RP4-mini-mu derivative in *Escherichia coli, Salmonella typhimurium, Klebsiella pneumoniae* and *Proteus mirabilis*. *Plasmid* **7**, 30–44.

van Gijsegem, F., Toussaint, A. and Schoonejans, E. (1985). *In vivo* cloning of the pectate lyase and cellulase genes of *Erwinia chrysanthemi*. *EMBO J.* **4**, 787–792.

Vanneste, J.L., Paulin, J. and Expert, D. (1990). Bacteriophage Mu as a genetic tool to study *Erwinia amylovora* pathogenicity and hypersensitve reaction on tobacco. *J. Bacteriol.* **172**, 932–941.

van Vliet, F., Silva, B., Van Montagu, M. and Schell, J. (1978). Transfer of RP4::Mu plasmids to *Agrobacterium tumefaciens*. *Plasmid* **1**, 446–455.

Vincent-Sealey, L. (1997). Investigation of the role of disulphide bond formation in the secretion and activity of virulence factors in *Erwinia carotovora* subsp. *carotovora*. PhD Thesis, Warwick University.

Vivian, A. and Gibbon, M. J. (1997). Avirulence genes in plant-pathogenic bacteria: signals or weapons? *Microbiology* **143**, 693–704.

Walker, D. S., Reeves, P. J. and Salmond, G. P. C. (1994). The major secreted cellulase, CelV, of *Erwinia carotovora* subsp. *carotovora* is an important soft rot virulence factor. *Mol. Plant–Microbe Interact.* **7**, 425–431.

Walters, K., Maroofi, A., Hitchin, E. and Mansfield, J. (1990). Gene for pathogenicity and ability to cause the hypersensitive reaction cloned from *Erwinia amylovora*. *Physiol. Mol. Plant Pathol.* **36**, 509–521.

Wandersman, C., Andro, T. and Bertheau, Y. (1986). Extracellular proteases in *Erwinia chrysanthemi*. *J. Gen. Microbiol.* **132**, 899–906.

Wandersman, C., Delepelaire, P., Letoffe, S. and Schwartz. M. (1987). Characterisation of *Erwinia chrysanthemi* extracellular proteases: cloning and expression of the protease genes in *Escherichia coli*. *J. Bacteriol.* **169**, 5046–5053.

Way, J. C., Davis, M. A., Morisato, D., Roberts, D. E. and Kleckner, N. (1984). New Tn*10* derivatives for transposon mutagenesis and for construction of *lacZ* operon fusions by transposition. *Gene* **32**, 369–379.

Wei, Z. M. and Beer, S. V. (1995). HrpL activates *Erwinia hrp* gene transcription and is a member of the ECF subfamily of sigma factors. *J. Bacteriol.* **177**, 6201–6210.

Wei, Z. M., Laby, R. J., Zumoff, C. H., Bauer, D. W., He, S. Y., Collmer, A. and Beer, S.V. (1992a). Harpin, elicitor of the hypersensitive response produced by the plant pathogen *Erwinia amylovora*. *Science* **257**, 85–88.

Wei, Z. M., Sneath, B. J. and Beer, S. V. (1992b). Expression of *Erwinia amylovora hrp* genes in response to environmental stimuli. *J. Bacteriol.* **174**, 1875–1882.

Wengenik, K. and Bonas, U. (1996). HrpXv, an AraC-type regulator, activates expression of five of the six loci in the *hrp* cluster of *Xanthomonas campestris* pv. *vesicatoria*. *J. Bacteriol.* **178**, 3462–3469.

Willis, J. W., Engwall, J. K. and Chatterjee, A. K. (1987). Cloning of genes for *Erwinia carotovora* subsp. *carotovora* pectolytic enzymes and further characterisation of the polygalacturonases. *Mol. Plant Pathol.* **77**, 1199–1205.

Wilmes-Riesenberg, M. R. and Wanner, B. L. (1992). Tn*phoA* and Tn*phoA'* elements for making and switching fusions for study of transcription, translation, and cell surface localisation. *J. Bacteriol.* **174**, 4558–4575.

Xiao, Y. X., Heu, S. G., Yi, J. S., Lu, Y. and Hutcheson, S. W. (1994). Identification of a putative alternative sigma factor and characterisation of a multicomponent regulatory cascade controlling the expression of *Pseudomonas syringae* pv. syringae Pss61 *hrp* and *hrmA* genes. *J. Bacteriol.* **176**, 1025–1036.

Yankosky, N. K., Bukanov, N. O., Gritzenko, V. V., Evtushenkov, A. N., Fonstein, M. and Debadov, V. G. (1989). Cloning and analysis of structural and regulatory pectate lyase genes of *Erwinia chrysanthemi* ENA49. *Gene* **81**, 211–218.

Yoder, M. D., Keen, N. T. and Jurnak, F. (1993a). New domain motif: the structure of pectate lyase C, a secreted plant virulence factor. *Science* **260**, 1503–1507.

Yoder, M. D., Lietzke, S. E. and Jurnak, F. (1993b). Unusual structural features in the parallel β-helix in pectate lyases. *Structure* **1**, 241–251.

Yoshida, A., Izuto, M., Ito, K., Yoshiyuki, K. and Izaki, K. (1991a). Cloning and characterisation of the pectate lyase III gene of *Erwinia carotovora* Er. *Agric. Biol. Chem.* **55**, 933–940.

Yoshida, A. Ito, K., Kamio, Y. and Izaki, K. (1991b). Purification and properties of pectate lyase III of *Erwinia carotovora* Er. *Agric. Biol. Chem.* **55**, 601–602.

Yoshida, A., Matsuo, Y., Kamio, Y. and Izaki, K. (1992). Molecular cloning and sequencing of the extracellular pectate lyase II gene from *Erwinia carotovora* Er. *Biosci. Biotech. Biochem.* **56**, 1596–1600.

Zink, R. T. and Chatterjee, A. K. (1985). Cloning and expression in *Escherichia coli* of pectinase genes of *Erwinia carotovora* subsp. *carotovora*. *Appl. Environ. Microbiol.* **49**, 714–717.

Zink, R. T., Kemble, R. J. and Chatterjee, A. K. (1984). Transposon Tn5 mutagenesis in *Erwinia carotovora* subspecies *carotovora* and *Erwinia carotovora* subspecies *atroseptica*. *J. Bacteriol.* **157**, 809–814.

Zink, R. T., Engwall, J. K., Mcevoy, J. L. and Chatterjee, A. K. (1985). RecA is required in the induction of pectin lyase and *carotovoricin* in *Erwinia carotovora* subsp. *carotovora*. *J. Bacteriol.* **164**, 390–396.

13 Molecular Genetic Methods in *Paracoccus* and *Rhodobacter* with Particular Reference to the Analysis of Respiration and Photosynthesis

M Dudley Page[1] **and R Elizabeth Sockett**[2]

[1] *Department of Biochemistry, University of Oxford, South Parks Road, Oxford, UK;*
[2] *Institute of Genetics, University of Nottingham, Queens Medical Centre, Nottingham NG7 2UH, UK*

◆◆◆

CONTENTS

Paracoccus and Rhodobacter

◆◆◆◆◆◆ I. AIMS

This chapter will highlight how molecular genetic work has been and is being used to enhance understanding of electron transport in *Paracoccus denitrificans* and aspects of electron transport and photosynthesis in *Rhodobacter sphaeroides* and *Rhodobacter capsulatus*. This approach uses selective examples of molecular genetic work on photosynthesis and respiration in these bacteria. This is deliberately restrictive and does not include important biochemical and biophysical work in these fields, plus extensive work on photosynthesis and respiration in other *Rhodobacter* species. For reviews of these important areas, which are beyond the purpose of this chapter, the reader is referred to a number of reviews (Scolnik and Marrs, 1987; Donohue and Kaplan, 1991; Ghozlan *et al.*, 1991;

Harms and van Spanning, 1991; Steinrücke and Ludwig, 1993; Williams and Taguchi, 1995; Baker *et al.*, 1998).

◆◆◆◆◆◆ II. INTRODUCTION

Paracoccus and *Rhodobacter* species are Gram-negative rod-shaped bacteria that are commonly isolated from soil and water environments. They are phylogenetically "close relatives" and form a separate group within the alpha proteobacteria (Ludwig *et al.*, 1993; Katayama *et al.*, 1995). They have been extensively studied as model systems for, and possible progenitors of, respiration and photosynthesis in eukaryotes. The aerobic electron transport chains of *P. denitrificans* and *R. sphaeroides* spp. resemble those found in mitochondria. Counterparts of complexes I (NADH dehydrogenase), II (succinate dehydrogenase), III (the cytochrome bc_1 complex) and IV (the aa_3-type cytochrome oxidase) are all present. These similarities led to the proposal that the eukaryotic mitochondrion originated as a symbiotic bacterium related to *P. denitrificans* (John and Whatley, 1975; Whatley, 1981). In a similar way, the evolution of the basic photosynthetic reaction centre, light-harvesting pigments and chlorophyll and carotenoid-synthesizing enzymes in chloroplasts of eukaryotes are thought to have links to the photosystems of *Rhodobacter*, although the eukaryotic oxygen-evolving capacity came separately (Meyer, 1994; Lockhart *et al.*, 1996).

Paracoccus and *Rhodobacter* species are extremely nutritionally versatile, growing heterotrophically on a wide variety of carbon sources, both aerobically and anaerobically with nitrate, nitrite or nitrous oxide as terminal electron acceptor. *P. denitrificans* can also grow autotrophically with either hydrogen or thiosulphate as energy source, and with the reduced one-carbon compounds methanol and methylamine which are oxidized, via formaldehyde, to carbon dioxide. *Rhodobacter* sp. grow photoheterotrophically on organic acids such as succinate, they also grow photoautotrophically on carbon dioxide, and anaerobically in the dark on glucose using dimethyl suphoxide as the terminal electron acceptor The nutritional versatility of these bacteria is due to their ability to assemble a considerable number of alternative respiratory enzymes and small electron transport proteins. Such versatility allows the generation of mutant strains, which can be rescued by growth in alternative conditions, in which the gene products being mutagenized are not required. This very useful feature has allowed basic studies on the gene products required for electron transport and synthesis of photosynthetic pigments in these bacteria.

Until relatively recently the focus of much research effort in *Paracoccus* and *Rhodobacter* has been on the isolation and analysis of respiratory or photosynthetic complexes, and studies of their prosthetic groups or metal centres by physical techniques (e.g. UV/visible spectroscopy, EPR and MCD spectroscopy for metal centres, redox potentiometry). Many or most of the proteins in these complexes are synthesized naturally at high levels and this, combined with the relative ease of growing the bacteria in large culture volumes, has allowed their purification and study, including amino acid sequence determination, without the need for cloning and over-expression . There has consequently been some delay in applying molecular genetic techniques in these organisms. It is remarkable that for a number of electron transport proteins: *Paracoccus* cytochromes c_{550} and cd$_1$ nitrite reductase, methylamine dehydrogenase, and *Rhodobacter* cytochrome c', (Timkovich and Dickerson, 1976; van Spanning *et al.*, 1990a; Chen *et al.*, 1992; Chistoserdov *et al.*, 1992; Fülop *et al.*, 1995; Tahirov *et al.*, 1996; Baker *et al.*, 1997), crystallization and collection of X-ray diffraction data have preceded cloning and sequencing of the structural gene(s).

This "biochemical advantage" has also meant that in *Paracoccus* and *Rhodobacter*, molecular genetics has predominantly been used as a tool rather than an end in itself, although genome mapping and sequencing projects are being undertaken with a view to characterizing the overall genetic make-up of these organisms. Many techniques used in these organisms have been adapted from those used in related genera such as *Pseudomonas* and *Rhizobium*. In recent years more genetic techniques have been applied due to increasing interest in the regulation of gene expression in these bacteria, coupled with a drive from crystallographic data for more sophisticated manipulations of protein function *in vivo*.

◆◆◆◆◆◆ III. BACTERIAL STRAINS USED FOR MOLECULAR GENETIC WORK

The standard *P. denitrificans* strain for molecular genetic studies is Pd1222, an undefined nitroso-guanidine-derived mutant of Pd1103, a rifampicin- and spectinomycin-resistant derivative of the type strain NCIMB 8944; Pd1222 lacks the major host restriction system responsible for the low conjugation frequencies observed for NCIMB 8944 (de Vries *et al.*, 1989; see below). Strains lacking both major and minor restriction systems, and strains lacking both these systems and the host DNA modifying system mentioned above have also been isolated. The closely related *Thiosphaera pantotropha* (Robertson and Kuenen, 1983) has recently been reclassified as *P. pantotrophus* (Rainey *et al.*, 1999). Furthermore, it has been shown that some strains of *P. denitrificans* (including a supposed type strain, DSM65) are in fact *P. pantotrophus* (Rainey *et al.*, 1999), so care must be taken in selecting a strain on which to work. *Thiobacillus versutus*, also closely related to *P. denitrificans*, is now known as *P. versutus* (Katayama *et al.*, 1995). *R. sphaeroides* wild-type strains WS8, 8253 and 2.4.1 are most commonly used; these are amenable to genetic manipulation but are restriction and modification positive. Spontaneous mutant derivatives of these strains which are antibiotic resistant (to nalidixic acid or rifampicin) are commonly isolated for use in conjugation experiments (see IV.C) (Sistrom *et al.*, 1984). Strains of *P. denitrificans* and *Rhodobacter* that are deficient in DNA repair and recombination, and thus in homologous recombination, have recently been constructed; they may be useful for ensuring the stability of plasmid-borne constructs used for complementation or for the (over)expression of genes in the homologous hosts (de Henestrosa *et al.*, 1997; Mackenzie *et al.*, 1995)

◆◆◆◆◆◆ IV. VECTORS FOR USE IN *RHODOBACTER* AND *PARACOCCUS*

A. Cloning Vectors

The most commonly used vectors for complementation or for maintaining plasmid-encoded genes in *Paracoccus* and *Rhodobacter* are the mobilizable broad host range vectors based on the RK2 replicon (incompatibility group P1; Ditta *et al.*, 1980) and the RSF1010 replicon (complementation group IncQ; Barth and Grinter, 1974; Chapter 2, II). However, a new mobilizable broad-host-range replicon, pBBR1 from *Bordetella bronchoseptica*, has recently been reported, derivatives of which have proved useful in *P. denitrificans*; the incompatibility group of pBBR1 is unknown, but it is not IncP1, IncQ or IncW (Antoine and Locht, 1992).

The RK2-derived vectors which have been most widely used are pRK404 (Ditta *et al.*, 1985) and its derivatives pRK415 (Keen *et al.*, 1988) and pEG400 (Gerhus *et al.*, 1990). All three plasmids have multiple cloning sites, support blue/white screening, and are substantially smaller (at 10.6, 10.5 and 11.4 kb, respectively) than pRK290 (20 kb) (Ditta *et al.* 1980); making them much easier to manipulate. pRK290, pRK404 and pRK415 carry a tetracycline-resistance marker while in pEG400 this has been replaced by the R100.1 spectinomycin/streptomycin-resistance determinant. IncP1 plasmids are stable in *R. sphaeroides* as long as selective pressure is maintained, but in its absence plasmids are lost from about 8% of cells per generation (Davis *et al.*, 1988; DeHoff *et al.*, 1988; Brandner *et al.*, 1989). The tetracycline-resistance marker of pRK415 only gives *R. sphaeroides* resistance to 1 μg ml^{-1} of the antibiotic; because of this, optical filters must be used to protect the photosensitive tetracycline when bacteria containing vectors such as pRK415 are grown photosynthetically at high light intensities (100 W m^{-2}). To overcome this and other limitations of these vectors, a pRK415 derivative, pRKD418, has been constructed which carries both tetracycline and trimethoprim-resistance markers and an enlarged multiple cloning site (Mather *et al.*, 1995). The tetracycline-resistance determinant is only weakly selectable in *P. denitrificans* (confers resistance to only 0.1–0.3 μg ml^{-1}; de Vries *et al.*, 1989), but pRK415 can be maintained by the selection pressure of functional complementation itself in experiments with this organism (e.g. Page and Ferguson, 1995). The copy number for vectors based on pRK404 in *Rhodobacter* sp. has been estimated at 5–9 per host genome (Davis *et al.*, 1988; Brandner *et al.*, 1989). Vectors have been modified to include alternative resistance markers when required to complement already antibiotically marked strains. Markers commonly used in this way in both genera include the kanamycin-resistance markers from

Tn5 and Tn903 (e.g. Chandra and Friedrich, 1986) and the spectino-mycin/streptomycin-resistance markers from R100.1 and Tn1831 (Prentki and Krisch, 1984; Gerhus *et al.*, 1990; van Spanning *et al.*, 1990a). The R388 trimethoprim resistance marker (Ward and Grinstead, 1982), the R1033 gentamycin (Hirsch *et al.*, 1986) and the plasmid Sa gentamycin/kanamycin-resistance markers (Ward and Grinstead, 1982) have been used to a lesser extent. The uses of other markers in *Rhodobacter* and *Paracoccus* have been discussed by Donohue and Kaplan (1991) and by Steinrücke and Ludwig (1993) respectively.

Mobilizable IncP1 cosmid vectors have been used to construct *R. capsulatus* and *R. sphaeroides* genomic DNA libraries which when conjugated into mutant strains allow the identification of complementing cosmids and thus isolation of the gene(s) affected in those strains. pLAFR1 (Friedman *et al.*, 1982) has been used in the isolation of *R. capsulatus* nitrogen fixation and hydrogenase genes by complementation of Nif and Hup⁻ mutants, respectively (Avtges *et al.*, 1985; Colbeau *et al.*, 1986), pVK102 (Knauf and Nester, 1982) in the isolation of *R. sphaeroides* DNA complementing a ribulose 1,5-bisphosphate carboxylase–oxygenase regulatory mutant (Weaver and Tabita, 1983), and pLA2917 (Allen and Hanson, 1985) in the isolation of *R. sphaeroides* flagellar genes (Sockett and Armitage, 1991).

The RSF1010 derivatives pNH2 (Hunter and Turner, 1988) and pJRD215 (Davison *et al.*, 1987) have been applied in the same way as the IncP1 cosmid vectors described above. pNH2 was employed to construct a *R. sphaeroides* genomic DNA library which was used in the isolation of photosynthesis genes (Hunter and Turner, 1988), while pJRD215 was employed to construct a *R. capsulatus* genomic DNA library which was used to clone the gene for the regulatory protein RegA (Sganga and Bauer, 1992). It is surprising that RSF1010 derivatives have not enjoyed greater use in *P. denitrificans* and *Rhodobacter* sp., given that plasmids are available which have useful resistance markers and are of moderate size (e.g. pKT230 and pKT231; Bagdasarian *et al.*, 1981). Vectors based on RSF1010 have a copy number of 15–20 (Franklin, 1985).

The pBBR1 derivatives pBBR1MCS-2 (Kovach *et al.*, 1995), pBBR1MCSKm (Ueda *et al.*, 1996), and pBBR122 (MoBiTech GmbH, Göttingen) have recently proved useful in *P. denitrificans* (Ueda *et al.*, 1996; Page and Ferguson, 1997). All three are much smaller than the commonly used broad host range plasmids (5.15 kb for pBBR1MCS-2 and pBBR1MCSKm and 5.3 kb for pBBR122), greatly facilitating their manipulation, and pBBR1MCSKm and pBBR1MCS-2 both have the pBluescript multiple cloning site and support blue/white screening. pBBR1MCS has a copy number of 30–40 in *E. coli* (Antoine and Locht, 1992). The IncP1 vectors pJB3Km1 and pJB3Tc20 (Blatny *et al.*, 1997) have similar properties to those of the pBBR1 series and have also been used in *P. denitrificans*.

B. Suicide Vectors

The observation that plasmids based on the ColE1 and the pMB1 replicons are not maintained in *P. denitrificans* or *Rhodobacter* sp. unless integrated into the host chromosome has led to the routine use of mobilizable pMB1-based "suicide plasmids", for example pSUP202 (Simon *et al.*, 1983) for mutagenesis by conjugation into these bacteria (Chapter 2, III.C). Suicide vectors based on the R6K replicon have also been shown to be functional in *R. sphaeroides* and *P. denitrificans* (Penfold and Pemberton, 1992; Page and Ferguson, 1994); these vectors require the R6K-specified Pir protein for replication and can be maintained only in host strains which provide this protein *in trans* (e.g. *E. coli* strains containing integrated lambda *pir*; Taylor *et al.*, 1989; Penfold and Pemberton, 1992). For more details on their use, see the transposon mutagenesis and gene replacement sections below.

C. Plasmid transfer

In the 1980s efforts were made to establish efficient protocols for *Rhodobacter* transformation (reviewed in Williams and Taguchi, 1995), however these were largely unsuccessful. Electroporation of *P. denitrificans* and *R. sphaeroides* have been reported (Solioz and Bienz, 1990; Donohue and Kaplan, 1991), but again efficiency problems have meant that it has not been widely used. Plasmids are most often transferred to *Paracoccus* and *Rhodobacter* sp. by conjugation with *E. coli* (Chandra and Friedrich, 1986; Moore and Kaplan, 1989) (Chapter 2, V). Matings may be either biparental or triparental, i.e. with the *trans*-acting *tra* genes residing either in the genome of the *E. coli* donor (*E. coli* SM10 and S17-1; Simon *et al.*, 1983) or on a self-mobilizing "helper" plasmid which is not propagated in the recipient, e.g. pRK2013 [Figurski and Helinski, 1979]; pRK2073 [Leong *et al.*, 1982; Ditta *et al.*, 1985; Hunter and Turner, 1988] is carried by a second *E. coli* donor strain. Counter-selection, after conjugation, against donor *E.coli* is achieved by the use of rifampicin (for *P. denitrificans*) or nalidixic acid for spontaneously resistant "wild-type" strains of *R. sphaeroides*. Improved counter-selection in *Rhodobacter* conjugations has been achieved by treating exconjugants with T4 phage (attacks only the *E.coli*) or where possible by subjecting them to photosynthetic growth conditions.

Paracoccus denitrificans type strain NCIMB 8944 proved to be unsuitable for genetic studies due to difficulties in acheiving high transconjugant frequencies (Paraskeva, 1979; de Vries *et al.*, 1989; Fellay *et al.*, 1989). By carrying out conjugations in the presence of NTG, de Vries and co-workers (1989) isolated a number of mutants exhibiting an enhanced frequency of conjugational plasmid transfer; one of these, PD1222, is now the standard strain for molecular genetic studies. Furthermore PD1222 is resistant to rifampicin, which facilitates counter-selection of *E. coli* after conjugation. "Back-conjugation" has been used in the past to transfer plasmids from *P. denitrificans* to *E. coli* by (Steinrücke and Ludwig, 1993) for subsequent analysis; but more commonly, plasmid DNA is isolated directly from *P. denitrificans* and *Rhodobacter* sp. by alkaline lysis-based procedures as for *E. coli* (e.g. Birnboim and Doly, 1979) (Chapter 2, VI).

◆◆◆◆◆◆ V. GENE IDENTIFICATION AND CLONING STRATEGIES

A. Reverse Genetics

For the reasons discussed in the introduction, cloning of many *Paracoccus* and *Rhodobacter* structural genes encoding electron transport components, has been achieved by reverse genetics. Both screening genomic DNA libraries with degenerate oligonucleotide probes designed on the basis of known protein sequences (Raitio *et al.*, 1987; van Spanning *et al.*, 1990a) and PCR with degenerate primers (Richter *et al.*, 1994; Witt and Ludwig, 1997) have been employed.

A second strategy, again reliant on the availability of purified protein, is cloning genes by antibody screening of expression libraries in *E. coli*. Differences in codon usage have not proved a barrier to the expression of *Paracoccus* or *Rhodobacter* sp. genes in *E. coli*, but a suitable promoter must be provided (see VI.D.1) (Harms *et al.*, 1987; Kurowski and Ludwig, 1987; Steinrücke *et al.*, 1987; Shaw *et al.*, 1996).

B. Cloning Genes by Complementation

Some *Rhodobacter* sp. genes have been cloned by complementation of *E. coli* mutants, e.g. the *R. capsulatus hemH* and *R. sphaeroides hisI* genes (Kanazireva and Biel, 1995; Oriol *et al.*, 1996). Additionally, the *R. sphaeroides recA* gene was cloned via complementation of a *P. aeruginosa recA* mutant (Calero *et al.*, 1994). Complementation in the homologous host has been the most widely used method, as in other bacteria, and obviously it has been essential for more "unique" genes like those encoding photosynthesis. Isolation of two *R. capsulatus* DNA fragments which complemented a number of photosynthesis-deficient mutants led to sequencing of the structural genes for the three reaction centre subunits and the alpha and beta subunits of the light-harvesting complex I (Youvan *et al.*, 1984). Similar methods resulted in cloning of *R. capsulatus* cytochrome oxidase *ccoNOQP* genes and *R. sphaeroides* flagellar motor genes (Sockett and Armitage, 1991; Koch *et al.*, 1998). Although, as mentioned above, complementation of heterologous hosts has been widely used in gene cloning, DNA hybridization using heterologous gene probes has only been used to clone a few *Rhodobacter* and *Paracoccus* genes, e.g. *R. sphaeroides mob* genes (Palmer *et al.*, 1998); *P. denitrificans nirS* and *nosZ* genes (Hoeren *et al.*, 1993; de Boer *et al.*, 1994). This approach was precluded by codon bias in some cases.

C. Transposon Mutagenesis and Marker Rescue

Some caution is required when using transposon mutagenesis to clone genes required for photosynthetic growth or cytochrome biogenesis. Both processes require the participation of many metallo-protein complexes, the syntheses of which are attained by multi-enzyme pathways. It is very

Paracoccus and Rhodobacter

easy, for example, to obtain transposon mutants of *Rhodobacter* which are photosynthetically incompetent, but these mutants could have defects in pigment biosynthesis genes, genes encoding pigment-binding proteins, regulatory genes and so on. Genomic clustering of genes with related function is common in these bacteria, and that too serves as both a help and a hindrance when using transposons. Helpfully, identification of one gene can lead to the characterization of others with related functions merely by sequencing upstream and downstream of the Tn insertion site. This was the case for the *P. denitrificans* NADH dehydrogenase gene *nqo1*; its identification, led to the cloning and sequencing of the genes for the other 13 subunits of the complex because these are all located in a single large gene cluster: *nqo7-nqo6-nqo5-nqo4-nqo2-urf2-urf1-nqo1-urf3-nqo3-urf4-nqo8-nqo9-urf5-urf6-nqo10-nqo11-nqo12-nqo13-nqo14* (Xu *et al.*, 1993a and references therein) Similarly for the denitrification gene cluster *nnr-norFEDQBC-nirJISECFD*, which was identified via the cloning of *nirS* (de Boer *et al.*, 1996 and references therein). Less helpfully, individual non-polar knock-outs of each gene may be required, after sequencing to ascertain function.

Some mutations in genes coding for the synthesis of bacteriochlorophyll, carotenoid or haem in *Rhodobacter* caused novel coloured phenotypes. Luckily the elegant "pre-molecular-genetic" work of June Lascelles (1960) had shown that the biosynthesis of bacteriochlorophyll and carotenoid was catalysed by a sequential array of enzymes and proceeded via a series of coloured intermediates. Thanks to this many mutants could be preliminarily identified by their pigmentation , or type of intermediate accumulating, and then the genes responsible cloned (for example Giuliano *et al.*, 1988; Bollivar *et al.*, 1994).

Mutagenesis with Tn5 and its derivatives has been widely used in *Paracoccus* and *Rhodobacter* (Chandra and Friedrich, 1986; Hunter 1988; Bell *et al.*, 1990). The Tn5 derivatives Tn5-*mob*, Tn5::*phoA* and mini-Tn5–*lacZ2* have also been used successfully (Moore and Kaplan 1989; Page and Ferguson 1994; Wodara *et al.*, 1994; see Chapter 4, X and XI). Unstable or temperature-sensitive RK2 derivatives have been used as mobilizable suicide vectors for Tn5 delivery in *R. sphaeroides* (Weaver and Tabita, 1983), but more recently suicide vectors based on the pMB1 replicon (see below), for example pSUP2021 (pSUP202::Tn5; Simon *et al.*, 1983); pRT733 (pJM701.3::Tn5*phoA*; Taylor *et al.*, 1989) and pUI800 (pSUP203::Tn5*phoA*; Moore and Kaplan, 1989), have been widely used. Tn5 integration is generally assumed to be random in *Paracoccus* and *Rhodobacter* sp., although there is no direct evidence that this is the case. One problem encountered during Tn5 mutagenesis of *P. denitrificans* has been vector co-integration (e.g. Page and Ferguson, 1994). Integration of two Tn5 elements per genome has also been observed (Page *et al.*, 1997), but Tn5-associated deletions have not.

Mini-Tn5::*lacZ2* (de Lorenzo *et al.*, 1990) was designed to construct random operon or gene fusions to identify promoters with certain regulatory properties. It is currently being applied for the identification of *P. denitrificans* genes expressed during anaerobic growth with nitrous oxide, as terminal electron acceptor (Baker and Ferguson, unpublished). Similarly,

Tn5::*phoA* mutagenesis has been used to identify genes for periplasmic proteins and membrane proteins with periplasmically orientated hydrophilic domains. Such proteins include reductase enzymes vital to alternative electron transport pathways in the bacteria (Moore and Kaplan, 1989), and components of energetic complexes like the bacterial flagellar motor (Sockett and Armitage, 1991).

A number of *Paracoccus* and *Rhodobacter* genes (including all the *P. denitrificans* c-type cytochrome biogenesis genes) have been cloned after segments of transposon-flanking genomic DNA were rescued from Tn5 or Tn5::*phoA* mutants using the Tn5 kanamycin resistance as a selectable marker. Tn5-inserted genes can be cloned into a broad host range vector and exchanged for the wild-type copy by a "reverse cloning" procedure (Klipp *et al.*, 1988; Schuddekopf *et al.*, 1993), but it is more usual to use Tn5-flanking DNA segments to probe a genomic DNA library (e.g. Sockett and Armitage, 1991; Page and Ferguson, 1995). If large DNA segments are cloned, wild-type genes adjacent to the Tn5-inserted one can be rescued (e.g. Bell *et al.*, 1990; Mittenhuber and Friedrich, 1991; Page *et al.*, 1997).

Antibiotic-resistance cassettes inserted by homologous recombination (see next section) can be rescued in the same way as Tn5; using a kanamycin-resistance cassette inserted in *mauC*, van der Palen *et al.* (1995) cloned the whole of the *P. denitrificans mauRFBEDACJGMN* methylamine utilization gene cluster by this "marker rescue" technique.

◆◆◆◆◆◆ VI. CHARACTERIZATION OF GENES AND GENE PRODUCTS

A. Gene Replacement

Homologous recombination is a versatile tool allowing the creation of marked or unmarked insertion or deletion mutations in one or more genes. For a more detailed discussion on gene replacement technology, see Chapter 10, II, III and IV. It has been used to establish the role of products encoded by many *Rhodobacter* and *Paracoccus* genes (by disruption followed by biochemical and energetic characterization of the mutant strain); to establish the role of individual subunits within multi-subunit complexes; to integrate gene-reporter fusions into the host genome; to replace host genes with exogenous ones; and to create a suitable genetic background for receipt of PCR-point-mutagenized genes. As gene replacement work progressed, especially in *Rhodobacter*, it sometimes gave hints of gene duplications, null mutations did not give null phenotypes, recombination was sometimes seen at "the wrong" sites (Hallenbeck and Kaplan, 1988). This later was found to be due to the presence of two chromosomes in the bacterium containing homologous gene copies, a novel but not unique feature.

The length of DNA required for homologous recombination to occur varies between organisms; in *P. denitrificans* successful recombination between plasmid-borne and genomic DNA has been achieved with as little as 350 bp of homologous DNA. The simplest form of insertional

Paracoccus and Rhodobacter

mutagenesis is the creation of a plasmid integrant strain via a single recombination between a cloned internal fragment of a gene and the chromosomal copy. This results in the formation of a mutant strain carrying one 5'-truncated and one 3'-truncated copy of the gene in question and has been used to inactivate genes in *R. sphaeroides* (Penfold and Pemberton, 1994) and in *P. denitrificans* (van Spanning *et al.*, 1995a). Plasmid integrant strains are unstable in the absence of selection, loss of the vector resulting in reversion.

Much more common is the use of double homologous recombination between a cloned gene which has been disrupted by insertion of an antibiotic-resistance cassette and the chromosomal copy of that gene to disrupt the chromosomal copy with loss of the vector. A key issue in double homologous recombination is differentiation between plasmid integrants (single crossover) and true gene replacements(double crossover). Haltia *et al.* (1989) utilized a pSUP202 derivative carrying a streptomycin resistance marker as the suicide plasmid to allow discrimination between *P. denitrificans* exconjugants retaining the plasmid in an integrated form and those which had eliminated it in a double crossover event. The suicide plasmids pGRPD1 and pRVS1 developed for use in *P. denitrificans* (van Spanning *et al.*, 1990a, 1991a) also carry a streptomycin-resistance marker. A similar strategy has been used for directed mutagenesis using the Ω-Sm cassette; in this case the suicide plasmid pARO181, carrying the kanamycin-resistance marker from Tn5 (Parke, 1990) was used (Page and Ferguson, 1997). A non-antibiotic marker, the *E. coli lacZ* gene expressed from the Tn*903 aph* promoter, has been used to identify plasmid integrant strains in *P. denitrificans*; these form blue colonies on media containing X-gal (Witt and Ludwig, 1997).

B. Disruption and Study of Multiple Genes

The nutritional versatility of *P. denitrificans* is reflected in its respiratory electron transport chain, which, while resembling in some respects that of the mitochondrion, is much more complex, with multiple oxidases, branch points and alternative routes for electron flow (van Spanning *et al.*, 1995b). Because of this complexity, a central strategy in the analysis of electron transport pathways in *P. denitrificans* has been the construction of strains in which the structural genes for several respiratory enzymes or electron transport proteins are simultaneously disrupted. There is a similar situation for respiratory electron transport in *Rhodobacter* sp., but additionally there are multiple bacteriochlorophyll (bchl)-binding proteins which are part of reaction centre and light-harvesting complexes. To individually characterize the spectral properties of these complexes, deletion strains missing several genes encoding bchl-binding proteins have been required (Jones *et al.*, 1992).

In some cases the disruption of single genes has yielded mutant strains with definite phenotypes. For example, disruption of the amicyanin structural gene, *mauC*, yielded a mutant strain incapable of growth with methylamine (van Spanning *et al.*, 1990b) and disruption of the cytochrome c_{551} structural gene, *moxG*, yielded a strain incapable of growth with methanol (van Spanning *et al.*, 1991a), confirming biochemical evidence that amicyanin and cytochrome c_{551} are the obligate electron acceptors for the methylamine and methanol dehydrogenases, respectively.

However, it has more often been found that disruption of single genes had little effect on the capacity for respiration of *P. denitrificans* with succinate, methanol or methylamine. For example, disruption of the cytochrome c_{550} structural gene, *cycA*, had essentially no effect on the growth rate of *P. denitrificans* under aerobic, anaerobic or methylotrophic conditions (van Spanning *et al.*, 1990a). This was unexpected since the available biochemical evidence indicated that cytochrome c_{550} was probably important during anaerobic and methylotrophic growth, conditions in which periplasmic electron transport is more active than during aerobic growth. The obvious conclusion was that one or more alternative electron carriers could substitute for cytochrome c_{550}. Similarly, disruption of the structural genes for cytochrome c_{553i} (*cycB*), and cytochrome c_{552} (*cycM*) had only very limited physiological effects (Ras *et al.*, 1991; van Spanning *et al.*, 1995b)

1. Identification of multiple genomic copies of genes

Mutagenesis of the aa_3-type cytochrome oxidase subunit I gene *ctaDI* (COI)gene by simultaneous deletion and insertion of a kanamycin-resistance cassette did not yield a cytochrome-aa_3-deficient strain; membranes prepared from the mutant contained spectrally normal cytochrome aa_3 at levels similar to those found in the parental strain, and it oxidized TMPD at wild-type rates. Southern blotting analysis verified that deletion of *ctaDI* had been achieved, but also showed that there was a second region of *P. denitrificans* DNA which hybridized strongly to a *ctaDI* probe. This region was cloned and sequence analysis showed that it contained a second copy of *ctaDI*, which was termed *ctaDII* (Raitio *et al.*, 1990). Furthermore, when construction of a mutant strain lacking the aa_3-type cytochrome oxidase was achieved by interrupting the subunit II structural gene COII (*ctaC*), the viability of the mutant strain confirmed the presence in *P. denitrificans* of a second terminal oxidase (Ludwig, 1992; and see above). It was clear that analysis of the electron transport chain of *P. denitrificans* would require the construction and analysis of mutant strains in which two or three genes had been disrupted.

2. Multiple gene deletion using marked and unmarked mutations

The disruption of multiple genes is generally achieved by a protocol of mutagenesis of a gene by insertion of an antibiotic-resistance cassette followed by its replacement with a second copy of the gene carrying an unmarked internal deletion. A second gene can then be mutated by insertion of the same antibiotic-resistance cassette, and the process repeated to produce strains mutated in two, three or more genes (see below). This procedure yields a mutant strain which has the same susceptibility to

Paracoccus and Rhodobacter

antibiotics as the parental strain rather than one carrying several antibiotic resistance markers. This means that integrants must be isolated, using an antibiotic resistance or other marker on the vector, as an intermediate stage, and then allowed to undergo a second recombination event leading to elimination of the vector and creation of the desired mutation (resolution). However, the spontaneous resolution of integrants occurs at a rather low frequency (e.g. 5×10^{-4}; van Spanning *et al.*, 1991a; see below). Thus it is highly advantageous, although not essential (Gong *et al.*, 1994) to have a method of positively selecting for strains which have undergone a double crossover event. In *Rhodobacter* suicide plasmids using the counter-selectable *sacB* gene (Chapter 4, VI) (Gay *et al.*, 1983) (encodes levan sucrase which gives suicide on sucrose) have been employed to create single or multiple mutations by recombination (Sabaty and Kaplan, 1996). This system is reportedly functional in *P. denitrificans* but it has yet to be exploited in this organism.

In *P. denitrificans* an alternative method has been developed by van Spanning *et al.* (1991a) with the construction of the mobilizable suicide vector pRVS1; this is now routinely used for construction of unmarked mutations in this organism. pRVS1 carries a streptomycin-resistance marker and the *E. coli lacZ* gene expressed from the Tn5 *kan* promoter, which is active in *P. denitrificans*. It was first used as follows: An unmarked deletion was created in *cycA* the cytochrome c_{550} gene, by removal of an internal fragment, filling-in and re-ligation, and this was cloned in pRVS1. The recombinant plasmid was then introduced by conjugation into the *P. denitrificans* mutant PD2121, in which *cycA* had been disrupted by insertion of a kanamycin-resistance cassette. Exconjugants which had integrated the construct after a single recombination event were selected as kanamycin- and streptomycin-resistant and were blue on plates containing X-gal ($20 \; \mu g \; ml^{-1}$) due to expression of β-galactosidase. These cells were scraped from plates and resuspended in liquid medium, diluted to 10^5 cells per millilitre, spread on plates without any antibiotic, and allowed to form colonies. At this stage a small proportion of the integrants (about 1 in 5×10^4) underwent a second recombination event eliminating pRVS1. The colonies were then harvested from the plates, resuspended in liquid medium, diluted to 10^5 per millilitre, and spread on to media without any antibiotics but containing X-gal at $80 \; \mu g \; ml^{-1}$. This concentration of X-gal inhibited the growth of integrant strains, presumably due to the toxicity of the indolyl derivative formed; under these conditions, cells retaining integrated pRVS1 formed tiny blue colonies, while those which had eliminated the plasmid formed large white colonies which were easily isolated. Southern blotting analysis confirmed the introduction of the unmarked mutation into the *P. denitrificans* genome.

The multiple unmarked mutation approach has revealed the remarkable ability of different *c*-type cytochromes of *P. denitrificans* to substitute for one another in facilitating electron transport from the various dehydrogenases to the terminal oxidases. A series of double mutants carrying unmarked mutations in the genes for cytochrome c_{550} and cytochrome c_{552} (*cycA cycM*), cytochrome c_{550} and cytochrome c_1 (*cycA fbcC*) or cytochrome c_{552} and cytochrome c_1 (*cycM fbcC*) were all able to grow with methanol and methylamine. Only when the genes for all three c-type cytochromes were disrupted ($cycA$ $cycM$ $fbcC$) was the mutant strain unable to utilize these substrates (de Gier et $al.$, 1995; unpublished data cited in van Spanning et $al.$, 1995b).

This approach has also settled the question of the number of terminal oxidases assembled by $P.$ $denitrificans$; up to five have been reported or suggested to exist on the basis of spectroscopy of whole cells and/or membranes (Ludwig, 1987; Poole, 1988). De Gier et $al.$ (1994) constructed a mutant strain lacking the aa_3-type cytochrome oxidase by unmarked deletion of both $ctaDI$ and $ctaDII$; this strain was then made the recipient in a third round of mutagenesis to create the strain ($\Delta ctaDI$ $\Delta ctaDII$ $qoxB::kan$), lacking both the aa_3-type cytochrome oxidase and the quinol oxidase. The fact that this mutant strain was viable indicated that $P.$ $denitrificans$ possessed a third terminal oxidase; this was purified from the triple mutant strain and shown to be a second cytochrome c oxidase, cytochrome cbb_3 (de Gier et $al.$, 1994, 1996). A triple mutant strain lacking both the aa_3-type and cbb_3-type cytochrome oxidases $\Delta ctaDI$ ($\Delta ctaDII$ $ccoNO::kan$) was then constructed; this was unable to oxidize cytochromes c (van der Oost et $al.$, 1995; de Gier et $al.$, 1996). This observation, taken together with the observation that a mutant lacking the QoxABCD quinol oxidase ($qoxB::kan$) was devoid of quinol oxidase activity, indicates that these three oxidases represent all the terminal oxidases assembled by $P.$ $denitrificans$ (de Gier et $al.$, 1994, 1996).

The replacement of marked mutations with unmarked ones is also necessary when it is important that the gene disruption should be non-polar, that is, it should not negatively affect the expression of downstream genes. This consideration is highly relevant to analysis of complex systems such as respiration and photosynthesis because of the clustering of functionally related genes. The value of replacing marked mutations with unmarked ones has been clearly documented for the $P.$ $denitrificans$ methanol oxidation (mox) gene cluster, $moxFJGIR$. Of these genes, $moxF$ and $moxI$ code for the large (α) and small (β) subunits of MDH, respectively, while $moxG$ encodes the cytochrome c_{551} apoprotein. The $moxJ$ and $moxG$ genes were first disrupted by insertion of a kanamycin-resistance cassette; these were then used to construct strains in which both $moxJ$ and $moxG$ were disrupted by unmarked internal deletions. Both the $moxJ::kan$ and $moxG::kan$ and the $\Delta moxJ$ and $\Delta moxG$ strains were analysed. No MDH β subunit could be detected in either the $moxJ::kan$ or the $moxG::kan$ strains, but was present in both the $\Delta moxJ$ and $\Delta moxG$ strains, clearly indicating that insertion of the kanamycin resistance cassette in $moxJ$ or $moxG$ eliminated expression of $moxI$, while unmarked deletions in either gene did not. Similarly, cytochrome c_{551i} was present in the $\Delta moxJ$ strain but absent in the $\Delta oxJ::kan$ strain (van Spanning et $al.$, 1991b).

Of these two unmarked mutations, the $moxG$ unmarked mutation was created by excision of a 175 bp internal BamHI fragment followed by religation. This manipulation created a frameshift mutation and introduced a new stop codon 80 bp upstream of the original stop codon. In contrast, the $moxJ$ unmarked mutation was created by replacement of an internal 48 bp NarI fragment with a 60 bp fragment composed of parts of the polylinkers of pUC4K and pUC19. The central portion of *moxJ* was thus removed and replaced with a segment of DNA such that the coding region remained in frame and no new stop codons were introduced.

In general, the creation of an in-frame deletion and/or insertion in a gene is to be preferred over a mutation creating a frameshift mutation which results in premature termination of translation; in *E. coli* it has been demonstrated that while some (though not all) of the latter class of mutation are polar, the former are not (Adhya and Gottesman, 1978, and references therein; Armengod *et al.*, 1988). In the event, the Δ*moxG* mutation was not polar on *moxI* (van Spanning *et al.*, 1991b); nor were unmarked mutations in either the *P. denitrificans mauC* or *mauJ* genes polar on the downstream *mauG*, despite creating premature stop codons 52 bp upstream of the original *mauC* stop codon and 90 bp upstream of the original *mauJ* stop codon, respectively (van der Palen *et al.*, 1995). Frameshift mutations which do not lead to premature termination of translation are also non-polar (Adhya and Gottesman, 1978).

Insertion mutations can also be non-polar if the inserted segments of DNA provide new promoters for the transcription of downstream genes. This was clearly demonstrated by Moreno-Vivian *et al.* (1989) during analysis of the *R. capsulatus nifENX-orf4-orf5-nifQ* operon. Pairs of *nifE::kan, nifN:kan, nifX::kan* and *orf4::gm* mutant strains were constructed in which the antibiotic resistance cassettes were orientated such that the direction of transcription of the antibiotic resistance genes was either the same as the direction of transcription of the *nifENXQ* genes, or opposed it. To investigate the effects of these various mutations on the expression of *nifQ*, an internal fragment of *nifQ* was in each case substituted with a cartridge containing the promoterless *lacZYA* operon from *E. coli* and a tetracycline resistance marker. In all cases insertion of the antibiotic resistance cassettes such that the direction of transcription of the antibiotic resistance genes opposed that of the *nifENXQ* genes, expression of *nifQ::lacZ* was eliminated. When the cassettes were orientated such that the direction of transcription of the antibiotic resistance genes was the same as that of the *nifENXQ* operon, inducible expression of *nifQ–lacZ* from the *nifENXQ* promoter was replaced by constitutive expression from the *kan* or *gm* promoters.

The level at which downstream genes are expressed may be high or low and appears to be dependent both on the cassette used and on the context. In the system described above, the level of *nifQ–lacZ* expression was highly dependent on the cassette used; a kanamycin-resistance cassette created by an internal *XhoI* deletion of Tn5, and the gentamycin resistance cassette from R1033 supported only a low level of *nifQ–lacZ* expression, but a very high level of expression was obtained when an internal *XhoI* fragment from Tn5 was used as a kanamycin-resistance cassette. When the same Tn5-internal *XhoI* fragment was to used to disrupt *ccoN* in the *R. capsulatus ccoNOQP* gene cluster, however, CcoO and CcoP were only very weakly expressed (Thöny-Meyer *et al.*, 1994).

In some cases useful levels of expression of downstream genes are not obtained by this strategy. A *norQ::kan* mutant constructed during a study of the *P. denitrificans norCBQDEF* gene cluster could not be complemented by *norQ* alone, suggesting that the presence of the Tn*903* kanamycin cassette in *norQ* had a negative effect on the expression of the downstream *norD*, *norE* and *norF* genes, despite its being orientated such that the *kan* promoter could potentially support their transcription (de Boer *et al.*, 1995).

3. Gene deletions with simultaneous rescue

Gene deletion clearly presents a problem when the gene of interest is an essential one; often genes can be deleted from *Rhodobacter and Paracoccus* as they are metabolically versatile, but there are some exceptions. In such a case, strains must somehow be constructed to allow the deletion, sometimes by introducing genes encoding products with properties similar to, but distinguishable from, the essential one under study. Deletion of the *P. denitrificans* NADH dehydrogenase subunit structural genes *nqo8* and *nqo9* was found to be lethal. So as a preliminary to a site-directed mutagenesis study of these genes, Finel (1996) used homologous recombination simultaneously to delete the *P. denitrificans nqo8* and *nqo9* genes and to integrate the *E. coli ndh* gene, encoding the NADH dehydrogenase II of this organism, into the *P. denitrificans* genome. Expression and assembly of the *E. coli* Ndh allowed construction of a viable Δ*nqo8-nqo9* strain of *P. denitrificans* Borghese and co-workers (1998), studying the vital *atpNAGDC* operon of *Rhodobacter capsulatus*, which encodes the F_1 sector of the ATP synthase, took a different approach and used strains, pre-conjugated with plasmid containing the *atp* operon as recipients for chromosomal deletion of those genes using a phage gene transfer agent (GTA) (Marrs, 1974).

4. Use of deletion strains

Deletion strains *of Rhodobacter* and *Paracoccus* have been used for a variety of experiments in assigning function to genes encoding photosynthesis and respiration properties. Characterization of gene deletion mutants has provided evidence supporting structural predictions made for individual polypeptides of a multi-enzyme complex. The *P. denitrificans* NADH dehydrogenase subunit II polypeptide (Nqo2) is always found as part of the membrane-associated complex, but analysis of the *nqo2* gene sequence data indicates that Nqo2 is an entirely hydrophilic protein which is unlikely to interact directly with the cytoplasmic membrane. The observation that Nqo2 was found in the *P. denitrificans* cytoplasm when genes downstream of *nqo1* were disrupted strongly suggests that the Nqo2 structure prediction was correct and that in the wild-type complex Nqo2 is tightly bound by other, membrane-integral, components (Yano *et al.*, 1994).

Gene deletion has also been used to construct strains having a suitable genetic background for PCR with degenerate primers. As the structural genes for a number of bacterial oxidases, including the *P. denitrificans* cytochrome *aa*$_3$ (see above) and the best-characterized quinol oxidase,

441

E. coli cytochrome bo_3, were cloned and sequenced, it became clear that the majority were members of a superfamily of haeme–copper oxidases (Saraste, 1990). On the basis of this information, de Gier *et al.*, (1994) designed degenerate primers, to conserved regions of subunit I of the haeme–copper oxidases, with the aim of amplifying by PCR a segment of the structural gene for subunit I of the *P. denitrificans* quinol oxidase. To avoid the obvious problem of competing amplification of segments of the cytochrome aa_3 subunit I structural genes *ctaDI* and *ctaDII*, genomic DNA from the D*ctaDI* D*ctaDII* mutant (see above) was used as the template for PCR. This approach led to the cloning of the *P. denitrificans* quinol oxidase structural genes essentially simultaneously with Richter *et al.* (1994; see above).

As has been stressed, *Paracoccus* species contain multiple cytochromes and *Rhodobacter* species contain multiple chlorophyll- and carotenoid-containing protein complexes. The use of selectively deleted strains has allowed purification of specific complexes from a background devoid of other related pigments/cytochromes. For example, *P. denitrificans* oxidase mutants have been used for oxidase purification, an *fbcFBC* deletion mutant over-expressed the quinol oxidase, facilitating purification of this enzyme (Richter *et al.*, 1994), while the cbb_3-type oxidase (CcoNOQP) is the only oxidase expressed in a Δ*ctaDI* Δ*ctaDII* *cyoB::Km* mutant (de Gier *et al.*, 1996). In *R. sphaeroides* Jones and co-workers (1992) have used multiply deleted strains to express, purify and spectrally characterize, single photosynthetic complexes away from other complexes in the cell with which they interact. *R. sphaeroides* strains lacking combinations of the *puc*AB, *puf*BA, *puf*LM and *puh*A genes (which encode the protein components of light-harvesting II, light-harvesting I, reaction centre L and M, and reaction centre H subunits respectively) have been used in this way.

Similarly, gene deletion has been used to create a suitable genetic background for expression of site-directed mutant genes in the homologous host; recombination between mutant alleles introduced on plasmid vectors and the wild-type form of the gene must obviously be avoided. Gerhus *et al.* (1990) replaced the *P. denitrificans* *fbcC* gene with a kanamycin-resistance cassette as a preliminary to introducing site-directed mutant alleles; similarly, Yun *et al.* (1990) replaced the *R. sphaeroides* *fbcFBC* operon with a kanamycin-resistance cassette prior to complementation with the *fbcFBC* operon containing a mutant *fbcB* allele, while Farchaus and Oesterhelt (1989) replaced the *R. capsulatus* *puf*LMX gene region with a kanamycin resistance cassette as a preliminary to introducing site-directed mutant alleles of *puf*M and *puf*L (Gray *et al.*, 1990; Wachtveitl *et al.*, 1993). An alternative strategy is replacement of the wild-type gene with the site-directed mutant allele by homologous recombination (Gong *et al.*, 1994; Knäblein *et al.*, 1997).

Site-directed mutagenesis has been most often applied to determine the ligands to metal centres in electron transport components of *Rhodobacter* and *Paracoccus* (e.g. Yano *et al.*, 1994) or non-liganding residues essential to protein structure, (e.g. Liebl *et al.*, 1997). Site-directed mutagenesis has been used extensively to determine structure and function in the *Rhodobacter* photosynthetic apparatus (Goldman and Youvan,

1995). Site-directed mutants can however be useful for protein purification studies: Konishi *et al.* (1991) used site-directed mutagenesis of the *R. sphaeroides fbcC* gene to produce and study a truncated soluble cytochrome c^1 which lacked the C-terminal membrane anchor domain. This strategy is interesting because, in principle, the water-soluble domains of membrane-associated subunits of respiratory complexes should be more amenable than the native proteins both to study by a range of physical techniques and to crystallization.

C. Heterologous Expression of Genes in *E. coli*

I. Reasons and methods

While we have previously mentioned that many respiratory enzymes and electron carriers in *Paracoccus* and *Rhodobacter* are expressed at sufficiently high levels to allow their study without cloning and over-expression, regulatory proteins and the proteins involved in prosthetic group biosynthesis (e.g. haeme, bacteriochlorophyll) are not highly expressed in the parent organism. Expression and purification of the products of such genes in *E. coli* have allowed, for example, demonstration *in vitro* of the enzymatic activities of the bacteriochlorophyll biosynthesis gene products BchM of *R. capsulatus* and BchM, BchH, BchI and BchD of *R. sphaeroides* (Bollivar *et al.*, 1994; Gibson and Hunter, 1994; Gibson *et al.*, 1995), and DNA footprinting studies with the *P. denitrificans* transcription regulator MauR (Delorme *et al.*, 1997).

Furthermore, respiratory enzymes are generally complex, consisting of a number of different polypeptides and having a number of different prosthetic groups. While it is possible in many cases to deduce which subunits contain which prosthetic group on the basis of sequence data, the expression and analysis of individual polypeptides is a powerful tool for the assignment of a particular prosthetic group to a particular subunit. For example, the *P. denitrificans* NADH dehydrogenase has 14–15 subunits and contains non-covalently bound FMN and at least five EPR-visible iron–sulphur clusters (Yagi, 1993). Expression of the Nqo1, Nqo2, Nqo3, Nqo4, Nqo5 and Nqo6 gene products in *E. coli* has allowed the assignment of iron–sulphur clusters to Nqo1 ([4Fe–4S]), Nqo2 ([2Fe–2S]), Nqo3 ([4Fe–4S] and [2Fe–2S]) and possibly to Nqo6 (Yano *et al.*, 1994, 1995, 1996; Takano *et al.*, 1996). It also allows the properties of individual prosthetic groups or metal centres to be studied in isolation. *E. coli* appears to be able to assemble many, although not all, of the prosthetic groups found in *Paracoccus* and *Rhodobacter* sp. respiratory chain proteins. The expression of individual polypeptides in *E. coli* has helped define open reading frames within a complex gene region.

Using *E. coli* promoters, moderate levels of expression can be achieved. Placing the *P. denitrificans* electron transfer flavoprotein structural genes under the control of the *lac* promoter resulted in yields of 30 mg l^{-1} of this heterodimeric protein (Bedzyck *et al.*, 1993), while using the T7 promoter system to drive *P. pantotrophus pazS* expression yielded up to 80 mg l^{-1}

Paracoccus and Rhodobacter

pseudoazurin (Leung *et al.*, 1997). Placing the *P. denitrificans cycA* gene under the control of the *tac* promoter has recently resulted in yields of up to 10 mg l^{-1} cytochrome c_{550} (Koffenhoffer and Ferguson, unpublished; see below).

Differences in codon usage (Steinrücke and Ludwig, 1993; see above) have not proved a barrier to the expression of *Paracoccus* or *Rhodobacter* genes in *E. coli*; however, transcription from *R. sphaeroides* promoters is low or undetectable in *E. coli* (Hallenbeck and Kaplan, 1987; Gerhus *et al.*, 1990), so that a suitable promoter must be provided and often the exisiting GC-rich promoter region of the gene deleted to prevent secondary structure formation and expression problems. Deletion of an inverted repeat region upstream of the *P. pantotrophus pazS* gene resulted in a twofold higher yield of pseudoazurin when the gene was expressed in *E. coli* (Leung *et al.*, 1997), while high-level expression of MauR in *E. coli* was achieved only after PCR amplification of the minimal *mauR* gene, using a forward primer which introduced an *Nde*I site at the position of the *mauR* ATG start codon allowing cloning in an expression vector without any upstream DNA (Delorme *et al.*, 1997). A similar strategy has been employed for expression in *E. coli* of *P. denitrificans* Nqo1, Nqo2, Nqo3, Nqo4, Nqo5 and Nqo6 (Yano *et al.*, 1994, 1995, 1996; Takano *et al.*, 1996), *R. capsulatus* HupT (Elsen *et al.*, 1997), and bacterioferritin (Penfold *et al.*, 1996), and *R. sphaeroides* BchH and BchI (Gibson *et al.*, 1995).

Paracoccus and *Rhodobacter* proteins expressed in *E. coli* appear to fold correctly such that they are competent for binding prosthetic groups and/or other polypeptides, and/or are catalytically active (Grabau *et al.*, 1991; Bedzyk *et al.*, 1993; Armengaud and Jouanneau, 1995; Kanazireva and Biel, 1995; Oriol *et al.*, 1996; Pollock and Barber, 1997; Elsen *et al.*, 1997). When the *P. denitrificans* Nqo1 and Nqo2 were expressed in *E. coli*, they associated to form a homodimer and the [2Fe–2S] centre of Nqo2 was assembled; the [4Fe–4S] centre of Nqo1 was not assembled and it contained no FMN, but both could be reconstituted *in vitro* (Yano *et al.*, 1996). When the *R. sphaeroides* BchH, BchI and BchD were expressed separately in *E. coli*, and crude cell-free extracts of the three strains mixed, the three gene products assembled to form the chlorophyll biosynthesis enzyme magnesium chelatase (Gibson *et al.*, 1995).

2. Problems with expression in *E.coli*

Some problems have arisen with heterologous expression; the *R. sphaeroides* DMSO reductase, which has the same MGD prosthetic group as biotin sulphoxide reductase, was produced only in an insoluble inactive form (Hilton and Rajagopalan, 1996) when expressed in *E. coli*, and the *R. capsulatus* xanthine dehydrogenase cannot be expressed in an active form in *E. coli* because it does not produce the enzymes's molybdopterin co-factor (Rajagopalan, 1996).

The expression of *Paracoccus* and *Rhodobacter* c-type cytochromes in *E. coli* has also proved difficult, due to the apparent inability of this organism efficiently to perform the ligation of haeme to apocytochrome polypeptides (McEwan *et al.*, 1989; Grisshammer *et al.*, 1991). Most *E. coli* strains synthesize only very low levels of endogenous *c*-type cytochromes, although synthesis of *Paracoccus versutus* holocytochrome c_{550} at moderate levels (1–2 mg l^{-1}) has been observed in *E. coli* W3110 (Ubbink *et al.*, 1992); however, an *E. coli* K12 strain has recently been described which supports the assembly of high levels of both endogenous and exogenous *c*-type cytochromes (JCB712; Iobbi-Nivol *et al.*, 1994). The basis for this characteristic is unknown, but it may relate to upregulation of the cytochrome *c* biosynthetic genes and/or haeme biosynthesis (Sinha and Ferguson, 1998). High levels of expression and assembly have been achieved for *P. denitrificans* cytochrome c_{550} (up to 10 mg l^{-1}), and moderate levels for the semiapo form (having covalently bound *c*-type haeme but no haeme d_1) of the *P. pantotrophus* cytochrome cd_1 nitrite reductase (about 1 mg l^{-1}). The expressed protein contains only the haeme *c* centre because *E. coli* cannot elaborate the d_1 haeme, but it appears to be correctly folded because it can be reconstituted with d_1 haeme isolated from the native enzyme to 90% of the expected specific activity (Koffenhoffer and Ferguson, unpublished).

D. Gene Expression in *Paracoccus* and *Rhodobacter*

1. *Paracoccus* and *Rhodobacter* as heterologous hosts

Because of the limitations associated with the use of *E. coli*, *P. denitrificans* and *R. sphaeroides* have been used as heterologous hosts as a host for the expression of "each other's" gene products, especially for assembly of holo-*c*-type cytochromes. While expression of the *Rhodopseudomonas viridis cycA* gene in *E. coli* led to accumulation of membrane-bound pre-apocytochrome c_2 (Grisshammer *et al.*, 1991), its expression in *P. denitrificans* led to formation of the soluble periplasmic holoprotein (Gerhus *et al.*, 1993); similarly, when the *Desulfovibrio vulgaris* (strain Hildenborough) tetrahaeme cytochrome c_3 gene was expressed in *E. coli* only the apoprotein accumulated (Pollock *et al.*, 1989), but when it was expressed in *R. sphaeroides* the holoprotein was correctly assembled (Cannac *et al.*, 1991). It may that *P. denitrificans* and *R. sphaeroides* can more easily assemble exogenous cytochromes *c* because they themselves assemble a wide variety of endogenous ones.

Rhodobacter mutant strains have been used as heterologous hosts in the cloning of some photosynthesis genes from plants. The enzyme phytoene synthase (Psy) catalyses the formation of phytoene, an intermediate in the carotenoid biosynthesis pathway of both *Rhodobacter* and tomato (*Lycopersicon esculentum*). Cloning of a partial cDNA for a tomato phytoene synthetase, PSY2, was achieved by complementation of a *Rhodobacter capsulatus* phytoene synthase mutant, with a plasmid containing a cDNA fragment under the control of a plasmid promoter (Bartley and Scolnick, 1993). Recently, a *R. sphaeroides* mutant defective in an oxygen sensor protein was complemented with plasmid expressing a rat liver mitochondrial drug receptor protein (Yeliseev *et al.*, 1997). Other heterologous genes expressed in *R. capsulatus* include *E. coli* cytochrome b_{562}, rat cytochrome b_5 and the *Ascaris* haemoglobin domain 1, which have been used as probes of haeme export to the periplasm (Goldman *et al.*, 1996).

2. Regulated over-expression of homologous genes

Although genes can be readily cloned into these bacteria, regulating their level of expression, and to some extent achieving high levels of expression, remain problematic. Obviously workers wish to over-express and study proteins in homologous hosts (especially where metal–protein assembly may be difficult in a heterologous host); expression should be ideally regulatable to prevent lethality. There is at present no established "good" system for the regulated over-expression of homologous genes in *P. denitrificans* or *Rhodobacter*. In this section we will review what is available and give examples of how each system has been applied.

One problem is that the promoters used for gene expression in *E. coli* are inactive or only weakly active in *Rhodobacter* and *paracoccus*. The *tac* and P_L promoters are reportedly non-functional in *P. denitrificans* (unpublished results cited in Harms and van Spanning, 1991). The *lac* promoter is constitutively functional to some extent in *R. sphaeroides*, the level of expression of the *lac* operon cloned in a broad host range vector being judged to be 50% of the uninduced level measured in *E. coli*; the *lac* promoter was not IPTG inducible, so this figure represents its maximal activity (Nano and Kaplan, 1982). The *H. thermophilus* cytochrome c_{552} was expressed in *P. denitrificans* to a level sufficient for easy detection by haeme staining from a construct in which it was under the control of both the *lac* and *tac* promoters (Sambongi and Ferguson, 1994). The vector pRK415 has been successfully used for functional complementation of *P. denitrificans* *c*-type cytochrome biogenesis mutants with promoterless *c*-type cytochrome biogenesis genes; however the level of expression (presumed to be driven by the vector *lac* and/or *tetR* promoters) was low (Page, Tomlinson and Ferguson, unpublished results).

Rhodobacter sp. promoters appear to work well in *P. denitrificans* (Gerhus *et al.*, 1993), but not strongly enough to give high levels of over-expression. The *Pseudomonas putida* TOL plasmid *Pm* promoter is also functional but not strong in *P. denitrificans* (Mermod *et al.*, 1986). *Rhodobacter* or *Paracoccus* genes, cloned along with their own promoters in broad host range plasmids and reintroduced into the source organism, may be over-expressed to some extent as a consequence of the increased gene dosage. An *R. capsulatus* strain carrying the *cycA* gene cloned in pRK404 contained 3.5 to 8.0 times more cytochrome c_2 than the plasmidless host, in line with the copy number of this vector (Brandner *et al.*, 1989), while all three subunits of the cytochrome bc_1 complex were overexpressed several-fold when a *P. denitrificans* *fbcC::kan* mutant was complemented with the *fbcFBC* operon cloned in pEG400 (Gerhus *et al.*, 1990). Similarly, mannitol dehydrogenase was over-expressed 3.4-fold when the *mtlK* gene cloned in pRK415 was introduced into *R. sphaeroides* (Schneider and Giffhorn, 1994). One of the stronger endogenous promoters in *P. denitrificans* is the *cycA* promoter which supports the synthesis of approximately 20 mg L^{-1} cytochrome c_{550} during anaerobic growth. It is also regulated; cytochrome c_{550} is expressed to some extent during aerobic growth, but is induced 6.5-fold during anaerobic growth with nitrate as terminal electron acceptor and 10 to 25-fold during aerobic growth on methanol or methylamine, and to a lesser extent during growth on choline (Stoll *et al.*, 1996).

The *P. denitrificans cycA* promoter has been used to drive the expression of a *P. denitrificans ccmG–E.coli phoA* fusion, after it was found that the endogenous *ccmG* promoter known to be present in the construct was insufficiently active for alkaline phosphatase activity to be detected in cell extracts (Page and Ferguson, 1997), and to drive the expression of a *cycA–napC* fusion in *P. denitrificans* (Roldán *et al.*, 1997). In this case the periplasmic domain of the membrane-anchored tetrahaeme *c*-type cytochrome NapC was expressed as a soluble periplasmic protein by replacement of the 5′ region of *napC*, encoding the N-terminal membrane anchor, with the region of *cycA* encoding the cleavable signal sequence of cytochrome c_{550} plus ten amino acids from the N-terminus of the mature protein. The construct contained the *cycA* promoter and regulatory regions. The level of the soluble CycA–NapC fusion present in recombinant cells was higher than that of authentic NapC, but only by a factor of two- to threefold.

In *Rhodobacter* some endogenous promoters have been employed recently for over-expression but there is still nothing to match the inducibility of *lac* in *E. coli*. The *R. sphaeroides cycA* promoter has been used for the homologous over-expression of the chemotaxis signalling protein CheW (Hamblin *et al.*, 1997).

One of the most highly regulated *Rhodobacter* promoters studied is the *R. capsulatus nifHDK* promoter (Pollock *et al.*, 1988). This promoter is totally repressed during growth in the presence of ammonium (and/or oxygen) and is induced approximately 2600-fold during diazotrophic growth (i.e. anaerobically in the absence of a nitrogen source other than N_2); intermediate levels of activity are observed during anaerobic growth with various amino acids as the sole (poor) nitrogen source. These authors constructed an expression vector, pNF3, comprising the *nifHDK* promoter, a multiple cloning site and a spectinomycin/streptomycin resistance marker. Bauer and Marrs (1988) demonstrated regulatable expression of the homologous *pufQ* gene from this vector in *R. capsulatus*. Measurement of the strength of the *nifHDK* promoter using a promoterless *lacZ* gene as reporter suggested that the derepressed *nifHDK* promoter was as active as the *pufQ* promoter (Pollock *et al.*, 1988); however, it is possible that the requirements of anaerobiosis and nitrogen limitation for gene expression in this system may be a disadvantage in some applications.

Duport *et al.* (1994), noting that the *R. capsulatus* fructose utilization operon, *fruBKA*, was induced more than 100-fold in the presence of fructose, constructed a cassette comprising the *fruBKA* promoter, a multiple cloning site and a gentamycin-resistance marker suitable for cloning the gene of interest and subsequent cloning of the cassette in a broad host range vector of choice. The authors cloned the *R. capsulatus fdxN* gene into this cassette and then cloned the cassette into pRK404 to create a vector which they used to demonstrate fructose-dependent complementation of a *R. capsulatus fdxN* mutant. An assessment of the relative strengths of the *fruBKA* and *nifHDK* promoters using a promoterless *lacZ* gene as reporter suggested that the *fruBKA* promoter may be two- to threefold weaker than the *nifHDK* promoter, while measurement of FdxN suggested that it might be tenfold less active (Jouanneau *et al.*, 1995). Activation of the *fruBKA* promoter requires the addition of fructose; disappointingly, despite this simple requirement, the system was found not to be active in *R. sphaeroides* WS8 which does grow on fructose (Shah and Sockett, unpublished); it has yet to be tested in *P. denitrificans*.

Frey *et al.* (1988) have constructed pαΩ, an IncQ vector carrying the very strong T4 phage gene 32 promoters and its terminator. Measurements of promoter activity indicated that the gene 32 promoters were as strongly transcribed in *P. denitrificans* as in *E. coli*, but it has yet to be used experimentally in this organism. The IncQ vector pMMB66EH (Fürste *et al.*, 1986) also incorporates transcription terminators downstream of the multicloning site; efficient transcription termination can increase both the stability of the expression vector and the half-life of the transcribed mRNA, enhancing protein yield (Makrides, 1996, and references therein). More recently, the *R. capsulatus* xanthine dehydrogenase has been over-expressed in the homologous host by placing the structural genes (*xdhAB*) under the control of the Tn5 *Kan* promoter; the enzyme reportedly comprised at least 0.05% of total cell protein (Leimkuhler *et al.*, 1998). The Tn903 *kan* promoter, which is highly transcribed in *R. sphaeroides* may have similar potential (Donohue and Kaplan, 1991).

Clearly, the technology for gene (over)expression in *Paracoccus* and *Rhodobacter* is only beginning to be developed. In the *Rhizobiaceae* this problem has been approached by the construction of a synthetic consensus IPTG-inducible promoter cassette (Giacomini *et al.*, 1994). It seems probable that this promoter may be active in *Paracoccus* and *Rhodobacter*; however the consensus promoter sequence in *Paracoccus* and *Rhodobacter* is now sufficiently well defined (Cullen *et al.*, 1997; Baker *et al.*, 1998) to allow the construction of a similar synthetic promoter, optimized for expression in these genera, to be attempted.

E. Reporters of Gene Expression in *Rhodobacter* and *Paracoccus*

I. Reasons and methods

There has always been considerable interest in regulation of gene expression in *R. capsulatus* and *R. sphaeroides*; pigmentation and growth properties of the bacteria change radically in response to changes in oxygen tension and light intensity. Synthesis of Bchl, haem, carotenoids and pigment binding proteins are elaborately regulated in response to these stimuli (Bauer and Bird, 1996; Zeilstra-Ryalls *et al.*, 1998). The respiratory systems of *P. denitrificans* also respond to environmental stimuli (van Spanning *et al.*, 1995, and references therein; Baker *et al.*, 1998).

Additionally, reporters are required where direct assays of gene product levels are difficult. For example, although many *Rhodobacter* photosynthetic pigments can be quantified spectrophotometrically, there are problems of spectral overlap. *Paracoccus and Rhodobacter* cytochromes are particularly hard to quantify by haeme staining and for the majority of the regulatory proteins in these bacteria there are no stains or spectral assays. For these reasons, reporter gene technologies have been very important in characterizing levels of gene expression under various environmental conditions.

The *E. coli lacZ* gene has been widely used to construct both transcriptional (to promoterless *lacZ*) and translational (to 5'-truncated *lacZ*) fusions to genes of interest in *Paracoccus denitrificans* and *Rhodobacter* sp.; *lacZ* fusions of both types have been extensively used as reporters of gene expression (for example Zeilstra-Ryalls and Kaplan, 1995), while vectors carrying the promoterless *lacZ* gene have been used to identify promoters (Harms *et al.*, 1989), and to study the regulation of cloned promoters (Kutsche *et al.*, 1996).

Broad host range IncP1 and IncQ vectors (Chapter 2, III.c, IV and V) for the construction of transcriptional *lacZ* fusions include the IncP1 vectors pGD499 and pGD500 (Ditta *et al.*, 1985), pZM400 (Ma *et al.*, 1993), pMP220 (Spaink *et al.*, 1987), pRW2 (Lodge *et al.*, 1990); and the IncQ vectors pMP190 (Spaink *et al.*, 1987), the pML5-pML9 series (Labes *et al.*, 1990) and pHRP309 (Parales and Harwood, 1993). Broad host range vectors for the construction of translational *lacZ* fusions include the IncP1 vectors pGD969 (Ditta *et al.*, 1985), pPHU234, pPHU235 and pPHU236 (Hübner *et al.*, 1991), and the IncQ vectors pUI108 (Nano *et al.*, 1985), pUI523A and pUI533A (Tai *et al.*, 1988) and the pML103-pML107 series (Labes *et al.*, 1990). pCB303 is a specialized IncP1 transcriptional fusion vector developed for the analysis of bidirectional promoters and carries divergently orientated promoterless *lacZ* and *phoA* genes separated by a multiple cloning site (Schneider and Beck, 1987; see below for *phoA* fusions).

Of the vectors listed, pGD499, pGD500, pGD969, pPHU234/5/6, pCB303, pHRP309 and pZM400 and pML5 have been used or shown to function in *R. capsulatus* (Ditta *et al.*, 1987; Hübner *et al.*, 1991; Richaud *et al.*, 1991; Armstrong *et al.*, 1993; Parales and Harwood, 1993; Pollich and Klug, 1995; Kutsche *et al.*, 1996; Lang *et al.*, 1996; Pasternak *et al.*, 1996; Nickel *et al.*, 1997; Leimkuhler *et al.*, 1998); pPHU234/5/6, pCB303, pUI108, pUI523A/33A and pHRP309 have been used or shown to function in *R. sphaeroides* (Nano *et al.*, 1985; Tai *et al.*, 1988; Parales and Harwood, 1993; Pasternak *et al.*, 1996), and pGD969, a pMP220 derivative, and pRW2 (Spiro, 1992) have been used in *P. denitrificans* (Ueda *et al.*, 1996). Some of these vectors exhibit a "non-zero" basal level of *lacZ* transcription in *R. sphaeroides* and *P. denitrificans*; the standard solution is insertion of an omega antibiotic-resistance cassette (Fellay *et al.*, 1987) upstream of the promoter under study (e.g. Tai *et al.*, 1988; Varga and Kaplan, 1989; Dryden and Kaplan, 1993; Parales and Harwood, 1993).

One criticism often made of analyses of expression using gene fusions cloned in broad host range vectors is that the presence of multiple copies of the regulatory region under study could result in the titration of regulator proteins and thus to erroneous results. One approach which avoids this possibility is integration of gene–reporter fusions into the host chromosome by homologous recombination; integrant strains contain both the gene–reporter fusion and a wild-type copy of the gene and hence retain the prototrophic phenotype, while only two copies of the regulatory regions of interest are present (Hübner *et al.*, 1993; Masepohl *et al.*, 1993; Delorme *et al.*, 1997) An alternative approach is to integrate the gene–reporter fusion into the genome at random sites; Calero *et al.* (1994) used mini-Tn5 (Chapter 4, XI) as a vehicle for the random integration of various *recA–lacZ* fusions into *R. sphaeroides* chromosomal DNA; fusions were cloned into the unique *Not*I site of the transposon. As with the homologous recombination method, only two copies of the *recA* regulatory regions were present in the recombinant strains, and the *recA* gene itself remained intact.

2. Using reporters to identify regulators of gene expression

Many overlapping regulatory circuits control the regulation of expression of genes encoding electron transport components (for example see Zeilstra-Ryalls *et al.*, 1998); many of these have been dissected using reporter genes. In some cases, due to the highly conserved nature of regulatory systems across genera, gene fusions to heterologous promoters have reported activities of cognate regulators. This shows that the approach may be worth trying as new regulators are discovered in different genera by genomic sequencing. For example Spiro (1992) cloned the *E. coli* FNR-dependent *melR* promoter upstream of promoterless *lacZ* in a broad host range vector. When introduced into *P. denitrificans*, the pattern of β-galactosidase expression indicated that *P. denitrificans* possesses a transcriptional regulator which has similar DNA binding specificity to FNR and which responds to a similar physiological signal, i.e. anaerobiosis.

More often, a homologous promoter fusion is used, sometimes along with mutagenesis, to identify regulators of expression of that promoter. Kranz and Hazelkorn (1985) used a *nifH–lacZ* fusion cloned in a broad host range vector to screen *R. capsulatus* Tn5 Nif⁻ mutants for inability to express β-galactosidase under conditions which would normally de-repress the *nifH* promoter. They identified nine Tn5 mutants, found that they mapped to four loci, and identified four regulatory protein genes; *nifR1*, *nifR2*, *nifR3* and *nifR4*. They then constructed *lacZ* fusions to each of these to determine what environmental signal they responded to. Bauer and Marrs (1988) placed the putative *R. capsulatus* regulatory protein gene *pufQ* under the control of the tightly regulated *nifHDK* promoter on a broad host range vector, and used a *lacZ* fusion to the downstream *pufM* gene to measure directly the effect of *pufQ* transcription under various expression regimes. In this way they were able to show that bacteriochlorophyll levels in a *R. capsulatus* strain carrying this plasmid correlated with the level of *pufQ* expression, confirming that *pufQ* was a regulatory protein controlling bacteriochlorophyll biosynthesis Flory and Donohue (1997) made *lacZ* fusions to *R. sphaeroides* ctaD, coxII and cycFG, and observed co-ordinate induction on anaerobiosis.

An alkaline phosphatase fusion has recently proved useful in the analysis of cytochrome c_{550} expression in *P. denitrificans*. As previously described, disruption of the cytochrome c_{550} structural gene, *cycA*, does not affect methylotrophic or anaerobic growth of this organism (van Spanning *et al.*, 1990a); this was unexpected in view of biochemical evidence that cytochrome c_{550} was likely to be an important electron carrier during growth under these conditions. Using a *cycA–phoA* translational fusion, it was found that cytochrome c_{550} expression was markedly (6.5- to 25-fold) higher during anaerobic heterotrophic and methylotrophic growth than during aerobic heterotophic growth. This result strongly implies that cytochrome c_{550} is, as had been expected, important as an electron carrier under these growth conditions and that in the *cycA* mutant one or more other electron transport proteins can take over its role (Stoll *et al.*, 1996).

Other reporters used include the promoterless chloramphenicol acetyl transferase (*cat*) gene (Benning and Somerville, 1992; Penfold and Pemberton, 1994), The *xylE* gene, coding for catechol 2,3-dioxygenase has also been used as a reporter of gene expression in *R. sphaeroides* (Dryden and Kaplan, 1993; Xu and Tabita, 1994). XylE is also functional in *P. denitrificans*. (Frey *et al.*, 1988). At least one IncP and two IncQ promoter probe vectors based on *xylE* have been constructed (pHX300, Xu and Tabita, 1994; pVDX18, Konyecsni and Deretic, 1988; pHX200, Xu *et al.*, 1993b).

3. Reporters of protein localization and transmembrane topology

In *Rhodobacter* sp. and to a lesser extent in *Paracoccus*, a great deal of biochemical activity is centred in the cell membrane and periplasm, more than is usual for *E. coli*. In understanding, for example, the organization of the *Rhodobacter* photosynthetic apparatus in the invaginations of the cytoplasmic membrane, it is essential to know the topology of each gene product which goes to form it (for example Yun *et al.*, 1991). Both LacZ and PhoA are effective reporters of protein localization and transmembrane topology in *P. denitrificans, R. capsulatus and R. sphaeroides* as in *E. coli.*; all three of which exhibit negligible levels of endogenous alkaline phosphatase activity, and *Rhodobacter* sp. have no or low endogenous ß-galactosidase activity. Translational fusions to *lacZ* and *phoA* have been used to analyse the transmembrane topology of components of the *R. capsulatus* c-type cytochrome biosynthetic apparatus (Beckman *et al.*, 1992; Lang *et al.*, 1996; Goldman *et al.*, 1997; Monika *et al.*, 1997), cytochrome c_y (Myllykallio *et al.*, 1997) and PucC (LeBlanc and Beatty, 1996). They have also been used for FbcB and RdxA in *Rhodobacter sphaeroides* (Yun *et al.*, 1991; Neidle and Kaplan, 1992), and FbcB (Yun *et al.*, 1991).

Additionally, measurements of alkaline phosphatase activity after cell fractionation provide an indication as to whether a fusion protein is membrane bound with the alkaline phosphatase domain projecting into the periplasm or is a soluble periplasmic protein (e.g. Sockett and Armitage 1991; Lang *et al.*, 1996; Monika *et al.*, 1997). Alkaline phosphatase fusions have also been useful as reporters of apocytochrome c_2 translocation in *R. sphaeroides* (Varga and Kaplan, 1989; Brandner *et al.*, 1991), and as

Paracoccus and Rhodobacter

reporters of apocytochrome c_2 and apocytochrome c_{550} translocation in mutants of *R. capsulatus* and *P. denitrificans* defective in *c*-type cytochrome biosynthesis (Beckman *et al.*, 1992; Beckman and Kranz, 1993; Lang *et al.*, 1996; Page *et al.*, 1996).

◆◆◆◆◆◆ VII. A SUMMARY "WISH LIST" FOR *RHODOBACTER* AND *PARACOCCUS* GENETIC METHODS

We note that there are many materials and techniques already developed for use in other Gram-negative bacteria, particularly the pseudomonads and the *Rhizobiaceae*, which could be tried in *P. denitrificans* and *Rhodobacter* sp. There are, however, some remaining needs that should be addressed in the future. These include better broad host range vectors that can be conjugated but, that are smaller with more cloning sites; better expression systems with strong promoters whose activity can be induced other than by alteration in levels of nutrients. Development of such tools will allow more workers to take advantage of the unique physiology of these bacteria and to study gene expression in them.

◆◆◆◆◆◆ VIII. GENOMIC STUDIES AND FUTURE PERSPECTIVES

Both *Paracoccus* and *Rhodobacter* species have been subject to pulsed field gel electrophoretic analysis of their genomes. In *R. sphaeroides* this led to the surprising discovery that *R. sphaeroides* has two chromosomes, and established that bacteria may have more than one chromosome (Suwanto and Kaplan, 1989). Since then similar studies on *P. denitrificans* (Winterstein and Ludwig, 1998) have shown it to contain three chromosomes. Programmed sequencing of *R. capsulatus* and *R. sphaeroides* genomes is being carried out (Fonstein *et al.*, 1998, http://capsulapedia.uchicago.edu/; Choudhary *et al.* 1997, http://www-mmg.med.uth.tmc.edu/sphaeroides/), and the *P. denitrificans* genome is slowly being sequenced "piecemeal" by interested groups working on their own gene clusters. It is unfortunate that although all three bacteria have very interesting physiological properties, and potential applications, they are not pathogens, and primary bacterial sequencing funding has been directed at pathogenic bacteria. Information from the genomic sequencing projects has already contributed further to the understanding of photosynthetic genes in higher plants, for example a chlorophyll synthetase gene from *Arabidopsis* has been identified by its homology to a corresponding *Rhodobacter* gene (Gaubier *et al.*, 1995). Even the relatively few *Paracoccus* gene sequences have also contributed; Rasmusson and co-workers (1998) were able to identify a mitochondrial iron–sulphur protein encoding gene on the basis of its homology to a sequence encoded by *Paracoccus*. Workers in other fields are finding that homology searches with *Paracoccus* and *Rhodobacter* sequences can be doubly beneficial, as often not just a sequence but a fully characterized mutant phenotype is available for each gene, due to the ease of mutant rescue by alternative growth modes. What is not possible by mutagenesis (due to lethality) in the mitochondrion, chloroplast, cyanobacterium or eukaryotic cell, can be elucidated by working on corresponding *Paracoccus* and *Rhodobacter* strains (for example Bartley and Scolnick, 1993; Smith *et al.*, 1996; Yeliseev *et al.*, 1997). So, unusually for bacteria, as eukaryotic genomic information grows, work on *Paracoccus* and *Rhodobacter* assumes a greater, not lesser importance. The future for genetic work in these bacteria is bright!

Acknowledgements

RES and MDP thank all authors of unpublished work for permission to cite it.

RES thanks the Royal Society and BBSRC for support. MDP thanks Simon Baker for discussion and comments, and Prof. Stuart Ferguson and the Oxford Centre for Molecular Sciences for support; the OCMS is supported by the BBSRC, EPSRC and MRC.

References

Adhya, S. and Gottesman, M. (1978). Control of transcription termination. *Annu. Rev. Biochem.* **47**, 967–996.

Allen, L. N. and Hanson, R. S. (1985). Construction of broad-host-range cosmid cloning vectors: identification of genes necessary for growth of *Methylobacterium organophilum* on methanol. *J. Bacteriol.* **161**, 955–962.

Antoine, R and Locht, C. (1992). Isolation and molecular characterisation of a novel broad-host-range plasmid from *Bordetella bronchiseptica* with sequence similarities to plasmids from Gram-positive organisms. *Mol. Microbiol.* **6**, 1785–1799.

Armengaud, J. and Jouanneau, Y. (1995). Overexpression in *E. coli* of the *fdxA* gene encoding *Rhodobacter capsulatus* ferredoxin II. *Protein Exp. Purif.* **6**, 176–184.

Armengod, M.E., Garcia-Sogo, M. and Lambies, E. (1988). Transcriptional organisation of the *dnaN* and *recF* genes of *Escherichia coli* K-12. *J. Biol. Chem.* **263**, 12109–12114.

Armstrong, G. A., Cook, D. N., Dzwokai, M., Alberti, M., Burke, D. H. and Hearst, I.E. (1993). Regulations of carotenoid and bacteriochlorophyll biosynthesis genes and identification of an evolutionarily-conserved gene required for bacteriochlorophyll accumulation. *J. Gen. Microbiol.* **139**, 897–906.

Avtges, P., Kranz, R. G. and Haselkorn, R. (1985). Isolation and organization of genes for nitrogen fixation in *Rhodopseudomonas capsulata*. *Mol. Gen. Genet.* **201**, 363–369.

Bagdasarian, M., Lurz, R., Ruckert, B., Franklin, F. C., Bagdasarian, M. M., Frey, J. and Timmis, K. N. (1981). Specific-purpose plasmid cloning vectors. II. Broad host range, high copy number, RSF1010-derived vectors, and a host–vector system for gene cloning in *Pseudomonas*. *Gene* **16**, 237–247.

Baker, S. C., Saunders, N. F., Willis, A. C., Ferguson, S. J., Hadju, J. and Fulop, V. (1997). Cytochrome cd_1 structure: unusual heme environments in a nitrite reductase and analysis of factors contributing to beta-propellor folds. *J. Mol. Biol.* **269**, 440–455.

Baker, S. C., Ferguson, S. J., Ludwig, B., Page, M. D., Richter, O-M. H. and van Spanning, R. J. M. (1998). Molecular genetics of the genus *Paracoccus*: Matabolically versatile bacteria with bioenergetic flexibility. *Microbiol. Mol. Biol. Revs.* **62**, 1046–1078.

Barth, P. T. and Grinter, N. J. (1974). Comparison of the deoxyribonucleic acid molecular weights and homologies of plasmids conferring linked resistance to streptomycin and sulfonamides. *J. Bacteriol.* **120**, 618–630.

Bartley, G. E. and Scolnick, P. A. (1993). cDNA cloning of *PSY2* a second tomato gene encoding phytoene synthetase. *J.Biol. Chem.* **268**, 25718–25721.

Bauer, C. E. and Bird, T. H. (1996). Regulatory circuits controlling photosynthetic gene expression. *Cell* **85**, 5–8.

Bauer, C. E. and Marrs, B. L. (1988). *Rhodobacter capsulatus puf* operon encodes a regulatory protein (PufQ) for bacteriochlorophyll biosynthesis. *Proc. Natl. Acad. Sci. USA* **85**, 7074–7078.

Beckman, D. L. and Kranz, R. G. (1993). Cytochromes *c* biogenesis in a photosynthetic bacterium requires a periplasmic thioredoxin-like protein. *Proc. Natl. Acad. Sci. USA* **90**, 2179–2183.

Beckman, D. L., Trawick, D. R. and Kranz, R. G. (1992). Bacterial cytochromes *c* biogenesis. *Genes Dev.* **6**, 268–283.

Bedzyk, L. A., Escudero, K. W., Gill, R. E., Griffin, K. J. and Frerman, F. F. (1993). Cloning, sequencing, and expression of the genes encoding subunits of *Paracoccus denitrificans* electron transfer flavoprotein. *J. Biol. Chem.* **268**, 20211–20217.

Bell, L. C., Page, M. D., Berks, B. C., Richardson, D. J. and Ferguson, S. J. (1990). Insertion of transposon Tn5 into a structural gene of the membrane-bound nitrate reductase of *Thiosphaera pantotropha* results in anaerobic overexpression of periplasmic nitrate reductase activity. *J. Gen. Microbiol.* **139**, 3205–3214.

Benning, C. and Somerville, C. R. (1992). Identification of an operon involved in sulfolipid biosynthesis in *Rhodobacter sphaeroides*. *J. Bact.* **174**, 6479–6487.

Birnboim, H. C. and Doly, J. (1979). A rapid alkaline extraction procedure for screening recombinant plasmid. *Nucl. Acids Res.* **7**, 1513–1523.

Blatny, J. M., Brautaset, T., Winther-Lassen, H. C., Haugan, K. and Valla, S. (1997). Construction and use of a versatile set of broad-host-range cloning and expression vectors based on the RK2 replicans. *Appl. Environ. Microbiol.* **63**, 370–379.

Bollivar, D. W., Jiang, Z. Y., Bauer, C. E. and Beale, S. I. (1994). Heterologous expression of the bchM gene product from *Rhodobacter capsulatus* and demonstration that it encodes S-adenosyl-L-methionine:Mg-protoporphyrin IX methyltransferase. *J. Bacteriol.* **176**, 5290–5296.

Borghese, R., Crimi, M., Fava. L. and Melandri, B. A. (1998). The ATP synthase atpHAGDC (F₁). operon from *Rhodobacter capsulatus*. *J. Bacteriol.* **180**, 416–421.

Brandner, J. P., McEwan, A. G., Kaplan, S. and Donohue, T. J. (1989). Expression of the *Rhodobacter sphaeroides* cytochrome c_2 structural gene. *J. Bacteriol.* **171**, 360–368.

Brandner, J. P., Stabb, E. V. Temme, R. and Donohue, T. J. (1991). Regions of *Rhodobacter sphaeroides* cytochrome c_2 required for export, haeme attachment and function. *J. Bacteriol.* **173**, 3958–3965.

Calero, S., Fernandez de Henestrosa, A. R. and Barbe, J. (1994). Molecular cloning, sequence and regulation of expression of the *recA* gene of the phototrophic bacterium *Rhodobacter sphaeroides*. *Mol. Gen. Genet.* **242**, 116–120.

Cannac, V., Caffrey, M. S., Voordouw, G. and Cusanovich, M. A. (1991). Expression of the gene encoding cytochrome c_3 from the sulfate-reducing bacterium *Desulfovibrio vulgaris* in the purple photosynthetic bacterium *Rhodobacter sphaeroides. Arch. Biochem. Biophys.* **286**, 629–632.

Chandra, T. S. and Friedrich, C. G. (1986). Tn5-induced mutations affecting sulfur-oxidising ability (Sox). of *Thiosphaera pantotropha. J. Bacteriol.* **166**, 446–452.

Chen, L., Mathews, F. S., Davidson, V. L., Huizinga, E. G. and Vellieux, F. M. (1992). Three-dimensional structure of the quinoprotein methylamine dehydrogenase from *Paracoccus denitrificans* determined by molecular replacement at 2.8Å resolution. *Proteins* **14**, 288–299.

Chistoserdov, A. Y., Boyd, J., Mathews, F. S. and Lidstrom, M. E. (1992). The genetic organisation of the *mau* gene cluster of the facultative autotroph *Paracoccus denitrificans. Biochem. Biophys. Res. Commun.* 184, 1181–1189.

Choudhary, M., MacKenzie, C., Nereng, K., Sodergren, E., Weinstock, G. M. and Kaplan, S. (1997). Low-resolution sequencing of *Rhodobacter sphaeroides* 2.4.1^T: chromosome II is a true chromosome. *Microbiology* **143**, 3085–3099.

Colbeau, A., Godfroy, A. and Vignais, P. M. (1986). Cloning of DNA fragments carrying hydrogenase genes of *Rhodopseudomonas capsulata. Biochimie* **68**, 147–155.

Cullen, P. J., Kaufmann, C. K., Boewman, W. C. and Kranz, R. G. (1997). Characterisation of the *Rhodobacter capsulatus* housekeeping RNA polymerase: *in vitro* transcription of photosynthesis and other genes. *J. Bacteriol.* **272**, 27266–27273.

Davis, J., Donohue, T. J. and Kaplan, S. (1988). Construction, characterisation, and complementation of a Puf⁻ mutant of *Rhodobacter sphaeroides. J. Bacteriol.* **170**, 320–329.

Davison, J., Heusterspreute, M., Chevalier, N., Ha-Thi, V. and Brunel, F. (1987). Vectors with restriction site banks. V. pJRD215, a wide-host-range cosmid vector with multiple cloning sites. *Gene* **51**, 275–280.

de Boer, A. P. N., Reijnders, W. N. M., Kuenen, J. G., Stouthamer, A. H. and van Spanning, R. J. M. (1994). Isolation, sequencing and mutational analysis of a gene cluster involved in nitrite reduction in *Paracoccus denitrificans. Antonie van Leeuwenhoek*, **66**, 111–127.

de Boer, A. P. N., Reijnders, W. N. M., Stouthamer, A. H., Westerhoff, H. V. and van Spanning, R. J. M. (1995). The nitric oxide and nitrite reductase gene clusters of *Paracoccus denitrificans*. Abstract Beijerinck centennial conference "Microbial physiology and gene regulation: emerging principles and applications".

de Boer, A. P. N., van der Oost, J., Reijnders, W. N. M., Westerhoff, H. V., Stouthamer, A. H. and van Spanning, R. J. M. (1996). Mutational analysis of the *nor* gene cluster which encodes nitric-oxide reductase from *Paracoccus denitrificans. Eur. J. Biochem.* **242**, 592–600.

de Gier, J.-W., Lubben, M., Reijnders, W. N. M., Tipker, C. A., Slotboom, D.-J., van Spanning, R. J. M., Stouthamer, A. H. and van der Oost, J. (1994). The terminal oxidases of *Paracoccus denitrificans. Mol. Microbiol.* **13**, 183–196.

de Gier, J.-W., van der Oost, J., Harms, N., Stouthamer A. H. and van Spanning, R. J. (1995). The oxidation of methylamine in *Paracoccus denitrificans. Eur. J. Biochem.* **229**, 148–154.

de Gier, J.-W., Schepper, M., Reijnders, W. N. M., van Dyck, S. J., Slotboom, D. J., Warne, A., Saraste, M., Krab, K., Finel, M., Stouthamer, A. H., van Spanning, R. J. M. and van der Oost, J. (1996). Structural and functional analysis of aa_3-type and cbb_3-type cytochrome *c* oxidases of *Paracoccus denitrificans* reveals significant differences in proton-pump design. *Mol. Microbiol.* **20**, 1247–1260.

de Henestrosa, A. R. F., del Rey, A., Tarrago, R. and Barbe, J. (1997). Cloning and characterisation of the *recA* gene of *Paracoccus denitrificans* and construction of a *recA*-deficient mutant. *FEMS Microbiol. Lett.* **147**, 209–213.

DeHoff, B. S., Lee, J. K., Donohue, T. J., Gumport, R. I. and Kaplan, S. (1988). In vivo analysis of *puf* operon expression in *Rhodobacter sphaeroides* after deletion of a putative intercistronic transcription terminator. *J. Bacteriol.* **170**, 4681–4692.

Delorme, C., Huisman, T. T., Reijnders, W. N., Chan, Y. L., Harms, N., Stouthamer, A. H. and van Spanning, R. J. (1997). Expression of the *mau* gene cluster of *Paracoccus denitrificans* is controlled by mauR and a second transcription regulator. *Microbiology* **143**, 793–801.

de Lorenzo, V., Herreo, M., Jakubzik, U. and Timmis, K. N. (1990). Mini Tn5 transposon derivatives for insertion mutagenesis. *J. Bacteriol.* **172**, 6568–6572.

de Vries, G. E., Harms, N., Hoogendijk, J. and Stouhamer, A. H. (1989). Isolation and characterisation of a *Paracoccus denitrificans* mutants with increasedconjugation frequencies. *Arch Microbiol.* **152**, 52–57.

Ditta, G., Stanfield, S., Corbin, D. and Helinkski D. R. (1980). Broad host range DNA cloning system for Gram negative bacteria. *Proc. Natl. Acad. Sci USA* **77**, 7347–7351.

Ditta, G., Schmidhauser, T., Yakobson, E., Lu, P., Liang, X.-W., Findlay, D. R., Guiney, D. and Helinski, D. R. (1985). Plasmids related to the broad host range vector, pRK290, useful for gene cloning and for monitoring gene expression. *Plasmid* **13**, 149–153.

Ditta, G. Virts, E. Palomares, A. and Kim, C. H. (1987). The *nif*A gene of *Rhizobium meliloti* is oxygen regulated. *J. Bacteriol.* **169**, 3217–3223.

Donohue, T. J. and Kaplan, S. (1991). Genetic techniques in Rhodospirillaceae. *Meth. Enzymol.* **204**, 459–485.

Dryden, S. C. and Kaplan, S. (1993). Identification of cis-acting regulatory regions upstream of the rRNA operons of *Rhodobacter sphaeroides*. *J. Bacteriol.* **175**, 6392–6402.

Duport, C., Meyer, C., Naud, I. and Jouanneau, Y. (1994). A new gene expression system based on a fructose-dependent promoter from *Rhodobacter capsulatus*. *Gene* **145**, 103–108.

Elsen, S., Colbeau, A. and Vignais, P. M. (1997). Purification and *in vitro* phosphorylation of HupT, a regulatory protein controlling hydrogenase gene expression in *Rhodobacter capsulatus*. *J. Bacteriol.* **179**, 968–973.

Farchqus, J. W. and Oesterhelt, D. (1989). A *Rhodobacter sphaeroides Puf* L, M and X deletion mutant and its complementation in *trans* with a 5.3 kb *Puf* operon shuttle fragment. *EMBO J.* **8**, 47–54.

Fellay, R., Frey, J. and Krisch, H. (1987). Interposon mutagenesis of soil and water bacteria: a family of DNA fragments designed for *in vitro* insertional mutagenesis of Gram-negative bacteria. *Gene* **52**, 147–154.

Fellay, R., Krisch, H. M., Prentki, P. and Frey, J. (1989). Omegon-Km: a transposable element designed for *in vivo* insertional mutagenesis and cloning of genes in Gram-negative bacteria. *Gene* **76**, 215–226.

Figurski, D. H. and Helinski, D. R. (1979). Replication of an origin-containing derivative of plasmid RK2 dependent on a plasmid function derived *in trans*. *Proc. Natl. Acad. Sci. USA* **76**, 1648–1652.

Finel, M. (1996). Genetic inactivation of the H^+-translocating NADH:ubiquinone oxidoreductase of *Paracoccus denitrificans* is facilitated by insertion of the *ndh* gene from *Escherichia coli*. *FEBS Lett.* **393**, 81–85.

Flory, J. E. and Donohue, T. J. (1997). Transcriptional control of several aerobically induced cytochrome structural genes in *Rhodobacter sphaeroides*. *Microbiology* **143**, 3101–3110.

Fonstein, M., Nikolskaya, T., Kogan, Y. and Haselkorn, R. (1998). Genome encyclopedias and their use for comparative analysis of *Rhodobacter capsulatus* strains. *Electrophoresis* **19**, 469–477.

Franklin, F. C. H. (1985). Broad host range cloning vectors for Gram negative bacteria. In *DNA Cloning*, vol. I. *A Practical Approach* (D. M. Glover, ed.), pp 165–184. IRL Press, Oxford.

Frey, J., Mudd, E. A. and Krisch, H. M. (1988). A bacteriophage T4 expression cassette that functions efficiently in a wide range of Gram-negative bacteria. *Gene* **62**, 237–247.

Friedman, A. M., Long, S. R., Brown, S. E., Buikema, W. J. and Ausubel, F. M. (1982). Construction of a broad host range cosmid cloning vector and its use in the genetic analysis of *Rhizobium* mutants. *Gene* **18**, 289–296.

Fülop, V., Moir, J. W., Ferguson, S. J. and Hadju, J. (1995). The anatomy of a bifunctional enzyme: structural basis for reduction of oxygen to water and synthesis of nitric oxide by cytochrome cd_1. *Cell* **81**, 369–377.

Fürste, J. P., Pansegrau, W. M, Frank, R., Blöcker, H., Scholz, P., Bagdasarian, M. and Lanka, E. (1986). Molecular cloning of the plasmid RP4 primase region in a multi-host-range *tacP* expression vector. *Gene* **48**, 119–131.

Gaubier, P., Wu, H. J., Laudie, M., Delsney, M., Grellet, F. (1995). A chlorophyll synthetase gene from *Arabidopsis thaliana*. *Mol. Gen. Gent.* **249**, 58–64.

Gay, P., Lecoq, D., Steinmetz, M., Ferrari, E. and Hoch, J. A. (1983). Cloning structural gene sacB which codes for exoenzyme levansucrase of *Bacillus subtilis*. *J. Bacteriol.* **153**, 1424–1431.

Gerhus, E., Steinrücke, P. and Ludwig, B. (1990). *Paracoccus denitrificans* cytochrome c_1 gene replacement mutants. *J. Bacteriol.* **172**, 2392–2400.

Gerhus, E., Grisshammer, R., Michel, H., Ludwig, B. and Turba, A. (1993). Synthesis of the *Rhodopseudomonas viridis* holo-cytochrome c_2 in Paracoccus denitrificans. *FEMS Microbiol. Lett.* **113**, 29–34.

Ghozlan, H. A., Ahmadian, R. M., Frölich, M. and Kleiner, D. (1991). Genetic tools for *Paracoccus denitrificans*. *FEMS Microbiol. Lett.* **82**, 303–306.

Giacomini, A., Ollero, F.J., Squartini, A. and Nuti, M. P. (1994). Construction of multipurpose gene cartridges based on a novel synthetic promoter for high-level gene expression in Gram-negative bacteria. *Gene* **144**, 17–24.

Gibson, L. C. and Hunter, C. N. (1994). The bacteriochlorophyll biosynthesis gene, bchM, of *Rhodobacter sphaeroides* encodes S-adenosyl-L-methionine: Mg protoporphyrin IX methyltransferase. *FEBS Lett.* **352**, 127–130.

Gibson, L. C. D., Willows, R. D., Kannangara, C. G., von Wettstein, D. and Hunter, C. N. (1995). Magnesium-protoporphyrin chelatase of *Rhodobacter sphaeroides*: reconstitution of activity by combining the products of the *bchH, -I*, and -*D* genes expressed in *Escherichia coli*. *Proc. Natl. Acad. Sci. USA* **92**, 1941–1944.

Giuliano, G., Pollock, D., Stapp, H. and Scolnik, P. A. (1988). Complete DNA-sequence, specific Tn5 insertion map, and gene assignment of the carotenoid biosynthesis pathway of *Rhodobacter capsulatus*. *Mol. Gen. Genet.* **213**, 78–83

Goldman, B. S., Gabbert, K. K. and Kranz, R. G. (1996). Use of haeme reporters for studies of cytochrome biosynthesis and haeme transport. *J. Bacteriol.* **178**, 6338–6347.

Goldman, B. S., Beckman, D. L., Bali, A., Monika, E. M., Gabbert, K. K. and Kranz, R. G. (1997). Molecular and immunological analysis of an ABC transporter complex required for cytochrome *c* biogenesis. *J. Mol. Biol.* **268**, 724–738.

Goldman, E. R. and Youvan D. C. (1995). Imaging spectroscopy and combinatorial mutagenesis of the reaction centre and light harvesting II antenna. In *Anoxigenic Photosynthetic Bacteria* (R. E. Blankenship, M. T. Madigan and C. E. Bauer, eds), pp. 1257–1268. Kluwer Academic Publishers, Amsterdam.

Gong, L., Lee, J. K. and Kaplan, S. (1994). The Q gene of *Rhodobacter sphaeroides*: Its role in *puf* operon expression and spectral complex assembly. *J. Bacteriol.* **176**, 2946–2961.

Gray, K. A., Farchaus, J. W., Wachtvertl, J., Breton, J. and Oesterhelt, D. (1990).

Initial characterisation of site-directed mutants of tyrosine M210 in the reaction centre of *Rhodobacter sphaeroides*. *EMBO J.* **9**, 2061–2070.

Grisshammer, R., Oeckl, C. and Michel, H. (1991). Expression in *Escherichia coli* of *c*-type cytochrome genes from *Rhodopseudomonas viridis*. *Biochim. Biophys. Acta* **1088**, 183–190.

Hallenbeck, P. L. and Kaplan, S. (1987). Cloning of the gene for phosphoribulokinase activity from *Rhodobacter sphaeroides* and expression in *Escherichia coli*. *J. Bacteriol.* **169**, 3669–3678.

Hallenbeck, P. L. and Kaplan, S. (1988). .Structural Gene Regions Of *Rhodobacter sphaeroides* Involved In Co2 fixation. *Photosyn. Res.* **19**, 63–71

Haltia, T., Finel, M., Harms, N., Nakari, T., Raitio, M., Wikström, M. and Saraste, M. (1989). Deletion of the gene for subunit III leads to defective assembly of bacterial cytochrome oxidase. *EMBO J.* **8**, 3571–3579.

Hamblin, P. A., Bourne, N. A. and Armitage, J. P. (1997). Characterization of the chemotaxis protein CheW from *Rhodobacter sphaeroides* and its effect on the behaviour of *Escherichia coli*. *Mol. Microbiol.* **24**, 41–51.

Harms, N. and van Spanning, R. J. M. (1991). C^1 metabolism in *Paracoccus denitrificans*: Genetics of *Paracoccus denitrificans*. *J. Bioenerg. Biomembr.* **23**, 187–209.

Harms, N., de Vries, G. E., Maurer, K., Hoogendijk, J. and Stouthamer, A. H. (1987). Isolation and nucleotide sequence of the methanol dehydrogenase structural gene from *Paracoccus denitrificans*. *J. Bacteriol.* **169**, 3969–3975.

Harms, N., van Spanning, R. J. M., Oltmann, L. F. and Stouthamer, A. H. (1989). Regulation of methanol dehydrogenase synthesis in *Paracoccus denitrificans*. *Antonie van Leeuwenhoek* **56**, 46–50.

Hilton, J. C. and Rajagopalan, K. V. (1996). Molecular cloning of dimethyl sulfoxide reductase from *Rhodobacter sphaeroides*. *Biochim Biophys Acta* **1294**, 111–114.

Hirsch, P. R., Wang, C. L. and Woodward, M. J. (1986). Construction of a Tn5 derivative determining resistance to gentamycin and spectinomycin using a fragment cloned from R1033. *Gene* **48**, 203–209.

Hoeren, F. U., Berks, B. C., Ferguson, S. J. and McCarthy, J. E. (1993). Sequence and expression of the gene encoding the respiratory nitrous-oxide reductase from *Paracoccus denitrificans*. New and conserved structural and regulatory motifs. *Eur. J. Biochem.* **218**, 49–57.

Hübner, P., Willison, J. C., Vignais, P. M. and Bickle, T. A. (1991). Expression of regulatory *nif* genes in *Rhodobacter capsulatus*. *J. Bacteriol.* **173**, 2993–2999.

Hübner, P., Masepohl, B., Klipp, W. and Bickle, T.A. (1993). *nif* gene expression studies in *Rhodobacter capsulatus*: *ntrC*-independent repression by high ammonium concentrations. *Mol. Microbiol.* **10**, 123–132.

Hunter, C. N. (1988). Transposon Tn5 mutagenesis of genes encoding reaction centre and light harvesting polypeptides of *Rhodobacter sphaeroides*. *J. Gen Microbiol.* **134**, 1481–1489.

Hunter, C. N. and Turner, G. (1988). Transfer of genes coding for apoproteins of reaction centre and light harvesting LH1 complexes to *Rhodobacter sphaeroides*. *J. Gen Microbiol.* **134**, 1471–1480.

Iobbi-Nivol, C., Crooke, H., Griffiths, L., Grove, J., Hussain, H., Pommier, J., Mejean, V. and Cole, J. A. (1994). A reassessment of the range of *c*-type cytochromes synthesised by *E. coli* K-12. *FEMS Microbiol. Lett.* **119**, 89–94.

John, P. and Whatley, F. R. (1975). *Paracoccus denitrificans* and the evolutionary origin of the mitochondrion. *Nature* **254**, 495–498.

Jones, M. R., Fowler, G. J. S., Gibson, L. C. D., Grief, G. G., Olsen, J. D., Crielaard, W. and Hunter, C. N. (1992). Mutants of *Rhodobacter sphaeroides* lacking one or more pigment–protein complexes and complementation with reaction-centre LH1 and LH2 genes. *Mol. Microbiol.* **6**, 1173–1184.

Jouanneau, Y., Meyer, C., Naud, I. and Klipp, W. (1995). Characterization of an *fdxN* mutant of *Rhodobacter capsulatus* indicates that ferredoxin I serves as electron donor to nitrogenase. *Biochim. Biophys. Acta* **1232**, 33–42.

Kanazireva, E. and Biel, A. J. (1995). Cloning and overexpression of the *Rhodobacter capsulatus hemH* gene. *J. Bacteriol.* **177**, 6693–6694.

Katayama, Y., Hiraishi, A. and Kuraishi, H. (1995). *Paracoccus thiocyanatus sp. nov.*, a new species of thiocyanate-utilising facultative chemolithotroph, and transfer of *Thiobacillus versutus* to the genus *Paracoccus* as *Paracoccus versutus* comb. nov. with emendation of the genus. *Microbiology* **141**, 1469–1477.

Keen, N. T., Tamaki, S., Kobayashi, D. and Trollinger, D. (1988). Improved broad-host-range plasmids for DNA cloning in Gram-negative bacteria. *Gene* **70**, 191–197.

Klipp, W., Masepohl, B. and Pühler, A. (1988). Identification and mapping of nitrogen fixation genes of *Rhodobacter capsulatus*: Duplication of a *nifA-nifB* region. *J. Bacteriol.* **170**, 693–699.

Knäblein, J., Dobbek, H. and Schneider, F. (1997). Organization of the DMSO respiratory operon of *Rhodobacter capsulatus* and its consequences for homologous expression of DMSOR/TMAOR. *J. Biol. Chem.* **378**, 303–308

Knauf, V. C. and Nester, E. W. (1982). Wide host range cloning vectors: A cosmid clone bank of an *agrobacterium* Ti plasmid. *Plasmid* **8**, 45–54.

Koch, H. G., Hwang, O. and Daldal F. (1998). Isolation and characterisation of *Rhodobacter capsulatus* mutants affected in cytochrome cbb^3 oxidase activity. *J. Bacteriol.* **180**, 969–978.

Konishi, K., van Dores, S. R., Kramer, D. M., Crofts, A. R. and Gennis, R. B. (1991). Preparation and characterisation of the water-soluble haeme-binding domain of cytochrome c_1 from the *Rhodobacter sphaeroides* bc_1 complex. *J. Biol. Chem.* **266**, 14270–14276.

Konyecsni and Deretic (1988). Broad-host-range plasmid and M13 bacteriophage-derived vectors for promoter analysis in *Escherichia coli* and *Pseudomonas aeruginosa*. *Gene* **74**, 375–386.

Kovach, M. E., Elzer, P. H., Hill, D. S., Robertson, G. T., Farris, M. A., Roop, R. M. R. II and Peterson, K. M. (1995). Four new derivatives of the broad-host-range cloning vector pBBR1MCS, carrying different antibiotic-resistance cassettes. *Gene* **166**, 175–176.

Kranz, R. G. and Hazelkorn, R. (1985). Characterisation of *nif* regulatory genes in *Rhodopseudomonas capsulata* using *lac* gene fusions. *Gene* **40**, 203–215.

Kurowski, B. and Ludwig, B. (1987). The genes of the *Paracoccus denitrificans* bc_1 complex. *J. Biol. Chem.* **262**, 13805–13811.

Kutsche, M., Leimkuhler S., Angermüller, S. and Klipp, W. (1996). Promoters controlling expression of the alternative nitrogenase and molybdenum uptake system in *Rhodobacter capsulatus* are activated by NtrC. *J. Bacteriol.* **178**, 2010–2017.

Labes, M., Puhler, A. and Simon, R. (1990). A new family of RSF1010-derived expression and *lac*-fusion broad-host-range vectors for Gram-negative bacteria. *Gene* **89**, 37–46.

Lang, S. E., Jenney, F. E. and Daldal, F. (1996). *Rhodobacter capsulatus* CycH: a bipartite gene product with pleiotropic effects on the biogenesis of structurally different *c*-type cytochromes. *J. Bacteriol.* **178**, 5279–5290.

Lascelles, J. (1960). The synthesis of enzymes concerned in bacteriochlorophyll formation in growing cultures of *Rhodobacter sphaeroides*. *J. Gen. Microbiol.* **23**, 487–498.

LeBlanc, H. N. and Beatty, J. T. (1996). Toplogical analysis of *Rhodobacter capsulatus* PucC protein and effects of c-terminal deletion on LHII. *J. Bacteriol.* **178**, 4801–4806.

Leimkuhler, S., Kern, M. Solomon, P. S., Mc Ewan, A. G., Schwarz, G., Mendel, R. R. and Klipp, W. (1998). Xanthine dehydrogenase from *Rhodobacter capsulatus* is more similar to its eukaryotic counterparts than to prokaryotic molybdenum enzymes. *Mol. Microbiol.* **27**, 853–869.

Leong, S. A., Ditta, G. S. and Helinski, D. R. (1982). Heme biosynthesis in *Rhizobium*. Identification of a cloned gene coding for delta-aminolevulinic acid synthetase from *Rhizobium meliloti*. *J. Biol. Chem.* **257**, 8724–8730.

Leung, Y.-C., Chan, C., Reader, J. S., Willis, A. C., van Spanning, R. J. M., Ferguson, S. J. and Radford, S. E. (1997). The pseudoazurin gene from *Thiosphaera pantotropha*: analysis of upstream putative regulatory sequences and overexpression in *Escherichia coli*. *Biochem. J.* **321**, 699–705.

Liebl, U., Sled, V., Brasseur, G., Ohnishi, T. and Daldal, F. (1997). Conserved non-ligating residues of the *Rhodobacter capsulatus* Rieske iron–sulfur protein of the bc^1 complex are essential for protein structure, properties of the [2Fe-2S] cluster, and communication with the quinone pool. *Biochemistry* **36**, 11675–11684.

Lockhart, P. J., Larkum, A. W. D., Steel, M. A., Waddell, P. J. and Penny, D. (1996). Evolution of chlorophyll and bacteriochlorophyll: The problem of invariant sites in sequence analysis. *Proc. Natl Acad. Sci. USA* **93**, 1930–1934

Lodge, J., Williams, R., Bell, A., Chan, B. and Busby, S. (1990). Comparison of promoter activities in *Escherichia coli* and *Pseudomonas aeruginosa*: use of a new broad-host-range promoter-probe plasmid. *FEMS Microbiol. Lett.* **67**, 221–226.

Ludwig, B. (1987). Cytochrome *c* oxidase in prokaryotes. *FEMS Microbiol. Revs.* **46**, 41–56.

Ludwig, B. (1992). Terminal oxidases in *Paracoccus dentrificans*. *Biochem. Biophys. Acta.* **1101**, 195–197.

Ludwig, W., Mittenhuber, G. and Friedrich, C. G. (1993). Transfer of *Thiosphaera pantotropha* to *Paracoccus denitrificans*. *Int. J. Syst. Bacteriol.* **43**, 363–367.

Ma, D., Cook, D. N., O'Brien, D. A. and Hearst, J. E. (1993). Analysis of the promoter and regulatory sequences of an oxygen-regulated *bch* operon in *Rhodobacter capsulatus* by site-directed mutagenesis. *J. Bacteriol.* **175**, 2037–2045.

Mackenzie, C., Chidmbaram, M., Sodergren, E. J., Kapan, S. and Weinstock, G. M. (1995). DNA repair mutants of *Rhodobacter sphaeroides*. *J. Bacteriol.* **177**, 3027–3035.

Makrides, S. C. (1996). Strategies for achieving high-level expression of genes in *Escherichia coli*. *Microbiol. Rev.* **60**, 512–538.

Marrs, B. (1974). Genetic recombination in *Rhodopseudomonas capsulata*. *Proc. Natl. Acad. Sci. USA* **72**, 971–973.

Masepohl, B., Angermüller, S., Hennecke, S., Hubner, P., Moreno-Vivian, C. and Klipp, W. (1993). Nucleotide sequence and genetics analysis of the *Rhodobacter capsulatus* ORF6nifUSVW region. *Mol. Gen. Genet.* **238**, 369–382.

Mather, M. W., McReynolds, L. M. and Yu, C. A. (1995). An enhanced broad-host-range vector for Gram-negative bacteria: avoiding tetracycline phototoxicity during the growth of photosynthetic bacteria. *Gene* **156**, 85–88.

McEwan, A. G., Kaplan, S. and Donohue, T. J. (1989). Synthesis of *Rhodobacter sphaeroides* cytochrome c_2 in *Escherichia coli*. *FEMS Microbiol. Lett.* **59**, 253–258.

Mermod, N., Ramos, J. L., Lehrbach, P. R. and Timmis, K. N. (1986). Vector for regulated expression of cloned genes in a wide range of Gram negative bacteria. *J. Bacteriol.* **167**, 447–454.

Meyer, T. E. (1994). Evolution of photosynthetic reaction centers and light harvesting chlorophyll proteins. *Biosystems* **33**, 167–175.

Mittenhuber, G., Sonomoto, K., Egert, M. and Friedrich C. G. (1991). Identification of the DNA region responsible for sulfur oxidising ability of *Thiosphaera pantotropha*. *J. Bacteriol.* **173**, 7340–7344.

Monika, E. M., Goldman, B. S., Beckman, D. L. and Kranz, R. G. (1997). A thiore-duction pathway tethered to the membrane for periplasmic cytochromes c bio-genesis; *in vitro* and *in vivo* studies. *J. Mol. Biol.* **271**, 679–692.

Moore, M. D. and Kaplan, S. (1989). Construction of Tn*PhoA* gene fusions in *Rhodobacter sphaeroides*: Isolation and characterization of a respiratory mutant unable to utilize dimethyl sulfoxide as a terminal electron acceptor during anaerobic growth in the dark on glucose. *J. Bacteriol.* **171**, 4385–4394.

Moreno-Vivian, C., Hennecke, S., Pühler, A. and Klipp, W. (1989). Open reading frame 5 (ORF5), encoding a ferredoxin-like protein, and *nifQ* are cotranscribed with *nifE*, *nifN*, *nifX*, and ORF4 in *Rhodobacter capsulatus*. *J. Bacteriol.* **171**, 2591–2598.

Myllykallio, H., Jenney, F. E., Moomaw, C. R., Slaughter, C. A. and Daldal, F. (1997). Cytochrome C*y* of *Rhodobacter capsulatus* is attached to the cytoplasmic mem-brane by an uncleaved signal-sequence-like anchor. *J. Bacteriol.* **179**, 2623–2631.

Nano, F. E. and Kaplan, S. (1982). Expression of the transposable *lac* operon Tn951 in *Rhodopseudomonas sphaeroides*. *J. Bact.* **152**, 924–927.

Nano, F. E., Shepherd, W. D., Watkins, M. M., Kuhl, S. A. and Kaplan, S. (1985). Broad host range plasmid vector for the *in vitro* construction of transcrip-tional/translational *lac* fusions. *Gene* **34**, 219–226.

Neidle, E. L. and Kaplan, S. (1992). *Rhodobacter sphaeroides rdxA*, a homolog of *Rhizobium meliloti fixG*, encodes a membrane protein which may bind cytoplas-mic [4Fe-4S] clusters. *J. Bacteriol.* **174**, 6444–6454.

Nickel, C. M., Vandekerckhove, J., Beyer, P. and Tadros, M. H. (1997). Molecular analysis of the *Rhodobacter capsulatus* chaperone *dna*KJ operon. *Gene* **192**, 251–259.

Oriol, E., Mendez-Alvarez, S., Barbe, J. and Gennis, I. (1996). Cloning of the *Rhodobacter sphaeroides hisI* gene: unifunctionality of the encoded protein and lack of linkage to other *his* genes. *Microbiology* **142**, 2071–2078.

Page, M. D. and Ferguson, S. J. (1994). Differential reduction in soluble and mem-brane-bound c-type cytochrome contents in a *Paracoccus denitrificans* mutant partially deficient in 5-aminolevulinate synthase activity. *J. Bacteriol.* **176**, 5919–5928.

Page, M. D. and Ferguson, S. J. (1995). Cloning and sequence analysis of *cycH* gene *Paracoccus denitrificans*: the *cycH* gene product is required for assembly of all c-type cytochromes, including cytochrome c₁. *Mol. Microbiol.* **15**, 307–318.

Page, M. D. and Ferguson, S. J. (1997). *Paracoccus denitrificans* CcmG is a periplas-mic protein-disulphide oxidoreductase required for c- and aa₃-type cytochrome biogenesis; evidence for a reductase role *in vivo*. *Mol. Microbiol.* **24**, 977–990.

Page, M. D., Pearce, D. A., Norris, H. A. and Ferguson, S. J. (1997). The *Paracoccus denitrificans ccmA, B* and *C* genes: cloning and sequencing, and analysis of the potential of their products to form a haem or apo-c-type cytochrome trans-porter. *Microbiology* **143**, 563–567.

Palmer, T., Goodfellow, I. G., Sockett, R. E., McEwan, A. G. and Boxer, D. H. (1998). Characterisation of the *mob* locus from *Rhodobacter sphaeroides* required for molybdenum cofactor biosynthesis. *Biochim. Biophys. Acta* **1395**, 135–140.

Parales, R. E. and Harwood, C. S. (1993). Construction and use of a new broad-host-range *lacZ* transcriptional fusion vector, pHRP309, for Gram-negative bac-teria. *Gene* **133**, 23–30.

Paraskeva, C. (1979). Transfer of kanamycin resistance mediated by plasmid R68.45 in *Paracoccus denitrificans*. *J. Bacteriol.* **139**, 1062–1064.

Parke, D. (1990). Construction of mobilizable vectors derived from plasmids RP4, pUC18 and pUC19. *Gene* **93**, 135–137.

Pasternak, C., Chen, W., Heck, C. and Klug, G. (1996). Cloning nucleotide

Paracoccus and Rhodobacter

sequence and characterization of *rpo*D encoding the primary sigma factor of *Rhodobacter capsulatus*. *Gene* **176**, 177–184.

Penfold, R. J. and Pemberton, J. M. (1992). An improved suicide vector for construction of chromosomal insertion mutations in bacteria. *Gene* **118**, 145–146.

Penfold, R. J. and Pemberton, J. M. (1994). Sequencing, chromosomal inactivation, and functional expression in *Escherichia coli* of *pps*R, a gene which represses carotenoid and bacteriochlorophyll synthesis in *Rhodobacter sphaeroides*. *J. Bacteriol.* **176**, 2869–2876.

Penfold, C. N., Ringeling, P. L., Davy, S. L., Moore, G. R., McEwan, A. G. and Spiro, S. (1996). Isolation, characterisation and expression of the bacterioferritin gene of *Rhodobacter capsulatus*. *FEMS Microbiol. Lett.* **139**, 1434–1439.

Pollick, M. and Klug, G. (1995). Identification and sequence analysis of genes involved in late steps in Cobalamin (Vitamin B12) synthesis in *Rhodobacter capsulatus*. *J. Bacteriol.* **177**, 4481–4487.

Pollock, V. V. and Barber, M. J. (1997). Biotin sulfoxide reductase. Heterologous expression and characterisation of a functional molybdopterin guanine dinucleotide-containing enzyme. *J. Biol. Chem.* **272** 3355–3362.

Pollock, D., Bauer, C. E. and Scolnik, P. A. (1988). Transcription of the *Rhodobacter capsulatus nif*HDK operon is modulated by the nitrogen source. Construction of plasmid expression vectors based on the *nif*HDK promoter. *Gene* **65**, 269–275.

Pollock, W. B. R., Chemerika, P. J., Forrest, M. E., Beatty, J. T. and Voordouw (1989). Expression of the gene encoding cytochrome c3 from *Desulfovibrio vulgaris* in *E.coli*. *J. Gen. Microbiol.* **135**, 2319–2328.

Poole, R. K. (1988). Bacterial cytochrome oxidases. In: Anthony, C. (ed.), *Bacterial Energy Transduction*, pp. 231–291, Academic Press, London.

Prentki, P. and Krisch, H. M. (1984). *In vitro* insertional mutagenesis with a selectable DNA fragment. *Gene* **29**, 303–313.

Rainey, F. A., Kelly, D. P., Stackebrandt, E., Burkhardt, J., Hiraishi, A., Katayami, Y. and Wood, A. P. (1999). A re-evaluation of the taxonomy of *Paracoccus denitrificans* and a proposal for the creation of *Paracoccus pantotrophus* comb. NOV. *Int. J. Syst. Bacterial.* (In press).

Raitio, M., Jalli, T. and Saraste, M. (1987). Isolation and analysis of the genes for cytochrome oxidase in *Paracoccus denitrificans*. *EMBO J.* **6**, 2825–2833.

Raitio, M., Pispa, J. M., Metso, T. and Saraste, M. (1990). Are there isoenzymes of cytochrome *c* oxidase in *Paracoccus denitrificans*? *FEBS Lett.* **261**, 431–435.

Rajagopalan, K. V. (1996). Biosynthesis of the molybdenum cofactor. In *E.coli and Salmonella*. 2nd edn (F. C. Neidhardt, R. Curtis, J. L. Ingraham, E. C. C. Lin, K. B. Low and B. Magasanik, eds), pp. 674–679. American Society for Microbiology, Washington, DC.

Ras, J., Reijnders, W. N. M., van Spanning, R. J. M., Harms, N., Oltmann, L. F. and Stouthamer, A. H. (1991). Isolation, sequencing, and mutagenesis of the gene encoding cytochrome c_{553i} of *Paracoccus denitrificans* and characterisation of the mutant strain. *J. Bacteriol.* **173**, 6971–6979.

Rasmusson, A. G., Heiser, V. Irrgang, K. D., Brennicke, A. and Grohmann, L. (1998). Molecular characterisation of the 76 kDa iron–sulphur protein subunit of potato mitochondrial complex I. *Plant Cell Physiol.* **39**, 373–381.

Richaud, P., Colbeau, A., Toussaint, B. and Vignais, P. (1991). Identification and sequence analysis of the *hup*R1 gene which encodes a response regulator of the NtrC family required for hydrogenase expression in *Rhodobacter capsulatus*. *J. Bacteriol.* **173**, 5928–5932.

Richter, O.-M. H., Tao, J.-S., Turba, A. and Ludwig, B. (1994). A cytochrome ba_3 functions as a quinol oxidase in *Paracoccus denitrificans*. *J. Biol. Chem.* **269**, 23079–23086.

Robertson, L. A. and Kuenen, J. G. (1983). *Thiosphaera pantotropha* gen. nov. sp. nov., a facultatively anaerobic, facultatively autotrophic sulfur bacterium. *J. Gen. Microbiol.* **129**, 2847–2855.

Roldán, M. D., Sears, H. J., Cheeseman, M. R., Ferguson, S. J., Thomas, A. J., Berks, B. C. and Richardson, D. J. (1998). Spectroscopic characterisation of a novel multi heme *c*-type cytochrome widely implicated in bacterial electron transport. *J. Biol. Chem.* **273**, 28785–28790.

Sabaty, M. and Kaplan, S. (1996). *mgpS*, a complex regulatory locus involved in the transcriptional control of the *puc* and *puf* operons in *Rhodobacter sphaeroides* 2.4.1. *J. Bacteriol.* **178**, 35–45.

Sambongi, Y. and Ferguson, S. J. (1994). Synthesis of holo *Paracoccus denitrificans* cytochrome c_{550} requires targeting to the periplasm whereas that of holo *Hydrogenobacter thermophilus* cytochrome c_{552} does not. *FEBS Lett.* **340**, 65–70.

Saraste, M. (1990). Structural features of cytochrome oxidase. *Q. Rev. Biophys.* **23**, 331–336.

Schneider, K. and Beck, C. F. (1987). New expression vectors for identifying and testing signal structures for initiation and termination of transcription. *Meth. Enzymol.* **153**, 452–461.

Schneider, K. H. and Giffhorn, F. (1994). Overproduction of mannitol dehydrogenase in *Rhodobacter sphaeroides*. *Appl. Microbiol. Biotechnol.* **41**, 578–583.

Schuddekopf, K., Hennecke, S., Liese, U., Kutsche, M. and Klipp, W. (1993). Characterization of *anf* genes specific for the alternative nitrogenase and identification of *nif* genes required for both nitrogenases in *Rhodobacter capsulatus*. *Mol. Microbiol.* **8**, 673–684 .

Scolnick, P. A. and Marrs, B. L. (1987). Genetic research with photosynthetic bacteria. *Annu. Rev. Microbiol.* **41**, 703–726.

Sganga, M. W. and Bauer, C. E. (1992). Regulatory factors controlling photosynthetic reaction center and light-harvesting gene expression in *Rhodobacter capsulatus*. *Cell* **68**, 945–954.

Shaw, A. L., Hanson, G. R. and McEwan, A. G. (1996). Cloning and sequence analysis of the dimethyl sulfoxide from *Rhodobacter capsulatus*. *Biochim. Biophys. Acta* **1276**, 176–180.

Simon, R., Priefer, U. and Pühler A. (1983). A broad host range mobilization system for *in vivo* genetic engineering: transposon mutagenesis in Gram negative bacteria. *Bio/Technol.* **1**, 784–791.

Sinha, N. and Ferguson, S. J. (1998). An *E.coli* ccm deletion strain substantially expresses Hydrogenobacter thermophilus cytochrome c552 in the cytoplasm. *FEMS Microbiol. Lett.* **161**, 1–6.

Sistrom, W. R., Macaluso, A. and Pledger, R. (1984). Mutants of *Rhodopseudomonas sphaeroides* useful in genetic analysis. *Arch. Microbiol.* **138**, 161–165.

Smith, C. A., Suzuki, J. Y., Bauer, C. E. (1996). Cloning and characterization of the chlorophyll biosynthesis gene *chlM* from *Synechocystis* PCC 6803 by complementation of a bacteriochlorophyll biosynthesis mutant of *Rhodobacter capsulatus*. *Plant Mol. Biol.* **30**, 1307–1314.

Sockett, R. E. and Armitage, J. P. (1991). Isolation, characterization, and complementation of a paralysed flagellar mutant of *Rhodobacter sphaeroides*. *J. Bacteriol.* **173**, 2786–2790.

Solioz, M. and Bienz, D. (1990). Bacterial genetics by electric shock. *Trends Biochem. Sci.* **15**, 175–177.

Spaink, H. P., Okker, R. J. H., Wijffelman, C. A., Pees, E. and Lugtenberg, B. J. J. (1987). Promoters in the nodulation region of the *Rhizobium leguminosarum* Sym plasmid pRL1JI. *Plant Mol. Biol.* **9**, 27–39.

Spiro, S. (1992). An FNR-dependent promoter from *Escherichia coli* is active and anaerobically inducible in *Paracoccus denitrificans*. *FEMS Microbiol. Lett.* **98**, 145–148.

Steinrücke, P. and Ludwig, B. (1993). Genetics of *Paracoccus denitrificans*. *FEMS Microbiol. Rev.* **104**, 83–118.

Steinrücke, P., Steffens, G. C. M., Panskus, G., Buse, G. and Ludwig, B. (1987). Subunit II of cytochrome oxidase from *Paracoccus denitrificans*. *Eur. J. Biochem.* **167**, 431–439.

Stoll, R., Page, M. D., Sambongi, Y. and Ferguson, S. J. (1996). Cytochrome c_{550} expression in *Paracoccus denitrificans* strongly depends on growth condition; identification of promoter region for *cycA* by transcription start analysis. *Microbiology* **142**, 2577–2585.

Suwanto A. and Kaplan S. (1989). Physical and Genetic mapping of the *Rhodobacter sphaeroides* 2.4.1. genome, presence of two unique circular chromosomes. *J. Bacteriol.* **171**, No 11, p5850–5859.

Tai, T.-N., Havelka, W. A. and Kaplan, S. (1988). A broad-host-range vector system for cloning and translational *lacZ* fusion analysis. *Plasmid* **19**, 175–188.

Tahirov, T. H., Misaki, S., Meyer, T. E., Cusanovich, M. A., Higuchi, Y. and Yasuoka, N. (1996). High-resolution structures of two polymorphs of cytochrome *c′* from the purple phototrophic bacterium *Rhodobacter capsulatus*. *J. Mol. Biol.* **259**, 467–479.

Takano, S., Yano, T. and Yagi, T. (1996). Structural studies of the proton-translocating NADH-ubiquinone oxidoreductase (NDH-1). of *Paracoccus denitrificans*: Identity, property, and stoichiometry of the peripheral subunits. *Biochemistry* **35**, 9120–9127.

Taylor, R. K., Manoil, C. and Mekalanos, J. J. (1989). Broad-host-range vectors for delivery of Tn*phoA*: use in genetic analysis of secreted virulence determinants of *Vibrio cholerae*. *J. Bact.* **171**, 1870–1878

Thöny-Meyer, L., Beck, C., Preisig, O. and Hennecke, H. (1994). The *ccoNOQP* gene cluster codes for a *cb*-type cytochrome oxidase that functions in aerobic respiration of *Rhodobacter capsulatus*. *Mol. Microbiol.* **14**, 705–716.

Timkovich, R. and Dickerson, R. E. (1976). The structure of *Paracoccus denitrificans* cytochrome c_{550}. *J. Biol. Chem.* **251**, 4033–4046.

Ubbink, M., van Beeumen, J. and Canters, G. W. (1992). Cytochrome c_{550} from *Thiobacillus versutus*: cloning, expression in *Escherichia coli*, and purification of the heterologous holoprotein. *J. Bacteriol.* **174**, 3707–3714.

Ueda, S., Yabutani, T., Maehara, A. and Yamane, T. (1996). Molecular analysis of the poly(3-hydroxyalkanoate). synthase gene from a methylotrophic basterium, *Paracoccus denitrificans*. *J. Bacteriol.* **178**, 774–779.

van der Oost, J., Schepper, M., Stouthamer, A. H., Westerhoff, H. V., van Spanning, R. J. M. and de Gier, J. W. L. (1995). Reversed electron transport through the bc_1 complex enables a cytochrome *c* oxidase mutant ($\Delta aa_3/cbb_3$). of *Paracoccus denitrificans* to grow on methylamine. *FEBS Lett.* **371**, 267–270.

van der Palen, C. J. N. M., Slotboom, D.-J., Jongejan, L., Reijnders, W. N. M., Harms, N., Duine, J. A. and van Spanning, R. J. M. (1995). Mutational analysis of *mau* genes involved in methylamine metabolism in *Paracoccus denitrificans*. *Eur. J. Biochem.* **230**, 860–871.

van Spanning, R. J. M., Wansell, C. W., Reijnders, W. N. M., Oltmann, L. F. and Stouthamer, A. H. (1990a). Mutagenesis of the gene encoding amicyanin of *Paracoccus denitrificans* and the resultant effect on methylamine oxidation. *FEBS Lett.* **275**, 217–220.

van Spanning, R. J. M., Wansell, C., Harms, N., Oltmann. L. F. and Stouthamer, A. H. (1990b). Mutagenesis of the gene encoding cytochrome c_{550} of *Paracoccus denitrificans* and analysis of the resultant physiological effects. *J. Bacteriol.* **172**, 986–996.

van Spanning, R. J. M., Wansell, C. W., Reijnders, W. N. M., Harms, N., Ras, J., Oltmann, L. F. and Stouthamer, A. H. (1991a). A method for introduction of unmarked mutations in the genome of *Paracoccus denitrificans*: construction of strains with multiple mutations in the genes encoding periplasmic cytochromes c_{550}, c_{551i} and c_{553i}. *J. Bacteriol.* **173**, 6962–6970.

van Spanning, R. J. M., Wansell, C. W., de Boer, T., Hazelaar, M. J., Anazawa, H., Harms, N., Oltmann, L. F. and Stouthamer, A. H. (1991b). Isolation and characterisation of the *moxJ*, *moxG*, *moxI*, and *moxR* genes of *Paracoccus denitrificans*: Inactivation of *moxJ*, *moxG* and *moxR* and the resultant effect on methylotrophic growth. *J. Bacteriol.* **173**, 6948–6961.

van Spanning, R. J. M., de Boer, A. P. N., Reijnders, W. N. M., de Gier, J.-W. L., Delorme, C.O., Stouthamer, A. H., Westerhoff, H. V., Harms, N. and van der Oost, J. (1995a). Regulation of oxidative phosphorylation: The flexible respiratory network of *Paracoccus denitrificans*. *J. Bioenerg. Biomembr.* **27**, 499–512.

van Spanning, R. J., de Boer, A. P., Reijnders, W. N., Spiro, S., Westerhoff, H. V., Stouthamer, A. H. and van der Oost (1995a). Nitrite and nitric oxide reduction in *Paracoccus denitrificans* is under the control of NNR, a regulatory protein that belongs to the FNR family of transcriptional activators. *FEBS Lett.* **36**, 151–154.

Varga, A. R. and Kaplan, S. (1989). Construction, expression, and localization of a CycA::PhoA fusion protein in *Rhodobacter sphaeroides* and *Escherichia coli*. *J. Bacteriol.* **171**, 5830–5839.

Wachtveitl, J., Laussermair, E., Mathis, P., Farchaus, J. W. and Oesterhelt, D. (1993). Probing the donor side of reaction centres: site-directed mutants of tyrosine L1620 *Rhodobacter sphaeroides* and *Rhodopseudomonas viridis*. *Biochim. Soc. Trans.* **21**, 43–44.

Ward, J. M. and Grinstead, J. (1982). Physical and genetic analysis of the IncW group of plasmids R388, 5a and R7k. *Plasmid*, **7**, 239–250.

Weaver, K. E. and Tabita, F. R. (1983). Isolation and partial characterisation of *Rhodopseudomonas sphaeroides* mutants defective in the regulation of ribulose bisphosphate carboxylase/oxygenase. *J. Bacteriol.* **156**, 507–515.

Whatley, F. R. (1981). The establishment of mitochondria: *Paracoccus* and *Rhodopseudomonas*. *Ann. NY Acad. Sci.* **361**, 330–340.

Williams, J. C. and Taguchi, A. K. W. (1995). Genetic manipulation of purple photosynthetic bacteria. In *Anoxigenic Photosynthetic Bacteria* (R.E. Blankenship, M. T. Madigan and C. E. Bauer, eds), pp. 1029–1065. Kluwer Academic Publishers, Amsterdam.

Winterstein, C. and Ludwig, B. (1998). Genes coding for respiratory complexes map on all three chromosomes of the *Paracoccus denitrificans* genome. *Arch. Microbiol.* **169**, 275–281

Witt, H. and Ludwig, B. (1997). Isolation, analysis and deletion of the gene coding for subunit IV of cytochrome *c* oxidase in *Paracoccus denitrificans*. *J. Biol. Chem.* **272**, 5514–5517.

Wodara, C., Kostka, S., Egert, M., Kelly, D. P. and Friedrich, C. G. (1994). Identification and sequence analysis of the *soxB* gene essential for sulfur oxidation of *Paracoccus denitrificans* GB17. *J. Bacteriol.* **176**, 6188–6191.

Xu, H. H. and Tabita, F. R. (1994). Positive and negative regulation of sequences upstream of the form II cbb CO_2 fixation operon of *Rhodobacter sphaeroides*. *J. Bacteriol.* **176**, 7299–7308.

Xu, X., Matsuno-Yagi, A. and Yagi, T. (1993a). DNA sequencing of the seven remaining structural genes of the gene cluster encoding the energy-transducing NADH-quinone oxidoreductase of *Paracoccus denitrificans*. *Biochemistry* **32**, 968–981.

Paracoccus and Rhodobacter

Xu, H. H., Viebahn, M. and Hanson, R. S. (1993b). Identification of methanol-regulated promoter sequences from the facultative methylotrophic bacterium *Methylobacterium organophilum* XX. *J. Gen. Microbiol.* **139**, 743–752.

Yagi, T. (1993). The energy-transducing NADH–quinone oxidoreductase (NDH-1). of *Paracoccus denitrificans*. *Biochim. Biophys. Acta* **1101**, 181–183.

Yano, T., Sled, V. D., Ohnishi, T. and Yagi, T. (1994). Expression of the 25-kilodalton iron–sulfur subunit of the energy-transducing NADH–ubiquinone oxidoreductase of *Paracoccus denitrificans*. *Biochemistry* **33**, 494–499.

Yano, T, Yagi, T., Sled, V. D. and Ohnishi, T. (1995). Expression and characterisation of the 66-kilodalton (NQO3). iron–sulfur subunit of the proton-translocating NADH–quinone oxidoreductase of *Paracoccus denitrificans*. *J. Biol. Chem.* **270**, 18264–18270.

Yano, T., Sled, V. D., Ohnishi, T. and Yagi, T. (1996). Expression and characterisation of the flavoprotein subcomplex composed of 50-kDa (NQO1). and 25-kDa (NQO2). subunits of the proton-translocating NADH–quinone oxidoreductase of *Paracoccus denitrificans*. *J. Biol. Chem.* **271**, 5907–5913.

Yeliseev, A. A., Krueger, K. E. and Kaplan, S. (1997). A mammalian mitochondrial receptor functions as a bacterial oxygen sensor. *Proc. Natl. Acad. Sci. USA* **94**, 5105–5106.

Youvan, D. C., Bylina, E. J., Alberti, M., Begusch, H. and Hearst J.E. (1984). Nucleotide and deduced polypeptide sequences of the photosynthetic reaction-center, B870 antenna, and flanking polypeptides from *R. capsulata*. *Cell* **37**, 949–957.

Yun, C.-H., Beci, R., Crofts, A. R., Kaplan, S. and Gennis, R. B. (1990). Cloning and DNA sequencing of the *fbc* operon encoding the cytochrome bc_1 complex from *Rhodobacter sphaeroides*. Characterisation of *fbc* deletion mutants and complementation by a site-specific mutational variant. *Eur. J. Biochem.* **194**, 399–411.

Yun, C.-H., van Doren, S. R., Crofts, A. R. and Gennis, R. B. (1991). The use of gene fusions to examine the membrane topology of the L subunit of the photosynthetic reaction centre and of the cytochrome b subunit of the cytochrome bc_1 complex from *Rhodobacter sphaeroides*. *J. Biol. Chem.* **266**, 10967–10973.

Zeilstra-Ryalls, J. and Kaplan, S. (1995). Regulation of 5-aminolevulinic acid biosynthesis in *Rhodobacter sphaeroides*. *J. Bacteriol.* **177**, 2760–2768.

Zeilstra-Ryalls, J., Gomelsky, M., Eraso, J. M., Yeliseev, A., Ogara, J. and Kaplan, S. (1998). Control of photosystem formation in *Rhodobacter sphaeroides*. *J. Bacteriol.*, **180**, 2801–2809.

14 Sporulation in *Bacillus subtilis*

D. H. Green and S. M. Cutting
School of Biological Sciences, Royal Holloway University of London, Egham, Surrey, UK

◆◆

CONTENTS

List of Abbreviations

PCR	polymerase chain reaction
MCS	multiple cloning site
RT	room temperature
RNAP	RNA polymerase
rbs	ribosome binding site
BGSC	*Bacillus* Genetic Stock Center

◆◆◆◆◆◆ I. INTRODUCTION

Spore formation in *Bacillus subtilis* has been extensively studied over the last 30 years as a model for cell differentiation. Classical genetic analysis has been used to identify genes which are essential to the developmental process and with the advent of gene cloning in the mid-1970s a multitude of sporulation genes have been cloned and characterized at the molecular level. The result of this work has revealed an elegant, yet complex, programme of orchestrated gene expression which drives the physiological and morphological changes that result in the formation of a mature spore. In the late 1980s PCR was developed, which has now radically changed the way in which genes are manipulated. It is now relatively simple to clone genes "at will" providing a much faster pace to genetic analysis. More importantly, in 1997 the complete genome of *B. subtilis* became

Sporulation in Bacillus subtilis

available. The genome sequencing project has revealed that of the total number of genes found on the chromosome more than 30% are of unknown function. This suggests that there are still more sporulation genes to be discovered. The purpose of this chapter is to outline some of the methods that are currently used in the *B. subtilis* community for gene analysis. While we have not described all, we believe that we have presented a simplified description of a number of "streamlined" techniques that are in common use today. We should point out that considerable *B. subtilis* methodology is available in *Molecular Biological Methods for Bacillus* published by John Wiley & Sons Ltd (Harwood and Cutting, 1990).

◆◆◆◆◆◆ II. IDENTIFICATION OF SPORULATION GENES

Most sporulation genes have been identified by mutation in the last 20–30 years using classical methods for mutagenesis (e.g. NTG chemical mutagenesis or UV irradiation of growing cells). Identifying developmental mutants (*spo* or sporulation mutants) has been simplified by the fact that mutants which fail to sporulate, or are blocked at an intermediate stage of development, are readily identified on sporulation agar plates. Colonies able to form intact spores are pigmented brown (Pig⁺) after prolonged incubation at RT or 37°C while *spo* mutants appear translucent (due to extensive cell lysis) and lack the spore-associated pigment (Pig⁻). Typically, *spo* mutants are then characterized further by a number of established techniques. Electron microscopy is used to define the stage of blockage in the development. For example, a stage II mutant (*spoII*) is unable to proceed beyond stage II and complete engulfment of the prespore chromosome which defines stage III (see Figure 1), thus microscopy reveals cells containing prespores but no forespores. In addition to microscopy, a number of biochemical tests can be used to measure enzymes or chemicals that are synthesized at unique stages of spore formation (e.g. glucose dehydrogenate, GDH, at stage III and dipicolinic acid, DPA, at stage IV; see Nicholson and Setlow, 1990).

In addition to classical mutagenic procedures transposon-mediated insertional mutagenesis has successfully been used to identify new *spo* genes (Sandman *et al.*, 1987). This method utilizes the Tn917 transposon. Under appropriate conditions Tn917 can transpose and disrupt *spo* genes which, in turn, as mentioned above are readily identified by their colony phenotype (Pig⁺/⁻). The advantages of using Tn917 have been that: (i) the transposon is marked with an antibiotic-resistance marker facilitating rapid genetic mapping; and (ii) DNA sequences flanking the transposon can be cloned by a process of chromosome walking.

Procedures for mutagenesis whether using classical methods or using Tn917 are only suitable for loci having an observable phenotype when disrupted. It is perhaps not surprising that many genes encoding structural components of the spore are redundant and do not give a definitive sporulation phenotype (e.g. the *cot* genes which encode the spore

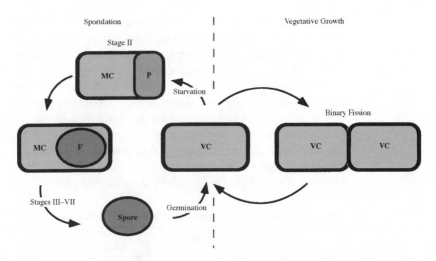

Figure 1. *B. subtilis* life cycle. *B. subtilis* under permissive growth conditions will ordinarily divide by binary fission (vegetative growth). Under conditions of nutrient limitation, the cell enters an irreversible developmental phase that results in the formation of a dormant endospore (for review see Errington, 1993). Initially, the genome is duplicated and one of the daughter chromatids is partitioned by a polar septum, forming the prespore (Stage II, or 2 h post-sporulation onset). The prespore is then engulfed, forming a "cell-within-a-cell" (termed the forespore) which lies within the mother cell (Stage III/3 h). Later stages (to Stage VII/7–8 h) entail the developing endospore becoming increasingly resistant to solvents, UV irradiation, heat and lysozyme, as the chromosome is condensed, and a thick cortex and proteinaceous coat are laid down on to the forespore. Ultimately, the mother cell is lysed and the endospore is released (Stage VII). The developmental cycle is completed when the endospore germinates under the appropriate environmental conditions and reinitiates vegetative growth. MC, mother cell; P, prespore; F, forespore; VC, vegetative cell.

coat proteins, and the *ssp* genes which encode the small acid-soluble proteins which encase the forespore chromosome). Generally, there are two strategies that have been used to identify phenotypically redundant genes. First, reverse genetics uses biochemical procedures to purify the protein (e.g. the spore coat proteins; Donovan *et al.*, 1987). Next, the amino-terminal sequence of the protein is determined using sequential Edman degradation, and finally, this sequence is used to design oligonucleotide probes (often these are "guessmers") to identify and clone the structural gene from libraries of *B. subtilis* fragments cloned in *Escherichia coli* plasmids (e.g. pUC18/19 or pBR322). A second, and more recent strategy is based on the fact that *spo* genes are transcribed by RNAP associated with one of five sporulation-specific sigma factors. The five sigma factors bind to core RNAP at different times (as well as in different compartments of the developing cell) and facilitate temporal and spatial gene expression. It is now possible to identify genes whose transcription is under the control of specific sigma factors. To achieve this, the ORF encoding a particular sigma factor is first cloned under the control of the IPTG-inducible *spac* promoter (*Pspac*; Yansura and Henner, 1984) using either an autonomously replicating episome or an integrative plasmid (see Table 1). Next, a library of transcriptional fusions (these are usually carried on a phage vector such as SPβ) is introduced into cells containing the inducible sigma factor gene and recombinants selected on a suitable antibiotic-resistant agar plate. Colonies which are Lac⁺ only in the presence of IPTG must contain genes whose expression is under the control of the inducible sigma factor. The *spo–lacZ* reporter gene can then be recovered and the *spo* gene identified by nucleotide sequencing across the boundary of the *spo–lacZ* junction. This strategy has proven successful in identifying a number of genes controlled by RNAP associated with σ^H (Jaacks *et al.*, 1989), σ^E (Beall *et al.*, 1993) and σ^G (Bagyan *et al.*, 1996).

Table 1. Vectors containing inducible sigma factor genes

	Plasmid[a]	Comments[b]
σ^H	pJOH7d	Integrative plasmid. *spoOH* fused to *Pspac* (Jaacks *et al.*, 1989)
σ^F	pSDA4	Autonomously replicating plasmid. *spoIIAC* fused to *Pspac* (Shazand *et al.*, 1995)
σ^E	pDG180	Autonomously replicating plasmid. *spoIIGB*[c] fused to *Pspac* (Frandsen and Stragier, 1995)
σ^G	pDG298	Autonomously replicating plasmid. *spoIIIG* fused to *Pspac* (Sun *et al.*, 1989)

[a]pJOH7d carries the *cat* gene (chloramphenicol resistance, 5 μg ml⁻¹) and pSDA4, pDG180 and pDG298 the *kan* gene (neomycin resistance, 2.5 μg ml⁻¹) for selection in *B. subtilis*. Selection for the latter three plasmids can also be made using phleomycin (0.1 μg ml⁻¹).
[b]Integrative plasmid: cannot replicate autonomously and introduction into recipient cells requires a single "Campbell-type" recombination into the chromosome.
[c]A modified form of *spoIIGB* devoid of the 5′-leader that encodes the pro-sequence of σ^E is fused to *Pspac* to ensure that a constitutively active form of σ^E is synthesized.

◆◆◆◆◆◆ III. GENETIC MAPPING

Having identified a sporulation mutation, the next step is to identify the genetic locus. For loci identified using either reverse genetics or IPTG induction to identify genes under the control of a particular sigma factor (see above) the task of mapping is straightforward. The *B. subtilis* genome has been sequenced in its entirety and is available to all researchers. Thus, clones identified using either of these methods can be partially sequenced and the complete DNA sequence retrieved from a computer database (see section VIII).

However, where the gene has been identified by classical procedures (chemical or UV mutagenesis) then the mutant allele must be methodically mapped on the chromosome. This can present a time-consuming task and early success depends upon the availability of genetic markers near to the mutant allele. Methods for genetic mapping have been described in detail elsewhere (Cutting and Vander-Horn, 1990; Azevedo *et al.*, 1993; Cutting and Youngman, 1993) and require two stages:

- **Stage I: Long-range mapping.** The mutation is assigned an approximate chromosomal location using PBS1-mediated transduction using a collection of auxotrophic or antibiotic-resistance markers spread around the chromosome. Mapping is continued until linkage is observed to one or two of these markers and is sufficient to place the mutation within a 300 kb region (which is the size of the chromosomal DNA that can be packaged in PBS1 phage heads).
- **Stage II: Fine structure mapping.** DNA-mediated transformational crosses are made to position the mutation to genetic markers in the vicinity of the mutation. This is an empirical task and is continued until linkage is observed between mutation and marker. Linkage can only arise between markers separated by no more than 30 kb of DNA since only short strands of chromosomal DNA can enter competent *B. subtilis* cells. At high linkages (greater than 70%), genetic distance is almost linear with 80% and 90% co-transformation frequencies representing a distance of 2 kb and 1 kb respectively.

With the completion of the *B. subtilis* genome sequencing project it should now be possible to identify rapidly the complete physical and partial genetic map of the chromosomal region to facilitate fine structure genetic mapping. Once high linkages are obtained it should be possible to determine whether the mutation is allelic to a known sporulation gene or represents a new locus. It is now probably simpler and less time consuming to clone the region of interest (from the wild-type strain using the PCR) and perform *in trans* complementation analysis by creating a merodiploid. A suitable system would be placement of DNA at the *amyE* locus using the vector pDG364 as described below in section VI. Conversion of the merodiploid cells to a Spo⁺ phenotype should be sufficient to define this as the genetic locus. The one requirement using this procedure is that the mutation is recessive! If the construction of merodiploids is not possible, then an alternative strategy would be "chromosome walking" from a cloned marker in the immediate proximity of the mapped mutation. This procedure is explained in more detail below (section IV). Some recent examples of sporulation mutations that have required exhaustive mapping are *bofB*, allelic to *spoIVFA* (Cutting *et al.*, 1990, 1991), *spoIIP*, allelic to *csfX* (Londono-Vallejo and Stragier, 1995) and *bofC* (Gomez and Cutting, 1997).

◆◆◆◆◆◆ IV. GENE CLONING

A variety of methods have been used to clone spo genes in B. subtilis. The most successful have used phage vectors (see also Chapter 3). The two most widely used phages are f105 and Spb. Derivatives of these temperate phages have been constructed which allow rapid construction of phage libraries of B. subtilis DNA. Transduction of a spo mutant with the phage library followed by selection for the Spo+ phenotype (chloroform-resistant colonies) allows rapid identification of clones. Both f105 and Spb have been engineered to allow retrieval of the cloned DNA from the phage genome to an E. coli plasmid and with the monopoly of PCR-based techniques this should now prove even simpler. Techniques for using the f105 and Spb phage cloning systems have been described extensively (Errington, 1990; Chapter 3, VII.D) and it is not necessary to repeat them here.

A. Chromosome Walking

At the completion of a programme of genetic mapping it should prove possible to establish the approximate position of a gene (defined by the mutation) within the existing genetic map of the chromosome. If plasmid clones are available for any markers in the proximity of the mutation then they can be used for a programme of chromosome walking (as shown in Figure 2). Initially, cloned DNA should be subcloned into an integrational plasmid and used to isolate DNA either to the right or left of the gene of

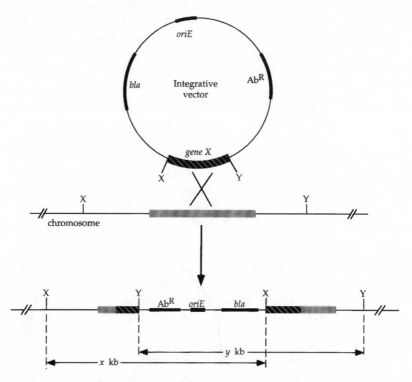

Figure 2. Chromosome walking. Schematic diagram of chromosome walking. An integrational plasmid containing a segment of *B. subtilis* DNA (gene X) is introduced into competent cells by DNA-mediated transformation with selection for the antibiotic resistance marker carried by the plasmid (Ab^R). The plasmid will integrate into the *B. subtilis* chromosome by a single crossover "Campbell-type" recombination event between homologous DNA shared between the plasmid and the chromosome. To clone DNA to the left or right of the cloned segment, chromosomal DNA from the recombinant is digested with restriction endonuclease X (to clone to the left) or Y (to clone to the right). The cleaved DNA is ligated at a low concentration (less than 10 ng ml^{-1}) to favour recircularization of plasmid monomers and transformed into *E. coli* with selection for Ap^R (*bla*, carried by the plasmid). Intact plasmids will now contain additional chromosomal DNA which extends out to sites X or Y. This procedure is dependent upon the availability of restriction sites in the plasmid DNA (sites used usually lie in the MCS). Note that there is a limit to how much chromosomal DNA can be cloned, thus, a restriction site over 5 kb from the point of insertion is unlikely to be cloned. Finally, walking to the left or right of the original marker will depend on the position of the restriction sites relative to the plasmid origin of replication which must be excised together with the *B. subtilis* DNA to enable reconstitution of the plasmid molecule.

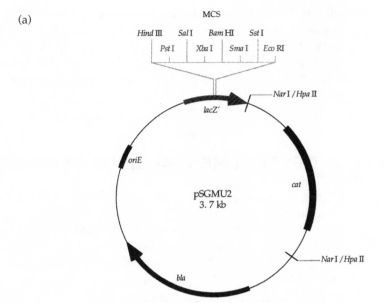

(a)

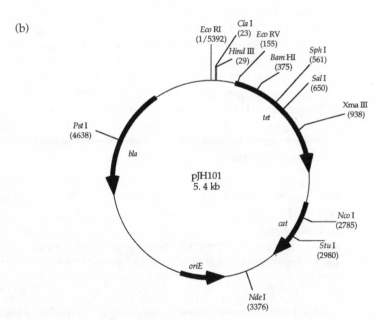

(b)

Figure 3. pSGMU2 and pJH101. Integrational vectors for the introduction of cloned DNA into the *B. subtilis* chromosome by Cambell-type recombination of homologous sequences on the vector and chromosome. Both vectors contain an *E. coli* origin of replication (*oriE*) that permits replication in *E. coli* but not in *B. subtilis*. (a) pSGMU2 (Errington, 1990) is based on pUC13 with the addition of a *cat* cassette (selectable Cm^R resistance in *Bacillus*). Cloning of DNA into this vector (in *E. coli*) is aided by the MCS and blue–white selection conferred by disruption of the *lacZ'* reading frame. (b) pJH101 (Ferrari *et al.*, 1983) offers a number of unique restriction endonuclease sites for cloning and a selectable marker (Cm^R encoded by the *cat* gene) in *B. subtilis*. Nucleotide positions of the restriction sites are shown in brackets.

interest. Integrational plasmids are unable to replicate autonomously in *B. subtilis* but carry a replicative origin for *E. coli*. These plasmids also carry an antibiotic-resistance marker (Ab^R) which can be selected for in *B. subtilis* (e.g. Cm^R, chloramphenicol resistance; Sp^R, spectinomycin resistance; Nm^R, neomycin resistance). Two such plasmids are pSGMU2 and pJH101 (see Figure 3).

◆◆◆◆◆◆ V. DEVELOPMENTAL GENE EXPRESSION

One of the most powerful tools for analysis of sporulation genes has been the analysis of gene expression. In principle, this entails fusing the gene of interest to a reporter gene and measuring synthesis of the reporter protein *in vivo*. The most widely used reporter gene is the *E. coli lacZ* gene which encodes β-galactosidase. Two types of *lacZ* fusion can be constructed:

- **Transcriptional.** The reporter gene is inserted downstream of the sporulation promoter. The reporter can be inserted as a cassette at any position within the *spo* gene so long as it is downstream of the sporulation-specific promoter. To facilitate expression of the cassette the *lacZ* ORF is fused in frame, at its 5′ end to the rbs, leader sequence and normally the first eight codons of a strongly expressed sporulation gene (normally from the *spoVG* or *spoIIAC* genes). This type of construct is the most widely used since any effects of gene expression due to translational control are removed.
- **Translational.** The reporter gene is fused, in frame, with the *spo* gene ORF creating a gene that synthesizes a protein chimera comprising *N*-terminal end of the *B. subtilis* protein and *lacZ*. Expression of *lacZ* is thus under both transcriptional and translational control.

A number of complete "systems" have been designed for both cloning and construction of reporter gene fusions. The phage φ105 has been exploited for the rapid conversion of cloned genes to *spo–lacZ* fusions and allowing construction of both translational and transcriptional (using the *spoIIAC* rbs and leader sequence) fusions (Errington, 1986). Tn917 has been exploited for mutagenesis, cloning and construction of transposon-tagged genes to *lacZ* transcriptional fusions (using the *spoVG* rbs and leader sequence; Youngman, 1990). The Tn917-based system allows transfer of the fusion to the SPβ temperate phage which, like φ105, allows easy transfer of the fusion to alternate genetic backgrounds (Chapter 3, V.E).

For genes which have not been cloned using φ105, SPβ or the Tn917-based systems a number of vectors exist that allow straightforward construction of *spo–lacZ* transcriptional fusions:

- **pDG268** (Figure 4a) contains the *lacZ* gene fused at its 5′ end to the translational leader sequence (rbs and first eight codons) of the *spoVG* gene (Stragier et al., 1988). The gene of interest (part or all of the gene, but not the transcriptional termination signals!) is inserted into a MCS preceding *lacZ*.

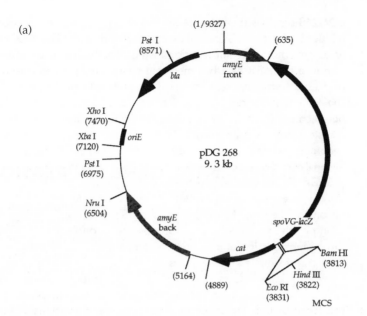

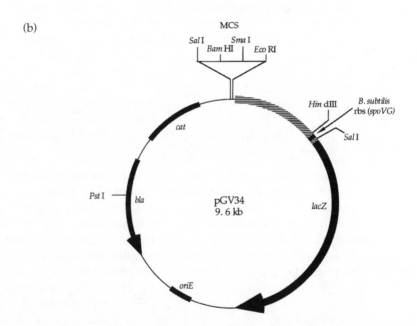

Figure 4. pDG268 and pGV34. Integrational vectors used for creating transcriptional fusions in *B. subtilis* by either double-crossover or single Cambell-type recombinational events, respectively. Both vectors contain an *E. coli* origin of replication (*oriE*) and only replicate in *E. coli*. (a) pDG268 contains *lacZ* fused at its 5′ end to the translational leader sequence (rbs and first 8 codons) of the *spoVG* gene (Stragier *et al.*, 1988). This vector integrates at the *amyE* locus via the front and back regions of *amyE* located on the vector, generating recombinants which are Cm^R and Amy⁻. Nucleotide positions of the restriction sites are shown in brackets. (b) pGV34 contains immediately following the MCS *B. subtilis* chromosomal DNA upstream of and including the *spoVG* rbs (Youngman, 1990).

pDG268 contains the *cat* gene (encoding Cm^R) as well as the front and back portions of the *amyE* gene allowing insertion of the *spo–lacZ* fusion at the *amyE* gene following linearization of the recombinant plasmid and introduction into *B. subtilis* by DNA-mediated transformation and selection for Cm^R (pDG268 in this respect, is similar to pDG364; see section VI). Since the *spo–lacZ* gene fusion will insert at the *amyE* gene by a double crossover between the front and back portions of *amyE* contained within the plasmid, transformants will be Amy⁻. Chromosomal DNA prepared from such a recombinant can then be introduced into other genetic backgrounds facilitating analysis of genetic dependence, etc.

- **pGV34** (Figure 4b) contains the *lacZ* gene fused at its 5' end to the rbs of the *spoVG* gene (Youngman, 1990). As with pDG268 (above) transcriptional fusions can be made by cloning into a MCS preceding the *lacZ* gene. pGV34 is an integrative vector which cannot replicate autonomously in *B. subtilis*, so following introduction of plasmid into competent *B. subtilis* cells by DNA-mediated transformation the *lacZ* fusion will integrate at the sporulation locus using a single "Campbell-type" recombination between homologous sequences carried on the vector and chromosome. The resulting integration produces two copies of the sporulation locus, one wild-type and the other fused to *lacZ*. This vector can also be used to introduce *spo–lacZ* fusions on to the SPβ prophage (Youngman, 1990).

A. Temporal

Initial characterization of gene expression involves measurement of β-galactosidase synthesis during sporulation driven from the sporulation promoter. This enables the time of initiation of gene expression to be determined. Two methods are used to induce sporulation, the resuspension method and the exhaustion method. Most laboratories working with *B. subtilis* use one of these two methods and each have their own advantages and are described in detail by Nicholson and Setlow (1990).

I. Resuspension method

This method relies on inducing sporulation in an essentially "unnatural" way. Cells are grown in a casein-based growth medium and at the mid-logarithmic phase of growth resuspended in an equal volume of a minimal medium and incubation continued. The time of resuspension represents T_0 and defines the initiation of the developmental process. At hourly intervals thereafter samples are withdrawn and assayed for β-galactosidase synthesis driven from the *spo* gene promoter. The precise time during mid-logarithmic growth that cells can be resuspended is relatively flexible between the range OD_{600} 0.2–0.8. Therefore, a number of different samples can be grown and resuspended at the same time which simplifies sampling later on.

2. Exhaustion method

This procedure more closely resembles the "assumed" natural conditions that promote spore formation. Cells are grown in a rich growth medium (DSM, Difco sporulation medium) and cell growth monitored (OD_{600nm}) at frequent intervals. A graph is plotted on two-cycle log paper of cell growth and the point at which growth is no longer exponential is defined as T_0, this can be extrapolated from the tangent. At intervals thereafter samples are withdrawn and assayed for β-galactosidase synthesis driven from the *spo* gene promoter. Generally, the exhaustion method produces greater levels of β-galactosidase activity. One reason is that cells reach a much higher cell density during sporulation using the exhaustion method (up to an OD_{600} ~2.5, compared with 1–1.5 using the resuspension method). While the temporal progress of sporulation is the same as when using the resuspension method, spores produced with the exhaustion method appear to be more robust with thicker spore coats (when viewed by electron microscopy). The one significant drawback with the exhaustion method is the necessity of taking numerous optical density readings to establish T_0; different samples will most likely not have the same time for T_0 making subsequent sampling of multiple cultures problematic. With both the resuspension and exhaustion methods extrapolating the point at which gene expression commences is made more accurate by taking frequent samples (≤30 min).

B. Dependence

The analysis of *spo*-directed β-galactosidase in developmental mutants has provided an important tool for characterizing sporulation gene expression. In this method a *spo–lacZ* fusion is introduced into cells containing a *spo* mutation and β-galactosidase synthesis measured during sporulation against a *spo*[+] control. If the *spo* mutation encodes a gene product that is required for *spo–lacZ* expression then no β-galactosidase synthesis will occur. Obvious examples would be regulatory genes encoding DNA-binding proteins or sigma factors. It is imperative that isogenic strains are used for dependence analysis since significant differences occur between different strains. For this purpose a collection of *spo* mutants in an appropriate strain background is invaluable. The recommended and most defined genetic backgrounds are those of the *spo*[+] strains 168 *trpC2* , PY79 and JH642 *trpC2 pheA1* all of which are available from the BGSC (see section VIII).

As the genetic dependence of more and more genes is characterized a picture has emerged as to how *spo* genes are regulated during spore formation. In brief, five sigma factors, σ^H, σ^F, σ^E, σ^G and σ^K bound to core RNAP direct the transcription of five developmental regulons (for a review, see Errington, 1993). The action of σ^F and σ^G is confined to the prespore/forespore compartment of the sporulating cell while σ^E and σ^K are restricted to the mother cell. In addition, at least four DNA-binding proteins, AbrB, SpoIIID, SpoVT and GerE (Kunkel *et al.*, 1989; Strauch *et al.*, 1989; Zheng *et al.*, 1992; Bagyan *et al.*, 1996) co-regulate expression of subsets of genes, either positively or negatively. To provide an initial characterization of a new developmental gene the first step would be to construct a *spo–lacZ* fusion and analyse gene expression during sporulation. The time at which the gene is initially expressed will provide a clue as to which sigma factor is directing transcription (as shown in Table 2). Next, the *spo–lacZ* fusion is introduced into a collection of developmental mutants blocked in sigma factor production. These strains should preferably contain null (insertion and deletion) mutations in each of the *sig* genes which encode the five sporulation-specific sigma factors, σ^H (*sigH* or *spoOH*), σ^F (*sigF* or *spoIIAC*), σ^E (*sigE* or *spoIIGB*), σ^G (*sigG* or *spoIIIG*) and σ^K (*sigK* or *spoIVCB/spoIIIC*). If gene expression is blocked in any of these mutants then clearly the appropriate sigma factor is required. However, the situation is more complicated for the four sigmas (σ^F, σ^E, σ^G and σ^K), because the *sig* genes themselves are dependent upon transcription by another (preceding) sigma factor, and in some cases intercompartmental signals are also required for their activation. The interdependence and cross-regulation of sigma factor gene expression has now been clearly defined (Losick and Stragier, 1992) and is summarized for use in dependence analysis in Table 2.

One problem sometimes encountered in the analysis of *spo*-directed β-galactosidase synthesis is in defining the block in gene expression, for example, when expression of a *spo–lacZ* fusion in a mutant background is reduced to 10% of that in a *spo⁺* strain. A number of factors may be involved in allowing these residual levels of gene expression:

- **Strain background**. As mentioned above it is essential that all strains be isogenic since considerable strain differences are now known to exist, and in some cases strains have been identified with mutations in more than one *spo* loci.
- **Oligosporogeny**. Many *spo* mutants allow a percentage of intact, heat-resistant, spores to be produced while retaining their defective allele. Presumably,

Table 2. Genetic dependence[a]

	Time[a]	sigH	sigF	sigE	sigG	sigK	spoIIID	spoVT	gerE
σ^H	*T0–1*	−							
σ^F	*T0.5–1.5*	−	−						
σ^E	*T1–3*	−	−	−			+/−		
σ^G	*T2–3*	−	−	−	−			+/−	
σ^K	*T3.5–*	−	−	−	−	−			+/−

[a]The table shows the approximate time periods during sporulation when the sigma factor proteins are synthesized and active. Gene expression controlled by a particular sigma factor is dependent upon the synthesis of other sigma factors as indicated in the table (−) and also a particular gene may be also co-regulated (+/−) by the SpoIIID, SpoVT or GerE DNA-binding proteins as shown. Genetic dependence should be established by: (i) analysis of the time of initiation of gene expression; (ii) introducing a *spo–lacZ* fusion into isogenic strains containing null alleles in each of the sigma factor genes (*sig*) and examining *spo*-directed β-galactosidase synthesis; and (iii) analysis in the *spoIIID, spoVT* or *gerE* genes if appropriate.

the mutations are "leaky", e.g. in a minority of sporulating cells the requirement for the defective gene product can be overcome (perhaps due to redundancy or epistasis). Whatever the mechanism for oligosporogeny this background can be as high as 1% in some mutants which will generate measurable background levels of β-galactosidase.

Most laboratories working on *spo* gene expression replicate the analysis of the synthesis of β-galactosidase from reporter genes many times using the same strains, thus establishing trends rather than accurate definitions of what constitutes a block in gene expression. It may be reasonable to state that levels of β-galactosidase synthesis below 10% can be considered a block in gene expression. Other factors which should be considered when analysing gene expression are:

1. **Forespore-specific gene expression**. Some genes are expressed only in the forespore chamber (e.g. σ^G-controlled genes) so in determining enzyme activity it is necessary to treat samples with lysozyme to allow entry of the β-galactosidase substrate ONPG to the spore interior.
2. **Dual recognition**. Some genes whose expression is controlled by σ^G are also recognized by σ^F, e. g. *bofC* (Gomez and Cutting, 1997), since they share the same promoter. Temporal analysis of gene expression may reveal two peaks (biphasic gene expression) where σ^G replaces σ^F in the RNAP holoenzyme complex.
3. *B. subtilis* contains a cryptic *lacZ* gene which provides basal levels of β-galactosidase activity. During analysis of *spo–lacZ* gene expression an isogenic *spo*⁺ strain containing no *spo–lacZ* fusions should be sampled in parallel and basal levels subtracted from those found in samples containing the *spo–lacZ* fusion.

C. *In Vivo* Determination of Sigma Factor Recognition

As further proof of sigma factor specificity *in vivo*, expression of a gene of interest can be induced by synthesis of the sporulation-specific sigma factor during vegetative growth. Four plasmids (see Table 1) have been constructed where the structural gene for the sigma factors, σ^H, σ^F, σ^E and σ^G are fused downstream of and under the control of the IPTG-inducible *spac* promoter, *Pspac* (Yansura and Henner, 1984). Cells containing a transcriptional fusion of the gene of interest to a reporter gene (*lacZ*) are transformed with plasmid DNA containing the IPTG-inducible sigma factor gene. Cells are grown in a suitable rich medium (LB or 2 XYT) to an OD_{600} ~ 0.2–0.5 and then the culture is divided into equal portions. To one, IPTG (final concentration of 1 mM) is added and incubation continued. Synthesis of β-galactosidase only in cells induced with IPTG demonstrates sigma factor specificity. Note that since expression of *spoIIIG* which encodes σ^G is controlled by RNAP associated with σ^F and because *spoIIIG* autoregulates, to avoid any problems associated with competition or inhibition when inducing σ^G, cells should contain a chromosomal disruption of the *spoIIIG* gene. Where possible, sigma factor recognition and the transcriptional start site should then be confirmed by primer extension analysis of mRNA isolated from induced and uninduced cells containing the

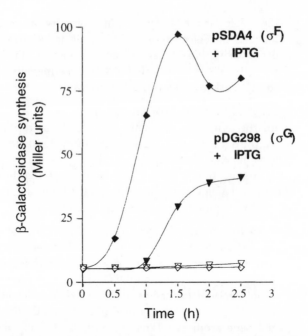

Figure 5. Induction of *spo* gene expression during vegetative cell growth. *bofC*-directed β-galactosidase synthesis was determined in *spoIIIGΔ1* cells containing the *bofC-lacZ* and either pSDA4 (*Pspac-IIAC*) or pDG298 (*Pspac-IIIG*). Cultures were grown in 2XYT medium with phleomycin (0.1 μg ml⁻¹) and at an OD₆₀₀ of 0.3 the culture was split and IPTG added to one portion (final concentration 1 mM) and growth of both cultures continued. Samples were removed and assayed for β-galactosidase activity at the indicated times after IPTG induction. Symbols: pSDA4 – IPTG (◊), pSDA4 + IPTG (♦), pDG298 – IPTG (▽), pDG298 + IPTG (▼).

plasmid, and the start site compared to that obtained from sporulation-specific mRNA.

Figure 5 shows this type of analysis performed on the *bofC* developmental gene (Gomez and Cutting, 1997). From genetic dependence studies using a *bofC-lacZ* transcriptional fusion, *bofC* expression was predicted to be under the control of Eσ^G or Eσ^F + Eσ^G since expression was dependent on both the *spoIIAC* and *spoIIIG* genes which encode σ^F and σ^G respectively. Induction of either σ^F or σ^G during vegetative cell growth was sufficient to promote *bofC*-directed β-galactosidase synthesis demonstrating that the *bofC* promoter was recognized by both Eσ^F and Eσ^G.

D. Protein Localization and Cell-specific Gene Expression

Recently techniques to detect the subcellular localization of gene products have led to a greater understanding of the roles of proteins involved in the control of sporulation. While immunofluorescence microscopy has been successfully employed, by far the most popular technique is the use of gene fusions to the Green Fluorescent Protein (GFP) of *Aequorea victoria*

(Webb *et al.*, 1995; Lewis and Errington, 1996; Barak *et al.*, 1996; Lin *et al.*, 1997; Kohler and Marahiel, 1997). GFP emits green light under illumination by a blue light source. As no metabolic substrates need to be introduced into the cells (as is the case with the *lux* system), GFP can be used in living, intact cells. GFP is always used as a translational fusion. To obtain reliable protein localization data it is important that the fusion does not disrupt the normal localization signals of the protein under study and if possible the activity of the protein as a whole. Usually the fusion is made to the C-terminal part of the protein. With the appropriate equipment, very high resolution for subcellular localization can be obtained; for example a fusion to the 3′ end of the *spoIIE* gene permitted detection of the SpoIIE–GFP fusion specifically in the septum between the prespore and the mother cell (Barak *et al.*, 1996; Wu *et al.*, 1998).

◆◆◆◆◆◆ VI. *IN TRANS* COMPLEMENTATION ANALYSIS

In trans complementation analysis is an important tool for characterizing sporulation mutations. Traditionally, *trans* complementation analysis has been used for: (i) determination of the dominance/recessive nature of mutations, which can provide important insights into gene function; and (ii) assigning multiple alleles to complementation groups (cistrons). In addition, complementation of a mutation with the wild-type allele can be used to complete a programme of gene mapping. For example, following extensive fine structure mapping of a new mutation it is often possible to identify candidate genes in which the wild-type allele could lie. With the completion of the genomic sequence of the *B. subtilis* chromosome, in principle, it is now possible to rapidly clone any putative gene using PCR. To facilitate *in trans* complementation analysis the wild-type gene must be placed extrachromosomally or at a position unlinked to the gene of interest. Rapid cloning into phage-based vectors is too cumbersome and time consuming for rapid analysis. The method of choice is to create a merodiploid at the *amyE* locus (Shimotsu and Henner, 1986). To introduce a gene at this position (found at 26° on the chromosome) the gene of interest is first cloned into pDG364 (Figure 6) at the MCS. The MCS is arranged such that flanking each side are the front and back portions of the *amyE* gene together with a selectable marker (the *cat* gene encoding resistance to Cm^R; 5 μg ml^{-1}). Plasmid DNA is linearized using restriction sites found in the pDG364 backbone (e.g. *Pst*I) and introduced into the mutant by transformation. Selection is made for Cm^R and transformants screened for production of amylase. Amy⁻ transformants will arise from introduction of the gene of interest by a double crossover recombination between the front and back portions of *amyE* carried on pDG364, as shown in Figure 6. Integration at the *amyE* locus is stable and chromosomal DNA from the recombinant can be prepared and used for further *trans* complementation analysis if necessary.

Merodiploids constructed at the *amyE* locus have been used to rapidly assign mutations to the *ftsH* locus (Lysenko *et al.*, 1997) and for analysis of dominance/recessivity of *spoIVF* alleles (Cutting *et al.*, 1991).

◆◆◆◆◆◆ VII. SELECTABLE MARKERS

One of the advantages of genetic analysis in *B. subtilis* is the large number of antibiotic-resistance markers that are available. The most widely used genes are *cat* (chloramphenicol resistance; Cm^R), *erm* (resistance to erythromycin and lincomycin; Er^R or MLS^R), *spc* (spectinomycin resistance; Sp^R), *neo* (resistance to neomycin; Nm^R), *tet* (tetracycline resistance; Tc^R) and *ble* (resistance to phleomycin or bleomycin; Pm^R) (see Table 1 in Chapter 2). These markers have enabled insertional mutagenesis of target genes and the tagging of transposons or phages. It is theoretically possible to construct strains with multiple antibiotic resistance markers (up to four). The logistics of performing this have been simplified using a series of plasmids (Steinmetz and Richter, 1994) which enable conversion of one chromosomal-borne antibiotic resistance gene to an alternative resistance gene. For example, a gene marked by the *cat* gene (*gen::cat*) can be converted to a *spc* gene (*gen::spc*) by DNA-mediated transformation of *cat*-containing cells with plasmid pCm::Sp followed by selection for Sp^R. pCm::Sp should be linearized before transformation (although most plasmid molecules will be linearized by entry into the cell), and contains front and back portions of the *cat* gene flanking the *spc* gene. Homologous recombination between these portions of the *cat* gene and the chromosomal-borne *cat* gene enable insertion of *spc* and simultaneous inactivation of the *cat* gene (transformants will therefore be Sp^R Cm^S). The ability to convert antibiotic-resistance markers enables rapid strain constructions and provision of reliable chromosomal markers. Plasmids facilitating the conversion of Cm^R to Er^R, Nm^R, Sp^R and Tc^R, and the conversion of Er^R to Cm^R, Nm^R, Pm^R and Sp^R are available from the BGSC (see section VIII below).

◆◆◆◆◆◆ VIII. AVAILABLE RESOURCES

A. Strains

The *Bacillus* Genetic Stock Center (BGSC) provides *Bacillus* strains free of charge to academic laboratories upon request to: Dr Daniel Zeigler, BGSC, Department of Biochemistry, The Ohio State University, Columbus, OH 43210, USA. Tel: 614-292-1538, Fax: 614-292-6773; e-mail: dzeigler@magnus.acs.ohio-state.edu.

B. Nucleotide and Protein Databases

Data generated from the international *B. subtilis* genome sequencing project can be found at the following world wide web sites:

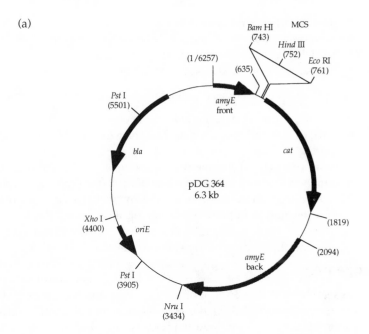

(a)

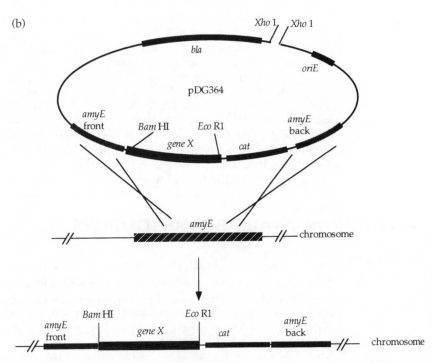

(b)

Figure 6. *In trans* complementation analysis using the *amyE* gene. (a) Map of pDG364 showing the multiple cloning site, *cat* gene and front and back portions of the *amyE* gene. Restriction sites that can be used for linearization are indicated; nucleotide positions are noted in brackets. (b) Schematic diagram showing the double-crossover recombinational event that generates a merodiploid using pDG364.

- Subtilist. http://www.pasteur.fr/Bio/Subtilist.html
- Micado site. http://locus.jouy.inra.fr/cgi-bin/genmic/madbase_home.pl
- NRSUB. http://acnuc.univ-lyon1.fr/nrsub/nrsub.html
- International Sequencing Project in Japan. http://bacillus.tokyo-center.genome.ad.jp

Related *B. subtilis* sequence data can be found at the following web sites:

- *B. subtilis* Protein Index. http://pc13mi.biologie.uni-greifswald.de
- Expasy's *B. subtilis* protein data base. http://expasy.hcuge.ch/cgi-bin/lists?subtilis.txt
- NCBI's eubacterial genome site. http://www.ncbi.nlm.nih.gov/Entrez/Genome/eub_g.html.
- The author's laboratory also lists the known *B. subtilis* sporulation genes and their function at: http://www.bl.rhbnc.ac.uk/staff/cutting

Acknowledgements

Research in this laboratory is supported by grants from the Medical Research Council (MRC), Biotechnology and Biological Sciences Research Council (BBSRC) and the Royal Society.

References

Azevedo, V., Alvarez, E., Zumstein, E., Damiani, G., Sgaramella, V., Erhlich, S. D. and Serror, P. (1993). An ordered collection of *Bacillus subtilis* DNA segments cloned in yeast artificial chromosomes. *Proc. Natl. Acad. Sci. USA* **90**, 6047–6051.

Bagyan, I., Hobot, J. and Cutting, S. M. (1996). A compartmentalized regulator of developmental gene expression in *Bacillus subtilis. J. Bacteriol.* **178**, 4500–4507.

Barak, I., Behari, J., Olmedo, G., Guzman, P., Brown, D.P., Castro, E., Walker, D., Westpheling, J. and Youngman, P. (1996). Structure and function of the *Bacillus* SpoIIE protein and its localisation to sites of sporulation septum assembly. *Mol. Microbiol.* **19**, 1047–1060.

Beall, B., Driks, A., Losick, R. and Moran, C. P. (1993). Cloning and characterization of a gene required for assembly of the *Bacillus subtilis* spore coat. *J. Bacteriol.* **175**, 1705–1716.

Cutting, S., Oke, V., Driks, A., Losick, R., Lu, S. and Kroos, L. (1990). A forespore checkpoint for mother cell gene expression during development in *B. subtilis. Cell* **62**, 239–250.

Cutting, S., Roels, S. and Losick, R. (1991). Sporulation operon *spoIVF* and the characterization of mutations that uncouple mother-cell from forespore gene expression in *Bacillus subtilis. J. Mol. Biol.* **221**, 1237–1256.

Cutting, S. M. and Vander-Horn, P. B. (1990). Genetic Analysis. In *Molecular Biological Methods for* Bacillus (C. R. Harwood and S. M. Cutting, eds), pp. 27–74. John Wiley & Sons, Chichester, UK.

Cutting, S. M. and Youngman, P. (1993). Gene transfer in gram-positive bacteria. In *Methods for General and Molecular Bacteriology* (P. Gerhardt, R. G. E. M, W.A. Wood and N.R. Krieg, eds), pp. 348–364. American Society for Microbiology, Washington, DC.

Donovan, W., Zheng, L., Sandman, K. and Losick, R. (1987). Genes encoding spore coat polypeptides from *Bacillus subtilis. J. Mol. Biol.* **196**, 1–10.

Errington, J. (1986). A general method for fusion of the *Escherichia coli lacZ* to chromosomal genes in *Bacillus subtilis*. *J. Gen. Microbiol.* **132**, 2953–2966.

Errington, J. (1990). Gene cloning techniques. In *Molecular Biological Methods for Bacillus*. (C. R. Harwood and S. M. Cutting, eds), pp. 175–220. John Wiley & Sons, Chichester, UK.

Errington, J. (1993). *Bacillus subtilis* sporulation: regulation of gene expression and control of morphogenesis. *Microbiol. Rev.* **57**, 1–33.

Ferrari, F. A., Nguyne, A., Lang, D. and Hoch, J. A. (1983). Construction and properties of an integrable plasmid for *Bacillus subtilis*. *J. Bacteriol.* **154**, 1513–1515.

Frandsen, N. and Stragier, P. (1995). Identification and characterization of the *Bacillus subtilis spoIIP* locus. *J. Bacteriol.* **177**, 716–722.

Gomez, M. and Cutting, S. (1997). *bofC* encodes a putative forespore regulator of the *Bacillus subtilis* σ^K checkpoint. *Microbiology* **143**, 157–170.

Harwood, C. R. and Cutting, S. M. (1990). *Molecular Biological Methods for* Bacillus. John Wiley & Sons, Chichester, UK.

Jaacks, K. J., Healy, J., Losick, R. and Grossman, A. D. (1989). Identification and characterization of genes controlled by the sporulation-regulatory gene *spoOH* in *Bacillus subtilis*. *J. Bacteriol.* **171**, 4121–4129.

Kohler, P. and Marahiel, M.A. (1997). Association of the histone-like protein Hbsu with the nucleoid of *Bacillus subtilis*. *J. Bacteriol.* **179**, 2060–2064.

Kunkel, B., Kroos, L., Poth, H., Youngman, P. and Losick, R. (1989). Temporal and spatial control of the mother-cell regulatory gene *spoIIID* of *Bacillus subtilis*. *Gene Develop.* **3**, 1735–1744.

Lewis, P.J. and Errington, J. (1996). Use of green fluorescent protein for detection of cell-specific gene expression and subcellular protein localisation during sporulation in *Bacillus subtilis*. *Microbiology UK* **142**, 733–740.

Lin, D.C.H., Levin, P.A. and Grossman, A.D. (1997). Bipolar localisation of a chromosome partition protein in *Bacillus subtilis*. *Proc. Natl. Acad. Sci. USA* **94**, 4721–4726.

Londono-Vallejo, J.-A. and Stragier, P. (1995). Cell–cell signaling pathway activating a developmental transcription factor in *Bacillus subtilis*. *Gene Develop.* **9**, 503–508.

Losick, R. and Stragier, P. (1992). Crisscross regulation of cell-type-specific gene expression during development in *B. subtilis*. *Nature* **355**, 601–604.

Lysenko, E., Ogura, T. and Cutting, S. (1997). Characterization of the *ftsH* gene of *Bacillus subtilis*. *Microbiology* **143**, 971–978.

Nicholson, W. L. and Setlow, P. (1990). Sporulation, germination and outgrowth. In *Molecular Biological Methods for Bacillus*. (C. R. Harwood and S. M. Cutting, eds), pp. 391–450. John Wiley & Sons, Chichester, UK.

Sandman, K., Losick, R. and Youngman, P. (1987). Genetic analysis of *Bacillus subtilis spo* mutations generated by Tn917-mediated insertional mutagenesis. *Gentics* **117**, 603–617.

Shazand, K., Frandsen, N. and Stragier, P. (1995). Cell-type specificity during development in *Bacillus subtilis*: the molecular and morphological requirements for σE activation. *EMBO J.* **14**, 1439–1445.

Shimotsu, H. and Henner, D. J. (1986). Construction of a single copy integration vector and its use in analysis of regulation of the *trp* operon of *Bacillus subtilis*. *Gene* **43**, 85–94.

Steinmetz, M. and Richter, R. (1994). Plasmids designed to alter the antibiotic resistance expressed by insertion mutations in *Bacillus subtilis*, through *in vivo* recombination. *Gene* **142**, 79–83.

Stragier, P., Bonamy, C. and Karmazyn-Campbelli, C. (1988). Processing of a sporulation sigma factor in *Bacillus subtilis*: how morphological structure could control gene expression. *Cell* **52**, 697–704.

Strauch, M. A., Spiegelman, G. B., Perego, M., Johnson, W. C., Burbulys, D. and Hoch, J. A. (1989). The transition state transcription regulator *abrB* of *Bacillus subtilis* is a DNA binding protein. *EMBO J.* **8**, 1615–1621.

Sun, D., Stragier, P. and Setlow, P. (1989). Identification of a new σ-factor involved in compartmentalized gene expression during sporulation of *Bacillus subtilis*. *Gene Develop.* **3**, 141–149.

Webb, C.D., Decatur, A., Teleman, A. and Losick, R. (1995). Use of green fluorescent protein for visualisation of cell-specific gene-expression and subcellular protein localisation during sporulation in *Bacillus subtilis*. *J. Bacteriol.* **177**, 5906–5911.

Wu, L.J., Feucht, A. and Errington, J. (1998). Prespore-specific gene expression in *Bacillus subtilis* is driven by sequestration of SpoIIE phosphatase to the prespore side of the asymmetric septum. *Gene Dev.* **12**, 1371–1380.

Yansura, D. G. and Henner, D. J. (1984). Use of *Escherichia coli lac* repressor and operator to control gene expression in *Bacillus subtilis*. *Proc. Natl. Acad. Sci. USA* **81**, 439–443.

Youngman, P. (1990). Use of transposons and integrational vectors for mutagenesis and construction of gene fusions in *Bacillus* species. In *Molecular Biological Methods for Bacillus*. (C. R. Harwood and S.M. Cutting, eds), pp. 221–266. John Wiley and Sons, Chichester, UK.

Zheng, L., Halberg, R., Roels, S., Ichikawa, H., Kroos, L. and Losick, R. (1992). Sporulation regulatory protein GerE from *Bacillus subtilis* binds to and can activate or repress transcription from promoters for mother-cell-specific genes. *J. Mol. Biol.* **226**, 1037–1050.

Index

Page numbers in *italics* refer to figures and tables.

luciferase (*luc, lux*) genes, 116–17, 153
  in clostridia, 200
  Erwinia homologues, 388–9, 408
  in mycobacteria, 267–8
Lyme disease, 210
lysins, bacteriophage, 120–2
lysis
  alkaline, in plasmid isolation, 84–5
  lysin-mediated, 121–2
  method, chromosomal DNA preparation, 178
  phage-mediated, 98–9, 101
lysogenic conversion, 99
lysogeny, 99
lysozyme
  in electroporation, 26
  in plasmid isolation, 84–5
  in spheroplast/protoplast generation, 30, 31
LysR, 384

magnesium ions (Mg^{2+}), 21, 23
maltose-binding protein, 334
marine bacteria, detection of plasmids, 86–7
marker exchange mutagenesis, *Erwinia* exoenzyme
    genes, 372–3
marker genes
  in *Bacteroides*, 232
  in mycobacteria, 257–8
  see also selectable markers
marker rescue, 13–14, 284, 285
  in halophilic Archaea, 302–4
  in *Rhodobacter* and *Paracoccus*, 433–5
MarR, 382
mauC gene, 436–7
mecA/B genes, 10–11
mer (mercury resistance) genes, 257
merodiploids, 470, 481–2
metalloproteases, extracellular, 365
Methanobacteriales, 289
Methanobacterium, 297–8
Methanobacterium ivanvovii, 297–8
Methanobacterium thermoautotrophicum, 282, 298
methanochondroitin, 294–5
Methanococcales, 289
Methanococcus, 289–94
Methanococcus jannaschii, 280, 282, 290, 292
Methanococcus maripaludis, 290, 291–4
Methanococcus voltae, 290–2, 293, 294, 296
methanogenic Archaea, 280, 288–98
Methanomicrobiales, 289
Methanosarcina, 294–7
Methanosarcina acetivorans, 295
Methanosarcina barkeri, 297
Methanosarcina mazei, 295
mevinolin, 300, 302
microcins, 72
mini-transposons, 142–5
missing contact assay, 381
mitochondrion, 428
mitomycin C, 99
mob genes, 54, 78, 82
mop genes, 402–3
motility
  in *Helicobacter pylori*, 335
  in plant pathogens, 358

Mot⁻ mutants, *Erwinia*, 358, 402–3
mox genes, 437, 439–40
mpr gene, 258
mRNAs, archaeal, 279
mtlK gene, 446
Mu bacteriophage, 108–9, 136–7, *139*
  in Archaea, 292
  d derivatives, 153, 351–2, 374, 406
  in *Erwinia*, 109, 350–3, 405
  mini-, 157, 352, 374
  -P1 hybrids, 109, 352
MukBEF proteins, 72
multimer resolution, 70–1
murein hydrolase, 121
mutagenesis
  in *Bacillus subtilis*, 468–9
  in *Bacteroides*, 235–9
  in clostridia, 192–4
  in *Erwinia*, 349–56, 372–4
  gene disruption, *see* gene disruption/deletion
      mutagenesis
  in halophilic Archaea, 299–300
  in *Helicobacter pylori*, 335–8
  localized, *out* cluster, 395–6
  in methanogenic Archaea, 290, 295–8
  in mycobacteria, 261–2
  random insertional *see* random insertional
      mutagenesis
  in *Rhodobacter* and *Paracoccus*, 433–5, 436–43
  shuttle *see* shuttle mutagenesis
  signature-tagged, 157–8, 341, 407
  site-directed *see* site-directed mutagenesis
  in *Sulfolobus*, 307–9
  transposon *see* transposon mutagenesis
mutational cloning, 111–13
mutations
  analysis, 283–4
  conditional, 156
Mx4 phage, 106, 108
mycobacteria, 251–69
  cloning vectors, 258–60
  conjugation, 256–7
  construction of recombinants, 257–60
  electroporation, 25–7, 255–6, 261
  gene expression, 265–8
  genetic exchange, 254–7
  genomes, 253–4
  growth, 252, 253
  integration vectors, *100*, 118, 259
  plasmids, 258–9
  promoters, 266
  recombinational properties, 254–5
  reporter genes, 267–8
  resources, 268–9
  RNA polymerase, 265
  selectable markers, 257–8
  strains, 252–3
  terminators, 267
  transduction, 103–4, 256
  transformation, 255–6
  transposons/transposition, 260–2
mycobacteriophages, 262–5
  conditionally replicating, 261–2
  generalized transducing, 256
  reporter, 117

Index